# ON MARS

Exploration of the Red Planet, 1958–1978
The NASA History

Edward Clinton Ezell
Linda Neuman Ezell

*Introduction to the Dover edition by*
Paul Dickson

**DOVER PUBLICATIONS, INC.**
Mineola, New York

NASA maintains an internal history program for two principal reasons: (1) Publication of official histories is one way in which NASA responds to the provision of the National Aeronautics and Space Act of 1958 that requires NASA to "provide for the widest practicable and appropriate dissemination of information concerning its activities and the results thereof." (2) Thoughtful study of NASA history can help agency managers accomplish the missions assigned to the agency. Understanding NASA's past aids in understanding its present situation and illuminates possible future directions.

One advantage of working in contemporary history is access to participants. During the research phase, the authors conducted numerous interviews. Subsequently they submitted parts of the manuscript to persons who had participated in or closely observed the events described. Readers were asked to point out errors of fact and questionable interpretations and to provide supporting evidence. The authors then made such changes as they believed justified. The opinions and conclusions set forth in this book are those of the authors; no official of the agency necessarily endorses those opinions or conclusions.

*Bibliographical Note*

This Dover edition, first published in 2009, is an unabridged and slightly corrected republication of the work originally published in Washington in 1984 in the NASA History Series as NASA SP-4212 under the title *On Mars: Exploration of the Red Planet 1958–1978*. The color photos originally on pages 385 to 388 can now be found between pages 364 and 365 in the Dover edition. A new Introduction by Paul Dickson has been added to this edition.

*Library of Congress Cataloging-in-Publication Data*

Ezell, Edward Clinton.
    On Mars : exploration of the red planet, 1958—1978 : the NASA history / by Edward Clinton Ezell and Linda Neuman Ezell ; introduction to the Dover edition by Paul Dickson.
        p. cm.
    Originally published: Washington, D.C. : Scientific and Technical Information Branch, National Aeronautics and Space Administration, 1984; in series; NASA history series (NASA SP-4212).
    Includes bibliographical references and index.
    ISBN-13: 978-0-486-46757-3
    ISBN-10: 0-486-46757-0
    1. Viking Mars Program (U.S.) — History. I. Ezell, Linda Neuman. II. Title.

TL789.8 .U6V524 2009
629.43'543—dc22

                                                                                2008052610

Manufactured in the United States of America
Dover Publications, Inc., 31 East 2nd Street, Mineola, N.Y. 11501

# Contents

# Tables

# Introduction to the Dover Edition

It seems so long ago that the first Viking images came back from Mars, and for many people—myself included—it was the moment in the pre-Internet Dark Ages that we began to understand that the analog age was about to give way to a digital one. What was hard to grasp during the summer of 1976, was that pictures from the surface of a body millions of miles away were being sent back in color in digital bits and bytes. Before the year was over more than 1,500 images came back to earth from the surface of the Red Planet, which was, according to those images, reddish brown. These were not the first digital images from space, but they were the ones that captured our imagination.

For the world, that first image (page 382, top) that came back from the Viking moments after its landing on the *Chryse Planitia* (Golden Plains) on July 20, 1976 was a singular moment. The picture was not as spectacular as those that would follow—it simply showed the Viking footpad on the right and vertical lines on the left caused by dust settling out of the atmosphere from disturbance by the retro-engines of the lander. But, when it showed up on the pages of our newspapers and magazines it was many things to people. America was celebrating its Bicentennial and this seemed to present a mark for the country as it moved into its third century. For NASA and the scientific community, the Viking would prove to be its most important and successful unmanned mission to date and pave the way for planetary missions to come. It also underscored the fact that humans could explore distant bodies efficiently, effectively, and economically from Pasadena, California, home of the Jet Propulsion Laboratory.

After that first image, came the second which was in color (page 385 and color section after page 364) and then some truly spectacular views including the first panoramic view (pages 382 and 383). As a member of the team put it: "Mars had become a place. It went from a word, an abstract thought, to a real place."

Of course, there was more to Viking than the images, the two 1,300 pound (600 kilogram) landers were packed with experiments designed to probe and sample the Martian surface. Each lander was powered by two radioisotope thermoelectric generators to convert heat to electricity from the radioactive decay of Plutonium 238. The landers required a mere 70 watts of power, less than that needed by most conventional light bulbs.

Now, a third of a century after those landings, Edward and Linda Ezell's narrative of the Viking's history and the first era of Mars exploration (1958–1978) is as relevant and important today as it was when first published. They show us how this program was conceived, developed, and sold to Congress in the wake of a more ambitious program—Mars Voyager—which was not funded. Then, as now, the politics of getting support for a mission is an important matter, especially when other issues loom large and it is easier for politicians to put space exploration in the margins. The formal Mars Voyager program died, but the idea of a Mars probe continued "under the radar" and was reborn as Viking in the face of an expanding war in Southeast Asia. (The name Voyager was later revived as the name for a Grand Tour of the plants.)

*On Mars* is a story of science, engineering, politics, and teamwork told by writers who were there when the team at the Jet Propulsion Laboratory picked the landing spot for Viking. They tell the story with the drama and clarity that the Viking program deserves. It is, in the very best sense of the word, a great adventure story.

—PAUL DICKSON

# Foreword

It is difficult for me to believe that there is already a written history of the Viking program. After all, the Vikings landed on Mars in 1976, just a few years ago. The *Viking 1* lander, designed for a 90-day surface mission, actually transmitted science messages to Earth for seven years. That such an extended mission was even possible is a crowning tribute to the people of Viking, in government, industry, and academia. This history is largely the story of those people of Viking, how they faced and dealt, emotionally at times, with technical problems, with adversity in the form of budget and people troubles, and with the political world. It is all too easy to forget these things in the face of gleaming, sophisticated spacecraft successfully performing their miracles in orbit around and on the surface of Mars.

This is indeed the time to capture the Viking history, for the memories of those who spent almost a decade with the program will rapidly fade, most have already moved on to new ventures, and relevant documents will disappear. The Viking history is of interest not only as the story of how the first planetary landing came to fruition or of how the first in-situ search for evidence of extraterrestrial life came about, but as a lesson on how thousands of individuals performed as a coherent team to accomplish what some believed to be impossible.

NOEL W. HINNERS
*Associate Administrator for*
*Space Science, NASA, 1974–1979*
*Director, National Air and*
*Space Museum, 1979–1982*

# Introduction

For many members of the Viking flight team, the early morning hours of 20 July 1976 were the culmination of 8 years of intense activity. Several of the scientists had more than 15 years invested in preparations for the investigations that would begin once Viking safely landed on the surface of Mars. The focus of everyone's attention on this day was the *Viking 1* spacecraft in orbit around Mars. Across 348 million kilometers, the team maintained contact with the 3250-kilogram craft from the Jet Propulsion Laboratory (JPL) in Pasadena, California. JPL this night stood jewellike, its brightly lit buildings contrasting sharply with the darkened silhouette of the San Gabriel Mountains. Outside the Theodore von Kármán Auditorium, converted into a press center for the mission, mobile television vans were being readied to broadcast the news of Viking's success or failure.

While reporters prepared stories and visitors strolled over the grounds, members of the flight team could be seen on the closed-circuit television monitors as they sat in the half-light of the control room. Elsewhere, hundreds of engineers, scientists, technicians, and support crews were at work or waiting to go to work. At 1:52 a.m., PDT, the audio circuit on the JPL television came to life, and George Sands, associate project scientist and for the moment the "Voice of Viking," announced: "We have separation. . . . We have engineering data indicating separation. . . . Separation is being confirmed all along the line." Eighteen minutes 18 seconds earlier, the time it took the confirming radio signal to travel from Mars to Earth, the lander had separated from the orbiter.[1]

By 2 a.m., the noise that had been building up at the press center and in the visitor areas diminished. Mission control, a small, glass-walled room with men seated around a circular console watching data displayed on television screens, was being projected on monitors around the lab. "Beyond the controllers' desks and the consoles, through the glass walls of his office . . . [was] Jim Martin, a big man in a short-sleeved blue shirt." James Slattin Martin, Jr., had the bearing and appearance of a military man. His closely cropped iron gray hair added to the image and encouraged nicknames like the "Paratroop Colonel" and the "Prussian General." Many members of the Viking team would attest publicly that he had run a tight project, but even those who had cursed him under their breath over the years had to admit that the incredible performance of the spacecraft during its 11-month cruise toward Mars and the normal postseparation checkout of the lander indicated that all the discipline and hard work Martin had put them through had been worth it. With a billion dollars invested in two spacecraft, someone had to have a firm grip. As the lander in its protective aeroshell fell freely toward the surface thousands of kilometers below it, Jim Martin listened to the controllers reporting tersely and calmly on the latest electronic news.[2]

At 3 a.m., Albert R. Hibbs, a senior advanced missions planner at JPL, relieved George Sands in the commentator's booth. Hibbs, a veteran "voice" of many earlier unmanned spacecraft directed from Pasadena, had what one observer called "marvelous sense of theater." Smiling, Hibbs noted that the deorbit burn of the lander's eight small rocket motors had gone smoothly and the spacecraft had proper velocity. Impishly, he noted that it was also going the right direction.

At 2 p.m., everyone was still waiting. Hibbs reported: "So far, everything that is supposed to have happened . . . has happened and right on schedule. We are rapidly approaching the surface of Mars. . . ." As the craft followed its curved trajectory, Hibbs noted that it had only 11 340 kilometers to go.*

4:43:08 a.m. PDT. Less than 10 minutes to touchdown, 28 minutes to confirmation. Al Hibbs informed his audience that he and George Sands would talk the lander down, but neither they nor anyone else at the mission center had any control over the spacecraft at this point; they could only keep listeners posted on the latest news. Obeying only its preprogrammed onboard computers, the lander was "inexorably going to the surface. . . ." By now "the lander has felt the impact of the Martian atmosphere, although we won't know for 19 minutes."

4:53:14 a.m. PDT. Hibbs reminded the people at JPL that "Viking should be on the surface by now, one way or another." A steady volume of 18-minute-old data kept flowing into the control center. The Viking team

---

*Hibbs and most of his Viking teammates used the common English measurements (miles and feet), but the authors have used metric units in this book to conform with NASA requirements that the *systeme internationale d'unites* (SI) be used in all NASA publications.

*Viking Project Manager James S. Martin, Jr., works at his desk at Jet Propulsion Laboratory.*

watched each new data point with increasing interest. The flight path analysis group had devised a visual display that portrayed the predicted descent curve for the lander—a single line graph that measured the lander's altitude against time. That line, a gentle curve sweeping downward from left to right, ended at touchdown. Once the lander in its aeroshell reached about 244 000 meters, the upper limits of the Martian atmosphere, onlookers could watch on the TV monitors as the actual path of the lander (in the form of data points) was plotted against the predicted normal curve. That graph was the tangible link between the watchers gathered in Pasadena and the Viking spacecraft approaching Mars. The first data point was right on the "nominal" curve.

From mission control, a disembodied voice began calling out the velocity and altitude of the spacecraft. The descent progressed rapidly. At 98 707 meters, the spacecraft was traveling 4718 meters per second.

When *Viking 1* reached 60 960 meters, Hibbs suggested, ". . . we can now put out some of the instruments that cannot stand the temperature of entry—pressure and temperature sensors that have to stick out of the aeroshell." Calling attention to a second graph on the television screens, he said that the viewers could watch the gravity forces "build up on that graph. Very violent changes in the effective combination of Mars' gravity and atmosphere on the spacecraft." In just the few seconds that it had taken him to make that remark, the acceleration force had increased from 2.7 times to 5 times the normal Martian gravity. By the time the spacecraft reached 30,000 meters, the atmosphere was beginning to exert a braking effect, slowing the lander to only 3000 meters per second. The gravity forces continued to rise—6.8, then 8.4, the maximum force encountered. At 27 000 meters, the velocity dropped to 1820 meters per second. As the craft passed the 24 000-meter mark, Hibbs reported: "Well, we're coming down. We're coming down. It's a long period of glide; almost flat glide to get rid of some more of the speed before the parachute comes out." From mission control, the callout of the descent continued in a measured, emotionless tone. When the craft passed through an altitude of 22 800 meters, it was moving at 982 meters per second. The acceleration forces had been reduced to 0.8. At 5:09:50 a.m., the parachute deployed, slowing the craft even further, to 709 meters per second.

5:11:27 p.m. PDT. 1463 meters, 54 meters per second. At 1400 meters, the terminal descent engines started. At 5:12:07.1 a.m. PDT July 20, a voice in mission control called out, "Touchdown, we have touchdown!" A chorus of cheers rose for the event completed 19 minutes earlier on Mars. "We have several indications of touchdown." Mars local time was 4:13:12 p.m. when *Viking 1* landed on the surface.

Jim Martin, who had been watching the descent curve on his monitor, stood up abruptly. He shook hands with William H. Pickering, former director of JPL, and exchanged congratulations with his teammates who rushed into his office. But then he paused for a moment to take another look at the televised data, wanting to be very sure that it had actually happened. A critical event in the life of the Viking project had come to a successful conclusion. Controllers and support personnel who had been quietly doing their tasks let loose with a burst of backslapping, embracing, and hand-shaking. In the auditorium, a newly opened bottle of cold duck was passed around as NASA public affairs officers and news people shared ceremonial sips. *Viking 1* was safely down on Mars.[3]

Nick Panagakos, public affairs officer from NASA Headquarters who had for weeks been answering questions for the press, smiled and shook his head. Like many of his colleagues, he had been telling people that Viking would land safely. But now that it had actually happened, he found it hard to believe. As the team in the control room settled back down to prepare for the reception of the first pictures of the Martian landscape, many persons around the Jet Propulsion Laboratory reflected on Viking's amazing odyssey.

*   *   *

When NASA planetary investigators began planning the exploration of Earth's closest neighbor, basic elements in their strategy were dictated by common sense. The space agency planners proposed to visit the nearest bodies first—the moon, Mars, and Venus. They planned to conduct simple projects initially and progress to more complex ones. Flyby spacecraft would be sent to take photographs and measurements and, after such basic reconnaissance had been made, heavier and more sophisticated orbiting craft would be sent to the target of investigation. After more detailed evaluations of the environment had been completed, atmospheric probes—either hard-landers (spacecraft that would crash-land) or soft-landers—would be used for further study. Different bodies would require different instrumentation. Photography, for instance, would not be suitable for cloud-covered Venus; on Mars it would be an experiment with exciting potential. During the past two decades, this strategy—flyby, orbiter, lander—has become a formalized part of NASA's planetary exploration program.

Mars, because it is reasonably close to Earth, has been the subject of much scientific examination. The Viking project was begun by NASA in the winter of 1968 to make landed scientific investigation of biological, physical, and related phenomena in the atmosphere and on the surface of Mars. The desire to explore for possible life forms on the Red Planet was one of the earliest goals of scientists who became part of the United States space science program, stretching Viking's roots back to the early 1960s. While NASA's first attempts to land craft on Mars were successful, that success did not come without a struggle. Chapter 1 examines the reasons scientists wished to have a closeup look at Mars and describes the new opportunities that opened with the coming of space travel. As Chapter 2 indicates, the dream was not transformed into reality until new and reliable launch vehicles became available in the mid-1960s, but the scientific community began early to prepare for landed investigations of the planet. Modest flyby probes such as *Mariner 4*, using less powerful rockets than the later Viking's, provided new if discouraging information about Mars. Despite initial photographic evidence that did not encourage the search for life, a small group of biological scientists—who called themselves exobiologists—began to develop instrumentation that would serve as the prototypes for life detectors on spacecraft that might fly in the future. These activities are related in Chapter 3, while Chapter 4 deals with the plans for NASA's first Mars lander project. Called Voyager* and conducted by the Jet Propulsion Laboratory, this project was ambitious, perhaps too ambitious for the times. Expansion of the war in Vietnam and demands for federal funds for many sectors of the American economy began a period of budget problems for NASA. Voyager died for a complex series of reasons in late summer 1967.

While budgetary stringencies were to remain with NASA planners from that time on, enthusiasm for a Mars lander project also continued. The focus of that spirit shifted from JPL to the Langley Research Center. The aggressive team at the Virginia center entered the Mars game just in time to see Voyager terminated. Chapter 5 chronicles the Langley entry into the planetary spacecraft business. Chapter 6 tells the story of the Viking orbiter within the context of advanced Mariner Mars spacecraft. Jim Martin and his colleagues, realizing that the JPL people had mastered the flyby and orbiter trade, persuaded them to become part of the Viking team. As Chapter 7 indicates, the Viking lander demanded many new inventions. In addition to new and complicated mechanical systems, it also required closely knit managerial, technical, and scientific teams that could come together in a cohesive organization during the data-gathering and analysis phases of the mission.

Before collection of scientific information could begin, landing sites for the craft had to be chosen. Data obtained from the 1971 Mariner orbiter assisted the specialists in this task but, as Chapter 8 recounts, there was considerable debate over the best places to land, given both scientific interests and engineering constraints. Despite the time and energy given to site selection, Mars held some surprises for the Viking team. The first orbiter photographs, which the team hoped would certify the suitability of the preselected landing sites, showed extremely hazardous terrain. Site certification, described in Chapter 9, became a renewed search for suitable and safe areas on Mars. For nearly a month, the project members labored to find

---

*NASA used the name "Voyager" again later for another planetary program, in which two spacecraft investigated Jupiter in 1979 flybys.

a safe haven for the lander. Finding a site for the second lander was an equally time-consuming job.

In Viking, NASA's most complex unmanned space project to that date, were many stories of great human effort and some of personal sacrifice. But the scientific results were the payoff. To have proved the technological capability to design, build, navigate, and land a spacecraft on Mars was not enough. Chapters 10 and 11 outline the scientific results of the Viking investigations and examine some of the unresolved questions. As so often is true in new fields of inquiry, as many questions were raised as were answered. And as earlier investigations of Mars have shown, the latest hypothesis can be upset by later, more detailed data. The Epilogue, therefore, considers possible future explorations of the Red Planet within the context of NASA's goals and other national priorities. One adventure was completed, but the exploration had just begun.

This book is just one of many possible histories that could be written about the events surrounding the Viking project. It is the official history because it was commissioned and paid for by the National Aeronautics and Space Administration. The authors began work shortly before Viking was scheduled to land on 4 July 1976, and they were present in Pasadena while Jim Martin and his team searched for a landing site. Exposure to the site selection process allowed us to see key project personnel at work and begin to understand the many complexities of Viking. We decided very quickly that we could not tell all the stories that participants might like to have told. We also concluded that, to appreciate fully the accomplishments of the project, readers should be exposed to the Mariner flights to Mars and to other planned but unconsummated missions to send landers to another planet. Thus our book evolved. In ignoring certain aspects or in describing others only briefly, we have not intended to slight other important aspects of the Viking effort. There are just too many stories and too many participants for them all to be included in this single volume.

# ON MARS

# 1

# Why Mars?

Since the 16th century, learned men have recognized Mars for what it is—a relatively nearby planet not so unlike our own. The fourth planet from the sun and Earth's closest neighbor, Mars has been the subject of modern scientists' careful scrutiny with powerful telescopes, deep space probes, and orbiting spacecraft. In 1976, Earth-bound scientists were brought significantly closer to their subject of investigation when two Viking landers touched down on that red soil. The possibility of life on Mars, clues to the evolution of the solar system, fascination with the chemistry, geology, and meteorology of another planet—these were considerations that led the National Aeronautics and Space Administration to Mars. Project Viking's goal, after making a soft landing on Mars, was to execute a set of scientific investigations that would not only provide data on the physical nature of the planet but also make a first attempt at determining if detectable life forms were present.

Landing a payload of scientific instruments on the Red Planet had been a major NASA goal for more than 15 years. Two related projects—Mariner B and Voyager—preceded Viking's origin in 1968. Mariner B, aimed at placing a capsule on Mars in 1964, and Voyager, which would have landed a series of sophisticated spacecraft on the planet in the late 1960s, never got off the ground. But they did lead directly to Viking and influenced that successful project in many ways.

When the space agency was established in 1958, planetary exploration was but one of the many worthy projects called for by scientists, spacecraft designers, and politicians. Among the conflicting demands made on the NASA leadership during the early months were proposals for Earth-orbiting satellites and lunar and planetary spacecraft. But man in space, particularly under President John F. Kennedy's mandate to land an American on the moon before the end of the 1960s, took a more than generous share of NASA's money and enthusiasm. Ranger, Surveyor, and Lunar Orbiter—spacecraft headed for the moon—grew in immediate significance at NASA because they could contribute directly to the success of manned Apollo operations. Proponents of planetary investigation were forced to be content with relatively constrained budgets, limited personnel, and little publicity. But by 1960 examining the closer planets with rocket-propelled probes was technologically feasible, and this possibility kept enthusiasts loyal to the cause of planetary exploration.

There is more to Viking's history than technological accomplishments and scientific goals, however. Viking was an adventure of the human mind, adventure shared at least in spirit by generations of star-gazers. While a voyage to Mars had been the subject of considerable discussion in the American aerospace community since the Soviet Union launched the first Sputnik into orbit in 1957, man has long expressed his desire to journey to new worlds. Technology, science, and the urge to explore were elements of the interplanetary quest.

## ATTRACTIVE TARGET FOR EXPLORATION

Discussion of interplanetary travel did not have a technological foundation until after World War II, when liquid-fueled rockets began to show promise as a transportation system. Once rockets reached escape velocities, scientists began proposing experiments for them to carry, and Mars was an early target for interplanetary travel.

Mars fell into that class of stars the Greeks called *planetes,* or "wanderers." Not only did it move, but upon close observation it appeared to move irregularly. The early Greek astronomer Hipparchus (160–125 B.C.) recognized that Mars did not always move from west to east when seen against the constellations of fixed stars. Occasionally, the planet moved in the opposite direction. This phenomenon perplexed all astronomers who believed Earth to be the center of the universe, and it was not until Johannes Kepler provided a mathematical explanation for the Copernican conclusion that early scientists realized that Earth, too, was a wanderer. The apparent motion of Mars was then seen to be a consequence of the relative motions of the two planets. By the time Kepler published *Astronomia nova* (New astronomy), subtitled *De motibus stellae Martis* (On the motion of Mars), in 1609, Galileo was preparing his first report on his observations with the telescope—*Sidereus nuncius* (Messenger of the stars), 1610. (See Bibliographic Essay for a bibliography of basic materials related to Mars published through 1958.)

From 1659, when Christiaan Huyghens made the first telescopic drawing of Mars to show a definite surface feature, the planet has fascinated observers because its surface appears to change. The polar caps wax and wane. Under close scrutiny with powerful telescopes, astronomers watch Mars darken with a periodicity that parallels seasonal changes. In the 1870s and 1880s during Martian oppositions with Earth,* Giovanni Virginio Schiaparelli, director of the observatory at Milan, saw a network of fine lines on the planet's surface. These *canali,* Italian for channels or grooves, quickly became *canals* in the popular and scientific media. Canals would be

---

*Appendix A describes some of the orbital relationships between Earth and Mars.

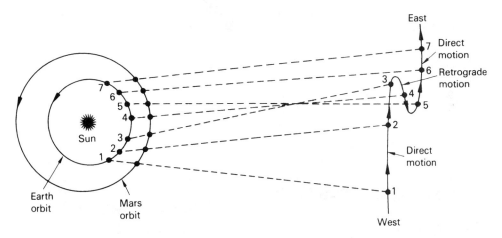

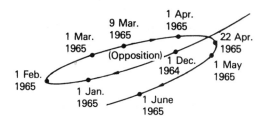

*The apparent motion of Mars. When Earth and Mars are close to opposition, Mars, viewed from Earth, appears to reverse its motion relative to fixed stars. Above, the simultaneous positions of Earth and Mars are shown in their orbits around the sun at successive times. The apparent position of Mars as seen from Earth is the point where the line passing through the position of both appears to intersect the background of fixed stars. These points are represented at the right. Below are shown the locations of Mars in the sky before and after the 1965 opposition. Samuel Glasstone,* The Book of Mars, *NASA SP-179 (1968).*

evidence of intelligent life on Mars. The French astronomer Camille Flammarion published in 1892 a 608-page compilation of his observations under the provocative title *La Planete Mars, et ses conditions d'habitabilité* (The planet Mars and its conditions of habitability). In America, Percival Lowell, in an 1895 volume titled simply *Mars,* took the leap and postulated that an intelligent race of Martians had unified politically to build irrigation canals to transport their dwindling water supply. Acting cooperatively, the beings on Mars were battling bravely against the progressive desiccation of an aging world. Thus created, the Martians grew and prospered, assisted by that popular genre science fiction. Percy Greg's hero in *Across the Zodiac* made probably the first interplanetary trip to Mars in 1880 in a spaceship equipped with a hydroponic system and walls nearly a meter thick. Other early travelers followed him into the solar system in *A Plunge into Space* (1890) by Robert Cromie, *A Journey to Other Worlds* (1894) by John Jacob Astor, *Auf zwei Planeten* (On two planets, 1897) by

3

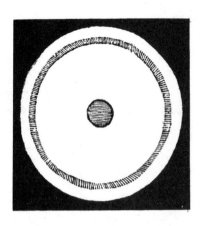
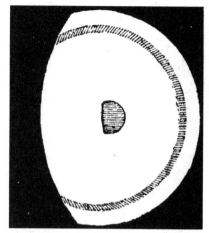

*These drawings of Mars by Francesco Fontana were the first done by an astronomer using a telescope. Willy Ley commented, "Unfortunately, Fontana's telescope must have been a very poor instrument, for the Martian features which appear in his drawings—the darkish circle and the dark central spot which he called 'a very black pill'—obviously originated inside his telescope." The drawing at left was made in 1636, the one at right on 24 August 1638. Wernher von Braun and Willy Ley,* The Exploration of Mars *(New York, Viking Press, 1956); Camille Flammarion,* La Planete Mars et ses conditions d'habitabilité *(1892).*

Kurd Lasswitz, H. G. Wells's well-known *War of the Worlds* (1898), and astronomer Garrett P. Serviss's *Edison's Conquest of Mars* (1898).[1] In "Intelligence on Mars" (1896), Wells discussed his theories on the origins and evolution of life there, concluding, "No phase of anthropomorphism is more naive than the supposition of men on Mars."[2] Scientists and novelists alike, however, continued to consider the ability of Mars to support life in some form.

Until the 1950s, investigations of Mars were limited to what scientists could observe through telescopes, but this did not stop their dreaming of a trip through space to visit the planets firsthand. Willy Ley in *The Conquest of Space* determined to awaken public interest in space adventure in the

*Christiaan Huyghen's first drawing of Mars (at left below), dated 28 November 1659, shows surface features he observed through his telescope. Of two later sketches, one of the planet as observed on 13 August 1672 at 10:30 a.m. (center below) shows the polar cap. At right below is Mars as observed on 17 May 1683 at 10:30 a.m. Flammarion,* La Planète Mars.

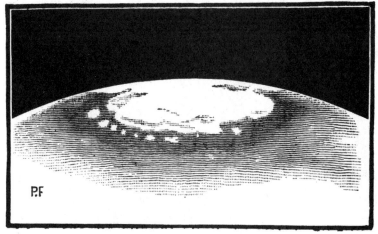

*Nathaniel E. Green observed changes in the southern Martian polar cap during opposition. The first sketch, at top, shows the polar cap on 1 September 1877, and the second, the cap seven days later. Flammarion,* La Planète Mars.

postwar era. His book was an updated primer to spaceflight that reflected Germany's wartime developments in rocketry. Ley even took his readers on a voyage to the moon. Considering the planets, he noted, "More has been written about Mars than about any other planet, more than about all the other planets together," because Mars was indeed "something to think about and something to be interested in." Alfred Russel Wallace's devastating critique (1907) of Percival Lowell's theories about life and canals did not alter Ley's belief in life on that planet. "As of 1949: the canals on Mars do exist," Ley said. "What they are will not be decided until astronomy has entered its next era" (meaning manned exploration).[3]

Ley's long-time friend and fellow proponent of interplanetary travel, Wernher von Braun, presented one of the earliest technical discussions describing how Earthlings might travel to Mars. During the "desert years"

5

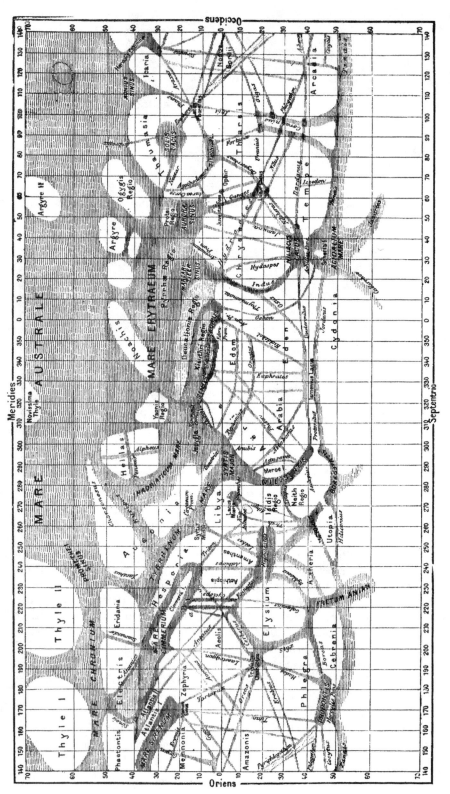

*Giovanni Schiaparelli's map of Mars, compiled over the period 1877–1886, used names based on classical geography or were simply descriptive terms; for example, Mare australe (Southern Sea). Most of these place names are still in use today. Flammarion, La Planète Mars.*

of the late 1940s when he and his fellow specialists from the German rocket program worked for the U.S. Army at Fort Bliss, Texas, and White Sands Proving Ground, New Mexico, testing improved versions of the V-2 missile, von Braun wrote a lengthy essay outlining a manned Mars exploration program. Published first in 1952 as "Das Marsprojekt; Studie einer interplanetarischen Expedition" in a special issue of the journal *Weltraumfahrt*, von Braun's ideas were made available in America the following year.[4]

Believing that nearly anything was technologically possible given adequate resources and enthusiasm, von Braun noted in *The Mars Project* that the mission he proposed would be large and expensive, "but neither the scale nor the expense would seem out of proportion to the capabilities of the expedition or to the results anticipated." Von Braun thought it was feasible to consider reaching Mars using conventional chemical propellants, nitric acid and hydrazine. One of his major fears was that spaceflight would be delayed until more advanced fuels became available, and he was reluctant to wait for cryogenic propellants or nuclear propulsion systems to be developed. He believed that existing technology was sufficient to build the launch vehicles and spacecraft needed for a voyage to Mars in his lifetime.

According to von Braun's early proposal, "a flotilla of ten space vessels manned by not less than 70 men" would be necessary for the expedition. Each ship would be assembled in Earth orbit from materials shuttled there by special ferry craft. This ferrying operation would last eight months and require 950 flights. The flight plan called for an elliptical orbit around the sun. At the point where that ellipse was tangent to the path of Mars, the spacecraft would be attracted to the planet by its gravitational field. Von Braun proposed to attach wings to three of the ships while they were in Mars orbit so they could make glider entries into the thin Martian atmosphere.*

The three landers would be capable of placing a payload of 149 metric tons on the planet, including "rations, vehicles, inflatable rubber housing, combustibles, motor fuels, research equipment, and the like." Since the ships would land in uncharted regions, the first ship would be equipped with skis or runners so that it could land on the smooth surfaces of the frost-covered polar regions. With tractors and trailers equipped with caterpillar tracks, "the crew of the first landing boat would proceed to the Martian equator [5000 kilometers away] and there . . . prepare a suitable strip for the wheeled landing gears of the remaining two boats." After 400 days of reconnaissance, the 50-man landing party would return to the seven vessels orbiting Mars and journey back to Earth.[5]

One item missing from von Braun's Mars voyage was a launch date. While he concluded that such a venture was possible, he did not say when he

---

*Earth's atmospheric pressure at sea level is 1013 millibars. From calculations made by A. Dollfus of the Paris Observatory in the 1950s, the mean Martian atmospheric pressure was determined to be about 85 millibars. The actual figure as determined by Viking measurements is 75 millibars.

expected it to take place. A launch vehicle specialist, von Braun was more concerned with the development of basic flight capability and techniques that could be adapted subsequently for flights to the moon or the planets. "For any expedition to be successful, it is essential that the first phase of space travel, the development of a reliable ferry vessel which can carry personnel into [Earth orbit], be successfully completed."[6] Thus, von Braun's flight to Mars would begin with the building of reusable launch vehicles and orbiting space stations. He and his fellow spaceflight promoters discussed such a program at the first annual symposium on space travel held at the Hayden Planetarium in October 1951, in a series of articles in *Collier's* in March 1952, and in *Across the Space Frontier*, a book published in 1952.[7] Two years later, however, von Braun concluded publicly that a major manned voyage to Mars was a project for the more distant future. As pointed out in an article entitled "Can We Get to Mars?":

> The difficulties of a trip to Mars are formidable. The outbound journey, following a huge arc [568 million kilometers], will take eight months—even with rocket ships that travel many thousands of miles per hour. For more than a year, the explorers will have to live on the great red planet, waiting for it to swing into a favorable position for the return trip. Another eight months will pass before the 70 members of the pioneer expedition set foot on earth again.[8]

Von Braun feared that it might "be a century or more" before man was ready to explore Mars.[9]

But five years later von Braun's response to an inquiry from the House Select Committee on Astronautics and Space Exploration indicated that his thinking had changed again. Gathering ideas for possible space activities, the House committee solicited opinions from the aerospace community and published its findings in *The Next Ten Years in Space, 1959–1969*. Von Braun considered "manned flight around the Moon . . . possible within the next 8 to 10 years, and a 2-way flight to the Moon, including landing, a few years thereafter." He believed it "unlikely that either Soviet or American technology will be far enough advanced in the next 10 years to permit man's reaching the planets, although instrumented probes to the nearer planets (Mars or Venus) are a certainty."[10]

A number of important technological and political events were instrumental in changing the rocket expert's thinking about American goals for space. Rocket technology had advanced considerably, as evidenced in the development of both American and Soviet intercontinental ballistic missiles. Soviet progress was forcefully impressed on the American consciousness by the orbital flights of *Sputnik 1* and *Sputnik 2* in the fall of 1957. Even as the Soviet Union stole a march on the Americans, von Braun and many others were busy defining and planning appropriate space projects for the United States.

Von Braun and his colleagues at the Army Ballistic Missile Agency in Huntsville, Alabama, had lost out to the Navy in September 1955 in the competition to launch an Earth satellite and had failed in their bid against the Air Force in November 1956 to be responsible for the development of intermediate range ballistic missiles. These setbacks prompted the managers of the agency to seek new justifications for the large launch vehicles they wanted to develop. Creating boosters that could be used for space exploration was the obvious answer. This goal was consistent with von Braun's long-time wish to see spaceflight a reality. In April 1957, Army Ballistic Missile Agency planners began to review United States missile programs in the light of known Soviet spaceflight capabilities and proposed a development strategy. The first edition of their sales pitch, "A National Integrated Missile and Space Vehicle Development Program," was issued on 10 December 1957. It reflected the post-Sputnik crisis:

> The need for an integrated missile and space program within the United States is accentuated by the recent Soviet satellite accomplishments and the resulting psychological intimidation of the West. . . . we are bordering on the era of space travel. . . . A review and revision of our scientific and military efforts planned for the next ten years will insure that provisions for space exploration and warfare are incorporated into the overall development program.[11]

The National Advisory Committee for Aeronautics (NACA) was also moving quickly in the wake of Sputnik. In an effort to define its role in the dawning space age, NACA's Committee on Aerodynamics resolved in November 1957 that the agency would embark upon "an aggressive program . . . for increased NACA participation in upper atmosphere space flight research." Subsequently, a Special Committee on Space Research under the direction of H. Guyford Stever, a physicist and dean at the Massachusetts Institute of Technology, was established "to survey the whole problem of space technology from the point of view of needed research and development and advise the National Advisory Committee for Aeronautics with respect to actions which the NACA should take."[12]

On 18 July 1958, the Working Group on Vehicular Program* of the Stever committee presented to NACA a revised edition of the Huntsville report on missile and space vehicle development. That document proposed an expanded list of possible goals for the American space program based on a phased approach to the development of successively more powerful launch vehicles. Those vehicles were divided into five generations:

First Generation—Based on SRBM boosters [short range]
Second Generation—Based on IRBM boosters [intermediate range]

---

*Members of the Working Group on Vehicular Program were W. von Braun, Chairman; S. K. Hoffman; N. C. Appold; A. Hyatt; L. N. Ridenour; A. Silverstein; K. A. Ehricke; M. W. Hunter; C. C. Ross; H. J. Stewart; G. S. Trimble, Jr.; and W. H. Woodward, Secretary.

Third Generation—Based on ICBM boosters [intercontinental]
Fourth Generation—Based on 1.5 million-pound-thrust [6.8-million-
newton] boosters
Fifth Generation—Based on 3 to 5 million-pound-thrust [13- to
22-newton] boosters.[13]

The planets, of course, were desirable targets for space exploration, but the realities of the emerging space race with the Soviet Union made the moon a more attractive goal politically for the late 1960s. In 1958, Stever's group did not think it would be possible to send a 2250-kilogram probe to Mars for at least a decade; it would be that long before the fourth-generation launch vehicle necessary for such a payload was ready. A manned mission to Mars or Venus was not projected to occur before 1977.

Implicit in the working group's timetable (table 1) was a gradual approach to space exploration. The proposed program was still ambitious, but it was increasingly apparent that scientific investigations in space would have to await new launch vehicles tailored to specific projects. It was technologically feasible to go to the moon and the planets, but the translation of feasibility into reality would require a national program and a new government agency to manage such activities.[14]

## OBJECTIVES IN SPACE

When the National Aeronautics and Space Administration (NASA), the new civilian space agency that superseded the National Advisory Committee for Aeronautics, officially opened its doors for business on 1 October 1958, a considerable body of knowledge could be grouped under the rubric *space science,* and many opinions were expressed about which aspects of space science should be given precedence for government monies. Scientists had been studying outer space for centuries, but observations made above Earth's filtering, obscuring atmosphere were a new step. Among the many disciplines that would benefit from using rockets in space were atmospheric research and meteorology, solar physics, cosmic ray study, astronomy, and eventually lunar and planetary investigation.

During most of the first half of the 20th century, professors had actively discouraged students from embarking on careers that would focus on the astronomy of the solar system, because most of the important information obtainable with existing equipment had been collected, digested, and published. Astronomy was described as "moribund"; it had "grown old from a lack of new data." Observations from space promised to change all that.

Before Sputnik, there were fewer than 1000 astronomers in the United States.[15] Budgets were tight, and research facilities were few. Until 1950, only 13 optical observatories with telescopes at least 914 millimeters in diameter had been built in the United States and, of these, 6 had been

*Table 1*
*Milestones of the Recommended U.S. Spaceflight Program,*
*July 1958*

| Item | Date | Event | Vehicle Generation |
|------|------|-------|--------------------|
| 1 | Jan. 1958 | First 20-lb [9-kg] satellite (ABMA/JPL) | I |
| 2 | Aug. 1958 | First 30-lb [14-kg] lunar probe (Douglas/RW/Aerojet) | II |
| 3 | Nov. 1958 | First recoverable 300-lb [140-kg] satellite (Douglas/Bell/Lockheed) | II |
| 4 | May 1959 | First 1500-lb [680-kg] satellite | II |
| 5 | Jun. 1959 | First powered flight with X-15 | |
| 6 | Jul. 1959 | First recoverable 2100-lb [950-kg] satellite | II and/or III |
| 7 | Nov. 1959 | First 400-lb [180-kg] lunar probe | II and/or III |
| 8 | Dec. 1959 | First 100-lb [45-kg] lunar soft landing | II and/or III |
| 9 | Jan. 1960 | First 300-lb [135-kg] lunar satellite | II and/or III |
| 10 | Jul. 1960 | First wingless manned orbital return flight | II and/or III |
| 11 | Dec. 1960 | First 10 000-lb [4500-kg] orbital capability | III |
| 12 | Feb. 1961 | First 2800/600-lb [1300/270] lunar hard or soft landing | III |
| 13 | Apr. 1961 | First 2500-lb [1100-kg] planetary or solar probe | III |
| 14 | Sept. 1961 | First flight with 1.5-million-lb [6.7-million-newton] thrust | IV |
| 15 | Aug. 1962 | First winged orbital return flight | III |
| 16 | Nov. 1962 | Four-man experimental space station | III |
| 17 | Jan. 1963 | First 30 000-lb [13 800-kg] orbital capability | IV |
| 18 | Feb. 1963 | First 3500-lb [1590-kg] unmanned lunar circumnavigation and return | IV |
| 19 | Apr. 1963 | First 5500-lb [2500-kg] soft lunar landing | IV |
| 20 | Jul. 1964 | First 3500-lb [1590-kg] manned lunar circumnavigation and return | IV |
| 21 | Sept. 1964 | Establishment of a 20-man space station | IV |
| 22 | Jul. 1965 | Final assembly of first 1000-ton [900-metric-ton] lunar landing vehicle (emergency manned lunar landing capability) | IV |
| 23 | Aug. 1966 | Final assembly of second 1000-ton [900-metric-ton] landing vehicle and first expedition to moon | IV |
| 24 | Jan. 1967 | First 5000-lb [2300-kg] Martian probe | IV |
| 25 | May 1967 | First 5000-lb [2300-kg] Venus probe | IV |
| 26 | Sept. 1967 | Completion of 50-man, 500-ton [450-metric-ton] permanent space station | IV |
| 27 | 1972 | Large scientific moon expedition | V |
| 28 | 1973/1974 | Establishment of permanent moon base | V |
| 29 | 1977 | First manned expedition to a planet | V |
| 30 | 1980 | Second manned expedition to a planet | V |

SOURCE: NACA, Special Committee on Space Technology, Working Group on Vehicular Program, "A National Integrated Missile and Space Vehicle Development Program," 18 July 1958, p. 6.

constructed before 1920 and 3 before 1900; in the 1950s another 6 were erected. But a boom occurred in the 1960s, when 28 new optical facilities were opened. Before the mid-1950s, only a handful of astronomers had more than very limited access to the large telescopes. One observer noted, "Not long ago, the study of the universe was the prerogative of a small group of men largely isolated from the rest of science, who were supported for the most part by private funds and were comfortable with projects that spanned decades."[16] Furthermore, astronomy had always been purely observational science with limited instrumentation. "Astronomers did not design experiments as physicists might; nor did they manipulate samples as chemists do." Faced with three major constraints—tight budget, lack of facilities, and the ever-present atmosphere through which they were forced to observe—astronomers saw few reasons for abandoning their 19th century ways. With World War II, change came to the field.

The war spawned radio astronomy and smaller, more sensitive instruments. Astronomers and their colleagues in other disciplines with whom they began to collaborate could detect, measure, and analyze wavelengths in the electromagnetic spectrum outside the visible range to which they had been limited. While radio astronomers probed the depths of the universe, finding among other phenomena radio galaxies more than a million times brighter than our own, a group of astronomers with highly sensitive equipment began to measure radiations and emissions from planetary atmospheres more accurately. In addition, the rocket, which could boost satellites and probes into space, promised to be another technological element that would open the way to a renaissance in astronomical research.[17]

In astronomical circles, the impact of the high-altitude rocket shots of the late 1940s was significant. Reacting to the first far ultraviolet spectra taken by V-2 rocket-borne instruments in October 1946, Henry Norris Russell, one of the most eminent astronomers of that generation, wrote, "My first look at one [rocket spectrum] gives me a sense that I [am] seeing something that no astronomer could expect to see unless he was good and went to heaven!"[18] Before the late 1950s, less than two percent of the astronomical community had been working in planetary studies. But experiments on board rockets and discussion of travel toward the moon, Mars, and Venus revived interest in the planets. Two "almost moribund fields—celestial mechanics and geodesy (the study of the size and shape of the earth)—were among the first to benefit from space explorations."[19]

American scientists were able to participate in this rocket-borne renaissance during the International Geophysical Year, 1 July 1957 through 31 December 1958, first suggested in 1950 by geophysicist Lloyd V. Berkner, head of the Brookhaven National Laboratory and president of the International Council of Scientific Unions. Originally Berkner saw this as a re-creation of the International Polar Years (1882, 1932), during which scientists from many nations had studied cooperatively a common topic—

the nature of the polar regions. The study proposed by Berkner would coincide with a period of maximum solar-spot activity, during which new instruments and rockets would be put to work to investigate widely many aspects of Earth science. Berkner's idea grew rapidly.

The National Academy of Sciences, a congressionally chartered but private advisory body to the federal government that attracted many of the nation's leading scientists, established the U.S. National Committee for the International Geophysical Year through its National Research Council. S. Fred Singer of the Applied Physics Laboratory, a member of the Council, had a strong interest in cosmic ray and magnetic field research, which led to his belief in using satellites as geophysical research platforms.[20] Singer proposed MOUSE—a Minimum Orbital Unmanned Satellite of the Earth—at the Fourth International Congress on Astronautics in Zurich in August 1953. A year later, at the urging of both Berkner and Singer, the International Scientific Radio Union adopted a resolution underscoring the value of instrumented satellites for observing Earth and the sun. Later that same month, September 1954, the International Union of Geodesy and Geophysics adopted an even more affirmative resolution. With momentum already established, the satellite proposal was presented to a Comité speciale de l'année géophysique internationale (CSAGI) planning meeting in Rome. After some maneuvering, the committee on 4 October 1954 adopted the following resolution:

> In view of the great importance of observations during extended periods of time of extra-terrestrial radiations and geophysical phenomena in the upper atmosphere, and in view of the advanced state of present rocket techniques, CSAGI recommends that thought be given to the launching of small satellite vehicles, to their scientific instrumentation, and to the new problems associated with satellite experiments, such as power supply, telemetering, and orientation of the vehicle.[21]

Two nations had the wealth and technology to respond to this challenge, the United States and the Soviet Union. During the next three years, the world scientific community watched the first leg of the space race, which culminated in the orbiting of *Sputnik 1* by the Soviets on 4 October 1957.[22]

After Sputnik's first success, it became increasingly clear that such a large-scale, cooperative scientific enterprise as the International Geophysical Year should not be allowed to die after only 18 months. Scientists from 67 nations had looked into a wide variety of problems related to Earth and the sun. To maintain the momentum behind those studies, Hugh Odishaw, executive director of the U.S. National International Geophysical Year Committee, and Detlev Bronk, president of the National Academy of Sciences, organized the Space Science Board in 1958. With many of the same members and staff that had worked on the international committee, the board was established to "stimulate and aid research, to evaluate proposed research, to recommend relative priorities for the use of space vehicles for scientific purposes, to give scientific aid to the proposed National Aeronautics and Space Agency, the National Science Foundation and the Department of Defense, and to represent the Academy in international cooperation in space research."[23] The Space Science Board had already held two meetings when NASA opened shop in the fall of 1958.

One of NASA Administrator T. Keith Glennan's first tasks was to pull together the many space-science-related activities that were scattered throughout the government. Launch vehicle development was managed by the Advanced Research Projects Agency of the Department of Defense. The Jet Propulsion Laboratory in Pasadena, California, and the Army Ballistic Missile Agency worked on the Explorer satellite project. Vanguard, another satellite venture, was directed by the Naval Research Laboratory. Many organizations, military and private, were already absorbed in the business of space exploitation. Besides carrying out existing projects and attending to the details of organization, NASA expanded its headquarters staff, acquired new field facilities, selected contractors, and sorted out its relationships with the Department of Defense and other government agencies. One participant in organizing the new agency's space science program recalled, "If anything stood out at the time, it was that everything seemed to be happening at once."[24] Within this context, scientists' proposals to send probes to Venus and Mars appeared to be very ambitious and certainly premature.

In April 1958 Abe Silverstein, a NACA veteran and associate director of the Lewis Propulsion Laboratory in Cleveland, went to Washington to participate in pre-NASA planning sessions and stayed on in a key position, director of the Office of Space Flight Development. Homer E. Newell, Jr., from the Naval Research Laboratory where he had been in upper-atmosphere research as superintendent of the Atmosphere and Astrophysics Division and science program coordinator for Project Vanguard, joined NASA on 18 October 1958, becoming Silverstein's assistant director for space sciences. Robert Jastrow, a Naval Research Lab physicist, and Gerhardt Schilling, a National Academy of Sciences staff member, were assigned to Newell's office. Jastrow immediately became immersed in plans for the future course of the space science program, and Schilling began studying ideas for lunar and planetary exploration.

America's space program was essentially two-sided; man-in-space was one dimension, space science the other. In the late 1950s, NACA's sounding rocket program and the Navy's Vanguard Project were the country's prime science activities, and those ventures were primarily "sky science," an examination of Earth-oriented phenomena from space. The only deep-space project in the works was the Air Force's yet-to-be-successful Pioneer probe. Since Administrator Glennan wanted to keep the growth of NASA's programs under control, Newell and his space science colleagues sought a gradual, rational expansion of existing science projects. Investigating the moon with unmanned spacecraft would obviously be more complicated and costly than near-Earth missions, so there was hesitancy to pursue a serious commitment to lunar science. Planetary studies seemed even further out of reach. According to Newell, Glennan was reluctant even to discuss planetary missions except in the framework of future planning.[25]

But the future came quickly. "Before Glennan left office NASA was engaged in space science projects that took in not only the earth and its environs, but also the moon and the planets, the sun, and even the distant stars," Newell remembered. Glennan, with some pride, turned over to his successor, James E. Webb, in 1961 "a well rounded program well under way."[26] Pressures for a broader space science program had come from several quarters—organized scientists (the Space Science Board), individual scientists (Harold C. Urey), and within the NASA fold (the Jet Propulsion Laboratory).

The Space Science Board's participation in planetary exploration discussions began in the summer of 1958 when Hugh Dryden, NACA's director, sought advice. The Air Force and the Jet Propulsion Laboratory had been promoting independently a planetary probe to Venus for 1959. Venus and Earth would be in their most favorable positions for such a mission, and it would be another 200 years before this particularly ideal opportunity came again. At the second meeting of the Space Science Board, 19 July 1958, Dryden asked the members to consider the wisdom of such an ambitious project. He feared that the mission as proposed was impractical because of limited time and a shortage of adequate tracking equipment for communications. Implicit in Dryden's hesitancy was the intent of NACA and the Eisenhower Administration to keep expansion of the space program in check.[27]

In response to Dryden's request for advice, Board Chairman Berkner established an ad hoc Committee on Interplanetary Probes and Space Stations. This group—chaired by Donald F. Hornig, professor of chemistry at Princeton University—considered two specific proposals for space projects, the first from Space Technology Laboratories of the Ramo-Wooldridge Corporation. Engineers proposed using a variant of the Air Force Thor intermediate range ballistic missile with an Able upper stage* (this two-stage launch vehicle had flown successfully in July 1958). Space Technology Lab's representatives advanced a concept for a 23-kilogram Venus probe plus the necessary tracking and communications equipment. The second suggestion came from Krafft Ehricke of the Astronautics Division of General Dynamics†, who proposed a considerably more complex mission. He wanted to use a yet-to-be-developed high-energy second stage with the Atlas intercontinental ballistic missile, which would be capable of delivering a 450-kilogram payload to the vicinity of Mars.[28]

---

*The Thor IRBM was developed by the Douglas Aircraft Company under contract (signed 27 December 1955) to the Defense Department, and the first strategic missile squadron was equipped with this IRBM on 1 January 1958. Douglas and STL collaborated to produce the Able second stage, based on components of the Vanguard launch vehicle.

†The Astronautics Division grew out of the Consolidated Vultee Aircraft Corporation (Convair).

Hornig's committee concluded that both proposals were technically feasible and furthermore believed that the time had come for action. The committee recommended unanimously to the Space Science Board that it was "urgently necessary to begin the exploration of space within the solar system with any means at our disposal if a continuing USA program of space science and exploration is to proceed at an optimum rate." Essential areas for study included:

1) The accurate determination of the astronomical unit [the distance between Earth and the sun].
2) Studies of the radiation environment
   a) High energy particles
   b) Low energy particles
   c) Gamma-rays
   d) X-rays
   e) Ultraviolet radiation
   f) Low frequency radiation
3) Measurements of electric and magnetic fields.
4) Study of radio propagation characteristics of outer space.
5) Study of the meteorite environment.
6) Study of the density, composition and physical properties of matter in space.

Some of these studies could be conducted with telescopes and spectrographs carried aloft by balloons, but most required close approaches or orbiting probes. The committee speculated that probes could possibly provide evidence of the existence of extraterrestrial life. Once an Atlas missile and a high-performance second stage were available, photographic or other viewing devices should be focused on the planets. The Hornig panel believed that "the most exciting experiments on both Venus and Mars seem to involve viewing devices, at least until it becomes possible to descend into their atmospheres." Since communications and tracking systems for such flights would require considerable development, the committee urged an early start on a planetary program.[29]

Hugh Odishaw—speaking for the Space Science Board in a special report to Glennan, Director Alan T. Waterman of the National Science Foundation, and Director Roy W. Johnson of the Advanced Research Projects Agency—underscored the committee's message. Since Thor-Able would be capable of transporting probes to the near planets, the board recommended "that a program aimed at launching a Mars probe during the 1961 conjunction [the time in the orbits of Earth and Mars when Mars disappears from Earth's view behind the sun] be immediately initiated." Odishaw also urged an early start on a high-performance second stage for Atlas "in order to provide a payload sufficient to carry out a more scientifically satisfying set of experiments on the planets Venus and Mars."[30] Instead of "resisting pressures" for early planetary exploration as requested by Hugh Dryden six months earlier, the Space Science Board strongly espoused such exploration in December 1958.

Individual scientists were also urging a broader space science program, and one of the most influential spokesmen for lunar and planetary studies was Harold Urey. Winner of the 1934 Nobel Prize in chemistry, Urey had a long, distinguished career behind him when in the early 1950s he turned his attentions to the origin of the solar system. In 1952, Yale University published his seminal book, *The Planets: Their Origin and Development.* In November 1958, Robert Jastrow traveled to the University of California in La Jolla to talk with this elder statesman of the space science community about the directions that NASA's space science program might take. Jastrow was converted to Urey's belief that the moon was a key element in unlocking the secrets of the universe, particularly for providing clues to the origin of the planets. Fascinated, Jastrow invited him to NASA Headquarters, where the scientist also convinced Homer Newell that a series of lunar projects should be undertaken. Newell noted years later that "the Ranger Project [a series of lunar probes] was in effect born on [that] day." As Jastrow set to work organizing an ad hoc Working Group on Lunar Exploration,* lunar enthusiasts had their foot in the door, and planetary advocates were not far behind.[31]

Within NASA, a major impetus for a larger space science program came from the staff of the Jet Propulsion Laboratory (JPL). Established in the summer of 1940 by the California Institute of Technology with contract funds provided by the U.S. Army Air Corps, JPL had over the years developed expertise in the fields of rocketry, instrumentation, telemetry, and tracking. After Sputnik, JPL joined the Army Ballistic Missile Agency in a successful partnership that launched *Explorer 1*, the first American satellite, on 31 January 1958 as part of the U.S. contribution to the International Geophysical Year. William H. Pickering, JPL director from 1954 to 1976 and a strong supporter of the American space program, wanted the United States, in the wake of Sputnik, to sponsor a space project that would outdistance the Soviet Union. His first proposal, "Red Socks," was for a seven-kilogram lunar payload. A major space first, according to Pickering, would be better for U.S. prestige than being the second nation to launch a satellite. While Red Socks never came about, the proposal was indicative of JPL's interest in projects other than Earth satellites.

Pickering had other aspirations as well. In a July 1958 letter to James R. Killian, presidential adviser for science and technology, the JPL director called for a significant role for his laboratory in the new space agency. Pickering urged that NASA "accept the concept of JPL as the national space laboratory. If this is not done, then NASA will flounder around for so long that there is a good chance that the entire program will be carried by the military." Instead of the space agency's being relegated to a position of

---

*Chaired by Jastrow, the working group included H. C. Urey, J. Arnold, F. Press, and H. Brown.

supporting research and developing scientific payloads, Pickering believed it could with JPL's guidance establish a realistic space program and maintain the civilian character that Eisenhower desired. "As you well know, one of the problems in the present space program is the multiplicity of committees and groups which are planning programs," Pickering reminded Killian. He believed that it was "essential for some competent group to be given a clear cut responsibility and told to draw up a realistic long term program which they can successfully complete on schedule." Only "if JPL does become the national space laboratory . . . does a complete experienced laboratory knowledgeable in all phases of the problem become the key asset of NASA."[32]

There was, however, a division of opinion at the Jet Propulsion Laboratory. Many of Pickering's colleagues believed that planetary investigation deserved more immediate attention than lunar goals. Whereas there were monthly opportunities for launching rockets to the moon, there were fewer such windows for trips to the planets. In 1958, the next launch opportunity for Mars would be October 1960 and the next practical chance for a Venus shot, December 1960–January 1961. Given these considerations, the JPL team after *Explorer 1*'s success began to look into possible planetary probe missions. One early example, undertaken at Pickering's request, was a design study for a 158-kilogram spacecraft that could be sent to Mars by a variant of the Army Ballistic Missile Agency's Jupiter intermediate range ballistic missile with two liquid-fueled upper stages. It quickly became apparent to the entire space science community that a more comprehensive study of possible planetary missions was needed.[33]

The space agency did want JPL to become a NASA field facility and began negotiations with the California Institute of Technology. But even before the contract between Cal Tech and NASA was signed, JPL staff members were discussing a long-range space program for the agency. A Silverstein memo suggesting that it begin thinking about future space projects had prompted the lab's actions. In Pasadena, the suggestion had been interpreted as a mandate—"a commission for JPL to plan a long range space program for NASA."[34]

John E. Froehlich, satellite project director at the lab, noted in the minutes of a 28 October 1958 meeting at JPL that he and his colleagues expected their study to "result in NASA's *major space program* but would not incorporate the entire national program." JPL, working jointly with the Army Ballistic Missile Agency, anticipated that this would be the working plan for the next five years, not just another proposal. The California team determined that NASA should concentrate on "putting up 'large payloads' for interplanetary research," not Earth-orbiting satellites.[35] Froehlich also recorded that the program must be "a compromise between a very conservative approach [and] a very wild, extravagant plan."[36]

A week later, JPL submitted to Silverstein a proposal to prepare a "Space Flight Program Study," the exact nature of which had been defined by Froehlich, Homer J. Stewart of Cal Tech, and Silverstein at NASA Headquarters. Once Glennan and Pickering concurred on JPL's interest in lunar and planetary studies, NASA agreed to the JPL study outline. On 18 November, the NASA Program Study Committee, composed of seven working group chairmen, began their task in earnest.[37] By executive order, President Eisenhower had the functions and facilities of JPL transferred from the Department of the Army to NASA on 3 December 1958.

Implicit in the Program Study Committee's work was a desire to influence launch vehicle and spacecraft development. At the end of the first five years, the necessary space vehicles had to be available for further work in space. "If this is not done, we will be entering the second five-year period doing what we are doing now—trying to fit available, but not entirely adequate, equipment to our program," Froehlich predicted. As a consequence, the study group and JPL's senior staff decided that the laboratory should concentrate its major energies on planetary goals while supervising others in the operation of lunar missions. As indicated in table 2, in which JPL launches are marked with asterisks, JPL planners considered two to three launches a year to be a comfortable maximum and Froehlich considered even that ambitious.

At a meeting with Homer Newell, John F. Clark, and Raymond Zavasky from headquarters, Director Pickering raised the issue of dividing planetary and lunar studies into two distinct fields. Newell saw two possibilities: JPL could "plan on doing the lunar work first and then later moving into deep space probes or go into deep space probes now with NASA finding some other agency or agencies to take on the lunar projects." Clark argued against any separation of lunar and planetary missions, stressing the similarities in guidance and communications requirements. Proposed near-misses (or *flybys* as they came to be called) of the moon and the planets would have analogous guidance requirements and should "accordingly be logical parts of a common program, while deep space probes would not necessarily have strict guidance requirements, and could themselves be a separate collection of projects." Although Pickering agreed to discuss these points while working on the laboratory's five-year plan, differences of opinion between JPL and NASA Headquarters were obvious.[38]

A 12 January 1959 meeting in Pasadena illustrated this growing divergence. Invited to discuss the progress of the evolving JPL-NASA study, the visitors from Washington included Abe Silverstein, Milton W. Rosen, Homer Newell, and Homer Stewart, who had been recruited from Cal Tech to Headquarters to direct the Office of Program Planning and Evaluation. After a few introductory remarks, Albert R. Hibbs of JPL described the missions portion of the study. The latest proposed lineup of flights (table 3) included a 1960 circumlunar mission and an escape toward Mars for a flyby of that planet. In 1961, JPL wanted to attempt a flight toward Venus, an escape out of the ecliptic (the plane about the sun in which all the planets

Table 2
Proposed Lunar and Deep Space Program, 1958

| Date | Mission | Payload Weight (kg) Required | Payload Weight (kg) Available | Launch Vehicle |
|------|---------|------|------|------|
| *1960* | | | | |
| * Aug. | Circumlunar | 159 | 230 | Titan |
| * Oct. | Two Mars flybys | 122 | 135 | Titan |
| *1961* | | | | |
| * Jan. | Two Venus flybys | 122 | 135 | Titan |
| May | Circumlunar | 159 | 230 | Titan |
| July | Lunar rough landing | 233 | 230 | Titan |
| * Sept. | Escape from Earth gravity | 120 | 135 | Titan |
| Nov. | Lunar satellite | 233 | 230 | Titan |
| *1962* | | | | |
| Feb. | Lunar rough landing | 233 | 230 | Titan |
| Apr. | Lunar satellite | 233 | 230 | Titan |
| * Aug. | Two Venus entries | 980 | 1360 | Juno V |
| * Nov. | Two Venus flybys | 161 | 135 | Titan |
| *1963* | | | | |
| Mar. | Lunar soft landing | 1810 | 2300 | Juno V |
| * June | Lunar soft landing Circumlunar with animal | 1810 | 2300 | Juno V |
| Aug. | Lunar soft landing | 1810 | 2300 | Juno V |
| * Oct. | Two Jupiter and two Mercury controlled flybys | 910 | 1360 | Juno V |

*JPL launches.
Source: J. D. McKenney, "Minutes of the Meeting of the NASA Program Study Committee . . . ," 15 Dec. 1958.

revolve), and a launch toward the moon that would produce a near-miss. Launches in 1962 would include orbiting lunar and Venus satellites, or perhaps a Venus entry probe and a Mars flyby. Lunar missions would occupy the following year with a circumlunar-return flight and a soft landing. Tentative goals for 1964 and 1965 were landings on Venus, another circumlunar-return, and a journey to Mars (1965). All these flights were by definition complete scientific exercises aimed at studying interplanetary space.

Pickering believed JPL's ambitious program was a sound one and would capture the interest and support of the scientific community. Since the recommended number of missions was limited to three to five a year, the director wanted each payload to be as advanced as possible. Toward that end, he wished to increase the laboratory's staff by 25 per cent. He also

Table 3

*JPL-Proposed Lunar and Planetary Missions, 12 January 1959*

| | Payload Number | Date | Mission | Scientific Package Weight (kg) | Gross Payload Required Weight (kg) | Gross Payload Available Weight (kg) | Nature of Measurements |
|---|---|---|---|---|---|---|---|
| Firm | 1 | 1 July 1960 | Circumlunar | 17 | 159 | 230 | Fields, atmosphere, photos of surface. |
| | 2 | 10 Oct. 1960 | Escape toward Mars | 14 | 161 | 135 | Interplanetary conditions, photos of Mars. |
| | 3 | 13 Oct. 1960 | Escape toward Mars | 14 | 161 | 135 | Interplanetary conditions, photos of Mars. |
| | 4 | 22 Jan. 1961 | Escape toward Venus | 14 | 161 | 135 | Interplanetary conditions, photos of Venus. |
| | 5 | 25 Jan. 1961 | Escape toward Venus | 14 | 161 | 135 | Interplanetary conditions, photos of Venus. |
| | 6 | Sept. 1961 | Escape out of ecliptic | 9 | 120 | 135 | Interplanetary conditions, measure A. U. |
| | 7 | Apr. 1962 | Lunar satellite | 23 | 233 | 230 | Gamma-rays, high-resolution mapping. |
| | 8 | 30 Aug. 1962 | Venus satellite | 1180[a] | 1770 | 1360 | Atmosphere, fields, surface nature. |
| | 9 | 2 Sept. 1962 | Venus satellite | 1180[a] | 1770 | 1360 | Atmosphere, fields, surface nature. |
| | 10 | 30 Nov. 1962 | Mars flyby | 14 | 190 | 135 | Atmosphere, photos, magnetic, and cosmic ray. |
| | 11 | 3 Dec. 1962 | Mars flyby | 14 | 190 | 135 | Atmosphere, photos, magnetic, and cosmic ray. |
| | 12 | June 1963 | Circumlunar & return | 1570[b] | 2300 | 2300 | Development test for Venus landing. |
| | 13 | 1963 | Lunar soft landing | 23 | 2300 | 2300 | Surface analysis, seismography. |
| | 14 | 1963 | Lunar soft landing | 23 | 2300 | 2300 | Surface analysis, seismography. |
| Tentative | 15 | 28 Mar. 1964 | Venus landing | 1100[b] | 2050 | ? | Weather, surface exploration. |
| | 16 | 1 Apr. 1964 | Venus landing | 1100[b] | 2050 | ? | Weather, surface exploration. |
| | 17 | Aug. 1964 | Circumlunar and return | 1570[b] | 2300 | 2300 | Manned flight. |
| | 18 | 20 Jan. 1965 | Circum-Mars and return | 2300[b] | 4500 | ? | Manned flight. |

[a]Including 1100-kg retrorocket.
[b]Including aerodynamic heating protection and aerodynamic controls or brakes, or both.
SOURCE: J. D. McKenney, "Minutes of the Meeting of the NASA Program Study Committee . . . ," 16 Jan. 1959.

contended that JPL should do nothing during 1959 that did not contribute directly to the development of deep space probes. In particular, it would be impossible to take on the direct technical supervision of NASA contracts in fields related to JPL projects. However, the JPL staff did expect to participate in NASA Headquarters committee activities and the like.

Abe Silverstein had in mind a different set of priorities when he looked at the rugged job NASA had ahead of it—managing an affordable but worthwhile national space program. He wanted JPL to be a part of NASA, to participate from the inside. He accepted the need for long-range planning, but NASA had to concentrate on the short run, on the creation of missions that would build congressional confidence so that legislators would support more ambitious projects for the years ahead. As a result, Silverstein was concerned with a different timetable, a launch and planning schedule for 1959. Long-range planning at this juncture could serve only as a guide. NASA did need to know where it was going, but Silverstein feared that JPL's five-year plan might take longer than five years to consummate and lock the agency on an unchangeable course.[39]

Obviously, NASA and JPL were looking at the future of spaceflight with different perspectives. NASA was still concerned with establishing its day-to-day activities and its short-term future. Working in Washington, Silverstein and his associates felt the often conflicting pressures from the White House, Capitol Hill, and the news media for a national space program that would at once surpass the Soviet Union's and be scientifically respectable without unbalancing the budget. Those pressures did not seem as important on the West Coast.[40]

JPL's plans were not only ambitious, they also reflected a difference in approach from that taken by Newell's space science office. Not unlike the von Braun team in Huntsville, Pickering's group thought of space probes in terms of their goals—the moon, Venus, Mars—while Newell's staff reflected the scientific community's concern with such topics as atmospheres; ionospheres; gravitational, magnetic, and electric fields; energetic particles; astronomy; biology; and environment. Likewise, Newell's suggestions to JPL for potential experiments for future missions reflected the disciplinary approach to space science taken during the International Geophysical Year.[41] JPL's goal-oriented study represented an engineer's way of looking at things. Neither view was better, both were necessary, but each had to accommodate the other, and that learning process would take years.

### NASA Long-Range Plans for Space Exploration

Not long after the meeting at JPL, NASA, spurred by pressure from Congress and the Space Science Board, was forced to do some long-range thinking of its own about the planetary exploration program. Two weeks into the new year of 1959 found Homer Stewart's Office of Program Planning and Evaluation working on a number of long-term questions. Besides looking into plans for the next year or two, Administrator Glennan wanted possible guidelines for the next 5 to 10 years.[42]

*Table 4*
*Influences on the Ten-Year Plan, 1960*

| JPL–Proposed Schedule | Goett Committee–Proposed Objectives[b] |
|---|---|
| Aug. 1960  Lunar miss (Vega) | 1. Man in space soonest—Project |
| Oct. 1960  Mars flyby (Vega) |    Mercury. |
| Jan.  1961  Venus flyby | 2. Ballistic probes. |
| June 1961  Lunar rough landing (Vega) | 3. Environmental satellite. |
| Sept. 1961  Lunar orbiter (Vega) | 4. Maneuverable manned satellite. |
| Aug. 1962  Venus orbiter (Vega) | 5. Manned spaceflight laboratory. |
| Aug. 1962  Venus entry (Vega) | 6. Lunar reconnaissance satellite. |
| Nov. 1962  Mars orbiter (Saturn 1) | 7. Lunar landing. |
| Nov. 1962  Mars entry (Vega) | 8. Mars-Venus reconnaissance. |
| Feb.  1963  Lunar orbit and return (Saturn 1) | 9. Mars-Venus landing. |
| June 1963  Lunar soft landing (Saturn 1) | |
| Mar.  1964  Venus soft landing (Saturn 1) | |

Ten-Year Plan[c]

| | |
|---|---|
| 1960: | First launching of meteorological satellite. |
| | First launching of passive-reflector communications satellite. |
| | First launching of Scout vehicle. |
| | First launching of Thor-Delta vehicle. |
| | First launching of Atlas-Agena B (DoD). |
| | First suborbital flight by astronaut. |
| 1961: | First launching of lunar impact vehicle. |
| | First launching of Atlas-Centaur vehicle. |
| | Attainment of orbital manned spaceflight, Project Mercury. |
| 1962: | First launching of probe to vicinity of Venus or Mars. |
| 1963: | First launching of 2-stage Saturn. |
| 1963-1964: | First launching of unmanned vehicle for controlled landing on moon. |
| | First launching of orbiting astronomical and radio astronomical laboratory. |
| 1964: | First launching of unmanned circumlunar vehicle and return to Earth. |
| | First reconnaissance of Mars or Venus, or both, by unmanned vehicle. |
| 1965-1967: | First launching in program leading to manned circumlunar flight and to permanent near-Earth space station. |
| Beyond 1970: | Manned lunar landing and return. |

[a]JPL, *Exploration of the Moon, the Planets, and Interplanetary Space*, ed. Albert R. Hibbs, JPL report 30-1 (Pasadena, 1959), pp. 95–114.
[b]NASA Hq., "Minutes of Meeting of Research Steering Committee on Manned Space Flight," 25–26 May 1959, p. 8.
[c]NASA Hq., Off. of Program Planning and Evaluation, "The Ten Year Plan of the National Aeronautics and Space Administration," 18 Dec. 1959, p. 10.

Stewart, one of the persons responsible for getting JPL's 5-year study under way, was charged with developing a 10-year master plan (1960–1970) for the agency. His recommendations, completed in December 1959, were influenced by two groups that were doing advanced planning at the time— the JPL NASA Study Program Committee and the Research Steering Committee on Manned Space Flight, chaired by Harry J. Goett of NASA's Ames Research Center. Stewart, borrowing from both these committees, secured balance among three important components of the space program—satellites, probes, and man-in-space.[43] The 10-year plan formalized the agency's goals for the 1960s (table 4).

The NASA Ten-Year Plan, presented by Associate Administrator Richard E. Horner, the number three official at NASA, to the House Committee on Science and Astronautics on 28 January 1960, established planetary missions as one of the firm goals of the space agency. The 1962 date for a probe to Venus or Mars and the 1964 photo-reconnaissance mission to Mars or Venus gave the JPL team something toward which to work. Many events would conspire to delay those flights, but exploration of the planets was securely part of the American space program.

# 2

# The Cart before the Horse:
# Mariner Spacecraft and Launch Vehicles

By August 1960 when Clarence R. Gates and his colleagues at Jet Propulsion Laboratory began studying plans for an interplanetary spacecraft called Mariner B, NASA's lunar and planetary program was taking the basic form it would have for a decade. Mariner B, designed to explore Mars and Venus and the space between, competed for both financial and manpower resources with several other space science projects. Lunar spacecraft—Pioneer, Ranger, Surveyor, and Prospector—were the main attraction, while Mariner and Voyager with their planetary objectives took second billing.[*][1] Lunar and planetary missions were arranged sequentially so that planners and scientists could progress from simple to complex tasks. Designers and engineers would likewise work on increasingly sophisticated spacecraft around a common chassis, or "bus," that could take successively more complex experiment packages into space. To meet these goals, NASA planned for the structured growth and development of several basic kinds of spacecraft. But spacecraft were only half the story.[2]

Reliable launch vehicles were essential to space exploration, and their lack had bedeviled the American space endeavor from the beginning. Reliability and payload capacity of the boosters (both proposed and in existence) defined the dimensions and possible use of each kind of spacecraft. While this relationship between launch vehicle and spacecraft was apparent in any space project, it had an especially negative effect on Mariner B.

## EVOLUTION OF UNMANNED SPACE EXPLORATION TO 1960

### Pioneer and Troublesome Launch Vehicles

Lunar exploration project Pioneer, America's bid in the early space competition, was approved in March 1958 under the initial direction of the Advanced Research Projects Agency, which assigned hardware develop-

---

*Lunar projects were given names related to terrestrial exploration activities; interplanetary projects were given nautical-sounding names that conveyed the impression of travel over great distances to remote lands.

ment to both the Air Force and the Army. But the two services each had a distinct approach to Pioneer, and the differences plagued the project from the start. On their first try, the Air Force team produced an unplanned pyrotechnic display when a Thor-Able launch vehicle exploded 77 seconds after liftoff from Cape Canaveral on 17 August 1958. *Pioneer 1*, launched on 11 October that year, was another disappointment; an early shutdown of the second stage prevented its attaining a velocity sufficient to escape Earth's gravity. After a 115 000-kilometer trip toward the moon and 43 hours in space, the probe burned up when it reentered Earth's atmosphere. The next month, *Pioneer 2*'s third stage failed to ignite; this spacecraft was also incinerated as it fell back to Earth. Meanwhile, the Army Ballistic Missile Agency and the Jet Propulsion Laboratory were working on a Pioneer lunar probe to be launched by a combination vehicle called Juno II, a Jupiter intermediate range ballistic missile with upper stages developed by JPL. A 6 December 1958 attempt to launch this four-stage rocket to the moon failed when the Jupiter first stage cut off prematurely. *Pioneer 3* reentered after a 38-hour flight.

*Pioneer 4*, the last of the series initiated by the Advanced Research Projects Agency, rose on its Juno II launch vehicle on 3 March 1959 and traveled without incident to the moon and beyond into an orbit around the sun, but without passing close enough to the moon for the lunar-scanning instruments to function. The U.S. attempt to beat the Soviet Union to the moon had already failed: *Luna 1*, launched 2 January, had flown by its target on 4 January. *Luna 2* next became the first spacecraft to land on another body in the solar system, crashing into the moon on 13 September 1959. *Luna 3*, launched 4 October, returned the first photographs of the moon's far side.

The U.S. effort continued to be less than successful. A sixth Pioneer lunar probe, a NASA-monitored Air Force launch, was destroyed when the payload shroud broke away 45 seconds after launch in November 1959. In 1960, two more NASA Pioneers failed, and the project died.* America's next entry was Ranger, NASA's first full-scale lunar project.[3]

*Ranger: Atlas-Vega versus Atlas-Agena*

The Ranger spacecraft—designed to strike the moon's surface after transmitting television pictures and gamma ray spectrometry data during descent—was one of the payloads planned for the Atlas-Vega launch vehicle. Atlas, an Air Force intercontinental ballistic missile developed by General Dynamics-Astronautics, had been selected by Abe Silverstein's Office of Space Flight Development for early manned orbital missions and deep space probes, and the decision had been based on several sound premises. If Atlas could be so adapted and if Thor and other intermediate-

---

*In 1965, NASA revived Project Pioneer with a new objective: to complement interplanetary data acquired by Mariner probes.

range ballistic missiles could be used for lightweight Earth satellites, then most of the funds NASA had earmarked for launch-vehicle development could be used for the development of a family of much larger liquid-propellant rockets for manned lunar missions. The space agency could purchase Atlas missiles from the Air Force and provide upper stages tailor-made for any particular mission, whether science in deep space or manned Mercury missions near Earth.

As defined in December 1958, three basic elements composed Atlas-Vega: (1) the Atlas missile, with its so-called stage and a half; (2) a modified Vanguard engine for the second stage; and (3) Vega, a new third stage under development at JPL. Vanguard was produced by General Electric. JPL's Vega would provide the extra thrust to reach the velocities necessary for planetary flights. According to the estimates, the combination would be able to place 2250 kilograms in a 480-kilometer Earth orbit or send approx-imately 360 kilograms to the moon. The first Atlas-Vega flight was optimis-tically scheduled for the fall of 1960.

On 17 December 1958 in Washington, representatives from NASA, the Advanced Research Projects Agency, the Army, and the Air Force consid-ered launch vehicle development and agreed that a series of versatile, increasingly powerful launchers was a desirable goal. However, NASA wanted its first new launch vehicle to be Atlas-Vega, while the Air Force favored the smaller Atlas-Agena. Since neither vehicle could meet the requirements of both organizations, NASA and the Air Force agreed to pursue their separate courses. Both approved Atlas-Centaur, a higher-energy rocket under development for future use, but only the space agency projected a need for the much larger Saturn.

Vega was the first element in NASA's proposal for "A National Space Vehicle Program," a document sent to President Eisenhower on 27 January 1959 specifying four principal launch vehicles—Atlas-Vega, Atlas-Centaur, Saturn I, and Nova (subsequently replaced by Saturn V). NASA began its hardware development program by contracting with General Dynamics, General Electric, and JPL for the production of eight Vega launch vehicles, being considered for Ranger flights to the moon and for a 1960 Mars mission. To send a spacecraft to Mars "with sufficient guidance capability and sufficient instrumentation to transmit information to the Earth, we need at least a thousand pounds [450 kilograms] of payload," Milton W. Rosen, chief of the NASA Rocket Vehicle Development Program, reminded senators during April 1959 hearings on the agency's 1960 budget. Vega was the first launcher in the NASA stable that had "such payload carrying capacity."[4]

Atlas-Vega, however, was not destined to fly to either the moon or the planets; a competitor blocked the way. The Air Force had been concealing a significant fact—Lockheed Missiles & Space Company had been develop-ing a much more powerful version of Agena, the B model.[5] The uprated Atlas-Agena B was unveiled in May 1959, almost instantly killing Atlas-

*An artist's concept of the Vega Mars probe as seen from the Martian moon Deimos was presented to the Senate Aeronautical and Space Science Committee on 7 April 1959.*

Vega. NASA began investigating the similarities between the two that spring, and in July the Civilian-Military Liaison Committee, established earlier to work out problems of mutual concern to NASA and the Department of Defense, ordered a review of the two systems. The committee's and NASA's findings agreed: one of the projects should be canceled. Since NASA was in no position to force the Air Force to terminate the somewhat more flexible Agena B, the agency conceded. On 7 December, Glennan telephoned JPL Director Pickering. All work on Vega would stop immediately.[6]

Glennan and his staff at NASA Headquarters were discomfited by Vega's cancellation. The duplicative project had not only cost them $17 million labeled for launch vehicle research, its cancellation had returned them to dependence on new Air Force rockets. JPL's unhappiness over losing Vega was compounded by dismay over NASA's new 10-year plan, which was clearly geared toward lunar rather than planetary activities.[7] Richard E. Horner, NASA associate administrator, wrote Pickering in December 1959 about the management's post-Vega thinking, discussing the recent transfer of the Army Ballistic Missile Agency in Huntsville, Alabama, to NASA (a transfer sought by NASA since October 1958) and Vega's cancellation. Although the cancellation was certainly "disturbing" and would "necessitate a major reorientation of the Laboratory work program," Horner believed that it would allow the entire NASA community to advance toward the agency's long-term objectives. Each NASA center working directly in space experimentation had been assigned "a major functional area of responsibility." The facility at Huntsville under the direction of Wernher von Braun was responsible for the development of launch vehicles and associated equipment. That organization would also control all launch-related activities to the point of orbital injection or some similar point in the trajectory of a probe. The Goddard Space Flight Center in Maryland would oversee the development and operation of Earth satellites and sounding-rocket payloads. Development and operation of spacecraft for lunar and interplanetary exploration was JPL's task. "It is pertinent to note here that the Administrator has decided that our efforts for the present . . . should be concentrated on lunar exploration as opposed to exploration of the planets," Horner added in his letter to Pasadena.[8]

Along with these clearly defined field assignments, major changes were taking place at NASA Headquarters. The former Office of Space Flight Development was divided into two directorates—the Office of Launch Vehicle Programs and the Office of Space Flight Programs.* Abe Silverstein would direct spaceflight, with JPL and Goddard reporting directly to him. Staff responsibility for launch vehicles would be directed by former Advanced Research Projects Agency specialist Maj. Gen. Don R. Ostrander, to whom the von Braun team would be accountable. These assignments were designed to establish clearer lines of responsibility for both administrative and functional purposes. (See charts in appendix G.)[9]

Within this new framework JPL, in carrying out its task of planning and executing lunar and planetary projects, would be in charge of mission planning, spacecraft development, experiments, mission operations, analysis of scientific data returned from space, and the publication of mission results. Since these activities could not possibly be carried out by JPL alone, headquarters "expected that a part of the developments will be contracted with industry and the Laboratory will assume the responsibility of monitoring such contracts," Horner noted. Pickering continued to resist such a role when he met with Silverstein a month later, but contracting for hardware development was agency policy. NASA would also exercise control over its field centers through annual program guidance documents written at headquarters. The Pasadena laboratory's independence was being curtailed as the men in Washington began to pull together a more centralized management system, but the relationship between headquarters and JPL was still not clearly defined.[10]

In December, going one step further in asserting headquarters' leadership, Silverstein outlined for JPL the space agency's plans for lunar and planetary missions for the next three years. Earlier that month the NASA Lunar Science Group, chaired by Robert Jastrow, had met to discuss proposals for lunar exploration. Harold Urey, Thomas Gold, Harrison Brown, and other scientists had agreed that a hard lunar landing, which by its crashing impact could help determine the nature of the moon's surface structure, would be an important first step. High-resolution pictures of the moon before impact would also be most important. Basing plans on the advice of the lunar group and the change in launch vehicles, Silverstein

---

*The distinction between programs and projects was first made clear by G. F. Schilling, Office of Space Science, late in 1959. *Programs* signified a related and continued series of undertakings geared toward understanding a broad scientific or technical topic; programs (e.g., examining the solar system) did not necessarily have foreseeable ends. *Projects* were the building blocks for programs and as such had limited objectives, limited duration (e.g., Project Mariner, Project Viking). While the space science personnel at NASA tended to maintain this distinction over the years, the concept was not as clearly observed in manned spaceflight, where the Apollo project grew so large it became a program.

advised Pickering that seven flights were planned through 1962. The first five would be launched by Atlas-Agena B for "lunar reconnaissance" in 1961–1962; two other spacecraft would be sent by Atlas-Centaur to Mars and Venus in 1962.[11] As part of an integrated lunar exploration program, the lunar spacecraft, Ranger, should also be capable of depositing an instrument package on the moon.

In late December, Homer Newell, Newell Sanders, Joseph A. Crocker, and Morton J. Stroller traveled to California to discuss how the projected flights fitted into the agency's long-range plans. Crocker explained that development should begin on four different spacecraft (designations in brackets indicate projects that emerged from this planning):

a. A spacecraft for use with the Agena on lunar work [Ranger],

b. a spacecraft for use with Centaur for planetary and lunar orbit, with perhaps a modification for soft landings [combination of Surveyor and Lunar Orbiter and Mariner B],

c. a spacecraft for use with Saturn on planetary work [Voyager] with some modifications, perhaps for instrumented landings of lunar rover vehicles [Prospector], and finally,

d. a spacecraft for use with the Saturn for unmanned circumlunar missions and return leading to perhaps some modifications for manned circumlunar missions and return.

Rather than be developed independently, the spacecraft would evolve, with more advanced spacecraft growing out of generation-to-generation experience.[12]

Pickering was still not fully reconciled to the moon-first priority laid down by Washington, believing that the limited opportunities for flights to the planets made it absolutely imperative that work begin immediately on planetary spacecraft. Newell and his colleagues relieved the director's anxieties somewhat by assuring him that there would be planetary flights "every time the near planets, Mars and Venus, were in optimum position." The JPL group was reminded, however, that the planetary program would be relying on the yet-to-be-developed Centaur launch vehicle for some time, until the more advanced Saturn family was ready.[13]

*Surveyor, Mariner, and the Centaur*

As headquarters directed, JPL personnel set about defining a lunar impact mission, but Atlas-Centaur-boosted spacecraft of the future were also an active concern. NASA hoped Surveyor, the first of these advanced craft, would allow a "tremendous stride forward in lunar exploration," since it would land softly on the moon, carrying a number of experiments,*

---

*The term *experiment*, as NASA uses it, refers to any exercise whose purpose is to gather scientific or engineering data (and also to the equipment used to perform an experiment). Few scientists would apply the term to some NASA experiments, e.g., photography of Earth from orbit.

including a surface sampler and an atmosphere analyzer. These instruments would provide scientists and designers information they needed to plan more sophisticated unmanned and manned landing missions. Mariner, the second spacecraft family to be powered by Atlas-Centaur, would be directed toward Venus and Mars. Two kinds of Mariner spacecraft were planned: an A model that would simply fly by those planets and a B model that could release a landing capsule toward Mars or Venus as the main bus flew by. A 1962 Mariner was expected to be launched toward Venus to measure the planet's surface temperature distribution, examine the atmosphere, and determine the extent of the magnetic field as it flew by.

Still later in the 1960s, two multipurpose spacecraft, Prospector and Voyager, atop mighty Saturn launch vehicles were to extend the scope of unmanned lunar and planetary exploration even further. Prospector was being designed to roam about the lunar surface as directed from Earth and examine the moon with a sophisticated array of instruments. Subsequent lunar rovers were to be used as logistic vehicles to marshal supplies for manned missions to the moon, or possibly as an early means of returning experiment samples. Voyager, too, was being designed with growth in mind. From the first missions in 1964 to either Venus or Mars with slightly larger landed payloads than the Mariner B capsule, Voyager was to grow larger and larger until a mechanized rover was sent to Mars or Venus. Prospector and Voyager represented the very distant future, but by the summer of 1960 JPL and NASA Headquarters were beginning to give serious attention to Surveyor and Mariner.[14] Both of these craft were scheduled for launch by Atlas-Centaur—the number two vehicle in NASA's plans—but development problems with the Centaur stage would seriously affect the timetable.

### CENTAUR: TROUBLESOME LAUNCH VEHICLE

One of the earliest plans for a U.S. probe to Mars was based on the Atlas-Centaur launch vehicle. In 1956, Krafft Ehricke of General Dynamics began to study high-energy second stages that might be used with the Atlas missile. In examining oxygen-hydrogen rocket stages, he had three objectives in mind—using the unexcelled thrust of Atlas, providing an upper stage with a maximum energy output for its weight, and developing a launch vehicle that could be used for several different kinds of mission. Three specific "important mission classes" were considered for this new vehicle:

> High-altitude satellites in the 8-hour, 12-hour and 24-hour orbits for the purpose of global surveillance, early warning, and global communication.

> Launchings of instrumented space probes to the lunar surface and into the inner solar system, primarily to Venus and Mars. . . .

> Establishment of a small manned orbital laboratory for a crew of three to inaugurate systematic preparations for deep space missions of manned spaceships.[15]

For several reasons, Ehricke and his associates settled on 13 500 kilograms for the weight of their proposed high-energy stage. This was close to the upper limit that the existing Atlas could boost, and a stage of this approximate weight would have about the same diameter as Atlas and a reasonable length. By October 1957, studies for the prototype Centaur were complete, and Ehricke took his ideas to the Advanced Research Projects Agency. The agency was intrigued and encouraged Ehricke's team to draw up a plan for a launch vehicle stage that used two Pratt & Whitney pump-fed engines rather than pressure-fed engines. On the basis of these discussions, General Dynamics submitted a proposal for a Mars probe in August 1958.

Ehricke noted that this particular suggestion for a flight to the Red Planet had been made because his team was "quite mission conscious and [wanted] to emphasize the importance of gaining an early capability to send probes to Venus and Mars in view of the infrequent intervals at which these missions [could] be flown." Some years are more favorable for planetary flights than others, and during advantageous years a rocket of given power can carry a much larger payload. Propitious opportunities for travel to Mars and Venus occur about every two years and generally last for about a month (appendix A). Unless the launch vehicle is unusually powerful, the geometry dictates a two-year delay once a launch window is missed. Separation between Mars and Earth at the time of closest approach varies from 55 million to 102 million kilometers over a cycle about 16 years long. (The most favorable opposition between 1970 and 1975 was in 1971, when the two planets were only 55.8 million kilometers apart.) Ehricke in 1958 looked toward a 1964 launch, to take a spacecraft past Mars in June 1965.

On 28 August 1958, the Advanced Research Projects Agency requested the Air Force Research and Development Command to oversee a contract with General Dynamics for the development of an upper stage for Atlas, to be propelled by oxygen and hydrogen. That stage, which was to weigh about 13 500 kilograms and have a diameter of about 3 meters, was to be powered by two engines capable of 67 000 newtons (15 000 pounds) of thrust each. Even though the effort required a major advance in the state of the art, an oxygen-hydrogen-powered stage appeared feasible. The resultant launch vehicle was intended to be a "space truck," bridging the gap between the less powerful Atlas-Agena and the much larger boosters of the future. Although a specific mission for the stage had not been defined, the first test flight was scheduled for January 1961, only 26 months after the contract with General Dynamics was signed.

Given the short development time, limited budget, and injunction against impinging on the military Atlas program, the government was expecting a great deal from General Dynamics, which was responsible for vehicle development and overall project integration, and Pratt & Whitney, which had a contract for building the oxygen-hydrogen engines. After considerable negotiation, NASA, the Advanced Research Projects Agency, and the Air Force agreed in the summer of 1959 to a compromise system of management. The Air Force named Lt. Col. John D. Seaberg Centaur project director and assigned him to the Ballistic Missile Division's offices at the Los Angeles Air Force Station. Seaberg had a strong background in the missile field and intimate knowledge of the relatively new technology surrounding liquid hydrogen, having worked on the Air Force's highly secret Suntan Project, which had sought to tame liquid hydrogen for use as an aircraft fuel. Seaberg reported directly to Milton Rosen, project director at NASA Headquarters. This arrangement became official on 1 July 1959, when responsibility for Centaur was shifted to the space agency.[16]

During the winter of 1959–1960, NASA established a Centaur Project Technical Team of specialists from the field centers, to undertake a thorough study of the project and recommend ways in which it might be best conducted.[17] Centaur had grown in importance to NASA since the cancellation of Vega and was rapidly becoming more than an austere research and development experiment. It was a probable answer to launching specific payloads. Centaur, with its much greater thrust and coast-restart capability, promised a major technological improvement over existing vehicles.[18]

In early 1960, NASA Headquarters and JPL conducted a series of studies to determine the most suitable launch vehicle for early Venus and Mars flyby missions. On 8 July 1960, a team from JPL gave Administrator

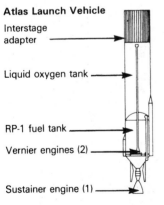

Nose fairing

Payload

Tank insulation panels

Liquid hydrogen tank

Liquid oxygen tank

Vernier engines (4)

Main engines (2)

**Centaur**

**Atlas Launch Vehicle**

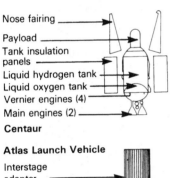

Interstage adapter

Liquid oxygen tank

RP-1 fuel tank

Vernier engines (2)

Sustainer engine (1)

Booster engine (2)

*Outlined at left are the major components of the proposed Atlas-Centaur two-stage launch vehicle for planetary probe missions. Below, the Centaur upper stage is nearly 10 meters tall and about 3 meters in diameter. General Dynamics/Astronautics, A Primer of the National Aeronautics and Space Administration's Centaur (San Diego, 1964).*

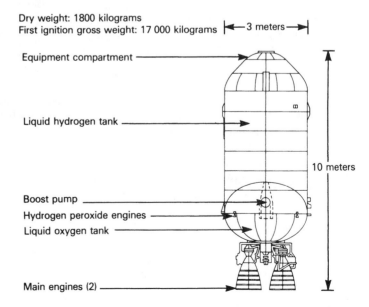

Dry weight: 1800 kilograms
First ignition gross weight: 17 000 kilograms   |— 3 meters —|

Equipment compartment

Liquid hydrogen tank

10 meters

Boost pump

Hydrogen peroxide engines

Liquid oxygen tank

Main engines (2)

Glennan a six-part briefing on the subject. Lab spokesman Robert J. Parks noted that the late 1960 Mars and early 1961 Venus launch windows would have to be ignored as NASA was "in no position to take advantage of them," but before 1970 there remained "exactly five opportunities to fire at Venus and four to fire at Mars." To make the best use of those, the proper order for developing spacecraft appeared to be "first planetary flybys, then planetary orbiters, and then the orbiter-landers, in which a part of the orbiting vehicle is detached and caused to enter the atmosphere and land on the planet relaying its information to the earth via the orbiter." Since Atlas-Centaur could not boost planetary orbiters (retrorockets would add considerably to the weight), JPL's 10-year flight schedule (see chart) called for using Centaur for flyby missions through 1964. In 1965, Saturn was to be used for planetary orbital experiments, leading to larger lander missions in 1967.[19]

The early flybys were important, since they would supply information about atmospheric and topographical conditions—data that would affect future landing craft. From the lab's point of view, the 1964 Venus and Mars opportunities were the big ones, and at least "three spacecraft developmental firings [were] required prior to . . . 1964." Repeating an increasingly familiar refrain, Parks told Glennan that after the first five Ranger launches, the planetary program would constitute "the major program activity of the Laboratory."[20]

Sending a spacecraft to either Venus or Mars depended on the availability of both Atlas-Centaur and sufficient funds. Atlas-Centaur was a big question mark, but nearly everyone was hopeful. Parks pointed out, however, that "FY61 fund limitations preclude developing and fabricating in time for a 1962 launching" a spacecraft meeting all the relatively severe requirements for a mission to Mars. Instead, JPL proposed a more modest spacecraft based on Ranger for a 1962 flight to Venus.

Although the small Ranger-class spacecraft would not be a true prototype of the 1964 Mariner, it would still provide an excellent early test. Assuming the availability of Atlas-Centaur in 1962, an 885-kilogram payload could be sent to Venus; 585 kilograms could be flown to Mars. Ranger weighed only 225 kilograms. Given the uncertain financial and launch vehicle situation, the JPL team favored sending the smaller craft to Venus in 1962, leaving the larger full-scale Mariner for the 1964 opportunity.[21]

Believing that Centaur would be ready on time, the Office of Space Flight Development disregarded JPL's advice. Headquarters planners in July 1960 proposed to launch a spacecraft designated Mariner A to Venus with Atlas-Centaur in 1962 after one test flight. Following a 1963 trial, a larger Mariner B, possibly with an instrumented lander, would be ready for Mars and Venus missions in 1964. JPL's austere 1962 super-Ranger was held in abeyance. Administrator Glennan approved the Mariner projects on 15 July 1960, just six days after he had approved three lunar Apollo feasibility studies.[22]

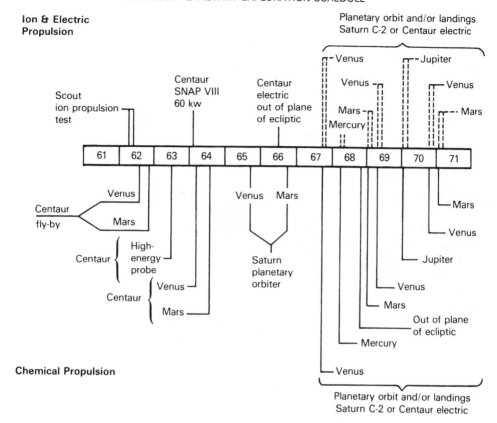

PROPOSED PLANETARY EXPLORATION SCHEDULE

*A proposed 10-year programming chart was shown to NASA Administrator Glennan at the 8 July 1960 planetary program briefing. The proposals for launch vehicle upper stages above the timeline would use nonconventional—ion and electric—propulsion. (SNAP stands for "system for nuclear auxiliary power"; SNAP VIII would produce 60 kilowatts of electrical power.) Proposals for upper stages shown below the timeline would use conventional—chemical—propulsion.*

## Planetary Mission Proposals

In August 1960, the Planetary Program Office at JPL began studying Mariner B, examining the feasibility of building a spacecraft capable of a variety of missions. Such a versatile craft using basic components with scientific instruments packed in modules promised lower production costs. A confidential "Mariner B Study Report" prepared in April 1961 concluded "that the Mariner B mission should involve a split capsule, in which the main body of the spacecraft passes by the planet and a small, passive capsule separates from the spacecraft and impacts the planet." Mariner B was expected to be used to investigate Mars and Venus.

In reviewing possible missions, Clarence Gates of JPL's Systems Division noted in JPL's study report that planners usually judged proposed spacecraft-borne experiments by three criteria:

(a) The experiment should be conservative and should be based to the maximum extent possible on previous experience, technology, and components; (b) the experiment should, in its own right, be significant in the contributions that it makes to technology and scientific knowledge; and (c) the experiment should be daring and imaginative, should take a substantial stride forward, and should bridge the gap between our present state of knowledge and the more distant future.[23]

Gates went on to point out that it was "rare for these considerations not to lead in diverse directions." In 1961, Mariner A typified a conservative approach with a high chance of success. That craft was fully attitude-stabilized, using the sun and Earth as references. Power was to be supplied by sun-oriented solar panels, with backup batteries. While the propulsion system could be operated for a midcourse correction maneuver, Mariner A had neither an approach nor a terminal guidance system; thus, it could not be expected to rendezvous reliably with specific celestial coordinates near the target planet. Mariner B, the next step, would be more advanced technologically, contributing to the design and development of the still more ambitious Voyager.

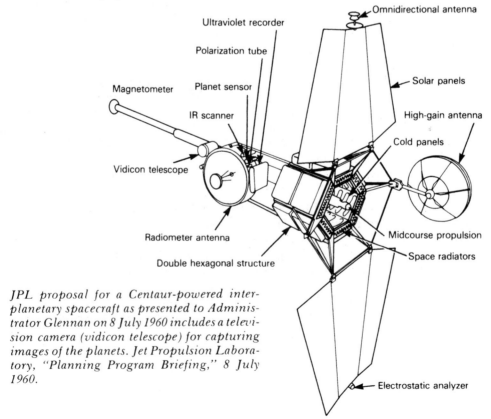

Omnidirectional antenna

Ultraviolet recorder

Polarization tube

Magnetometer    Planet sensor

Solar panels

IR scanner

High-gain antenna

Cold panels

Vidicon telescope

Radiometer antenna

Midcourse propulsion

Double hexagonal structure

Space radiators

*JPL proposal for a Centaur-powered interplanetary spacecraft as presented to Administrator Glennan on 8 July 1960 includes a television camera (vidicon telescope) for capturing images of the planets. Jet Propulsion Laboratory, "Planning Program Briefing," 8 July 1960.*

Electrostatic analyzer

Plans for Voyager called for a 1080-kilogram spacecraft with a several-hundred-kilogram capsule capable of surviving atmospheric entry and descent to the planetary surface. Among the technological accomplishments required before Voyager could fly in 1967 were: "(a) approach guidance which will place the spacecraft in desired relation with respect to the planet; (b) techniques for aerodynamic entry into a planetary atmosphere; and (c) propulsion systems for the addition of the relatively large velocity increments required by the planetary orbiters."[24] But between Mariner A and Voyager lay the largely undefined Mariner B.

Gates and his associates looked at four basic missions to determine the best way to bridge that technological gap. First was a proposal for a Mars flyby and return mission. While passing by the planet, the spacecraft would collect photographs and other scientific information and then return to Earth where a reentry package would be recovered, complete with developed photographs. The Instrumentation Laboratory at the Massachusetts Institute of Technology had studied such a planetary mission for the Air Force in 1958–1959,[25] and the Air Force had successfully recovered a 38-kilogram data capsule from the Earth-orbiting *Discoverer 13* on 10 August 1960, proving the recovery concept feasible. To the JPL planners, however, such a mission was "unattractive"; the quality and quantity of data that could be transmitted electronically to Earth from Mars was "entirely adequate."

A second mission under consideration was a flyby with more instrumentation than on Mariner A. Since this project seemed repetitive, something had to be done to improve its appeal. An approach guidance system would enable the craft to pass closer to Mars but would also increase the demands placed on the communications and power capabilities, which in turn would add unwanted weight. All additions to the weight of the basic craft would subtract from the scientific payload, but tradeoffs between different elements of the spacecraft became the norm.

A planetary orbiter was the third suggestion, but it would require a major new element, a retromaneuver package. Once a spacecraft reaches that point in its flight where the gravity of the target planet begins to attract it, a retrorocket must be fired to slow its speed so that it can go into orbit. Even if this equipment were available in time, its weight would probably increase the total beyond the predicted capability of Atlas-Centaur. Guidance technology necessary for such an orbital mission was another uncertainty.

A lander mission, the fourth consideration, would also require advanced propulsion and guidance technology that would not be ready by the early 1960s. Two other problems with a lander mission were protecting scientific instruments during entry into the Martian atmosphere and developing a communications link to operate from the Martian surface.

After studying the four missions, Gates and his colleagues made three suggestions:

One might (a) place the main body of the spacecraft in orbit around the planet and subsequently direct a small capsule to enter the atmosphere

37

and land upon the surface; (b) the main body of the spacecraft might be directed to go by the planet and place a small capsule in orbit about the planet; or (c) the main body of the spacecraft might be directed to fly by the planet and send a small capsule to enter the atmosphere of the planet and land upon its surface.

Of these, the JPL planners considered the flyby with capsule the most promising. An orbiter-lander capsule mission was too ambitious technically, and a flyby with orbiting capsule would produce no data beyond that obtained from a flyby. The split capsule concept was the most attractive proposal, and it became the basis for the first missions that would employ Mariner B spacecraft.[26]

While the staff at JPL had been studying Mariner B proposals, Wernher von Braun's Army missile group based at Huntsville, Alabama, had become part of NASA. Designated the George C. Marshall Space Flight Center effective 1 July 1960, the new center was to oversee the development of NASA's large launch vehicles. Colonel Seaberg subsequently reported directly to Hans Hueter, director of the Light and Medium Vehicle Office, as Centaur was also shifted from Air Force management to Marshall control.

In midsummer 1960, there was considerable confidence within NASA that Centaur could be made to work, and the Centaur Project Technical Team requested the purchase of four more Centaur stages beyond the six already on order. Later that year, however, Atlas-Centaur began giving the NASA team problems.[27] During the first test of Centaur's dual engines at Pratt & Whitney's Sycamore Canyon facility near San Diego in November, a procedural error by test personnel led to the ignition of only one engine. Unignited propellant from the second exploded, damaging both engines.[28] Only after two more explosions in January 1961 was the cause of the faulty ignition understood and the problem corrected.[29]

The explosions delayed the scheduled June test flight of Centaur until December, and all NASA and Department of Defense projects tied to Atlas-Centaur were also affected. The predicted payload capacity of the first Centaur was lowered as well. On 17 January, Edgar M. Cortright, assistant director for lunar and planetary programs, in response to the new limitations, recommended that "Surveyor and Mariner B missions . . . be reshaped to fit the expected Centaur performance but in such a way as to have growth capability." While the design of the two spacecraft was being scaled down to meet Centaur's reduced lift capacity, NASA Headquarters and JPL, during the winter months of 1961, began to worry about the 1962 Mariner A mission to Venus. The revised Centaur launch schedule seemed to rule out such a flight (table 5). Alternative missions would have to be devised for 1962, but NASA still hoped to use Mariner A for Venus flights in 1964 and 1965, reserving Mariner B for Mars investigations.[30]

Meanwhile, NASA and Space Technology Laboratories examined Able M, an Able upper stage that could be used with Atlas. Originally developed for lunar missions, Able was considered briefly in 1961 as a backup for a Mariner A flight.

By the second week of August, it was generally recognized that Centaur would not be ready in time for a 1962 launch to Venus.[31] Consequently, Oran W. Nicks of headquarters and Daniel Schneiderman of JPL got together to discuss their mutual problem. Nicks was fully informed on the status of Centaur, and Schneiderman had a detailed knowledge of Ranger. Together they became convinced that JPL's earlier proposal for an austere spacecraft built on the Ranger chassis deserved another look. "As the result of the optimism generated by Schneiderman during the discussion," Nicks approved JPL's study of an Atlas-Agena for such a mission.[32]

*Table 5*
*Centaur Launch Schedule as Modified in January 1961*

| Vehicle | Date | Mission | Orbit | Payload (kg) | |
|---------|------|---------|-------|--------------|---|
| 1 | Dec. 1961 | Vehicle test | — | | — |
| 2 | June 1962 | Vehicle test | — | | — |
| 3 | Oct. 1962 | Vehicle test | — | | — |
| 4 | Dec. 1962 | Vehicle test | 24-hr, 30° | | 45 |
| 5 | Feb. 1963 | Vehicle test | 24-hr, 30° | | 113 |
| 6 | Apr. 1963 | Vehicle test | 24-hr, 30° | | 113 |
| 7 | June 1963 | Vehicle test | Escape | Surveyor, | 340 |
| 8 | Aug. 1963 | Vehicle test | 24-hr, 30° | Advent, | 299 |
| 9 | Sept. 1963 | Spacecraft | Escape | Mariner, | 544 |
| 10 | Oct. 1963 | Vehicle test | 24-hr, 30° | Advent, | 299 |
| 11 | Nov. 1963 | Spacecraft | Escape | Surveyor, | 340+ |
| 12 | Dec. 1963 | Vehicle test | 24-hr, 30° | Advent, | 299 |
| 13 | Feb. 1964 | Mariner | Venus | | 544 |
| 14 | Feb. 1964 | Mariner | Venus | | 544 |
| 15 | Mar. 1964 | Advent | 24-hr equatorial | | 227 |
| 16 | Apr. 1964 | Surveyor | Lunar landing | | 952 |
| 17 | May 1964 | Advent | 24-hr equatorial | | 227 |
| 18 | June 1964 | Surveyor | Lunar landing | | 952 |
| 19 | July 1964 | Advent | 24-hr equatorial | | 227 |
| 20 | Aug. 1964 | Surveyor | Lunar landing | | 952 |
| 21 | Sept. 1964 | Advent | 24-hr equatorial | | 227 |
| 22 | Oct. 1964 | Mariner | Mars | | 635 (?) |
| 23 | Nov. 1964 | Mariner | Mars | | 635 (?) |
| 24 | Dec. 1964 | Surveyor | Lunar orbit | | 726 |

As revised 17 Jan. 1961, the Atlas-Centaur launch vehicle would have six test flights before a Surveyor lunar landing was attempted in June 1963. That mission would have been followed by a DoD Advent communications satellite launch and then a Mariner planetary flight. Planned as further tests of Centaur, these missions would have carried scientific payloads.
SOURCE: Edgar M. Cortright to Thomas F. Dixon, "Recommendations on the Centaur Program," 17 Jan. 1961.

In its political desire to beat the Soviet Union to a planetary shot, the United States wanted to launch probes to the planets in 1962 if at all possible and chose Venus as the most likely target, since flights to Earth's closest neighbor would require less powerful rockets. On 28 August 1961, JPL proposed a 1962 Venus mission based on an Atlas-Agena launch vehicle, using a hybrid spacecraft that combined features of JPL's lunar Ranger and Mariner A. This proposed spacecraft, called Mariner R, could carry about 11 kilograms of instruments. The 1962 project would not have a significant influence on the schedule for lunar Rangers, but a reallocation of launch vehicles would be required.[33]

*A Successful Flyby Mission*

On 30 August 1961, the NASA Office of Space Flight Development took three actions: it approved Mariner R, canceled Mariner A, and directed JPL to prepare Mariner B for a Centaur flight in 1964 to either Mars or Venus. In less than 11 months, the lab personnel designed, developed, procured, and modified components for, fabricated, tested, and launched two Ranger-derived Mariner R spacecraft. Trajectory calculations, launch operations, mission design, and ground support facilities also had to be readied on a crash schedule as launches were set for 22 July and 27 August 1962. The first spacecraft was destroyed by the range safety officer less than five minutes after launch when the Atlas stage became erratic. Quick measures corrected the launch vehicle checkout procedures and the computer's guidance program, allowing the second attempt to proceed as planned. On schedule at 2:53 a.m., Mariner R-2 rose from its pad at Cape Canaveral. For a few moments, new guidance troubles with Atlas intimated yet another failure, but the ground crew overcame the malfunction in time for the separation of the Agena stage. *Mariner 2* was off on a long and successful journey to Venus.[34]

Success was sorely needed. The first three Ranger missions had been outright failures, and *Ranger 4* had crashed uncontrolled onto the far side of the moon on 26 April 1962, returning no useful data. *Mariner 2*'s successful journey blunted the mounting criticism of the unmanned lunar and planetary program and took some of the bite out of the NASA-JPL investigation of Ranger shortcomings. At a 14 December *Mariner 2* press conference in Washington, the NASA administrator declared the flyby "an outstanding first in space for this country and for the free world. . . ." Despite the space-race jargon, he was correct: *Mariner 2* was "the most significant and perhaps the most spectacular of our scientific efforts to date."[35]

Telemetered signals transmitted a large quantity of scientific and engineering data from the Mariner spacecraft for 130 days. During that time, the probe reported on the interplanetary environment, supplied data on Venus as it flew past on 14 December, and relayed additional information on outer space until radio contact was lost on 3 January. During its lifetime, *Mariner 2* provided intelligence almost continuously on magnetic fields, cosmic dust, charged particles, and solar plasma. In addition, the infrared radiometers scanned the surface of Venus for 42 minutes when the spacecraft flew by at a distance of 35 000 kilometers, finding average temperatures to be about 415°C. The extremely high temperatures and an obscuring atmosphere did not make Venus a likely locale for extraterrestrial life, and exobiologists began to consider the Red Planet a more desirable target for their search.[36]

While *Mariner 2* was readied for its flight to Venus, the Centaur team continued to have difficulties that led to additional schedule slips. On 9 April 1962, NASA Headquarters once again revised Mariner plans. The B mission with its soft-landing capsule was postponed until the 1964 Mars launch opportunity, and the 1964 Venus mission became another Mariner R flight.[37]

### Scientific Organization and Payloads for Mars

Mariner B required the development of two kinds of experiments—those that would be carried on the flyby bus and those that would be landed on the planet's surface—but NASA had no general procedure for selecting scientific experiments for its missions. In April 1960, the Space Sciences Steering Committee was formed to bring together all the key people within the agency who had an interest in the space sciences. Reporting directly to Abe Silverstein, the committee, chaired by Homer Newell, recommended which projects should be undertaken and established working relations with outside scientists by forming a series of subcommittees. Headed by NASA personnel, these subcommittees had members and consultants from the scientific establishment, especially those associated with the Space Science Board of the National Academy of Sciences. By February 1961, there were seven discipline subcommittees—aeronomy, astronomy, bioscience, ionospheric physics, lunar science, particles and fields, and planetary and interplanetary science.[38]

Once the Space Sciences Steering Committee was in operation, Newell had some control over the advice that was given the agency about the kinds of missions it should fly. Thus, early in March 1961 he wrote Hugh Odishaw of the Space Science Board asking for suggestions for Mariner B experiments. Newell told Odishaw that present plans called for a planetary flyby and a planetary entry capsule. The main craft would come within 11 000–16 000 kilometers of Mars. If the mission was flown without the landing capsule, the probe could carry about 80 kilograms of scientific instruments. If an entry package was flown, instruments weighing about 23 kilograms could be landed, but it was uncertain how much weight the flyby half could support. Newell asked the Space Science Board to review "this problem and suggest a list of appropriate experiments."[39]

Odishaw responded with a report from several committees on 31 March. While the short notice prohibited an exhaustive reply, Odishaw noted that Mars missions had two desirable objectives—the study of the planet itself and the study of the interplanetary medium. Board scientists gave priority to "photographing the planet, determining atmospheric composition and conducting simple investigations of surface properties." And spacecraft experiments at the flyby distances should include study of the Martian magnetic field, radiation, aurora, airglow, and the like. After five days of briefings and discussions at JPL, the Space Science Board's Planetary Atmospheres Study Group developed a specific list of experiments for a lander mission:

*Spacecraft flyby*

> Radiation package
>
> Cosmic dust package
>
> Photographic equipment (1-km resolution)
>
> Magnetometer
>
> Infrared spectrometer
>
> Ultraviolet spectrometer

*Capsule*

> Television
>
> Temperature and pressure-measuring equipment operative during descent
>
> Radar altimeter
>
> Mass spectrometer
>
> Gas chromatograph

Odishaw added that it was "gratifying to note that the experiments planned by JPL for the Mariner B mission followed closely those recommended in the first interim report of the board's Committee on the Chemistry of Space and Exploration of the Moon and Planets, which was provided to NASA on February 1, 1959." The Space Science Board scientists, less enthusiastic about a probe that would study only the space between Earth and Mars, did recommend experiments for such a mission, but they clearly believed priority should be given the capsule-lander project.[40]

The summer of 1961 passed quickly, with planetary and unmanned space exploration taking a backseat to the accelerated manned lunar project Apollo. Yuri Gagarin's 12 April 1961 orbital mission galvanized American determination as the Soviet Union once again took the lead in space. On 26 May 1961, President Kennedy urged a joint session of Congress to commit the nation to landing and returning a manned expedition to the moon by the end of the 1960s.[41] Despite a sympathetic understanding of the plight of the space science community, Administrator James E. Webb, Glennan's successor, ordered the space agency's priorities to reflect the new national interest in reaching the moon. This change led to a reorganization of the agency.[42]

In 1961, "momentous decisions on both program and administrative matters [were] made in quick succession" at NASA, two of which left a lasting mark on the agency, as one historian put it. "One was the decision to strengthen NASA's general management by greatly strengthening the staff of the Associate Administrator, the other was the decision to reorganize NASA as a whole." The changes were effective 1 November 1961.[43]

Establishing an independent Office of Space Sciences under Homer Newell's direction was the key change for the unmanned planetary program (see chart in appendix G). Edgar Cortright became Newell's deputy, while Oran Nicks was named director of lunar and planetary sciences. Nick's organization included Charles P. Sonett, chief of lunar and planetary sciences, and N. William Cunningham, Fred Kochendorfer, and Benjamin Milwitsky, chiefs of Ranger, Mariner, and Surveyor offices. Orr E. Reynolds became director of the Bioscience Program Office, with Freeman H. Quimby serving as his chief of exobiology programs. Colonel D.H. Heaton began directing the Launch Vehicle and Propulsion Programs Office, with Commander W. Schubert and D. L. Forsythe as chiefs of the Centaur and Agena launch vehicle offices. This team would guide the lunar and planetary program until the next reorganization two years later.

During October and November 1961, Ford's Aeronutronic Division began work on a preliminary design for a Mariner B landing capsule as NASA personnel began examining tentative experiments for the spacecraft and capsule. From 64 original proposals, 8 experiments were chosen for the flyby bus and 10 for the capsule.[44] Changes in this payload were quick in coming, however. On 19 February 1962, Sonett informed Nicks that a cutback in Centaur payload weight, due to Defense Department changes associated with its Advent satellite, forced his staff to review again the list of proposed Mariner B experiments. Investigators had already been warned by Newell that their proposed scientific payloads would be subject to limitations placed on the overall payload by engineering constraints. "It now appears that we will have to exercise our options to hold off some of these people," Sonett wrote. "We intend to fund them, wherever possible, for backup research so as not to put them out of the program entirely."[45]

On 4 May 1962, Newell wrote the investigators whose experiments were being dropped. Power, telemetry, and weight considerations had become "critical due to factors connected with booster capability and spacecraft design. . . . In view of these conditions, the successful entry of the capsule into the Mars atmosphere hinges upon the restriction to very light, simple instrumentation and direct transmission to Earth rather than by use of a capsule-bus telemetry system." Most unfortunately, the limitations on capsule performance would apparently confine the landed experiments "to

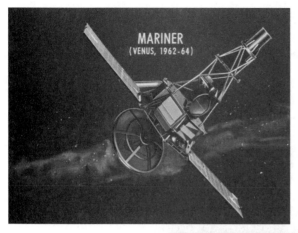

MARINER
(VENUS, 1962-64)

*Artist conceptions of Mariner space-craft were shown on slides in early 1962 Office of Space Sciences brief-ings on progress of the planetary program. Fabrication of Mariner R was scheduled for early 1962 com-pletion and design of Mariner B for mid-1962, with completed prototype in mid-1963. Voyager design and development was to begin in mid-1962.*

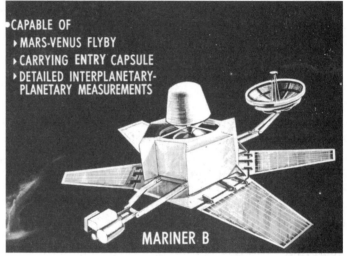

- CAPABLE OF
  - MARS-VENUS FLYBY
  - CARRYING ENTRY CAPSULE
  - DETAILED INTERPLANETARY-PLANETARY MEASUREMENTS

MARINER B

those intended to investigate the question of life and atmospheric composition." Nevertheless, NASA intended to develop a basic capsule design that would be flexible enough to permit investigators to fly more sophisticated experiments on subsequent missions to Mars.[46]

The uncertainty surrounding Centaur, both as to schedule and lift capacity, threw plans for Mariner B into a tailspin. The 1963 Mariner B test flight and 1964 Venus mission were canceled, and a 1964 test of the Mars version was added:[*][47]

| P[robe]-37 | Mariner R [*Mariner 1*] | 1962 Venus Mission |
| P-38 | Mariner R [*Mariner 2*] | 1962 Venus Mission |

*Because of some congressional confusion over the use of such terminology as Ranger A, Surveyor B, Mariner R, and the like, Nicks suggested that all published NASA documents use a clearer system—Ranger Lander, Mariner Mars (year), Surveyor Orbiter, etc. This nomenclature was adopted in materials intended for external use, but internally NASA continued to use the briefer alphabetical designations.

| P-40 | Mariner R | 1964 Venus Mission |
|------|-----------|--------------------|
| P-41 | Mariner R | 1964 Venus Mission |
| P-39 | Mariner B | 1st quarter 1964 test flight |
| P-70 | Mariner B | 1964 Mars Mission |
| P-71 | Mariner B | 1964 Mars Mission |
| P-72 | Mariner B | 2nd quarter 1965 test flight |
| P-73 | Mariner B | 1965 Venus Mission |
| P-74 | Mariner B | 1965 Venus Mission |

Continued problems with Centaur forced additional adjustments to the proposed Mariner timetable.[48] After 10 postponements of the first Atlas-Centaur launch, NASA tried again on 8 May 1962. Fifty-six seconds after liftoff, the vehicle exploded, and a week later the House Committee on Science and Astronautics began hearings to examine this troubled launch vehicle program. By late summer, the Office of Space Sciences—Nicks, Cortright, and Newell—had decided not to rely on Centaur for a 1964 Mariner B flight to Mars. Instead, they planned to use Atlas-Centaur in 1965 to send a B-class spacecraft to Venus, if the launch vehicle was ready then. The 1964 Mars B mission would be replaced by an Atlas-Agena-launched, lightweight spacecraft called Mariner C.[49]

During the fall of 1962, NASA personnel tackled various launch vehicle problems and studied their impact on the lunar and planetary probe program. On 7 September, 28 representatives from NASA Headquarters, Goddard Space Flight Center, and JPL met in Washington to take a new look at the relative merits of the proposed missions for the exploration of Mars during 1964. As they reviewed Mariners A, B, and R—their schedules, plans, and difficulties —Oran Nicks pointed to the problems with Centaur that had necessitated their using a spacecraft lighter than Mariner B. William G. Stroud, chief of the Aeronomy and Meteorology Division at Goddard, outlined his center's proposal for a planetary mission with a 210-kilogram spacecraft launched by an Atlas-Agena-Able. Stroud had in mind a hard lander equipped to measure the temperature, pressure, and composition of the Martian atmosphere and to detect life. Goddard's plan called for two launches in 1964 and three in 1965. In his turn, Robert Parks, now JPL's planetary program director, reviewed the lab's 1964 Mars proposal to send a 338-kilogram spacecraft launched by an Atlas-Agena on a flyby photographic mission. Similar in concept to the Venus Mariner R mission, the Mars flight would carry a television camera and an infrared spectrometer designed to detect organic molecules of the type produced by vegetation.

In the long sessions that followed these opening presentations, the specialists reviewed a number of important issues. Some of the major technical questions concerning the Goddard plan included: 1. Was it feasible to sterilize the capsule so that it would not contaminate the Martian environment? 2. Was the single 64-meter antenna to be built at Goldstone,

California, sufficient for communications with a capsule on Mars? 3. Could existing command and guidance systems provide the necessary accuracy needed to land a capsule? 4. Would a single biological experiment provide meaningful results? The JPL proposal also was scrutinized: 1. Was existing tape recorder technology adequate for storing and relaying television picture signals to Earth? 2. Could the infrared detector and its related filters be protected against long exposure to space environment? In studying these questions, it became obvious that the detection of life, whether by a landed detector or television pictures taken as the spacecraft flew past the planet, was a predominant theme of both proposals.[50]

Parks wrote to Nicks 13 days after their Washington meeting, "One point about which we all seem to be sincerely convinced is the . . . importance of the biology of Mars." This conviction had been reinforced from many scientific quarters, including the 1962 Iowa Summer Study Group sponsored by the Space Science Board of the National Academy of Sciences. This body enthusiastically supported the search for extraterrestrial life. Parks noted:

> Although the chances (1) that life does exist on Mars and (2) that importing earth life forms would distort or contaminate the study of Mars life (if it does exist) are both admittedly not great, it does appear quite important that we not take undue chances in this regard. The cost of not taking this chance is small. The only thing to be lost is a possible delay in obtaining the information relative to the basic physical information about the solar system that can be obtained only, or most quickly, by landings on Mars. The answers to a great many of these basic physics questions can be learned by measurements in interplanetary space, by flyby and landing measurements of Venus, and flyby measurements of Mars.[51]

Once having made clear his preference for an early flyby to Mars rather than a lander, Parks, like others concerned over the Russian challenge, suggested that NASA's Mars strategy would probably be influenced by the competition from the USSR. He wondered if the Soviet Union was likely to send a spacecraft to Mars that would contaminate the surface even though the USSR had indicated that it also had plans for sterilization. If it did land a spacecraft, was it likely to "scoop us in obtaining Mars biology data?" Though Parks believed that the Soviet Union might well risk contaminating Mars, he did not believe that would justify NASA's taking such a chance as well.* The state of the Soviet "scientific instruments and long range communications is behind ours and gives us a definite advantage in making these difficult and delicate measurements." Even if the United States did not land an instrumented package on Mars until much later, Parks determined that the U.S. could demonstrate its space exploration capabilities through flybys until a safe and sufficiently large lander could be developed.

*The Soviet Union launched its first spacecraft to Mars on 1 November 1962, but after traveling about 106 million kilometers the transmitters aboard *Mars 1* fell silent.

Some specific requirements had to be met before NASA attempted landing on Mars. In Park's view, total capsule sterilization was the first problem for designers at JPL. A second concern was for "well thought-out and well-tested biological instruments (the present state of development of biological sensing instruments for a planet is . . . considerably behind the requirement)." NASA would have to develop and thoroughly test an entry and landing capsule capable of carrying a number of biological and atmospheric experiments, in addition to the indispensable communications equipment. An approach and guidance control system was a fourth consideration. Also desirable was a communications link that used a flyby craft as a relay. Parks clearly favored flyby spacecraft on the first mission, to help find safe, biologically interesting landing sites for later missions. Many technical difficulties had to be resolved before landers could be sent to Mars and Venus. The people at Goddard, he contended, either did not understand the problem or were allowing enthusiasm to overshadow logic.[52] The JPL-Goddard dispute would continue for months, reflecting both a difference in approach to planetary exploration and a JPL concern over the Goddard staff's intrusion into what had been an exclusive preserve of the California laboratory. The continued problems with Centaur ultimately answered the flyby versus lander question.

Centaur was a genuine troublemaker for the Office of Space Sciences, since its two major projects, Surveyor and Mariner, were structured around it. The Centaur crisis came to a head at a mid-September 1962 meeting at the Marshall Space Flight Center. From the very beginning, Wernher von Braun and Marshall's top management had not favored Centaur and had accepted the project only reluctantly. Saturn was their primary mission. "Only a few crumbs which have fallen from the banquet table of thought and effort at MSFC have been given to Agena and Centaur," wrote the Agena program chief.[53] But beyond the problem of time and inadequate resources was von Braun's basic disagreement with the design approach of Centaur. Assigning Marshall the Centaur job had indeed been a serious error.

In September 1962, von Braun told Newell that the best lunar payload he could expect with the existing Centaur design was 810 kilograms. Projected Surveyor weights ranged from 1125 to 1260 kilograms, and similar weight problems would exist for Mariner B.[54] Von Braun wanted to cancel Centaur and use Saturn for Surveyor and Mariner and so recommended to the Senior Council of the Office of Space Sciences in August 1962. Brian O. Sparks, JPL deputy director, presented a similar recommendation to Newell on 13 September: "The performance schedule and funding problems associated with the Centaur program have finally reached the point where it appears that the Centaur vehicle will not be able to meet the requirements of the unmanned lunar and planetary programs of this country."[55] After reviewing all Centaur's technical faults, the team at JPL noted that the formally approved Centaur program "is totally intolerable, as it precludes any sensible Surveyor Project, completely obviates any timely contribution by Surveyor to the Apollo program and forces Mariner to continue indefinitely on Atlas-Agena with the attendant lack of confidence to achieve even minimal objectives."

This trend toward minimum goals should be reversed, JPL urged. "Rather than progressive reductions in spacecraft weight allowance during the development stage, a clear margin for weight increase is needed." Additional payload capacity could lead to enhanced spacecraft reliability through the use of redundant systems (a lesson learned from Ranger) and further hardware improvements, impossible with a smaller capacity launch vehicle. Greater reliability might also reduce the total number of launches required to achieve particular goals. Looking at all possible launch vehicle combinations, JPL specialists concluded that the Saturn C-1 combined with the Agena had several obvious advantages:

> (a) The C-1 development program appears to be on a sound basis and reasonably predictable. [The first Saturn C-1 test flight took place on 27 October 1961 (SA-1) and the second (SA-2) on 25 April 1962.]
> (b) Substantial performance margins above our minimum requirements can be confidently expected.
> (c) Substantial use of all stages is already programmed for other purposes.
> (d) No new stage development is required.
> (e) The resulting over-all funding requirements can be expected to be essentially the same as those now expected for the Centaur-based program.[56]

JPL planners anticipated that a Saturn-Agena could boost an 810-kilogram Mariner B, a significant increase over the 225–350 kilograms proposed for Mariner C. That meant "many of the current physical and weight constraints on these spacecraft [could] be relaxed, redundancy . . . added in key areas, and realistic mission flexibility . . . incorporated" into planetary space probes. Marshall could apparently ready the first planetary Saturn-Agena for a 1965 launch of Mariner B to Venus; a Mariner B mission to Mars on Saturn-Agena might also be feasible for 1966.[57]

NASA management in Washington—especially Homer Newell—reacted negatively to the suggestion that Centaur be replaced with Saturn-Agena. Instead, Newell concluded that Centaur needed a new home. At the end of September, the project was transferred to the Lewis Research Center in Cleveland, which had been under the direction of Abe Silverstein since November 1961.[58] "Although the Centaur development has been fraught with difficulties, many of them were of a management nature," Newell suggested. He admitted that the arguments advanced in favor of Saturn were attractive at face value, but "the development status of the Saturn was presented with somewhat disproportionate optimism, compared to the Centaur." Newell also believed that JPL critics were being overly optimistic since they were counting on the successful adaptation of an untested Saturn second stage and Agena stage "to provide an operationally suitable vehicle on a competitive time scale with Centaur." Nor was NASA's director of space sciences convinced that Saturn would be as economical as it had been portrayed. Newell and his associates were not ready to abandon Atlas-Centaur for a new steed.[59]

## A Review of Planetary Spacecraft for the 1960s

Although Centaur's future looked brighter at Lewis where Silverstein's enthusiasm was catching, the changes came too late for Mariner B, which was in jeopardy by the end of 1962. The longer Centaur was delayed the less likely it became that Mariner B would fly, especially since the next-generation spacecraft, Voyager, was being more precisely defined with each passing day. In December 1962, JPL informed headquarters that Mariner B–Centaur could not be launched in 1965 and proposed launching the mission in 1966 with Saturn-Agena. Oran Nicks wanted to continue the spacecraft's development with Atlas-Centaur, but he, too, noted that this would likely lead to technology that would not be used until the Voyager program. Perhaps, he suggested, a variation of the Mariner B capsule might be flown on the Voyager mission to Venus planned for 1967.[60]

More and more signs pointed to Mariner B's decline and Voyager's ascendancy. Independent Mariner B and Voyager programs would cost too much and, if Mariner B were flown, Voyager would surely be delayed, something no one at NASA wanted to see. In late December 1962, when Homer Newell asked Harry J. Goett, director of Goddard, for a plan for developing Mariner B's capsule, he requested that his specialists also consider possible Voyager applications for the hardware.[61]

At the outset of 1963, the proposed planetary science program consisted of three kinds of spacecraft. The first was Mariner C, the pared-down craft without a lander, which would be launched by Atlas-Agena, fly by Mars, and make a series of measurements, relaying them along with television images back to Earth. Uncertainty plagued Mariner B, the second spacecraft. It had been restructured and reoriented to take advantage of the 1966 Mars launch opportunity and, with a landing capsule, was to be launched by either Atlas-Centaur or Saturn-Agena. Third was the more ambitious Voyager, which was to send combination orbiter and lander spacecraft to Venus and Mars. The most likely time for Voyager's first flight was the 1967 Venus launch window. But the planetary program was to take some twists and turns that would alter the original plans. Mariner C, the 1964 Mars mission, would take on a vitality and distinct direction of its own. Mariner B would become a long-term project, transformed into a mission called Mariner Mars 66, inextricably entwined in the evolution of Voyager. Above all else, 1963 was to be the year in which Voyager, at least on paper, got off the ground.[62]

NASA learned some valuable lessons from Mariner B. First, it had been too ambitious for its time, representing too large a technological jump. The 1962 Venus flight and the revised 1964 mission to Mars made more sense, for they built upon the lunar experiences of Ranger. Second, launch vehicles

49

would continue to make advanced planning a chancy business at best, and launch scheduling would become nearly impossible. Atlas-Centaur would fly successfully only near the end of 1963. Then six more flights would be made before Centaur was considered operational and ready for the 30 May 1966 launch of Surveyor to the moon. No one within NASA had anticipated such delays when planetary flights with Centaur were first proposed in the early 1960s.[63]

Not all the Mariner B experiences had negative overtones, however. Mariner B gave the space agency and prospective experimenters an opportunity to define the investigations that could and should be performed on Mars, and variations of several of the experiments proposed in October 1961 would fly on later Mariners and ultimately on the Viking missions. Mariner B also forced the early study of such basic questions as spacecraft sterilization and aerodynamic entry into planetary atmospheres. Looking toward a 1964 landing mission, NASA seriously examined these topics much earlier than it might have otherwise, which was fortunate, because both entry and sterilization were extremely complex. Finally, Mariner B sparked theoretical and practical design work on devices for the detection of extraterrestrial life by scientists and engineers who were excited and challenged by the prospective search for life on Mars.

# 3

# The Search for Martian Life Begins: 1959–1965

The search for extraterrestrial life was a direct by-product of 20th century biochemists' quest for the origins of life on Earth. Instruments proposed by scientists to determine if there were detectable life forms or the organic matter necessary for such life forms to exist elsewhere in the solar system were based on the assumption that the laws governing the evolution of life on Earth are universally applicable, as are the laws of physics. When the Viking spacecraft was launched to Mars in 1975, they carried three biological experiments and a gas chromatograph–mass spectrometer, instruments with an intellectual and technological history reaching back to the early days of American space science. In fact, the development of life-detecting devices predates the availability of both spacecraft and launch vehicles.

As with many aspects of modern biology, the search for extraterrestrial life begins with Charles Darwin. His classic work, *On the Origin of Species* (1859), sparked considerable discussion of evolution, but it also led to speculation over the original source of life. In the 1860s, Louis Pasteur concluded that the spontaneous generation of microbes was not possible; all life on Earth came from preexisting life. What was the origin of those life forms? The Darwinian theory led many scientists to believe that the multiplicity of plant and animal species had a common source. In an 1871 letter, Darwin suggested that perhaps Earth's atmosphere, too, had evolved.

> It is often said that all the conditions for the first production of a living organism are now present, which could ever have been present. But if (and oh! what a big if!) we would conceive in some warm little pond, with all sorts of ammonia and phosphoric salts, light, heat, electricity, etc., present, that a protein compound was chemically formed ready to undergo still more complex changes, at the present day such matter would be instantly devoured or absorbed, which would not have been the case before living creatures were formed.[1]

In his speculation, Darwin rejected the premise that Earth's environment had always been static.

Most scientists disagreed with the theory that life on Earth had its beginnings in a prebiotic environment, until the idea was simultaneously revived in the 1920s by two biochemists, J. B. S. Haldane of Great Britain and Aleksandr Ivanovich Oparin of the Soviet Union. Haldane and Oparin independently asserted that although it was very unlikely for life forms or organic molecules to appear abiologically in an oxygen-rich atmosphere, such compounds could have appeared millions of years ago in a very different environment. They postulated that in a prebiotic, sterile era organic compounds of ever-increasing complexity accumulated in the seas and eventually by random combinations produced a living molecule. On the nature of that prehistoric atmosphere, Haldane and Oparin disagreed.

Haldane favored a combination of ammonia, carbon dioxide, water vapor, and little or no oxygen. Organic compounds were synthesized by energy from ultraviolet light. Gradually the evolutionary process produced more complex molecules capable of self-duplication. Oparin's primordial atmosphere consisted of methane, ammonia, water vapor, and hydrogen. According to his theory, an abundance of organic compounds in the seas, given enough time, would permit the formation of organic molecules that would be the foundation for yet more complex life forms. Despite their work, most other biochemists through the 1940s insisted on attempting to synthesize organic compounds in oxygen-rich environments. In the 1950s, the focus shifted to the production of amino acids.

As with improved astronomical instruments, new biochemical techniques, such as paper chromatography,* opened new doors. One door led to the study of amino acids, the building blocks of protein. Biochemists believed that amino acids might hold clues to the origin of life, since primeval forms of life were assumedly protein-centered. Melvin Calvin commented on the logic behind these early studies: "We had every reason to suppose that the primitive Earth had on its surface organic molecules." If one went further and postulated a "reducing," or oxygen-poor atmosphere, "most of the carbon was very largely in the form of methane or carbon monoxide, . . . the nitrogen was mostly in the form of ammonia, there was lots of hydrogen, and oxygen was all . . . in the form of water." Given these simple molecules, was it possible to create more complex ones in the laboratory? Calvin and several other scientists began to experiment with reduced atmospheres containing primarily carbon compounds.[2]

Stanley L. Miller, while pursuing his doctoral studies at the University of Chicago, was the first to produce amino acids in a reducing atmosphere. Working under Harold Urey, he developed a closed-system apparatus into which he introduced a mixture of methane, ammonia, water, and hydrogen. When subjected to a high-frequency spark for a week, milligram-quantities of glycine, alanine, and alpha-amino-*n*-butyric acid were pro-

---

*The process of separating a solution of closely related compounds by allowing a solution to seep through an absorbent paper so that each compound becomes absorbed in a separate zone.

duced. Apparently, he was on the right track. Miller reported his early results in *Science* magazine in May 1953.[3] Norman Horowitz, a biologist from the California Institute of Technology, commented: "This experiment on organic synthesis in simulated primitive earth atmosphere is the most convincing of all the experiments that have been done in this field."[4]

Six years later Miller and Urey reported further on the implications of their research. The absence of hydrogen in Earth's present atmosphere was a clue. They had begun their study assuming that cosmic dust clouds, from which presumably the planets had been formed, contained a great excess of hydrogen. "The planets Jupiter, Saturn, Uranus, and Neptune are known to have atmospheres of methane and ammonia," they noted, similar to primitive Earth's atmosphere. Given the lower temperatures and higher gravitational fields of these outer planets, time had not been sufficient for the excess hydrogen to escape. Miller and Urey held that Earth and the inner planets had "also started out with reducing atmospheres and that these atmospheres became oxydizing, due to the escape of hydrogen." Their production of amino acids in the laboratory indicated that before the development of an oxygen-rich atmosphere (the result of biological activity), the primitive environment was conducive to the formation of many different complex organic compounds. As soon as oxygen began to replace the hydrogen, experiments indicated that the spontaneous production of those compounds (amino acids) ceased.[5]

Miller's experience in the laboratory spurred further research, and with it speculation reappeared about the presence of life on other planets. As Miller and Urey pointed out in 1959, living matter does not require oxygen to grow and flourish; it was "possible for life to exist on the earth and grow actively at temperatures ranging from 0°C, or perhaps a little lower, to about 70°C. . . . Only Mars, Earth, and Venus conform to the general requirements so far as temperatures are concerned."[6] Because of the opacity of the heavy clouds on Venus, little could be deduced about the planet. Mars, on the other hand, had a clear atmosphere. Seasonal changes observed on the Martian surface suggested the possibility of vegetation.

The Red Planet became very important to the scientists searching for the origins of earthly forms of life. "If we find life on Mars, for example, and if we find that it is very similar to life on earth yet arose independently of terrestrial life, then we will be more convinced that our theories are right." Miller went on to argue:

> The atmosphere of Mars would have been reducing when this planet was first formed, and the same organic compounds would have been synthesized in its atmosphere. Provided there were sufficient time and appropriate conditions of temperature, it seems likely that life arose on this planet. This is one of the important reasons for the tremendous interest in finding out if living organisms are on Mars and why most of all we want to examine these organisms. We want to examine them in biochemical detail, and this would involve bringing a sample back to the

earth. What are the basic components of these organisms? Do they have proteins, nucleic acids, sugar? If they are completely different, then our theories about the primitive earth and the results of this experiment seem not at all convincing. If Martian organisms are identical to the earth's organisms in basic components, then there seems to be the possibility that some cross-contamination occurred between the earth and Mars. But, if Martian organisms have small but significant differences, then it would seem that theirs was probably an independent evolution, under the kind of conditions that we envision as those of the primitive earth.[7]

In 1959, Miller and Urey concluded, "Surely one of the most marvelous feats of the 20th-century would be the firm proof that life exists on another planet." They could have been addressing NASA when they added, "All the projected space flights and the high costs of such developments would be fully justified if they were able to establish the existence of life on either Mars or Venus."[8]

Especially significant for the search for extraterrestrial life were developments in the field of comparative biochemistry. Nobel-Prize-winning geneticist Joshua Lederberg told a Stockholm audience in the spring of 1959 that "comparative biochemistry has consummated the unification of biology revitalized by Darwin one hundred years ago." For many years, Lederberg noted, there had been a "pedagogic cleavage of academic biology from medical education." Lederberg cited two other specialists in the field in making his point: "Since Pasteur's startling discoveries of the important role played by microbes in human affairs, microbiology as a science has always suffered from its eminent practical applications. By far the majority of the microbiological studies were undertaken to answer questions connected with the well-being of mankind." By the late 1950s, however, research into the chemical and genetic aspects of the microbiological world led medical and biological investigators to realize that their work had much in common. "Throughout the living world we see a common set of structural units—amino acids, coenzymes, nucleins, carbohydrates and so forth—from which every organism builds itself. The same holds for the fundamental processes of biosynthesis and of energy metabolism."[9] This global perspective on the underlying unity of life on Earth, together with the common chemical origin of the planets, made it not unreasonable to postulate the possibility of life on other bodies in the solar system. Furthermore, the discovery of life elsewhere would give biological theory a long-sought universality. The origin of life studies and the work in comparative biochemistry formed the intellectual foundation that permitted respectable scientists to discuss the possible existence of extraterrestrial life.

## THE RISE OF EXOBIOLOGY AS A DISCIPLINE

As earth-bound biologists began to consider the existence—past or present—of life forms on other planets, two themes developed, detection and protection. How do you detect something whose nature and existence are unknown? How do you protect one planet from contamination by the biota of another? Detection and protection of life in the solar system were the subjects of considerable debate and investigation during the decade (1959-1968) that preceded the selection of biology experiments for Viking. Concern about possible contamination of other bodies by terrestrial organisms that might stow away aboard space probes got an impetus with the launch of Sputnik in 1957.

## Planetary Protection

Josh Lederberg was one of the first scientists to express publicly his worries about improperly sterilized spacecraft being the source of cosmic pollution. In 1961, he noted, "a corollary of interplanetary communication is the artificial dissemination of terrestrial life to new habitats."[10] His interest in planetary protection went back three years to the orbiting of *Sputnik 1*.

On his way back to the United States from a year as a Fulbright lecturer in Melbourne, Australia, Lederberg stopped to visit for a few days with Haldane, who was teaching in Calcutta. Lederberg recorded his recollections of a dinner party given on 6 November 1957, an evening on which another Soviet space spectacular seemed likely in celebration of the 40th anniversary of the Russian Revolution.

> The night of our arrival was the occasion of a lunar eclipse which was regarded as an important religious festival in Calcutta. It was also the occasion for a good deal of dinner table conversation. . . . Many members of the group were quite strongly pro-Soviet in their inclinations and they were almost gleeful at the prospect that the Soviet Union would follow up its October 4th triumph with another launch, perhaps even directed at the moon during the lunar eclipse. So, [we] even stayed up to see if there would be such a demonstration although we were well aware of the physical difficulties of arranging for something that could be visible from earth.* That occasion led me to think very sharply about the extent to which political motives would outweigh scientific ones in the further development of the space program. . . .[11]

When he returned to the Univeristy of Wisconsin where he was chairman of the medical genetics department, Lederberg circulated among the scientific community several editions of a memorandum expressing his concern over lunar and planetary contamination. His thoughts were subsequently formulated in a paper presented in May 1958 at the Satellite–Life Sciences Symposium, sponsored by the National Academy of Sciences, the

---

*Ironically, Jet Propulsion Laboratory proposed detonation of an atomic bomb on the lunar surface in response to the orbiting of Sputnik. William H. Pickering to Lee A. DuBridge, with summary of Red Socks proposal, 25 Oct. 1957, JPLHF 2-581.

American Institute of Biological Sciences, and the National Science Foundation, and in an article for *Science*.[12]

At the National Academy of Sciences, Lederberg's interest further stimulated concern over possible biological contamination in outer space. The Academy noted that improperly sterilized spacecraft might "compromise and make impossible forever after critical scientific experiments." Resolutions adopted in February 1958 by the Academy Council urged scientists "to plan lunar and planetary studies with great care" and called for the International Council of Scientific Unions "to encourage and assist the evaluation of such contamination and the development of means for its prevention." The Academy further intended to participate in the planning of "lunar or planetary experiments . . . so as to prevent contamination of celestial objects in a way that would impair the unique . . . scientific opportunities."[13]

An ad hoc Committee on Contamination by Extraterrestrial Exploration, formed by the International Council of Scientific Unions, met in May 1958 to draw up a code of conduct that would permit lunar and planetary exploration but at the same time prevent contamination. After being circulated throughout the scientific community, the proposed standards were adopted in October 1958. During the remaining months of 1958 and throughout 1959, the International Council of Scientific Unions' Committee on Space Research (COSPAR) and the U.S. Space Science Board continued to develop guidelines for the sterilization of space probes.[14]

The Space Science Board also expanded its activities into the field of life sciences in 1959 as the board members became interested in experiments that would investigate "the viability of terrestrial life forms under extraterrestrial conditions" and the implications of contamination.[15] The group's ad hoc committee on the subject, chaired by Lederberg, concluded that sterilization was technically feasible and that effective procedures could be developed, provided sufficient emphasis was given the problem. Toward that end, the Space Science Board sent suggestions to NASA and the Advanced Research Projects Agency on 14 September 1959. NASA Administrator Glennan assured the Space Science Board that the space agency had "adopted the general policy of sterilizing, to the extent technically feasible, all space probes intended to pass in the near vicinity of or impact upon the moon or planets."[16] Moreover, Abe Silverstein requested that JPL, Goddard Space Flight Center, and Space Technology Laboratories begin coordinated work on sterilization techniques.

While NASA Headquarters, its field centers, and contractors worked toward protecting the moon and planets from terrestrial microorganisms, the agency was studying more closely its participation in the life sciences.[17] To determine NASA's role in that field, Glennan established an ad hoc Bioscience Advisory Committee in July 1959. Chaired by Seymour S. Kety of the Public Health Service, the advisory board* reported 25 January 1960

---

*Other members included W. O. Fenn, D. R. Goddard, D. G. Marquis, R. S. Morison, C. T. Randt, and C. A. Tobias.

that life sciences had and would continue to have an important place in the American space program. The objectives of space research in this area were twofold—"(1) investigations of the effects of extraterrestrial environments on living organisms including the search for extraterrestrial life; (2) scientific and technologic advances related to manned space flight and exploration."[18] Kety and his colleagues also noted that existing space-related life-science activities were predominantly in applied medicine and applied biology. These activities were important, but support of more basic research in the biological, medical, and behavioral sciences was more crucial.

Besides supporting an Office of Life Sciences at NASA and arguing vigorously for the complete independence of life-science research from the military, the committee urged the space agency to search for extraterrestrial life on Mars. Kety and his colleagues recognized that a basic study of extraterrestrial environments would further man's understanding of the fundamental laws of nature. The origin of life and the possibility of its presence elsewhere in the universe were indeed challenging issues.

> For the first time in history, partial answers to these questions are within reach. Limited knowledge acquired over the past century concerning atmospheric and climatic conditions on other planets, the topographical and seasonal variety in color of the surface of Mars, the spectroscopic similarities . . . have suggested the presence of extraterrestrial environments suitable for life and permitted the formulation of hypotheses for the existence there of some forms of life at present or in the past.

The Kety committee believed that within the foreseeable future these hypotheses might be tested, indirectly at first by astronomical observations and by samplings taken mechanically from other planets, and finally by direct human exploration. The discovery of extraterrestrial life, or its absence, "will have important implications toward an ultimate understanding of biological phenomena."[19] Although these specialists believed that biological studies would "not be complete until the scientist himself is able to make meticulous investigations on the spot," they realized that manned missions to Mars belonged to the distant future.

As NASA went about establishing its Office of Life Sciences in the spring of 1960, the agency found itself with a 10-year plan that called for planetary missions in 1962 and 1964 and a recommendation from the Bioscience Advisory Committee to search for life. Given the scientific interest in Mars and the apparent feasibility of sending probes to that planet by the mid-1960s, it would have been difficult to argue against the idea. In August 1960, NASA authorized JPL to study spacecraft concepts for a mission to the Red Planet, a mission that would land a capsule on the surface and initiate the search for life beyond Earth. Although the Kety committee in 1959 and the Space Science Board's summer study at Iowa State University in 1962 both called for the biological investigation of Mars, a 1964 summer study sponsored by NASA and the Space Science Board was a further step in articulating the essential issues for exobiology as a field of inquiry.

*1964 Summer Study*

Professional biological interest in the search for life elsewhere in the universe had been growing for at least half a dozen years before the 1964 Summer Study gave exobiology the intellectual respectability needed to draw bright young scientists to the field. The "old-timers"—Lederberg, Colin Pittendrigh, and Wolf Vishniac, in their 30s and 40s—all had substantial and estimable careers in biology behind them before they launched into their quest for biota on Mars. Commenting on his early years in exobiology, Lederberg noted that his Nobel Prize for work on the genetics of bacteria had given him professional stability, which made it possible for him "to stay in a non-reputable game. Not disreputable, mind you, but non-reputable. It might have been very, very difficult otherwise and it would [have been] very hard for a capable young scientist who's had a lot of risks to take in his career to hitch it to something as uncertain as exobiology."[20] Gerald A. Soffen's experience is an example of the personal turmoil that could result from wishing to pursue the field of exobiology.

Jerry Soffen had begun his scientific career as a biologist. After earning a zoology undergraduate degree at the University of California in 1949, Soffen went on to study biology at the University of Southern California. Two books influenced the course of his subsequent career. One was A. I. Oparin's *The Origin of Life.* Soffen believed that Oparin was addressing himself to genesis—the origins of life, "the origins of me." Oparin's book started Soffen thinking about the beginnings of life, but Harold F. Blum's *Time's Arrow and Evolution* was even more influential. Blum's concept was simple and elegant—evolution conformed to the second law of thermodynamics. The universe's supply of energy is slowly diminishing, and all biological forms must adapt to lower, less satisfactory energy sources. Simple organisms present in a more primitive age when the oceans supplied them with a very rich nutrient broth had to develop more specialized and complex mechanisms for gathering energy (nutrients) as the ocean environment became less rich. Evolution is not a random process, since organisms must make orderly changes to survive in a changing world. This process leads to more complex, not simpler, organisms. Furthermore, organic evolution on Earth must be viewed as but a small part of the evolution of the entire universe.[21]

Soffen was so overwhelmed by the philosophical implications of Blum's work that he went to Princeton to do his doctoral work under Blum. During his doctoral studies, Soffen heard Stanley Miller summarize his investigations into the origins of life on Earth. As were many of his contemporaries, Soffen was taken by the brilliance and simplicity of Miller's theory. But the crucial factor for Soffen was the dawn of the era of spaceflight. Men could now reasonably talk of exploring the planets, and the search for life on other worlds was no longer just a dream. Soffen's interest in space exploration and the search for life on Mars brought him to another crossroads in his career while he was doing postdoctoral work at the New York State University School of Medicine in 1960.

Would Soffen pursue a safe, respectable career in biology studying mollusks, or would he gamble and undertake the study of exobiology, a new field not accepted as legitimate by many scientists? Soffen did not have fame or a Nobel Prize, as did Josh Lederberg, to give him academic security, and many professionals warned him against entering the new discipline. One physicist, Leo Szilard, told Soffen he was the wrong person from whom to seek advice. Instead, Soffen must ask himself what he wanted from life; no one else could decide the best course for him to follow. Soffen made his choice in 1961 when he joined the staff at the Jet Propulsion Laboratory, and he spent the next eight years managing the development of biological instruments, including exobiological detectors for spacecraft.[22] A wish to counter some of the professional risks associated with committing a career to exobiology was one of the reasons NASA convened the 1964 Summer Study at Stanford University.

After the usual staff work by Orr Reynold's Bioscience Programs Division, NASA got the summer study proposal moving by sending, in February 1964 over Homer Newell's signature, a letter to Chairman Harry Hess of the Space Science Board. Newell reminded Hess that "one of the prime assignments" of the space agency was "the search for extraterrestrial life," and he noted that the report of the Iowa City Summer Study of 1962 also described this undertaking as "the most exciting, challenging, and profound issue not only of the century but of the whole naturalistic move-ment."[23] There were those within and without the space science community who would question that priority, but even the most skeptical admitted that the discovery of life on a distant planet would have scientific, sociological, and theological implications of the first magnitude.[24]

Newell's letter set in motion a series of meetings between NASA and Space Science Board staff members. By mid-April, the board had readied its proposal for a summer study. Dean Colin Pittendrigh, professor of biology at Princeton, and Joshua Lederberg were appointed cochairmen of the study, and a distinguished group of scientists were named to the steering committee and the working group of participants for the June discussions (of the 37 persons who made up the core of the 1964 Summer Study, 9 would become key figures in the Viking Project). The summer meetings provided a much-needed forum where scientists could advise NASA as to what research they wanted the agency to support.

Some, Lederberg among them, had begun to worry about relations between the Space Science Board and NASA. Such sessions as the one in 1964 at Stanford were important decision-making exercises. But who would participate in such studies other than the interested and the enthusiastic, he mused? Thus, he viewed their reports as basically reputable, authoritative, and responsible endorsements, but also biased. While the views expressed that summer were generally those of proponents, the fact that they had been made publicly did achieve at least two things. First, the thinking of the participants who proposed a search for life on Mars had been sharpened, since their ideas were to be exposed to the critical evaluation of the larger scientific community. That is, those ideas became explicit targets for critical discussion. Second, the proposals had to be advanced in language that would permit broad discussion by legislators and laymen, as well. The study permitted NASA to discover how much scientific interest and support existed for the search for Martian life and to obtain the endorsement of the specialists for what the agency's advance planners wanted to do. Once a report with the Space Science Board–National Academy of Sciences imprimatur appeared, the space agency could move ahead.[25]

Those who participated in the 1964 Summer Study were believers and enthusiasts. Basic to their inquiries was a wish to know if life on Earth was unique. They could not prejudge the likelihood of life on other planets. While a speculation that it might exist was a relatively reasonable one, the biological community had no firm basis for assuming that other planets would be either fertile or barren. According to the 1964 summer conferees, "At stake in this uncertainty is nothing less than knowledge of our place in nature. It is the major reason why the sudden opportunity to explore a neighboring planet for life is so immensely important.[26]

Mars was a scientifically likely abode for life, the most Earthlike of all the planets. Although the Martian year was 687 days, the length of the day was "curiously similar to that of Earth, a fact that to a considerable degree ameliorates an otherwise very severe environment." The Red Planet had retained a tenuous atmosphere with surface pressures variously estimated from 10 to 80 millibars; the gaseous composition of that atmosphere was still a mystery in 1964. But scientists had concluded that oxygen was virtually nonexistent: "Oxygen has been sought but not detected; the sensitivity of measurement implies a proportion not greater than 0.1 per cent by volume." Water was also scarce. Water vapor had been measured spectroscopically with only traces detected in the atmosphere.

*Table 6*
*Physical Properties, Mars and Earth (1964)*

| Property | | Earth | Mars |
|---|---|---|---|
| Atmospheric pressure: | | 1000 millibars | 10–80 millibars |
| Gaseous composition: | oxygen | 20.00% | <00.1% |
| | carbon dioxide | .03% | 5–30% |
| | nitrogen | 78.00% | 60–95% |
| Water vapor: | | 3 g cm$^{-2}$ | 2x10$^{-3}$ g cm$^{-2}$ |

On Mars, surface temperatures overlapped the range on Earth. At some latitudes, daily highs of +30°C had been measured, and ranges of 100° within a 24-hour period were not unknown.[27]

But knowledge of the Martian surface had not progressed much beyond Lowell's observations at the beginning of the century. There was general agreement that the polar caps were frozen, but whether it was water or carbon dioxide was still a matter "of some controversy." Nor was there any understanding of a transport mechanism that could account for the seasonal alterations of the poles. "Our knowledge of what lies between the polar caps is limited to the distinction between the so-called 'dark' and 'bright' areas and their seasonal changes." The bright areas were generally believed to be deserts, with their "orange-ochre," or buff, appearance. The green color attributed to the darker regions was likely an optical illusion due to the contrast with the bright regions. Of biological interest were the seasonal changes in the dark areas. As was noted in the 1964 summer session report:

> In several respects they exhibit the kind of seasonal change one would expect were they due to the presence of organisms absent in the "bright" (desert) areas. In spring, the recession of the ice cap is accomplished by development of a dark collar at its border, and as the spring advances a wave of darkening proceeds through the dark areas toward the equator and, in fact, overshoots it 20° into the opposite hemisphere.[28]

The authors of *Biology and the Exploration of Mars* were quick to point out that the seasonal changes did not require the presence of living organisms. "Indeed, the question is whether the Martian environment could support life at all; and further, whether its history would have permitted the indigenous origin of life." Those were clearly two different questions.

One of the "more rewarding exercises" the summer study participants engaged in was the "challenge to construct a Martian ecology assuming the most adverse conditions indicated by present knowledge." That task posed no insuperable problems. Life forms could be conceived to exist with little or no oxygen. Some terrestrial organisms can survive freeze-thaw cycles of +30°C to -70°C. Others cope well with very low humidity, deriving their water supply metabolically. The intense ultraviolet radiation at the surface of Mars did not seem to be an insurmountable problem either, as some members of the study believed that organisms might exploit that radiation as an energy source. "The history of our own planet provides plenty of evidence that, once attained, living organization is capable of evolving adjustments to very extreme environments."[29]

Does life in fact exist on Mars?—this was a question of a different sort. That life forms *could* subsist on the planet was no kind of proof that life had actually emerged there. But the members of the study held that, "Given all the evidence presently available, we believe it entirely reasonable that Mars is inhabited with living organisms and that life independently originated there. However, it should be clearly recognized that our conclusion that the biological exploration of Mars will be a rewarding venture does not depend upon the hypothesis of Martian life." Two essential scientific questions should not be prejudged:

a. Is terrestrial life unique? The discovery of Martian life, whether extant or extinct, would provide an unequivocal answer.

b. What is the geochemical (and geophysical) history of an Earth-like planet undisturbed by living organisms? If we discover that Mars is sterile we may find answers to this alternative and highly significant question.[30]

## Scientific Aims of Martian Exploration

Having established that Mars was a worthy object of study, the summer study scholars addressed the precise aims of an investigation. "We approach the prospect of Martian exploration as evolutionary biologists." Whereas the emergence of organisms "was a chapter in the natural history of the Earth's surface," these scientists sought to test the generalized hypothesis that the evolution of life "is a probable event in the evolution of all planetary crusts that resemble" the Earth. Thus, they conceived the overall exploration of Mars "as a *systematic study of the evolution of the Martian surface and atmosphere* [italics in original text throughout unless noted otherwise]." Their aims in the summary were:

(1) determination of the physical and chemical conditions of the Martian surface as a potential environment for life,

(2) determination whether life is or has been present on Mars,

(3) determination of the characteristics of that life, if present, and

(4) investigation of the pattern of chemical evolution without life.[31]

As biologists, they had as "much interest as the planetary astronomers in a thorough study of the meteorology, geochemistry, geophysics and topography of Mars." Whatever the ultimate outcome of the search for life, its full meaning would be understood only within the broader context.

Four basic avenues of approach were suggested for the exploration of Mars, with the first three tasks ultimately leading to the fourth:

(a) laboratory work needed to develop techniques for planetary investigations and the knowledge needed to interpret their findings;

(b) Earth-bound astronomical studies of Mars;

(c) the use of spacecraft for the remote investigation of Mars; and

(d) a direct study of the Martian surface by landing missions.[32]

But by 1964, especially with the difficulties in planning Mariner B, it was apparent to all that defining lander payloads was a "complex and demanding task."

The planners needed more information about the structure of the Red Planet's atmosphere. Would parachutes work? Would retrorockets be necessary? They hoped *Mariner 3* and *4*, scheduled for launch in November 1964, would provide some answers on which spacecraft designers could base their plans. But even if complete knowledge for safely landing an instrumented package existed, the "principal design difficulty would remain: it concerns the problem of life detection. *What minimal set of assays will permit us to detect Martian life if it does exist?* A debate on this question for the past several years has yielded a variety of competing approaches." Each alternative was directed to monitoring some manifestation of life according to cues taken from terrestrial biology. An examination of life-detection concepts as they had evolved by 1964 provides an understanding of the problems facing the exobiologists, as well as the implied "Earth chauvinisms"[33] (a term popularized by Carl Sagan to describe the tendency to assume that living beings anywhere would be similar to those on Earth).

The very first grant NASA made in the area of biological science was to Wolf Vishniac for $4485 to develop "a prototype instrument for the remote detection of microorganisms on other planets." This money, awarded in March 1959 for work on what became known as Wolf Trap, initiated research in the field of life detection. Vishniac and his colleagues realized immediately that they faced a difficult task.[34]

Wolf Vladimir Vishniac was one of the pioneers in the search for extraterrestrial life. Born in Berlin in 1922, the son of Latvian parents who had fled the chaos of the Russian civil war, he was an associate professor of microbiology at the Yale University School of Medicine when he joined 18 other scientists* 19–20 December 1958 at the Massachusetts Institute of Technology to discuss the problems of detecting life on other planets and the possibility of contaminating those distant environments. The group, which took the name Panel on Extraterrestrial Life (or EASTEX, to distinguish it from a West Coast group led by Lederberg and called WESTEX, which met during 1959 and 1960 at Stanford University and JPL), was jointly sponsored by the National Academy of Sciences–National Research Council and the Armed Forces Committee on Bioastronautics. Melvin Calvin, professor of chemistry at the University of California at Berkeley, and Vishniac served as chairman and vice-chairman of EASTEX through 1961. At that first meeting in December 1958, one of the basic questions

---

*Dean Cowie, Carnegie Institute of Washington; Richard Davies, JPL; George A. Derbyshire, Space Science Board; Paul M. Doty, Thomas Gold, W. R. Sistrom, and Fred L. Whipple, Harvard; H. Keffer Hartline, Rockefeller Institute; Martin Kamen, University of California, San Diego; Cyrus Levinthal, Bruno B. Rossi, and A. Luria, MIT; E. F. MacNichol, Johns Hopkins; Stanley Miller, Columbia; John W. Townsend, Jr., NASA; Bruce H. Billings, Baird-Atomic, Inc.; Herbert Freeman, Servo-Mechanisms Laboratory; and Richard S. Young, Army Ordnance Missile Command.

addressed by the physicists and biologists was what kinds of life forms they might reasonably expect to find away from their own planet.[35]

Four basic hypotheses were advanced as to the nature of that life. One might find (1) living things that were essentially the same as those found on Earth; (2) life forms with the same chemistry but with peculiarities resulting from evolution in a different environment—both at the present and in the past—(3) organisms with a chemical base other than carbon (for example silicon, however unlikely that appeared in the "carbon chauvinistic" understanding of chemistry); or (4) very primitive life forms representing the initial steps along the evolutionary path. Two other distinct possibilities also existed—that life had evolved only on Earth and all the other planets were sterile, or that life had once flourished, or at least begun, on other planets only to succumb to environmental factors that precluded successful adaptation and evolution. In December 1958, few of the scientists gathered in Cambridge would have fervently backed one of these six possibilities over any other.

How does a scientist detect that which he is uncertain exists and whose form he is unsure of? Vishniac and his colleagues had to make some basic assumptions, and one of them was that life elsewhere would have a carbon base. Early in the 1960s, Vishniac in an interview said that scientists were "not acquainted with any forms of life except those that are carbon-based. It may be that carbon is indeed the only useful element that provides the structural basis for life, because of its chemical versatility." There was the possibility that other elements or combinations of elements might take on similar functions. "For instance, silicon-based life has been suggested—but silicon will not make as large and as stable compounds as will carbon. Compounds must be stable enough to . . . serve as structural units and to preserve some kind of continuity from generation to generation." Furthermore, a life-base compound must be reactive enough to permit metabolism to take place. "Carbon is particularly suited for that because it combines with itself, and with many other elements, perhaps to a greater extent than does any other element." Vishniac and others concluded that the simplest assumption was to say that life "always will be based on carbon. It may turn out that we are deluding ourselves—that we are simply limited in our imagination because of our limited experience." That was the constant intriguing possibility inherent in space research.[36]

Accepting the assumption regarding carbon, the exobiologists were still faced with defining life forms. What is life? What is a living thing? Three NASA authors who sought to analyze the life-detection problem wrote:

> The difficulties associated with assigning an unequivocal definition to the phenomenon of life lead one to utilize various approaches to a better understanding of the living state. From the standpoint of the problem of the detection of life on extraterrestrial bodies, it may be pertinent to list and scrutinize closely the criteria most commonly attributed to living

systems. *Thus the initial task of the exobiologist is to describe life in such a manner that tests can be devised that can demonstrate, unequivocally, the existence of extraterrestrial life.*[37]

These three scientists suggested five accepted manifestations of life: growth, movement, irritability, reproduction, and metabolism. Taken together, they provided an indication of living organisms, but the early students of exobiology had to determine which of these manifestations were primary to their search for living forms on other planets, "especially if those forms are exclusively microbes, as is suspected by some to be the case for Mars." A second factor to consider was the kind of detectors that might be sent to the planets. Given weight and size limitations, detectors that would test for the existence of microbiological life forms seemed more realistic than bulkier hardware created to locate larger organisms.[38]

When the exobiologists began developing life detectors, they built on the foundation provided by modern genetic theory, especially that relating to the cell as a living system. During the 1950s had emerged the revolutionary concept that the storage and transfer of basic biological information took place within the cell. A cell was "visualized as a society of macromolecules, bound together by a complex system of communication regulating both their synthesis and their activity."[39] If the cell could store and transmit biological information, it had to be able to reproduce and metabolize. Reproduction is the process which maintains biological information by its constant renewal. Metabolism has been characterized as "the fire that genetic material keeps going outside itself, to get the other material to work for it, in the service of its own distinctive goal: its own survival and replication."[40] Therefore, the minimum requirements of life can be represented as an interdependence among macromolecules, metabolism, and reproduction.

The exobiologists examined each of the three attributes to determine its relevance to the problem of detecting life. Many scientists working with Earth-bound experiments assigned top priority to reproduction. While there was certainly no argument that life could not exist very long without it, the exobiologist found it a difficult phenomenon on which to base an extraterrestrial experiment. It is a discontinuous process and "the reproductive rate varies enormously from species to species and, depending on environmental conditions, often within the species." Even at the macromolecular level, reproduction (replication) is often discontinuous in many life forms. With all the factors known to complicate observations of the reproduction of life on Earth, the detection of reproduction of life "in an exotic situation could be extremely difficult."[41]

Lederberg and others had proposed visual observations on Mars and Venus for microscopic and macroscopic life. But as with observations of reproduction, a living organism might not provide the scientist with motion or other visible clues during the short life span of an extraterrestrial experiment. The authors of the summer study report concluded that, as

65

attractive as the idea of visual observation was, "we can easily imagine circumstances in which this type of observation would be inconclusive."[42] A more reliable basis was needed.

Metabolism appeared the most promising attribute on which to base life-detection experiments, primarily because it was a continuous process. "Even life forms that are considered to be in a highly inactive state (e.g., bacterial spores and plant seeds) carry on measurable, albeit extremely low, rates of metabolism." Metabolism also could be measured in several ways (changes in pH or temperature, the evolution of gases). But after "lengthy discussions and deep deliberation," the exobiological community agreed that "a truly meaningful life detection program must be based on [several] fundamental attributes of life."[43] Scientists would not be convinced by negative answers from any single life detector. They wanted some direct visual inspection by television and a program that would land an automated biological laboratory (ABL). While not fully defined in 1964, the ABL would permit a number of chemical analyses and a variety of biological experiments. Plans included an onboard computer by which a variety of programmed assay sequences could be initiated, contingent on results of prior steps, and a sustained discourse between the computer and investigators on Earth. By remote control of their mechanical surrogate, the scientists on Earth could carry out investigations much as they would in their terrestrial laboratories. It was "in short an ambitious concept," but "realizable with current technology."[44]

## Mechanisms for Detecting Life on Mars

There was no shortage of life-detection concepts.[45] Speaking to this point at the beginning of the summer study on 15 June 1964, Lederberg compared the Mars life-search to the work that he and his colleagues normally did in their laboratories. In their everyday biochemical experiments, they were limited by approaches and hardware. Similarly, in the proposed exobiological studies, they needed to focus on the target and think about the best collective experiments for some years hence. The basic problem would come to deciding which instruments to develop. Scientists could quickly think of many experiments that might be done.[46] Once the redundant ideas were eliminated, a reasonable number of practical-looking concepts remained, among which were several that NASA had supported over the past several years. But translation of concepts into hardware was a challenge. In May 1963, NASA's Ames Research Center, Moffett Field, California, had been assigned the task of evaluating the many exobiology experiments. Ames had been serving as NASA's "in-house" life science research laboratory since the arrival of Richard S. Young in 1960, and in 1962 an Exobiology Division was established there. Hence, scientists at Ames were familiar with the issues the exobiologists were addressing their experiments to.[47]

*Mars Surface Television.* "The first thing man generally does in a new and strange environment is to look around." That was exactly what scientists wanted to do through one of the large Voyager-class landers, using television to view the topography immediately surrounding the craft. "There may be both geologic and biological surprises in the landscape. . . ." Television pictures would also permit the mission team to check out and monitor the condition of the lander. And not to be overlooked was the public-relations value of pictures as scientists and laymen alike shared a closeup view of Martian scenery.[48]

*Vidicon Microscopes.* A more sophisticated use of television cameras was the proposed microscope-television combination. Based on the suggestion of Joshua Lederberg, this idea was being pursued at his Instrument Research Laboratory at Stanford and in Gerald Soffen's facility at the Jet Propulsion Lab. "The detection of life by looking for it sounds elementary; however, this seemingly simple technique is extremely complex and involves numerous technical problems." Stanford and JPL scientists and instrument-makers were confronted by the difficulty and uncertainty in recognition and identification of microorganisms by microscope.[49] Beyond that, the large information return required to produce pictures of suitable quality appeared to be beyond computer capabilities projected for Mariner 1966. Although the Ames life-detection experiments team rejected the vidicon microscope for the Mariner flight, members of the summer study believed it had sufficient merit to be considered for a 1971 mission like Voyager.[50]

*Wolf Trap.* Wolf Vishniac originally developed this device in 1958–1960 to demonstrate the feasibility of automatic remote detection of the growth of microorganisms. He wanted to prove that such an instrument could be built, and having once committed himself to the experiment he seemed unable to set it aside for other ideas that might have been more fruitful. Defending this first exobiological instrument became part of Vishniac's promotional work on behalf of the Mars biology program.[51]

In a 1960 issue of *Aerospace Medicine*, Vishniac explained that microorganisms "are responsible for the major amount of turnover of matter on earth and . . . life of the higher plants and animals is inconceivable in [their] absence."[52] The object of Wolf Trap was the growth of Martian microbes, if they existed and could be trapped. At the heart of the instrument was a growth chamber with an acidity (pH) detector and light sensor; the former would sense the changes in acidity that almost inevitably accompany the growth of microorganisms, while the latter would measure the changes in the amount of light passing through the growth chamber. Microorganisms, such as bacteria, turn a clear culture medium cloudy (turbid) as they grow, and the light sensor would detect such changes. The pH measurement would complement the turbidity measurement, providing an independent check on growth and metabolism.

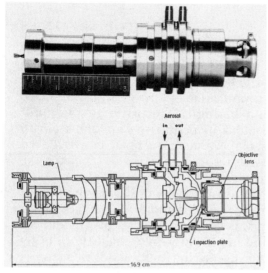

*A vidicon microscope, January 1965, being considered for future use in seeking extraterrestrial life had no moving parts. An aerosol for carrying particles was injected into the instrument and onto the impaction plate through a nozzle in the condenser lens. The objective lens and lamp were fixed in relation to the plane of focus. The sample was collected through a gas-operated aerosol aspirator.*

By mid-1963 Vishniac, with the assistance of C. R. Wilson and others, had progressed from a simple feasibility model to a more complex breadboard* design. A contract with Ball Brothers Research Corporation for the development of the second-generation instrument was let by the University of Rochester in 1961. Late in 1963, the Ames life-detection experiments team report noted several problems still unresolved, notably the likelihood of false signs of growth resulting from the sampling technique, and said the experiment probably could not be ready for 1966 but might be a 1969 candidate.[53]

*Multivator.* Conceived by Joshua Lederberg and worked out in prototype form by Elliott Levinthal and his assistants in the Instrumentation Research Laboratory at the Stanford School of Medicine, multivator was intended to be a miniature multipurpose biochemical laboratory in which a

---

*An assembly of parts used to prove the workability of a device or principle without regard to the final configuration or packaging of the parts.

*Model of Wolf Trap life detection device with cover removed.*

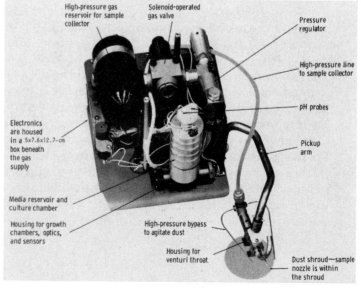

series of simple measurements could be made on samples of atmospheric dust. A variety of measurements was studied, and they all included testing a small sample of dust with a fluid reagent and reading out a simple optical or electrometric measurement. Lederberg and his associates originally hoped to cultivate Martian microorganisms in a defined culture medium, as in Wolf Trap, but they concluded that the brief communication times between a Mars lander and Earth monitoring stations would limit the opportunities of observing changes based on growth. Enzymatic activity might be a more realistic behavior to study. Thus, they began to concentrate on detecting the action of enzymatic phosphatase on phosphate containing chemicals that become fluorescent following removal of the phosphate grout. When enzymatic activity took place, the resulting glow would be determined by a detector, perhaps photoelectrically.[54] The Ames team evaluating the multivator in August 1963 decided that the instrument was maturing rapidly but that the experiments it would house would require "a great deal more effort" before they would be ready to be sent off on a mission to Mars.[55]

*Minivator.* A variant on the multivator concept, devised by Jerry L. Stuart of JPL, minivator had an improved sample-collection device. Driven by gas-powered turbine, the sample collector separated large and small particles by centrifugal action. Again, the instrument development was ahead of work on the experiments it would house. The Ames team assumed that the best features of the multivator and minivator would be combined.[56]

*Gulliver.* Named after Jonathan Swift's fictional traveler to strange places, the Gulliver instrument was the work of Gilbert V. Levin. After many years in the public health field, where he sought better methods for detecting bacterial contaminants in polluted water, Levin asked T. Keith Glennan, NASA's first administrator, if the agency would be interested in developing life-detection instruments for use on space probes. A contract for the work was let in 1961.[57]

Gulliver consisted of a culture chamber into which a sample of soil could be introduced. In the chamber was a broth whose organic nutrients were labeled with radioactive carbon. If microorganisms were put into the broth, they would metabolize the organic compounds, releasing radioactive carbon dioxide that could be trapped on a chemically coated film at the window of a Geiger counter. The radioactivity readings would be relayed to Earth by the spacecraft's radio transmitter. Gulliver had the virtue of being able to detect growth, as well as metabolism, since the rate of carbon dioxide production would increase exponentially with growing cultures.

Sample acquisition was the early Gulliver's unique feature. The instrument had a mechanism consisting of two 7.5-meter lengths of kite line wound around small projectiles in the manner of harpoon lines to prevent snagging. The string was coated with a sterile silicone grease to make it sticky. After the lander arrived on Mars, the projectiles would be

**Multivator Chambers**

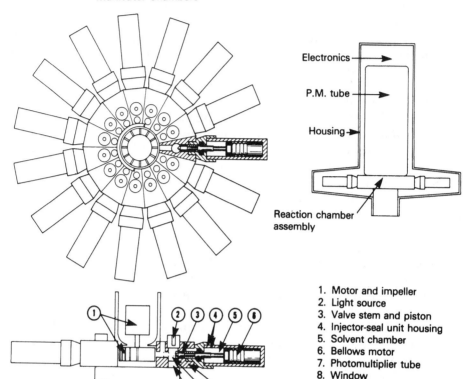

Electronics

P.M. tube

Housing

Reaction chamber
assembly

1. Motor and impeller
2. Light source
3. Valve stem and piston
4. Injector-seal unit housing
5. Solvent chamber
6. Bellows motor
7. Photomultiplier tube
8. Window
9. Reaction chamber
10. Reaction chamber unit block

*Multivator, above in January 1965, was a miniature laboratory for biochemical experiments on Mars. Gulliver III, at left, also in January, carried sticky strings and projectiles to be fired and retrieved with dust and loose particles to be tested for production of carbon dioxide indicating life.*

fired in mortar fashion and then reeled in together with adhering soil particles. After the lines were retrieved, Gulliver would be sealed and an ampule broken, releasing the sterile radioactive nutrients onto the samples.

The Ames life-detection experiments team gave Gulliver high marks because, unlike other experiments of the time, it had a sampling mechanism. But they also raised questions about the nature of the technique, since samples delivered to the growth and control chambers would not be identical. The chambers would contain a metabolic poison to serve as a check on chemically produced radioactive carbon that might otherwise be interpreted as signs of metabolism, and experimental control to prevent false results required a common sampling source. The Ames team concluded that sample acquisition might be a problem. It further noted that Gulliver was the most advanced experiment in terms of hardware development and the only one likely to be ready for flight in 1966.[58] Other life-detection concepts are listed in table 7.

Given the conclusion that no single life detector would be sufficiently accurate and conclusive in its results, an automated biological laboratory containing several experiments was the prudent choice. But before such sophisticated, expensive hardware was landed on Mars, a successful orbiter program was necessary; scientists and engineers needed more data regarding the planet's atmosphere (density and chemical composition) and surface. An orbiter's sustained seasonal observations would permit thorough evaluation of features considered suggestive of life and a better informed selection of landing sites for the laboratory.

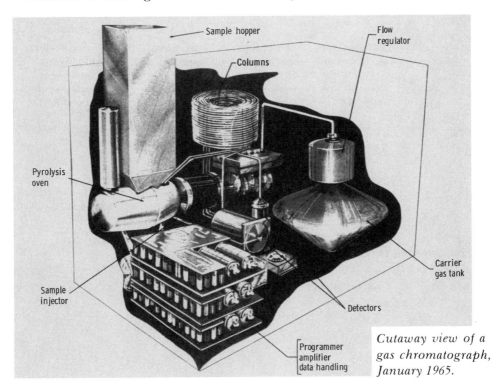

*Cutaway view of a gas chromatograph, January 1965.*

*Table 7*
*Ames Life-Detection Team Evaluation of Proposed Biology Instruments—*
*Development Status, 1963*

| Experiment | Status | Date Available | Manpower Support to Meet 1966 Date | Monetary Support to Meet 1966 Date |
|---|---|---|---|---|
| Vidicon microscope | Science—conceptual. Device—none. | ? | ? | ? |
| Wolf trap | Lab feasibility model. Engineering is conceptual. | April 1963 | Univ. of Rochester will need 1 Ph.D. & 4 techs. Subcontractor requirements unknown. | Double present funding. |
| Multivator | Science—functional feasibility. Device—conceptual. | ? / 1 Sept. 1963 | Sufficient available. Sufficient available. | Sufficient available. $10 000 for development of Mark II. $200 000 for flight hardware. |
| Minivator | Science—none. Device—flight-sized breadboard. | — / Now | — / 3 engineers. 4 technicians. | — / $200 000 for flight prototype. $40 000 for test & evaluation. |
| Gulliver | Advanced breadboard demonstrated. Ready to start work on prototype. | With proper funding and interface definition, 1 yr from contract award. | 10 persons required in engineering area | Between $250 000 and $350 000, depending on required experiment configuration |
| Optical rotation | Some functional feasibility demonstrated. | 14 mos from contract start. | — | $274 652 |
| "J" band | Science—functional feasibility. Device—conceptual. | 1 Aug. 1963. / 1 Aug. 1964 for flight prototype. | 2 scientists. 4 technicians. / 8-10 persons 1 yr. | $100 000 / $300 000–400 000 |
| Gas chromatograph | Feasibility breadboard. | Nov. 1964. | Additional: 4 engineers and 5 technicians. | $425 000 |
| Mass spectrometer | Conceptual. | May be ready 1966 launch date. | 2 assistants for Dr. Biemann and services of Consolidated Systems Corp. | $350 000 |

*Table 7*
*Ames Life-Detection Team Evaluation of Proposed Biology Instruments—*
*Development Status, 1963, Continued*

| Comments | Weight (kg) | Volume (cu cm) | Power Required (av/peak, watts) | Possible Lifetime | Sterilizable by 150°C for 24 hrs |
|---|---|---|---|---|---|
| Data rate requirements demand power available only with much larger boosters. Development of sample handling, methods for discrimination of biologicals requires more work. | Not defined | Not defined | 10 | Not defined | Yes |
| | 1.1–2.3 | 2460–3280 | 0.25/1 | 10-hr minimum | Yes |
| Depends on stability of phosphatase assay substrate. Can accomodate wide variety of biochemical experiments including some already proposed. | 1.4 | 1558 | 0.5/3–5 | Days–week | Yes |
| Science input lacking; accommodation similar to Multivator. | 2.3 | Not defined | 1–2/5–10 | 2 wks | Yes |
| | 3.2–5.4 | 4920–9840 | 2–3/4–5 | Not defined | Yes |
| | 2.4 | 2132 | 0.5/1.1 | Not defined | Yes |
| Sample acquisition and handling development not begun. | Not defined | Not defined | 2–3/10 | Weeks | Yes |
| | 3.04 | 3280 | ?/14.5 | Not defined | Yes |
| Support requirements appear to be underestimated by experimenter. | — | — | — | — | — |

SOURCE: Based on data presented in NASA, Ames Research Center, Life-Detection Experiments Team, "A Survey of Life-Detection Experiments for Mars," Aug. 1963, pp. 70-71.

When looking into automated biological laboratories, the summer study group had to consider how such advanced landers would be scheduled in relation to Mariner flights. Mariner flyby spacecraft were slated for launch in November 1964 by Atlas-Agena. Replacing the ill-fated Mariner B, Mariners E and F, approved in December 1963 for combination flyby and probe missions, were planned for 1966 (as Mariner 1966) if Atlas-Centaur were operational by that time. Thus, the members of the 1964 Summer Study preferred "a gradualistic approach" to the ultimate goals of landing a large automated laboratory on Mars and eventually returning samples for study. The scientific community favored exhausting all avenues of research, Earth-based observations and nonlanding missions, before committing itself to that big step.

However, the summer study members saw several "constraints to proceeding in a completely unhurried step-by-step fashion." Those included a "combination of celestial mechanics and the operational realities of space research." Preparation for flight required years of experimental design and spacecraft development and the coordination of effort among large numbers of persons in a wide range of disciplines. As individual scientists, accustomed to following their own idiosyncratic process of trial and error in designing laboratory experiments, they found the world of space research filled with tightly controlled schedules and very specific dos and don'ts. They noted further that the scientist was "plagued by the prospect of investing years of work only to encounter a mission failure or cancellation in which it is all lost—at least until a new opportunity arises, perhaps years hence." While the scientists might "chafe under these circumstances," it was the nature of the enterprise.

Added to the technological and scientific limitations was the small number of launch opportunities for flights to Mars. The "attempt to develop a systematic and gradualistic program is thus constrained to some extent by the fact that, while favorable opportunities occur in the 1969–1973 period, they will not return before 1984–1985." Therefore the summer study members argued for "a substantial program" that would exploit the Saturn launch vehicles during the 1969–1973 launch window. Explicit in their recommendations was concentration on activities that would lead to landings. "The first landing mission should be scheduled no later than 1973, and by 1971 if possible."[59]

## THE RESULTS OF MARINER 4

Whereas 1964 was a year of optimism for the burgeoning field of exobiology, 1965 was one of external criticism and reappraisal. New scientific information provided by the *Mariner 4* flyby mission altered perceptions of the Red Planet and raised serious questions about the search for life there. Criticism of NASA's exobiology program came from two quarters, members of the *Mariner 4* science team and scientists who were critical of the space program in general terms.

Variously known during its developmental phase as Mariner C, Mariner M, and Mariner 1964, *Mariner 4* was one of two spacecraft launched for Mars in 1964. Conceived in mid-1962 when NASA's advanced planners realized that the Centaur stage would not be ready for a 1964 mission, Mariner C was planned as a lighter Agena-sized spacecraft capable of a mission to Mars. As *Mariner 2* to Venus in 1962 had been a scaled-down Mariner A, the 1964 Mars craft was a revision of Mariner B without the lander.[60] Although smaller than either NASA or the scientists would have preferred, it would provide the first photographs of Mars, an exciting prospect. From November 1962 when the Project Approval Document was signed to liftoff of the two craft in November 1964, this first Mars mission was a challenging exercise. Constant battles against growing payload weights and difficulties with perfecting scientific instruments added a hectic air to preparations for the 1964 flights.[61]

As the launch date approached, trouble seemed to be the key word. *Mariner 3* was launched toward Mars about midday on 5 November. After a short delay while Agena circuits and relays were retested, the launch went normally, but an hour later telemetry indicated that while the scientific instruments were on there was no indication of power from the solar panels. Quickly the launch team determined that the cylindrical fiberglass nose fairing designed to protect the spacecraft during its initial ascent had failed to separate from it. Efforts to break the spacecraft free were frustrated when its circuits went dead after the batteries were drained. As *Mariner 3* blindly headed out into space, destined to enter solar orbit, NASA and contractor personnel searched for the cause of the problem and a quick solution before the 25th, the scheduled date for the second launch.[62]

Working around the clock for 17 days, a composite team from Lewis Research Center, Lockheed Missiles & Space Company, and JPL modified the nose fairing and produced a flawless launch of the second spacecraft on 28 November.[63] Everything went according to plan with *Mariner 4*. The Agena D separated from the Atlas at an altitude of 185 kilometers and went into a parking orbit. After coasting for more than 30 minutes, the Agena engine fired again and Mariner was on the path to Mars. With only 45 minutes elapsed since liftoff from Cape Canaveral, *Mariner 4* separated from Agena and continued its journey through space alone.

It took seven and a half months to travel the 525 million kilometers to Earth's neighbor. The 260-kilogram spacecraft began its brief encounter with the planet on 14 July 1965. Among other measurements, the vidicon television system during a 25-minute sequence took 21 full pictures and a fraction of a 22d of the Martian surface at distances of 10 000 to 17 000 kilometers. After being stored overnight on a tape recorder, the images were transmitted to Earth the next day. For eight and a half hours, JPL received

Mariner 4, *above, is prepared for a center-of-gravity test at Jet Propulsion Laboratory. At right, the spacecraft starts on its way from Kennedy Space Center on 28 November 1964.*

bits of electronic data that would be reconstructed into visual images. The pictures revealed a heavily cratered Mars[64]

What could one learn from 21½ pictures of 1 percent of the Martian surface taken from an average distance of 13 000 kilometers? For *Mariner 4*, expectations helped color perceptions. On 11 January 1965, Robert B. Leighton, principal investigator for the television experiment and professor of physics at the California Institute of Technology, had written Glenn A. Reiff, Mariner project manager, commenting that the *Mariner 4* pictures would "be of enormous interest to the scientific community and the public at large," but proper interpretation of those pictures was as important as their initial acquisition.[65] From the outset, NASA and JPL officials had carefully informed the public that Mariner would not produce pictures of sufficient resolution to detect plant or animal life, but while reporters told their audiences that "the pictures are not expected to resolve the mystery of life on Mars," they would usually add such phrases as "but may answer long standing questions about the 'canals' of the red planet," hinting that *Mariner 4*'s photography might indeed be spectacular.[66]

Before and during the flight, scores of articles about *Mariner 4*, the 1964 Summer Study, exobiology, Voyager, and other aspects of the exploration of Mars appeared in the American press.[67] Most carried the caveat that the 20-some photos would be equivalent to the best telescopic views of the moon from Earth and that "even the broadest earth river would not be visible at such a distance," but writers argued that it might still be possible to view the irrigated bands along the canals if any existed on Mars.[68] *Mariner 4* would not necessarily detect life, but the scientific community hoped it would provide additional insights into the likelihood of Martian biology. David Hoffman of the *New York Herald Tribune* commented on this dichotomy in an article on 14 July, the day the pictures were taken: "In what almost amounts to a non sequitur, NASA says the photo mission is not designed to answer 'the question of life on Mars.' But only to 'shed light on the possibility of extraterrestrial life.'"[69]

For believers in Martian canals and for scientists dedicated to the extraterrestrial life search, the pictures were disappointing. In a 29 July 1965 statement, the Mars television team led by Leighton summarized their first thoughts on the significance of the photographs: "Man's first close-up look at Mars had revealed the scientifically startling fact that at least part of its surface is covered with large craters. Although the existence of Martian craters is clearly demonstrated beyond question, their meaning and significance is, of course, a matter of interpretation." Their opinion was that the craters led "to far-reaching fundamental inferences concerning the evolutionary history of Mars and further enhances the uniqueness of Earth within the solar system." Seventy craters were clearly visible in photos 5 through 15, and they ranged in diameter from 4.8 to 120 kilometers. NASA specialists noted that it seemed likely that there were both larger and smaller craters in addition to those discerned in the photos. The rims of the craters appeared to rise as much as 100 meters above the surface, and the interiors seemed to descend to several hundred meters. The number of large craters was closely comparable to the densely cratered upland areas of the moon. They added that no Earth-like features, such as mountain chains, great valleys, ocean basins, or continental plates, were identifiable in the small region sampled by *Mariner 4*. And certainly no canals were seen.

From the pictures, the TV team thought some fundamental inferences could be drawn:

1. In terms of its evolutionary history, Mars is more Moon-like than Earth-like. Nonetheless, because it has an atmosphere, Mars may shed much light on early phases of Earth's history.

2. Reasoning by analogy with the Moon, much of the heavily cratered surface of Mars must be very ancient—perhaps two to five billion years old.

3. The remarkable state of preservation of such an ancient surface leads us to the inference that no atmosphere significantly denser than the present very thin one had characterized the planet since that surface was formed.

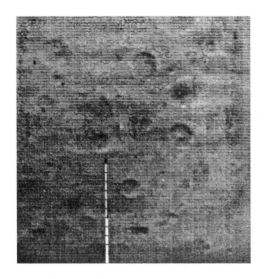

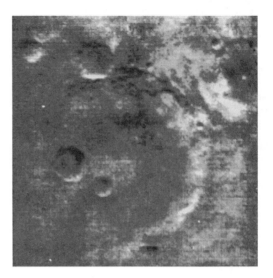

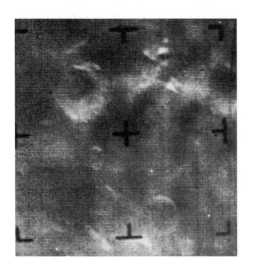

Mariner 4 *revealed a heavily cratered Mars, more like the moon than like Earth. Photos taken 14 July 1965, just before the closest approach of 9700 kilometers, were radioed back as digital data. At top left, Mare Sirenum, bordering on Atlantis. Above, Atlantis between Mare Sirenum and Mare Cimmerium. At left, bright region, northwestern Phaethontis. Below at the White House 31 July 1964, JPL Director William Pickering shows* Ranger 8 *photo of the moon to President Johnson. NASA Associate Administrator for Space Science and Applications Homer E. Newell is with him. Behind the president are Dr. Donald F. Hornig, special assistant to the president for science and technology, and Dr. Edward C. Welsh, executive secretary, National Aeronautics and Space Council.*

Similarly, it is difficult to believe that free water in quantities sufficient to form streams or to fill oceans could have existed anywhere on Mars since that time. The presence of such amounts of water (and consequent atmosphere) would have caused severe erosion over the entire surface.

4. The principal topographic features of Mars photographed by Mariner have not been produced by stress and deformation originating within the planet, in distinction to the case of the Earth. Earth is internally dynamic giving rise to mountains, continents, and other such features, while evidently Mars has long been inactive. The lack of internal activity is also consistent with the absence of a significant magnetic field on Mars as was determined by the Mariner magnetometer experiment.

5. As we had anticipated, Mariner photos neither demonstrate nor preclude the possible existence of life on Mars. The search for a fossil record does appear less promising if Martian oceans never existed. On the other hand, if the Martian surface is truly in its primitive form, the surface may prove to be the best—perhaps the only—place in the solar system still preserving clues to original organic development, traces of which have long since disappeared from Earth.[70]

The fifth point notwithstanding, the findings of the TV team were a genuine blow to the exobiologists. Leighton, Cal Tech astronomers Bruce C. Murray and Robert C. Sharp, and JPL television experts Richard K. Sloan and J. Denton Allen presented an official report in the 6 August 1965 issue of *Science*, restating the same basic conclusions. The apparent absence of water over hundreds of millions of years, the very thin atmosphere, and extremely low temperatures were strong arguments against the hypothesis for life put forward during the 1964 Summer Study. New tabular data for the physical properties of Mars are shown in table 8.

*Table 8*
*Physical Properties of Mars: Mariner 4 Findings*

| | Earth | Mars (1964 Summer Study) | Mars 1 | Mars 2 | Mars 3 |
|---|---|---|---|---|---|
| | | | (alternative *Mariner 4* figures) | | |
| Atmospheric pressure | 1000 millibars | 10–30 millibars | 4.1–5.7 | 4.1–6.2 | 5.0–7.0 |
| Gaseous composition of atmosphere: | | | | | |
| oxygen | 20% | <0.1% | — | — | — |
| carbon | 0.03% | 5–30% | 100% | 80% | 50% |
| nitrogen | 78% | 60–95% | — | 20%[a] | — |
| argon | trace | trace | — | — | 50% |
| Temperature range | 58°C to −88°C | +30°C ±50°C | −93°C ±20°C | −98°C ±25°C | −103°C ±20°C |

[a]Nitrogen plus argon.
SOURCE: NASA, *Mariner-Mars 1964: Final Project Report*, NASA SP-139 (Washington, 1967), pp. 321–22.

No matter which of the alternative atmospheric estimates from *Mariner 4* readings one chose, the possibility for life, past or present, seemed diminished.[71]

### External Criticism of the Search for Life on Mars

Criticism of the American space program, latent for several years, burst forth in 1963–1965. The two most prominent fault-finders were Barry Commoner, a microbiologist at Washington University, St. Louis, and critic-at-large-of-scientific priorities; and Philip H. Abelson, a physicist and the editor of *Science*. Both scientists, long-time critics of Apollo's lunar goals, extended their remarks to the exploration of Mars.

Commoner attacked the search for extraterrestrial life in June 1963 on the eve of Abelson's appearance before the Senate Astronautical and Space Science Committee. The committee was seeking "Scientists' Testimony on Space Goals," and Commoner noted that Abelson was the only witness expected to express reservations about the nation's priorities in space research. Of the 10 who were scheduled to testify, all but Abelson had a direct financial interest in the space program.* While Abelson attacked Apollo specifically, Commoner was upset by the argument that the extraterrestrial life search was "the most exciting, challenging and profound issue. . . that has characterized the history of Western thought for 300 years." Believing the possibility of life on other planets was extraordinarily low, he thought that such rhetoric was "a weak prop for the serious decision given its profound economic and social consequences."[72]

Scientists Commoner and Abelson did not agree with NASA's scientific goals. Simply put, they would have preferred to spend Apollo and Mariner-Voyager dollars on other investigations. They were also worried about the "social consequences" of space research in a world that was underfed and potentially revolutionary. In a September 1963 speech to the American Psychological Association, Abelson said there were no predictable economic advantages to be derived from the exploration of the moon or Mars, arguing that "the half of the world that is undernourished could scarcely be expected to place a higher value on landing on the moon than on filling their stomachs."[73]

The exobiologists were accustomed to defending their work on scientific grounds, but they were understandably perplexed when they were criticized in a manner that combined scientific disagreements and differences in opinion over social and economic priorities. Lederberg and others were reasonably certain of Commoner's political motivations, but they were not sure of his scientific views, as he diverged from the origin-of-life hypothesis that underpinned the search for extraterrestrial life. Was it

---

*The other nine were S. Ramo, H. C. Urey, P. Kursh, C. S. Pittendrigh, F. Seitz, L. V. Berkner, L. A. DuBridge, M. Schwarzschild, and H. H. Hess.

scientist Commoner or social-critic Commoner who opposed the extraterrestrial search?[74]

Abelson was even more difficult to understand. He was a long-time student of the extraterrestrial life question. In 1960, he had advised NASA that of all the near planets only Mars was a likely abode for life, but that the risk of contaminating the planet with Earth life-forms precluded our going there. By the next year, however, he was arguing persuasively that no place in our solar system other than Earth could support life as we know it. Thus, his editorial in the 2 February 1965 issue of *Science*, "the voice of American science," was particularly telling: "In looking for life on Mars we could establish for ourselves the reputation of being the greatest Simple Simons of all time."[75] Using the latest scientific information from *Mariner 4*, Abelson built a case against future expeditures of tax dollars to look for life on Mars; he was convinced life did not exist there. For a mixture of scientific and political motives, he effectively used *Science* as a forum for the scientifically based denunciations of NASA's goals.[76]

### 1964 Summer Study Revisited; or "Postscript: October 1965"

Against this background of scientific and political criticism, the discouraging new information provided by *Mariner 4* posed serious questions for those who believed that there might be life on Mars and that continuation of the search was respectable and worthwhile. Joshua Lederberg later looked back on October 1965 as a bleak time for exobiology. With most of the scientific community in agreement with a *New York Times* editorial saying that "Mars is probably a dead plannet," only a few "diehards" (Lederberg's decription of his associates of the 1964 Summer Study) refused to give up and accept Mars as a barren world.[77] In a postscript to *Biology and the Exploration of Mars*, those diehards held that, "during the interval between publication in March 1965 of the Summary and Conclusions of our Study and the appearance of this volume, our knowledge of Mars has been raised to an entirely new level by the success of the Mariner IV mission."[78] Lederberg and 25 other "desperate" persons met in late October 1965 to discuss the impact of *Mariner 4* on their proposed search for life: "The essence of our position was, and still is, the immense scientific importance of evaluating the uniqueness of life on Earth; of discovering facts that will permit more valid inference of its abundance in the Universe; and the fact that the new space technology allows us to obtain empirical evidence on the frequency with which living organization and its precursors emerge in the evolutionary history of planets." Even with the new Mariner data in hand, the scientists still thought "that life, even in essentially terrestrial form, could very well have originated on Mars and have survived in some of its contemporary micro-environments." While finding life clinging to the side of an inactive volcano or at the edge of some warm spring on Mars would be difficult, it was not totally unreasonable to expect.[79]

There was another justification for going to the Red Planet. The summer study participants believed it was "important to re-emphasize . . .

a major aspect of our position that critics have unaccountably missed; we sought to emphasize 'that our conclusion that the biological exploration of Mars will be a rewarding venture does not depend upon the hypothesis of Martian life.'" Throughout their deliberations, they had cast their questions in the broad context of the general evolutionary process in nature. "Our position is . . . fully justified even if life has not emerged there; but we will again be misunderstood if that emphasis is taken to mean we believe the chance of discovering fully fledged life is negligible."[80]

At the end of 1965, the scientists who believed that looking for life on Mars was a respectable enterprise faced those who were equally devoted to the proposition that such an exercise was foolishness of the gravest order. Voyager, with its goal of placing automated biology laboratories on Mars, would become the focus of the two groups' debate. Voyager would be scrutinized because of costs and general disenchantment with the space program, but the central issue would continue to be the validity of searching for life on the Red Planet. To that issue, scientists could bring only informed speculations. *Mariner 4* had provided only clues. No one could yet say with certainty that Mars was lifeless. And the search continued.

# 4

# Voyager: Perils of Advanced Planning, 1960–1967

Voyager was an advanced mission concept first considered in the spring of 1960 when the NASA staff was beginning to define its long-range plans for lunar and planetary missions.[1] In their semiannual (1 April–30 September 1960) report to Congress, the agency managers reported that preliminary mission studies were under way for a second planetary series. The Voyager orbiters were to be designed to orbit Venus and Mars and were to be "phased in time and capabilities with the Saturn launch vehicle." Orbits of the planets for long periods would make possible excellent investigations of their environments, and landing capsules would be able to provide information on the lower atmospheres and surfaces.[2] In designing the Voyager spacecraft, NASA engineers and scientists hoped to use new data gleaned from the Mariner flights—information that would help them design Voyager's scientific instruments to answer the proper questions and solve technological problems posed by Voyager's large size.

Unfortunately, the real world of politics, with too many projects competing for too few federal dollars, is seldom as neat as planners hope. For the Voyager proponents, the real world was an unhappy one. Delays on the Atlas-Centaur launch vehicle during the early 1960s prompted many changes in the Mariner project, which in turn delayed the acquisition of information about the Martian environment essential to the designers of Voyager. But Kennedy's decision to mount a full-scale assault on the moon was an even bigger blow for the supporters of unmanned space exploration. Once the manned Apollo decision had been made, the Marshall Space Flight Center and its industrial contractors concentrated on the preparation and production of Saturns for the lunar missions. Launch vehicles for space science projects would become available only after the top-priority goal had been met. From the start, Voyager was by definition a second-class project. As Congress became restive over the increased expenditures for Apollo, monies originally marked for space science and Voyager were reallocated to help pay for the moon program. Added to this were a costly war in Vietnam and the domestic troubles of the late 1960s. All post-Apollo missions proposed by the space agency faced reduced appropriations, which put

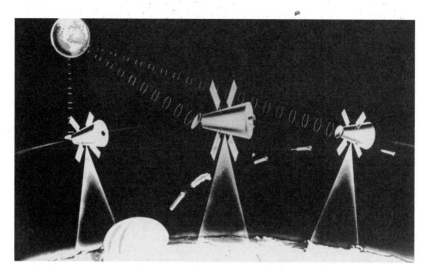

One of the first conceptual views of Voyager, above, was published in the NASA-Industry Plans Conference, July 28–29, 1960. The artist concept below was described by Edgar M. Cortright during March 1961 NASA hearings before Congress: "This spacecraft, weighing about 2,400 pounds [1090 kilograms], would be designed to orbit the target planet and to inject a several-hundred-pound capsule capable of surviving atmospheric entry and descent. . . . Thus the orbiting spacecraft would observe the planet and its atmosphere . . ., while the landing capsule would make detailed measurements during descent and on the ground. . . . Numerous . . . developments are required to accomplish this difficult but fascinating and distinctly realistic mission, which may well include among its rewards the discovery of extraterrestrial life." Senate Committee on Aeronautical and Space Sciences, NASA Scientific and Technical Programs, hearings, 87th Cong., 1st sess., 28 Feb., 1 Mar. 1961.

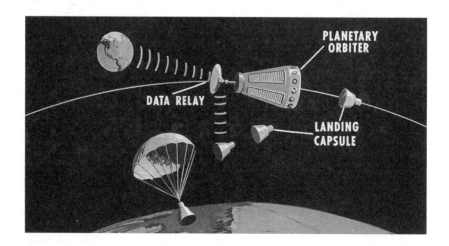

Voyager in deep fiscal trouble by summer 1967. A request in August from the Manned Spacecraft Center in Houston for proposals to study manned missions to Mars was the last bit of bad luck. Congress rebelled and terminated all Voyager work.

At first glance, it would appear that the Voyager project of the 1960s, like Mariner B and Mariner 66, was just another project that never progressed beyond the drafting table, but it was more than that. Voyager, with thousands of man-hours of work behind it, performed by dozens of specialists and costing many millions of dollars, helped to refine an understanding of the best approaches for a combination orbiter-lander investigation of Mars. Upon the solid foundations laid by Voyager personnel, the Viking team that followed them could construct a successful mission. The story of Voyager's troubles is essential to an understanding of Viking's accomplishments.

## ORIGINS OF VOYAGER

For the duration of the Voyager project, there were two distinct perspectives of the enterprise—one view from NASA Headquarters and another from the Jet Propulsion Laboratory in Pasadena. As with Ranger and Mariner, Voyager was initially a JPL undertaking, with nearly all the early study and design done in the California lab. In contrast, JPL had contracted out to industry for the design and development of Surveyor, the lunar soft-lander. This difference may have been indicative of the Pasadena team's bias for planetary missions but, for whatever reason, the team had a particular attachment to Voyager. JPL staffers had very specific ideas about how Voyager should be developed (orbiters first, with the addition of landers much later) and managed (loosely knit organization of delegates from various laboratory divisions). Furthermore, JPL wanted to conduct the total project within the walls of the laboratory. The West Coast planners favored small "manageable" undertakings, while NASA Headquarters called for centralized management under one responsible individual, with centers assuming a supervisory role over industrial contractors. As Voyager became a pet project with headquarters managers, the differences between JPL and Washington became obvious. In Pasadena, JPL personnel muttered about pencil-pushers who had no understanding of the problems of engineering the nuts and bolts of a Mars-bound spacecraft, and not uncommon in the nation's capital were exasperated remarks about the single-mindedness and independence found at JPL. While these differences were not responsible for the cancellation of the project, they made the work of Donald P. Hearth, responsible for Voyager at headquarters, and Donald P. Burcham, Voyager manager at JPL, more difficult. From the beginning, even Voyager's most optimistic supporters saw trouble ahead for the planetary spacecraft.

JPL planners began to study Voyager-class missions in 1961 to determine more clearly what flights with what size spacecraft would be a reasonable step beyond Mariner B. In May 1962, the laboratory's Planetary Program Office commissioned a study of advanced missions and spacecraft. In addition to Voyager with flights to Venus and Mars, a second kind of advanced spacecraft was examined—Navigator, which would explore the sun, comets, Mercury, and Jupiter and require still more powerful launch vechicles. Under the direction of Philip K. Eckman, the advanced planetary spacecraft study group, with representatives from all the technical divisions of JPL, examined large orbiter missions for Voyager because it believed that too little was known about the Martian and Venusian atmospheres to permit the development of spacecraft landing systems for either planet. One of the most important results of this initial phase of the advanced study was the determination of "the maximum orbiter-spacecraft payload." One member recalled that the group had been "hard pressed to come up with an in-orbit payload in excess of 500 pounds [230 kilograms] of instruments" for the "ideal" payload.[3] The group's work was the subject of three days of discussions by JPL and NASA representatives in early November 1962 (table 9).

Five men participated in the November Voyager review: Donald Hearth and Andrew Edwards, Jr., from headquarters; and Peter N. Haurlan, manager of the JPL Voyager study, Philip Eckman, and Robert J. Parks from JPL. Hearth, with NASA since 1962, was chief of Advanced Programs in the Lunar and Planetary Program Office and the key headquarters representative at the winter meeting. He had been an aeronautical research engineer at the Lewis Flight Propulsion Laboratory (of NACA) in Cleveland in the 1950s and a project engineer for Marquardt Corporation, where he had managed research related to hypersonic ramjets and similar advanced power plants. Hearth believed that fiscal 1963 activities were "proceeding along logical lines" and that JPL was doing a good job. However, he was disturbed by the postponement of work on landers, as preliminary research was necessary for comparison studies of alternative missions. Hearth preferred to push ahead with a total mission study, refining the details as new information about the planets became available.

A more pressing concern, according to Hearth, was the work load the Pasadena laboratory was assuming. "It appears to me that JPL is planning on doing too much in-house starting in 1964. Their plans for bringing in contractors next year looks good; however, I question the relative in-house and out-of-house level." Providing some overlap (with the JPL effort) from contractors appeared advisable, and Hearth expanded his thoughts on the subject in a memorandum to Oran Nicks:

1.) JPL (Haurlan) did not have complete information on Voyager expenditures thus far in FY63.
2.) JPL should have conducted mission capability comparisons (even on just a preliminary basis) earlier in the committee activity.

*Table 9*
*Highlights of Advanced Planetary Spacecraft*
*Group Investigations, 1962*

Missions Considered

   Flyby—very short duration.

   Planetary orbiter—longer duration but does not permit examination of planetary surface.

   Direct landing—"most exciting" mission, but technological requirements for such a mission are quite severe.

   Other—sample return, flyby or orbiter with landing capsule, flyby with multiple capsules, etc.

   *Conclusion*—Advanced orbiter appears most feasible in period under study, 1966–1973.

General Mission Objectives

   Acquire sufficient environmental information to permit confident design of large landing vehicle, both manned and unmanned.

   Permit biological examination of the near planets.

   Investigate planetary atmospheres.

   Study planetary geology.

Major Technological Problem Areas

   Launch facility limitations—not enough launch pads for quick turnaround required by launch window schedule.

   Tracking system limitations—deep space network too limited to permit communication with multiple spacecraft.

   Spacecraft power limitations—need to improve both solar-cell and radioisotope-thermoelectric-generator technology.

   Sterilization—need to develop techniques for sterilization and develop hardware that can survive sterilization process.

   Flexibility—need to develop capability to incorporate new knowledge from one mission into the next, even with short interval between planetary opportunities.

SOURCE: JPL, "Advanced Planetary Spacecraft Study Report," vol. 1, EPD-139, 28 Dec. 1962, pp. II-1 to II-8, V-1 to V-2.

3.) Haurlan and [Eckman] did not have definite schedules for committee activities . . . [and] schedule charts were not available. Between the three of us, we made up such a chart during my visit.
4.) JPL is thinking of doing more of the Voyager job in future years in-house than is reasonable.

Trying to maintain greater control over the progress of Voyager, Hearth asked the study group to provide NASA with monthly reports, quarterly project reviews, and all back minutes of advanced planetary spacecraft study group meetings. Hearth made a final point that would have discomfitted the team in Pasadena. From "the current situation," it appeared likely that JPL could manage Voyager in future years, but there was the chance that NASA Headquarters might decide otherwise. "If another NASA Center or if a strong industry contractor [was] to manage the project," Hearth thought that "they should be brought into the project *now* because the studies being conducted this year will establish the system design concept to be followed in future years."[4]

NASA had been considering broader industry participation in the Voyager project since early 1962,[5] and eight companies had active, internally sponsored concept studies in progress:

| | |
|---|---|
| Continuous study during the last 9–12 months | AVCO<br>General Electric |
| Study during the last few months | Douglas, Santa Monica<br>Convair, San Diego<br>Convair, Fort Worth |
| Study just starting | Lockheed, Sunnyvale<br>North American Aviation,<br>  Space and Information<br>  Systems Div.<br>Space Technology Laboratories |

In addition to the JPL exercise that would cost $700 000, Hearth recommended to Nicks that headquarters fund two industrial contractors ($75 000 each) to conduct mission and predesign studies. From their findings, two systems would be selected for further study.

Industrial participation would have four advantages according to Hearth. First, "it would provide a 'check' on the JPL results. This is important since a decision will, presumably, be made this year which will determine the approach to a system involving many millions of dollars." Second, NASA would have a wider base of "*funded* Voyager studies" in the event that Voyager management did not go to JPL. Third, by investing $150 000, NASA "would provide encouragement to the management" of numerous companies by demonstrating that NASA was "serious about Voyager" and that a substantial part of the task would be assigned to industry. Finally, contracts with industry would allow NASA to direct the studies "along lines desired" by the agency, and Hearth had no doubt that considerably more than $75 000 would be expended by each company in its studies. "In addition, the agency would gain an "early insight into the firm's capability for Voyager."[6]

Whereas Hearth had planned to contract with AVCO or General Electric for this short-term study, with a more elaborate preliminary design project in fiscal 1964, the lure of money brought a number of other contractors onto the scene.[7] The original plan for a six-month contract starting 1 January was replaced by a 5 March 1963 competitive request for proposals for a formal design study.[8] With an eye on a 1967 launch to Venus, Hearth decided that he could not afford to sacrifice six or seven months on a preliminary exercise. He told the NASA senior management at a February briefing for Administrator Webb that it would be difficult to meet the next Venus launch opportunity less than four and a half years away, but the undesirable alternative was to wait six years to launch the first planetary spacecraft (Mars 1969) "having the mission capability and scientific return possible with Voyager." Hearth believed strongly that they should set June 1967 as their goal.[9]

*Request for Proposals—Voyager*

Aiming for 1967 and 1969 launches to Venus and Mars, the NASA Headquarters staff decided to spend about $200 000 in 1964 on two contracts to examine mission and predesign aspects of a Voyager flight.[10] "Request for Proposal No. 10-929, Voyager Design Studies" was sent to 21 companies 5 March 1963. Potential contractors were to summarize their cost and scientific proposals—based on NASA's statement of work defining the projected studies—for developing an advanced spacecraft to perform "orbiter/lander missions to Mars and Venus from 1967 through 1975."[11] This Voyager-class spacecraft, launched by a Saturn booster, would be capable of more difficult missions than Mariner, carry more scientific instrumentation, collect and return more data, and have a longer operational lifetime.

Two contractors would be given six months to recommend their design concepts. Their proposals would consider both the orbiter and the lander and evaluate landers that could be released both before and after achievement of planetary orbit. Flight weight was set at 2700 to 3175 kilograms, the planetary payload for the Saturn IB booster, but smaller craft (1800 kilograms) would also be examined in case the Air Force Titan IIIC launch vehicle were employed instead. Growth of subsequent Voyager craft to weights as great as 27 000 kilograms was another area of study. Spacecraft in the heaviest class could be sized to fit the Saturn V, called the Advanced Saturn. Don Hearth was the technical director for this phase of the Voyager investigations.[12]

A total of 37 industrial organizations was represented at the Voyager preproposal briefing at NASA Headquarters on 11 March 1963, where delegates had the opportunity to ask questions before they finalized their proposals, due on the 25th.[13] Of the 13 companies submitting proposals, 10 were judged acceptable. A technical evaluation team* met on 27 March to begin the selection process. Using an elaborate formula, the team decided that the Missile and Space Vehicle Division of General Electric, Valley

---

*D. Hearth, chairman, B. C. Lam, A. Edwards, E. A. Gaugler, F. D. Kochendorfer, P. N. Haurlan, and L. E. Richtmyer.

Table 10
Ranking of Contractors Bidding on 1963 Voyager Study

| Contractor | Overall Rank | Composite Score (of possible 600) | Total Estimated Cost (cost rank) | Fee Requested | Overhead Rate | Man Hours | G & A* Rate | Computer Time |
|---|---|---|---|---|---|---|---|---|
| Missile and Space Div., General Electric | 1 | 524.5 | $125 000 (6) | 8.0% | 120% | 6 100 | 10.5% | $ 9 000 |
| Research and Advanced Development Div., AVCO | 2 | 443.4 | 144 546 (7) | 7.0 | 105 | 9 131 | 8.0 | 13 200 |
| Missiles & Space Co., Lockheed Aircraft Corp. | 3 | 406.5 | 122 315 (5) | 7.0 | 80 | 8 530 | 6.5 | 3 500 |
| Space Technology Laboratories | 4 | 358.6 | 169 189 (8) | 8.5 | 103 | 10 850 | 9.9 | ------- |
| Space and Information Systems Div., North American Aviation Inc. | 5 | 337.8 | ------------- | ------- | ------- | ------- | ------- | ------- |
| Aeronutronics Div., Ford Motor Co. | 6 | 334.4 | 96 109 (1) | 0.0 | 131 | 4 284 | 0.0 | ------- |
| Martin Marietta Corp. | 7 | 332.6 | 186 505 (9) | 7.0 | 102 | 19 184 | 16.9 | ------- |
| Aerospace Div., Boeing Co. | 8 | 301.9 | ------------- | ------- | ------- | ------- | ------- | ------- |
| McDonnell Aircraft Corp. | 9 | 276.4 | 98 939 (3) | 12.0 | 80 | 7 080 | 6.6 | ------- |
| Astronautics Div., General Dynamics Corp. | 10 | 265.7 | 99 944 (4) | 7.0 | 47 | 7 335 | 7.17 | ------- |
| International Telephone & Telegraph | 11 | ------ | 97 916 (2) | 0.0 | 125 | 9 480 | 12.9 | ------- |

----- = data not available.

*General and administrative; expenses such as executive salaries.

SOURCE: Donald Heath and Andrew Edwards to Carl M. Grey, "Technical Evaluation of Proposals Received in Response to RFP No. 10-929," 2 Apr. 1963; Heath, note of conversation with Grey and R. W. Loud, 4 Apr. 1963; and Grey, "Technical Evaluation of Proposals," 4 Apr. 1963.

Forge Space Technology Center, Pennsylvania, was the clear first choice. While other companies were competitive from a cost standpoint, only AVCO Corporation,* Lockheed Missiles & Space Company, and Space Technology Laboratories, Inc., had submitted technically acceptable proposals. After careful scrutiny, the evaluation team favored awarding the second contract to AVCO. Although AVCO's "proposal was not as smooth and as well organized as the Lockheed proposal, it did demonstrate a better understanding of the scope of the technical study."[14] The two contractors were notified of their selection in early April.

*Contractor Proposals*

Despite public and congressional scrutiny of Voyager, the contractor studies were conducted as planned during the summer months of 1963.[15] General Electric secured the services of 20 distinguished scientists to review the company's progress and "suggest modifications which would increase the overall scientific value of the program." Several familiar names were on the list—Melvin Calvin, Joshua Lederberg, Wolf Vishniac, Carl Sagan, Harold Urey.[16] Scientific community and industry were working together for their mutual benefit.

A host of technical questions were being examined by the contractors, as the following list sent to the Voyager project managers at AVCO and General Electric indicates:

1. What can Voyager do scientifically that Mariner B cannot do?

2. How large a Mars lander is required for long lifetime (one month or more)?

3. If a relay orbiter is employed, what is the trade-off between lander data rate, science payload weight, and lifetime?

4. How does Martian lander performance (data rate, science payload, and lifetime) and weight compare with and without a relay orbiter?

5. What are the problems associated with the use of a high-gain directional antenna on the Martian surface?

6. Can such an antenna be designed and developed for a '69 mission (without undue risk) based upon what we currently know and expect to learn about the planetary surface in the near future?

7. If the answer to question 6 is no, what type of additional scientific data is required?

8. Will a Voyager lander and *relay* plus science orbiter system be capable of obtaining the type of data indicated by the answer to question 7?

9. Once these data are obtained, how much time will be required for the design and development of a high-gain antenna (for landers) for use in a flight mission?

---

*Reentry research was a strong point with AVCO, since it had worked with the Air Force in 1956 and 1957 to develop the heat-sink reentry vehicle.

10. How heavy must a Martian lander be if it is to use a high-gain antenna?

11. Is an orbiter really necessary? For science? For communications relay? If an orbiter is necessary, must it be used simultaneously with a lander? Can basic orbiter spacecraft be designed to be modified efficiently from a science plus relay orbiter to a pure relay orbiter?

12. What is the trade-off between orbiter science weight and lander science weight?

13. What is the *minimum* weight of a pure relay orbiter as a function of data rate? Of a pure science orbiter?

14. How does the weight of a science plus relay orbiter compare to a relay orbiter?

15. Can an orbiter be designed (with or without minor modifications) to perform both Mars and Venus missions? How much of a weight penalty results from designing an orbiter for both Mars and Venus compared to an orbiter for only Mars?

16. What are the critical technology problems associated with Voyager? What is the development time and cost? What will flight units cost? On what experience is this based?

17. Starting with the 1969 opportunity, what type of Voyager program for Mars is possible with the changing energy requirements between 1969 and 1975?[17]

Before any hardware could be developed for Voyager missions to the near planets, all these many complex technical issues had to be resolved by NASA and its contractors. Time, however, was an issue of equal importance. By early fall 1963, no one at the space agency still considered a 1967 launch to Venus practical, and a mission to Mars in 1969 seemed even less likely.[18]

Growing friction between Hearth's office at headquarters and JPL's Advanced Planetary Spacecraft Study Group was another negative factor. The study group continued to stress the orbital portion of Voyager's mission and exclude the lander from its research. During the second phase of its study, which paralleled the AVCO and General Electric contracts, the team in Pasadena turned its attention to orbiter missions in the 2700- to 3175-kilogram class and during a third phase examined the technical aspects of joining and later separating an orbiter and lander. However, the work did not include studies of the lander itself. In fact, the engineers at JPL were growing increasingly skeptical about the desirability of an orbiter-lander spacecraft.[19]

Since NASA Headquarters had assumed control of Voyager, the laboratory managers had become resentful over the allocation of Voyager work and responsibility among the NASA centers. A memorandum for internal use only at JPL recorded that the laboratory had been directed by NASA

Headquarters to terminate its Advanced Planetary Spacecraft Study (APSS) as of September 1963. Later analysts explained JPL's perceptions of the controversy:

> Many factors probably played a role in this decision; one of these was the reporting of recent Mars observations, indicating that the surface pressure was much less than had been previously estimated, making the problem of successful entry and descent more difficult. Another reason appears to be budgetary considerations. *A third reason, though never publicly expressed, may have been related to certain political questions related to the future of the Laboratory and whether or not it was to be directly involved in planetary landing missions.* The fourth and most pressing reason was the initiation of the Mariner 1966 project and the lack of available manpower to support APSS work concurrently.[20]

Early in 1963, three JPL scientists—Lewis D. Kaplan, Guido Munch, and Hyron Spinrad—had revealed new data about Mars that had serious implications for proposed Mars landing studies. The new estimate for the surface level atmospheric pressure was 10–40 millibars, or one-third the previously estimated pressure.[21] Homer Newell called a special colloquium for 1 October 1963 to discuss the subject. As Newell later told members of the Senate Committee on Aeronautical and Space Sciences, a dozen or so planetary astronomers could not agree on the best figure, and their estimates ranged from 10 to 115 millibars. While *Mariner 4* would resolve this issue in the summer of 1965, the uncertainty did cause concerns—though not insurmountable ones—for the Voyager team in the interim. AVCO and General Electric were given an extra month to examine the implications of the lower pressure for their proposed landers.[22]

The JPL budget was not an inconsiderable issue. The space science budget was tied to the shuffling of the Mariner flights during 1963. As Oran Nicks pointed out to the Administrator in a 1963 year-end review, several flights planned for 1964 had been eliminated, including the Venus missions that would have duplicated the successful *Mariner 2*. Turning to the Mars aspect of the planetary program, Nicks told Webb that the two Mariner B flights planned for 1966 had been scrapped because of "recent budget problems for Fiscal Years 1964 and 1965." Mariner B with its small, biologically oriented landing capsule had begun to compete for Voyager dollars. Instead, a reincarnation of Mariner B—Mariner 1966—with a lighter and less expensive capsule had been scheduled for two Centaur-powered flights in 1966 to determine the constituents of the Martian atmosphere and obtain more accurate measurements of the surface pressure. While there was still time to prepare for a Venus mission in 1967, the fiscal 1965 budget crunch seemed to preclude such a flight. If Voyager funds were cut back or dropped entirely from the 1965 budget, a planetary mission would not be possible before 1971.[23]

JPL's contention that the lab's future was inextricably bound to NASA politics over what center would manage the agency's planetary projects, had a hollow sound to it, as did claims about manpower shortage. Hearth and his associates in the headquarters Advanced Programs and Technology Office were the first to acknowledge the crucial and central role that JPL had played in the NASA planetary program, but in a late summer memo Hearth told Nicks that JPL was using Voyager as a hostage to induce the agency to increase its manpower levels. "As you know, JPL has been going through a detailed evaluation of their personnel assignments as a result of their current man-power ceiling." It appeared to Hearth that JPL would not be submitting a proposed project development plan for Voyager or the cost and schedule information that headquarters needed. Apparently, the lab would "dissolve the Advanced Planetary Spacecraft Study Committee which essentially [would] terminate the current Voyager activity at JPL."[24] Simply put, the managers in Pasadena had decided not to work on Voyager during 1964. This did not quite agree with JPL's position that the laboratory had been "directed to terminate its APSS work."

Hearth was sure it would mean trouble for the project if JPL were to use Voyager to garner more job slots, but he argued that without Pasadena's assistance his office would be crippled. "In addition, we cannot propose a program without a center ready and willing to accept project management." Although he could delay his Voyager recommendations to the NASA managers for six months while his team selected another center or for one year while they waited for JPL, either of those delays would "jeopardize the chance for a 1969 Voyager launch." Hearth frankly felt that JPL was being "short-sighted" and would be left "without significant programs in another 2 to 3 years without Voyager." But he also had an inkling that some people at NASA Headquarters also wanted to delay Voyager. "Obviously, NASA management may decide to defer Voyager indefinitely," but he did not want that to happen without their having "all the technical and scientific facts available."[25]

Hearth presented the Voyager case at a December 1963 planetary program briefing for Administrator Webb. Summarizing first the Mariner program to date, he noted that the revised figures for the Mars atmospheric pressure, coupled with budget problems, had led to the termination of Mariner B. To survive a hard landing, a capsule would have to weigh at least 360 kilograms, and Atlas-Centaur could not be expected to deliver more than about 225 kilograms. The new Mariner 1966 would use a chassis like *Mariner 4*'s to transport a small atmosphere probe to Mars. Turning to Voyager, Hearth discussed the JPL, AVCO, and General Electric concepts as they had emerged during the April to October study.[26]

Engineers for AVCO and GE had studied Mars and Venus missions, with AVCO giving Venus greater attention, but it was obvious to both contractors that Mars was NASA's primary target. General Electric recommended two identical landers carried aboard a single orbiter bus. Primary communications from the landers to Earth would be via a relay in the orbiter, with secondary links directly from the landers. Solar cells and batteries would be used to power the orbiter, while radioisotope thermo-electric generators would provide both electricity and heat for the lander. Having concentrated basically on Mars missions, the General Electric engineers emphasized "biological and geophysical-geological experiments," recommending Syrtis Major (10°N., 285° long.) as a landing site for one lander and Pandorae Fretum (24°S, 310° long.) for the second. These were two of the more interesting areas for biological exploration. The appearance of Syrtis Major did not change much with the seasons. Its boundaries "are sharp and stable, and it is one of the darkest areas of the planet." Pandorae Fretum did change with the seasons, the dark color developing in spring, deepening with summer, and becoming light in the fall for the duration of winter. While the choice of these sites would eliminate close examination of the polar regions and the "darkening wave," they considered their choices the best ones "in view of the high priority of the life detection [experiments] and the eventual requirements for choosing sites for manned landing missions."[27] GE would wait until after the first successful landings to define future sites, but AVCO made the proposals in table 11.[28]

General Electric proposed a rather ambitious series of scientific investigations, considering the weight limits on instrumentation for both the orbiter (98 kilograms) and the lander (70 kg). Biological instruments would easily constitute a third of the payload projected for the lander. AVCO Corporation's landed science payload was greater (91 kg), but the proposed orbital instrumentation was less (61 kg). In either case, the weight was substantially more than the 23 kg of experiments that could have been landed with a Mariner B-class capsule. During more favorable Mars launch

*Table 11*
*AVCO Proposals for Missions to Mars, 1963*

| Launch Opportunity | Lander | Landing Site | Latitude | Longitude |
|---|---|---|---|---|
| 1969 | 1 | Solis Lacus | 28°S | 90° |
| | 2 | Syrtis Major | 15°N | 286° |
| 1971 | 1 | South Polar Cap | 83°S | 30° |
| | 2 | Mare Cimmerium | 18°S | 235° |
| | 3 | Lunae Palus | 15°N | 65° |
| | 4 | Aurorae Sinus | 15°S | 50° |
| 1973 | 1 | Propontis | 45°N | 185° |
| | 2 | Elysium | 25°N | 210° |
| 1975 | 1 | North Polar Cap | 78°N | 220° |
| | 2 | Nepenthes-Thoth | 25°N | 225° |

opportunities—1971 and 1973—larger scientific packages could be landed using the same orbiter and launch vehicles.

Besides the weights of the landers (GE, 657; AVCO, 762), the major difference between the two contractors' approaches was the number of landers; one for AVCO and two for GE. AVCO's lander was encapsulated before launch for sterility and for protection during the descent. The blunt body of the aeroshell would protect the lander during entry and slow the descent. A parachute, deployed when the aeroshell and heatshield were discarded, would slow the craft further. At impact, the lander would be protected by aluminum crush-up pads (touchdown velocity 12 meters per second). After a relatively hard landing, the craft would roll and tumble until it came to a stop, and six petals, which when closed protected the internal parts, would open and erect the lander and raise it off the ground. AVCO also planned to use radioisotope thermoelectric generators to provide electricity. General Electric's capsules by comparison were much simpler. They consisted of "moderately blunt sphere cones," which entered point downward instead of blunt end down as with the AVCO approach. General Electric proposed to use rockets, tip bars, and explosive anchors to orient the cone once it was on the surface.

Hearth told Webb at the December briefing that "the areas of agreement were quite significant even though the studies were conducted independently and separately of one another." Both contractors called for similar scientific capabilities, and "they agreed quite well on cost and what the prime technical problems and development problems" were. But would NASA underwrite Voyager missions to the planets beginning in 1969?[29]

### Mariner 1966 and Advanced Mariner

Hearth's attempt to sell the NASA management on a 1969 Mars Voyager was unsuccessful. The administrator decided that the resources required—manpower and dollars—made it too ambitious for a 1969 mission. He preferred to defer the first Voyager launch until 1971. With the first manned lunar landings accomplished, the space agency would be under less political and financial pressure, and Voyager could proceed. To fill the gap between the 1964 Mariner C flyby and the 1971 Voyager orbiter-lander, NASA's planetary program staff proposed to add a 1968–1969 Advanced Mariner to the schedule to supplement a Mariner 1966 Mars atmospheric capsule mission.[30]

A Mariner 1966 mission would "make maximum use" of Mariner 1964 technology.[31] Plans called for a nonsurviving atmospheric capsule that would crash onto the Martian surface after it had relayed its scientific data. But not everyone favored the concept, since it added new technological problems in several areas—planetary atmosphere entry dynamics, communication links between a flyby craft and capsule, and sterilization. NASA planners began discussing a 1966 capsule in January 1964, and it quickly became apparent that JPL did not favor the idea.[32]

*Table 12*
*Voyager System Weights from 1963 Contractor Studies*

| System | General Electric (kg) | AVCO (kg) |
|---|---|---|
| **Orbiter** | | |
| Structure | 190 | 147 (includes thermal |
| Harnessing | 48 | ---- control) |
| Power supply | 99 | 209 |
| Guidance and control | 103 | 84 |
| Communications | 131 | 128 |
| Thermal control | 40 | --- |
| Propulsion (dry) | 212 | 209 |
| Diagnostic instrumentation | 13 | --- |
| Payload (scientific) | 98 | 61 |
| | 934 | 838 |
| **Lander** | 2 landers | 1 lander |
| Heatshield | 41 | 204 (includes structure) |
| Structure | 181 | 95 (adapter sterile can) |
| Retardation | 72 | |
| Thermal control | 41 | |
| Power supply | 51 | 136 |
| Orientation | 26 | |
| Communications | 65 | 91 |
| Payload deployment | | 145 (touchdown and |
| and installation | 25 | deployment) |
| Spin and separation | 19 | |
| Retrorocket | 45 | |
| Adapter and radiator | 21 | |
| Payload (scientific) | 70 | 91 |
| | 657 each | 762 |
| | 657 | |
| **Fuel** | | |
| Orbit insertion and midcourse | 939 | 1361 |
| **TOTAL** | 3187 | 2961 |

SOURCE: General Electric Co., Missile and Space Div., Valley Forge Space Center, "Voyager Design Study," vol. 1, "Design Summary," 15 Oct. 1963; and AVCO Corp., Research and Advanced Development Div., "Voyager Design Studies," vol. 1, "Summary," 15 Oct. 1963, p. 111. All metric conversions are to the nearest kilogram.

By mid-March, Hearth told Oran Nicks that he was compelled to recommend eliminating the capsule from the proposed Mariner 1966 mission. JPL, understaffed and unenthusiastic, would not support the project if it included a capsule, and it was too late to assign the "entry probe" to another center. Considering the technical risks of the capsule, Hearth had to yield in face of the laboratory's intransigence.

NASA's fiscal 1965 budget would not support the Mariner 1966 project either. The $5.25 billion approved by Congress was $195 million less than

*Table 13*
*Experiments Recommended for Voyager 1969 in 1963 Contractor Studies*

| General Electric | AVCO |
|---|---|
| Orbiter | Orbiter |
| Biological | Biological |
| Television survey | Infrared spectra of surface |
| Infrared spectrum survey | |
| Geophysical-geological | Geophysical-geological |
| Stereo-television mapping | Television mapping |
| Magnetic field survey | Magnetic field survey |
| Charged particle flux survey | Radio absorption (lander to orbiter) |
| | Spectral albedo. |
| Atmospheric | Atmospheric |
| Ionosphere profile | Infrared radiometry of surface |
| Infrared emission | |
| Space environment | |
| Micrometeoroids | |
| Magnetic fields | |
| Landers (2) | Lander (1) |
| Biological | Biological |
| Growth | Biological detection |
| Metabolic activity | Microscopic examination of soil |
| Existence of organic molecules | Chemical structure of soil |
| Existence of photoautotroph | |
| Turbidity and pH changes | |
| Microscopic characteristics (TV) | |
| Organic gases | |
| Macroscopic forms (TV) | |
| Surface gravity | |
| Geophysical-geological | Geophysical-geological |
| Surface penetrability | Television mapping |
| Soil moisture | Magnetic field |
| Seismic activity | Solar optical absorption |
| Surface gravity | |
| Atmospheric | Atmospheric |
| Temperature | Temperature |
| Pressure | Pressure |
| Density | Density |
| Composition | Composition |
| Altitude | Wind velocity |
| Light level | |
| Electron density | |

SOURCE: General Electric Co., Missile and Space Div., Valley Forge Space Center, "Voyager Design Study," vol. 1, "Design Summary," 15 Oct. 1963, p. 2-2; and AVCO Corp., Research and Advanced Development Div., "Voyager Design Studies," vol. 1, "Summary," 15 Oct. 1963, p. 9.

the agency had requested. Administrator Webb announced that NASA would maintain the momentum and direction of its programs despite the loss of anticipated funds, while meeting its lunar goals. Although the decision did not "involve the transfer to manned space flight of funds from space science," those programs would "require some adjustments." Mariner 1966, however, was doomed. According to the news release issued at NASA Headquarters, "the combination of a heavy workload at the Jet Propulsion Laboratory, the short lead time available, and the importance of applying our resources to a major advance beyond the limited Mariner" made it "unwise" to undertake a Mars mission in 1966 with the current Mariner spacecraft. Development of a spacecraft "with much greater scientific promise for launch to Mars in 1969" was being initiated.[33]

Canceling the 1966 capsule called for changes in Mariner 1964 and the Advanced Mariner (Mariner 1969). Hearth recommended flying the 1964 Mariner on an occultation trajectory—the spacecraft would fly behind Mars as viewed from Earth. A radio signal would be transmitted as the craft approached the planet, and that signal would be blocked as the craft passed behind it. Analysis of the behavior of the radio signal could determine more precisely the composition and density of the Martian atmosphere.*[34]

At the loss of the 1966 Mars mission in July 1964, Hearth called for an immediate study of the capsule for the 1969 Mariner. Early study was essential if Nicks' Lunar and Planetary Programs Division was to coordinate its plans effectively with Orr Reynolds' Bioscience Programs Division, which was working toward a 1 August 1964 deadline for a proposal for a "minimum acceptable" biological lander payload for 1969. Hearth believed that should sufficient information be "obtained over the next three years on the Martian atmosphere, . . . a survivable biological lander is possible in 1969." He also thought that a lander mission was "preferable over an orbiter mission although the orbiter will be given careful study."[35] Hearth explained this in detail for Nicks because he did not believe that JPL could handle the entire Advanced Mariner mission, even if industrial contractors were used. The problem as Hearth saw it was choosing a NASA center to assist JPL. To assign Mariner 1969 to one organization and Voyager to

---

*"If all other factors producing apparent motion of the spacecraft were accounted for (e.g., the actual motion of the spacecraft, the motion of the deep space stations on the rotating Earth, the lengthening of the transit time of the signal, and the refractivity of the Earth's lower atmosphere), the remaining unexplained changes in the radio signal could be attributed to refraction by the atmosphere of Mars. (For a successful experiment, it was necessary to account for the total change in frequency or phase of the signal due to all causes other than refraction by the Martian atmosphere to an accuracy of at least one part in $10^{11}$.) Since the geometry obtained from the estimated trajectory is known, the measured changes could be used to estimate the spatial characteristics of the index of refraction (or refractivity) in the electrically neutral atmosphere and electrically charged ionosphere of Mars. Thus, by measuring and then analyzing the changes in the characteristics (frequency, phase, and amplitude) of the radio signals from the spacecraft, it was hoped to learn more about the composition, density, and scale height of the Martian atmosphere." NASA, *Mariner-Mars 1964: Final Project Report*, NASA SP-139 (Washington, 1967), pp. 316–17.

another would be unwise, because "the missions and spacecraft are too closely related." For Hearth, the only solution was to assign another center the responsibility for some portion of either Mariner 1969 or Voyager. "It is logical that this be the capsule. There is no question that such an arrangement will be difficult, to say the least," but he could see no alternative. Three centers could possibly assist JPL with its planetary work—Goddard, Langley, or Ames. Because of their earlier interest in the landing capsule for Mariner 1966, Hearth recommended the Ames Research Center team at Moffett Field, California.[36]

## MISFORTUNES OF VOYAGER

During the financial belt-tightening related to the fiscal 1965 budget, there was growing pressure from Congress, the Bureau of the Budget, and the White House to hold down costs. Congressional concerns became particularly strong following the failure of *Ranger 6* to transmit any of its prescribed 3000 pictures of the lunar surface before it crashed into the moon on 2 February 1964. The representatives on Capitol Hill told Webb and his associates that no more failures would be tolerated.

### Phased Project Planning

Joseph Karth, acting chairman of the House Subcommittee on NASA Oversight, was particularly bothered by the apparent weakness of the managerial chain between NASA Headquarters and Jet Propulsion Laboratory. Karth and other congressmen were rightly worried, since JPL was responsible for several key projects in addition to Ranger—Lunar Surveyor and the planetary Mariners, with Voyager likely to be the lab's next big project. Over the years, Karth and his staff had seen instances of JPL management resistance or reluctance to accept organizational and procedural changes recommended by NASA Headquarters. The *Ranger 6* failure gave everyone—congressmen, NASA managers, JPL staffers—the opportunity to reflect on the need for better program management in general and closer liaison between NASA managers and the California Institute of Technology-JPL team in particular.[37] One of the tools Administrator Webb chose to strengthen his managerial control over all new projects was Phased Project Planning.* This scheme played an especially important role in the subsequent life and death of Voyager.

In mid-July 1964, Associate Administrator Robert C. Seamans, Jr., advised that all "new projects should be planned on a phased basis with successful contracts for advanced studies, program definition, prototype design, and flight hardware and operations." Phased development would

---

*It has been noted that Phased Project Planning bears remarkable resemblance to the Air Force approach to systems management—conceptual phase, definition phase, acquisition phase, operation phase—as set forth in the Air Force Systems Command's 375 manual series. Arnold A. Levine, *Managing NASA in the Apollo Era*, NASA SP-4102 (Washington, 1983)

permit projects to "evolve in an orderly manner with maximum realism."[38] Voyager was one of about half a dozen new projects on which the headquarters staff experimented with the new procedure months before the official guidelines were promulgated in October 1965.

After nearly three years of advanced Voyager studies by JPL and others, the NASA managers took the initial steps in December 1964 to place the planetary project on the phased track to a 1971 mission to Mars. The decision came after four months of hectic conferences in Washington, during which Mariner-Mars 69 was approved (12 August), agonized over (September through October), and terminated (20 November). The prolonged debate was the result of Homer Newell's belief that a 1969 mission was necessary to satisfy the scientific community and Congress but, knowing that fiscal year 1966 funds for both Mariner-Mars 1969 and Voyager 1971 were not likely to be appropriated, NASA finally canceled Mariner-Mars 69 in an attempt to preserve Voyager. No one was happy with the compromise.[39] Donald F. Hornig, President Johnson's special assistant for science and technology, was dismayed over the loss of yet another Mars launch window in 1969. Seamans assured him that, if at all possible, some kind of flight, perhaps a Mars flyby that would test the basic 1971 Voyager without a lander, would be attempted in 1969. Still, the associate administrator noted that the money for Voyager was going to be tight. Four flights, two in 1971 and two in 1973, were expected to cost $1.25 billion. With that kind of price tag, a 1969 mission might have to be dropped in favor of less expensive test flights.[40]

While various persons continued to express unhappiness about the loss of another Mars opportunity, Seamans signed the project approval document for Voyager on 16 December 1964. During that same week, Don Hearth, slated to become Voyager project manager at headquarters, submitted his suggestions for the Voyager office in Washington.[41] Voyager was officially on its way. The first external step was the announcement on 15 January 1965 of requests for proposals from industrial contractors to work under JPL's direction on the preliminary design, phase IA of the phased program.[42]

The 22 January proposers conference at JPL was attended by 113 representatives from 28 companies. Three months later, after an elaborate source selection process, three firms were selected to make 90-day preliminary design studies: the Aerospace Division of the Boeing Company, Seattle; the Missile and Space Division of General Electric, Valley Forge; and TRW Space Technology Laboratories, Redondo Beach, California.[43] As the contractors began their work, Seamans, Newell, and other top NASA managers went to the Congress to explain Phased Project Planning, their hopes for Voyager, and their projections for its cost. The fiscal 1966 appropriations hearings proved as difficult as those of the preceding year.

President Johnson on 25 January 1965 recommended a $7.114-billion space budget for fiscal 1966. Of this amount, NASA would receive $5.26

billion, the Department of Defense $1.6 billion, the Atomic Energy Commission $236 million, the Weather Bureau $33 million, and the National Science Foundation $3 million. Of the NASA request, $43 million was earmarked for Voyager. Associate Administrator Seamans labeled the budget austere, but he said that the chances of landing Apollo's first crew on the moon on schedule were still good. He said that the $43 million, to be spent on further defining the Voyager orbiter and lander, would allow the agency to meet its milestones—a Mars flyby test of the spacecraft in 1969 and complete missions in 1971 and 1973.[44]

In testimony before Congress, Seamans, Newell, and Cortright explained phased planning and its applications to Voyager. Such planning gave design engineers the chance to refine project details incrementally, while the agency's managers maintained the big picture with all its critical milestones clearly delineated. Implicit in phased project planning was the assumption that the process would allow choice of the best technological alternatives. But phased planning was a double-edged management tool. By clearly delineating important decision points, it could be used to force the redirection or termination of a project. For Voyager it did both.[45]

As the contractors worked on the first phase (3 May to 30 July), several factors came to the attention of NASA managers that affected the execution of phase IB, an in-depth study of the lander. Once again, the agency was called on to tighten its programmatic belts; the budget request for $5.26 billion yielded an appropriation of $5.175 billion for fiscal year 1966.

*Table 14*
*NASA Budget Summary, Fiscal 1963 to 1966*
(in billions)

| Year | Budget Request | Authorization | Appropriation |
|------|------|------|------|
| 1963 | $3.7873 | $3.7441 | $3.6741 |
| 1964 | 5.7120 | 5.3508 | 5.1000 |
| 1965[a] | 5.4450 | 5.2275 | 5.2500 |
| 1966 | 5.2600 | 5.1904 | 5.1750 |

SOURCE: NASA, "Back-up Book—FY 1976, Hearings," sec. 6.
[a]Includes $141 million supplemental request; the appropriation includes a supplement of $74.5 million.

Voyager, as a new start, was vulnerable, but other projects such as the adaptation of the Centaur to the Saturn IB were also at risk, since such development diverted money away from the completion of the Saturn V, Apollo's powerful booster.

The unfavorable budget was trouble enough without the additional bad news brought by the radio occultation experiment aboard *Mariner 4*.[46] The Martian atmosphere was much less dense than previously estimated.

All proposals for landing capsules had to be thrown out as new aerodynamic analyses were performed based on the much lower pressure range (4–7 millibars, rather than the earlier estimates of 10–30 millibars).

The latest Mariner findings also jeopardized use of the Saturn IB launch vehicle, on technical grounds, adding to its financial difficulties. Given the 3000-kilogram weight limit for the spacecraft, much of the scientific payload would have to be sacrificed to provide the lander with additional means for slowing its descent through the thin Martian atmosphere. No matter which approach to the problem was taken—larger aeroshell, braking rockets, larger parachutes—it would mean too much weight for the Saturn IB. The larger Saturn V could provide the extra booster power, but it was seemingly too powerful and too costly to be realistic.

## Voyager Capsule Advisory Group

As early as March 1965, Oran Nicks and the Lunar and Planetary Program Office had begun plans for a Voyager Capsule Coordination Group to control studies being conducted at JPL and at the Ames and Langley Research Centers.[47] After preliminary meetings at which the centers exchanged information on their capsule activities, Homer Newell set up a panel of experts* to advise Don Hearth, Nicks, and the space science office on two basic questions:

> 1. Is the Martian atmosphere and surface sufficiently well known at this time to permit the design of a survivable capsule to be included in the 1971 operational Voyager mission, or will the design of such a capsule have to be based upon the results of a non-survivable atmospheric probe and/or other measurements to be made during the 1969 opportunity?
>
> 2. If the Voyager Program is to proceed on the basis of a survivable capsule in 1971, what general size and type of capsule should be selected?[48]

The concern at NASA Headquarters over the safe landing of Mars capsules was not totally spontaneous. For a number of months, this topic had been discussed throughout the U.S. space community. During the American Astronautical Society Symposium on Unmanned Exploration of the Solar System in early February 1965, the disagreements over priorities in Mars exploration bubbled to the surface. Some of the symposium participants wanted the 1969 atmospheric probe reinstated. Alvin Seiff, chief of the Ames Vehicle Environment Branch, was the leading proponent of an 11-kilogram Mars atmospheric probe. Others thought that 1971 was too early for a landing.

Implicit in this disagreement was a difference of opinion about the kinds of landers to be used and the best time to land the first life detectors. Whereas Seiff and his colleagues at Ames favored hard-landers, or "crash-

---

*J. E. Naugle, chairman, P. Tarver, E. Levinthal, U. Liddle, J. Hall, O. Reynolds, C. Goodman, R. F. Fellows, F. Johnson, H. M. Schurmeier, C. F. Capen, L. Lees, and G. Munch.

ers," Langley designers wanted soft-landers. Between them were men like Gil Levin, who wanted to get on with biological investigations at the very earliest opportunity, and Temple W. Neumann, program engineer for the NASA-sponsored automated biology laboratory being developed at the Aeronutronics Division of Philco. Neumann told the symposium participants that a biology laboratory could be hard-landed as part of a 1971 Voyager mission without prior detailed mapping of the Martian surface.[49] He, too, was ready to proceed.

Bruce Murray, a planetary astronomer at the California Institute of Technology and chairman of the Cal Tech-JPL planetary exploration study group, argued for a more evolutionary approach. At the Denver symposium, Murray remarked on the need for large-scale photographic mapping of Mars before landers could be safely deposited on the surface. Finding a satisfactory site, landing a craft there, and interpreting the biological instrument results would require a great deal of work and several hundred times more photographs than the 20 or so expected from *Mariner 4*.[50]

Gil Levin, the father of the biological sampler Gulliver, put his finger on another recurring concern when he noted that the Soviet Union would probably beat the United States to a Mars landing. In addition to capturing yet another first in the international space sweepstakes, Levin feared that the Soviet Union would contaminate the Martian surface. He reported that the Soviet Academy of Sciences did not appear to have an interest in completely sterilizing its spacecraft, putting the American program in an awkward position. The NASA team wanted to reach Mars ahead of the USSR so it could be certain of examining an undisturbed, uncontaminated planet, but NASA needed more time to develop its own sterilization techniques.

Levin's remarks were sparked by Homer Newell's statement that only rugged experiments and small capsules that could withstand existing sterilization procedures would be flown at first. Initial studies had indicated that the larger and the more complicated the lander, the greater the technical difficulties of sterilization. Components and assemblies had to be developed that could withstand sterilization temperatures ($135°$–$150°C$) and still perform satisfactorily after months in the cold void of space. By early 1965, the Josh Lederberg–Elliott Levinthal team at Stanford was realizing that the biggest problem facing the multivator life detector was the creation of chemical compounds that would not be rendered useless when heated to such extreme temperatures. On the other hand, the radioisotopes used in Gulliver were not heat-labile (subject to breaking down when heated). Levin was ready to send a Gulliver to Mars, but other experimenters needed more time.[51]

Amid the controversy over the timing and nature of Mars capsules and landers, the formation of the Voyager capsule advisory group was a prudent act, as the initial scientific results from *Mariner 4* confirmed. Turning to the questions posed by Newell when he established the panel, group chairman John Naugle reported at the end of August that new observations, including the Mariner occultation experiment, indicated that the lower limit for the surface pressure was in the region of 10 millibars. Furthermore, "in view of the agreement between the ground based and occultation studies, it appeared to the group that . . . the information that could be obtained from a 1969 atmospheric probe would not warrant its inclusion in the Voyager program." The 1969 atmospheric mission was eliminated once and for all.[52]

The new atmospheric data raised questions of equal significance about the possibilities of safely landing a capsule in 1971. At NASA Headquarters, Newell and his associates decided to postpone the scheduled request for proposals on the preliminary design of landing capsules until "the implications of the apparent low Martian surface pressure are determined."[53] While delaying the next step of the phased project plan gave the NASA managers time to think, it also helped to blunt the momentum necessary to the survival of such projects.

*Saturn IB–Centaur vs. Saturn V*

After several weeks of study, accompanied by many leaks to the news media, NASA Headquarters officials announced in mid-October 1965 that development of the Saturn IB–Centaur would be terminated and that Voyager would be launched with the 33 360-kilonewton (7.5-million-pound-thrust) Saturn V booster.[54] The decision had a number of cascading results. First, since Saturn V was not scheduled to fly until 1967 and the early production was assigned to Apollo, there would be no 1969 Voyager test flight. The 1971 lander mission would have to be delayed until 1973, and the 1971 flight opportunity would be dedicated to an orbiter mission without a lander. Second, morale suffered. Within NASA and contractor circles, people were discouraged by another two-year postponement. Congressional and press reactions were equally gloomy. But more telling was the effect Saturn V had on the space science budget. Total costs for a rescheduled project based on the large Saturn soared, and some estimates ran as high as $2 billion. Greater costs in a period of tightening agency budgets did not argue well for the survival of Voyager.[55]

Since Voyager planners had resisted the use of the Saturn V launch vehicle for several years, the switch came as an unpleasant shock to many. During 1964, JPL had commissioned General Electric to study the possible use of the Air Force Titan IIIC or NASA's Saturn V in place of the Saturn IB–Centaur.[56] In evaluating these and other studies, Ed Cortright concluded that the Titan IIIC-Centaur launch vehicle would not be powerful enough. Whereas Saturn IB–Centaur could boost a 2700-kilogram payload, Titan-Centaur could lift only 1270 kilograms. NASA planners were also hesitant to use the Titan because it was an Air Force booster. In addition to pursuing the basic principle of not becoming involved with too many different launch vehicles, Webb, Dryden, and Seamans—after their experiences with Atlas-Centaur—wished to stay clear of military boosters. And, although using the Titan IIIC would have saved about $10 million per Voyager launch, the dollars spent on Titan would have diverted money from the development of the Saturn family while purchasing an inadequate launch vehicle for Voyager.[57]

Whereas the Titan IIIC-Centaur combination was not powerful enough for Voyager, Saturn V was too powerful. In February 1963, Don Hearth had told Webb that the 18 000-kilogram payload capacity of Saturn V was 6.6 times that needed for first Voyager flights. "In addition," he noted, "we recognize that Apollo will place heavy demands on the Advanced Saturn launch vehicle during the time period of interest for Voyager."[58] By mid-1965, Saturn V was still too big for Voyager, unless two were flown at the same time, but the desire to keep that launch vehicle in production beyond the first lunar missions made it appear more practical for use in the planetary program.

The Saturn IB-Centaur combination was considered a diversionary project by many managers, diverting monies that could be used for the larger booster. Seamans wrote White House officials in late 1965 to that effect: " . . . the development cost of combining Centaur with Saturn IB would peak in FY 1966, 1967, 1968, while relatively little vehicle development effort is required to use Saturn V." Although the first flight of the advanced launch vehicle was still two years away, Seamans noted that "the projected cost of one additional Saturn V for 1971 and later Voyager flights is probably about equal to two Saturn IB Centaurs." As the year ended, the NASA managers believed that Saturn V was "a technically feasible and economic vehicle for Voyager [launching two spacecraft on one vehicle], with as great a probability of mission success as separate launches of smaller vehicles."[59]

Management's acceptance of Saturn V was not enthusiastically received throughout the agency. Newell's "space science people were sort of horrified at the thought of using Saturn Vs."[60] There was no absolute certainty that two spacecraft could be launched by one of the big boosters at about the same cost as two Saturn IB-Centaur combinations. There was surely less flexibility. If budgets tightened further, at least one flight could be made at each opportunity with the smaller vehicle. With Saturn V, two very large spacecraft were required for each launch.

At JPL and elsewhere, the launch vehicle switch was viewed with some suspicion. JPL staffers "felt Headquarters used the finding of [new data on the Martian atmosphere] as a rationalization for concepts they were already 'enamored of' such as out-of-orbit landing and mammoth scientific payloads, without adequately considering either the feasibility of some reasonable alternatives or the effects at the project level." There was also the belief that Webb had decided to force Saturn V on Voyager to maintain the Saturn production line and keep the Marshall Space Flight Center team "happy and working." Many persons at the project working level were afraid that headquarters did not understand how disruptive the decision could be for Voyager.[61]

Angered and dismayed because it had not been properly consulted about the decision, the JPL team believed there were several explicit reasons for not using Saturn V. Although launch vehicle cost was usually a small part of a planetary mission cost, the team feared use of the Saturn V would make the program too costly because increased payload capability would "escalate the cost of the spacecraft." It also would be too big a technological leap over the Mariners. And it might lead to a program "too big for JPL to handle alone or perhaps even to oversee."[62]

If the change from Saturn IB–Centaur to Saturn V was bad news in Pasadena, the cancellation of the 1971 Voyager mission was worse. On 22 December 1965, a little more than two months after the October launch vehicle decision, Homer Newell's office notified JPL that there would be no 1971 mission. On the 22d, NASA announced publicly that Voyager would not fly until 1973. To replace the 1971 orbital Voyager, the agency planned a 1967 flight to Venus using the *Mariner 4* backup spacecraft modified for this new purpose. In 1969, a pair of heavier Mariner-class craft would be launched by Atlas-Centaur boosters. In 1973, after passing up the 1971 opportunity, two identical Voyager craft would be launched to Mars by a single huge Saturn. According to this plan, both spacecraft would orbit Mars and release large landing capsules that would search for evidence of Martian life. Under the revised phased plan, capsule procurement would begin in late 1966 or early 1967.[63]

The 22 December 1965 decision was more than just another delay; it was the death knell for Voyager. In a published interview, Hearth admitted that work on Voyager spacecraft would "go on a low back burner basis for the next year and a half to two years before [it was picked] up again."[64] JPL would continue design work on landing capsules with support from Langley and Ames, but the next phase of the procurement cycle would be delayed "for some time."

The immediate reason for canceling the 1971 flight was a lack of funds. NASA had hoped to obtain $150 million in the 1967 budget with which to start hardware development for Voyager, but the Bureau of the Budget slashed the $5.6-billion overall request to $5.012 before it went to Capitol Hill. Since Apollo and Surveyor were reaching critical periods in their maturation, the planetary program took the greatest cuts. Voyager was allocated only $10 million. As Webb subsequently informed Sen. Clinton P. Anderson, chairman of the Committee on Aeronautical and Space Sciences, "The President specifically rejected the initiation of the Voyager program in the FY 1967 budget. In his consideration of the requirements of the space program for FY 1967, the President specifically included limited funds to permit continued study of the Voyager system aimed toward a 1973 Mars landing mission."

Looking back, Homer Newell concluded that NASA could not have managed two large programs simultaneously—there was just not enough money for the moon and the planets.[65]

For the next 22 months, Voyager continued at a reduced pace. The paperwork multiplied for all concerned, but the avalanche of correspondence and reports was misleading, for the agency's money and enthusiasm went elsewhere. Some dollars were reprogrammed to begin work on the 1967 Mariner Venus flyby and the twin 1969 Mariner Mars flybys. NASA could finally fall back on Mariner missions launched by Atlas-Centaur, since that vehicle was approaching flight readiness. To all concerned, it was apparent that in times of tight budgets it was easier to rely on existing and proved hardware, like Mariners, than to take the step up to more advanced technology.[66]

*Mission Guidelines and Management Assignments*

From January through September 1966, the JPL Voyager team under Don Burcham's direction prepared more than a dozen Voyager project estimates. Each of these lengthy documents detailed alternative missions and the technological and scientific tradeoffs required to execute a planned series of four Voyager flights for 1973, 1975, 1977, and 1979. These estimates were gigen to the JPL managers, the Voyager capsule advisory group, and the space science office team during a series of reviews from July through October. In mid-September, Voyager Project Estimate-14 was presented to Newell and his staff. This document, called a "feedback VPE" because it included many space science office recommendations, was approved in a revised set of Voyager project guidelines sent to JPL by Newell on 5 October. But some of the modifications of the plan upset JPL. The big change was that headquarters wanted the lab to examine the pros and cons of launching two orbiter-lander combinations that carried different—rather than identical—experiment payloads, with the possibility of a direct-entry landing instead of delivering the lander from an orbiting bus.[67]

In an attempt to secure approval for the development of the capsule systems (phase B of the procurement plan), JPL managers made their VPE-14 presentation to Associate Administrator Seamans on 17 October 1966.[68] But before any action was taken on phase B, considerable discussion on the best management arrangement for Voyager had to take place during the winter months. When finally signed on 27 January 1967, the project approval document for phase B called for a Voyager Program Management Office to parallel the Lunar and Planetary Program Office within Newell's Office of Space Science and Applications. Like Apollo, the Mars project had grown enough in size, duration, and cost to be called a program.[69]

Other changes proposed in the approval document were more significant, and from the JPL point of view revolutionary. Von Braun's Marshall Space Flight Center would be established as the management organization for both the Voyager spacecraft and the Saturn V launch vehicle. JPL and

Langley would work together on the development of lander systems and report to Marshall. This plan was never executed because a disaster in the Apollo program diverted NASA's attention from planetary missions. On 27 January, the day the project approval document was signed, a flash fire killed three astronauts during a test of the Apollo 204 spacecraft. The tragedy profoundly unsettled the American space program. As the agency investigated the awesome fire, Webb decided in early February to delay assigning responsibility for Voyager to Huntsville. In the interim, the administrator approved the creation of a Voyager Program Office in Newell's organization and a Voyager Interim Project Office in Pasadena. Oran Nicks would be program director and Hearth his deputy and acting project manager. The California office would be abolished once the project was assigned to another center.[70]

In discussing these changes with Webb and Seamans, Newell remarked that the transfer of project management from JPL to the Interim Project Office had been made because the next nine months were critical in preparing Voyager for its 1973 launch date. He also noted that they must "continue to draw heavily upon the existing project management team in JPL during the transition." Hearth's team of 77 persons began operation in a downtown Pasadena bank building on 20 March 1967.[71]

In Washington, meanwhile, Seamans, Newell, Cortright, and Nicks were explaining the agency's Voyager decisions to Congress. After the Apollo fire, the congressmen tended to be even sharper in their questioning, and they no longer accepted as readily the rationale of a race with the Soviet Union for first place on Mars. Representative Karth questioned the wisdom of assigning Voyager tasks to different organizations. Pursuing rumors that JPL was being deprived of Voyager management so that Marshall would be certain to have an adequate workload in the post-Apollo period, Karth asked if the split in responsibility had come about "as a result of certain" NASA centers running out of work for the future. He did not really expect the NASA officials to answer such a question in the affirmative, but he confessed that the new arrangement appeared suspect after "some 5 or 6 years of experience with the Voyager program." Ed Cortright responded that it would not be in the government's interest to enlarge JPL, a contractor, at a time when the agency's centers were likely to be cut back, especially when Marshall had personnel available from a phased-down Saturn program and Langley had pertinent, valuable skills developed from its management of Lunar Orbiter.[72]

Several years later, a Harvard Graduate School of Business Administration team studied the Voyager management shift and, while reflecting something of a JPL bias, questioned NASA's judgment:

> . . . as of the middle of 1967, the Voyager Program had an unusual and complex management structure. Much of the actual work was still being done at JPL, which was technically a contractor associated with OSSA [Office of Space Science and Applications], even though its official role

109

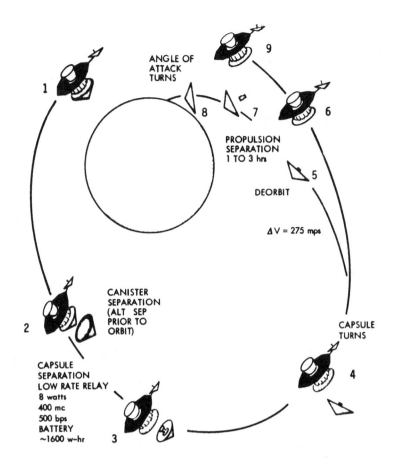

*The September 1966 JPL Voyager Project Estimate-14 briefing gave a profile of the planned orbital operations of the Voyager spacecraft. 1—The Voyager craft approaches the point of insertion into orbit of Mars. 2—After orientation of the orbiter (capsule), the lander in its aeroshell (canister) separates from the orbiter. 3—The orbiter and lander continue around the planet. 4, 6, 9—The orbiter is turned to achieve the attitude for communications with Earth. 5—A retrorocket impulse alters the velocity of the lander by 275 meters per second, causing it to deorbit. 7, 8—One to three hours after deorbit, the propulsion unit on the lander canister is released and the canister is oriented for final approach to the Mars surface. VPE-14 Project Study, September 1966.*

was much reduced. Two [Office of Advanced Research and Technology] centers, Ames and Langley, were involved in capsule work with Langley being given responsibility for the capsule bus system. Kennedy and Marshall, two [Office of Manned Space Flight] centers, were also on board. . . . On top of this structure was the [Voyager Interim Project Office] an arm of OSSA but staffed from the centers, and of course there was the program office at Headquarters in Washington.[73]

## Voyager Terminated

The viability of the new management arrangement became a topic only for conjecture because Voyager was canceled in 1967 (see appendix B for a summary of Voyager project highlights, 1966–1967). The cancellation was only one of a series of interlocking circumstances, which taken together remind us that 1967 was an unhappy year for the United States at home and abroad. Foremost among the problems facing the nation was the war in Southeast Asia. More than a half million Americans were on military duty in Vietnam. By 1967, nearly 25 000 had died in a conflict that was costing taxpayers at home $2 billion monthly. With each new expenditure in Vietnam, the Johnson administration was faced with a growing budgetary deficit, which forced the president to reduce nondefense expenditures and raise taxes. If no other factors had conspired to undermine the planetary projects NASA wanted to pursue, the cost of the Vietnam war alone would have diminished the chances for a big Mars mission. But other factors did also conspire against Voyager.

*JPL engineers also outlined their plan for landing Voyager. At about 6100 meters, the craft would be traveling 140–335 meters per second, depending on the density of the Martian atmosphere. To slow the lander canister, braking rockets would fire. At about the same time, the inertial guidance system and the radar altimeter would be activated to control the final approach. At a slant range of 610 meters, the lander would be pyrotechnically separated from the aeroshell. By this time, the craft would have slowed to 45–105 meters per second. At 25 meters from the surface, it would stabilize at 1.5 meters per second by firing the terminal descent engines. The engines would shut down at 3 meters to prevent undue alteration of the terrain. At touchdown, the lander would be traveling 3–8 meters per second. VPE-14 Project Study, September 1966.*

SLANT RANGE = 2000 ft
SEP AEROSHELL
VEL 150-350 fps
DEPLOY LEGS

SLANT RANGE MARK 20,000 ft
ALT 11,000 - 20,000 ft
VEL 450-1100 fps
(MACH 0.5 TO 1.20)
IGNITION
INERTIAL ATT. STAB.
RADVS ACQUIRE
SWITCH TO RADVS

ALT = 80 ft
BEGIN 5 fps
DESCENT

ALT = 10 ft
CUT-OFF
VEL ≈ 5 fps

TOUCHDOWN
VEL ≈ 10-25 fps
CONTINUE LOW RATE RELAY
FOR ABOUT 5 MIN

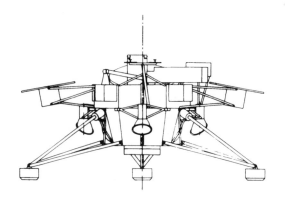

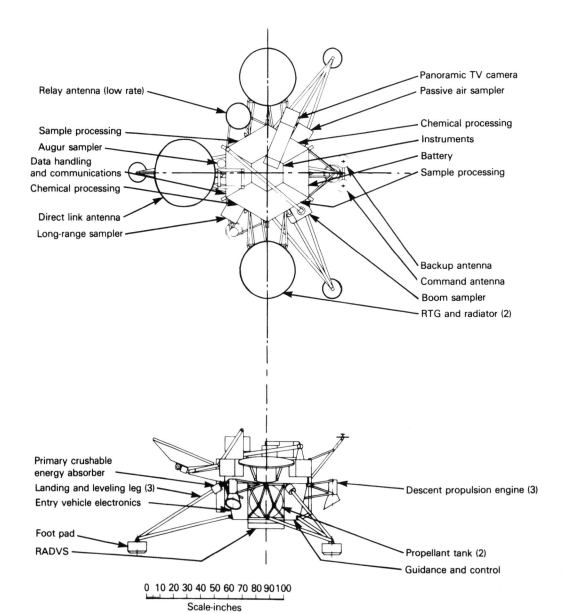

Relay antenna (low rate)

Sample processing

Augur sampler

Data handling
and communications

Chemical processing

Direct link antenna

Long-range sampler

Panoramic TV camera

Passive air sampler

Chemical processing

Instruments

Battery

Sample processing

Backup antenna

Command antenna

Boom sampler

RTG and radiator (2)

Primary crushable
energy absorber

Landing and leveling leg (3)

Entry vehicle electronics

Foot pad

RADVS

Descent propulsion engine (3)

Propellant tank (2)

Guidance and control

0 10 20 30 40 50 60 70 80 90 100

Scale-inches

There were the growing costs of Apollo, escalated further by the fire. As one reporter deduced, "The explosive spacecraft fire that killed three Apollo astronauts . . . may seriously delay unmanned spacecraft space projects as well as those involving man." The Apollo setbacks would cost more money—money that had been earmarked for Voyager and other planetary projects. The Office of Space Science and Applications had asked for $695 million for 1968 (an increase of $88 million over 1967) to provide funds for Voyager ($71.5 million). Now, noting that the orbiter-lander project had been "on NASA's back burner for about three years as a result of one budgetary crisis after another," the newspapers reported that the proposed 1973 landing date was "no longer realistic in view of the added costs likely to be imposed as a result of the Apollo accident."[74]

A secondary budget problem for Voyager was growing cost projections within the program itself. In House and Senate hearings, NASA representatives were questioned about the total estimated cost for Voyager. Sen. Margaret Chase Smith of Maine asked Webb for his best total cost figure. He responded with $2.2 billion for research and development through fiscal 1977. On top of that were "administrative operations costs—that is the salaries of our civil service personnel," as well as $40 million for facilities and $55 million for two additional 64-meter radar tracking antennas for the Deep Space Network, which could be used for other projects, too.[75] Voyager's growing price tag and the general record of NASA's cost predictions prompted Representative Karth to lecture the space agency's managers, noting that over the years, when project failures and budget overruns had occurred, NASA had used a by now too familiar excuse—youth and inexperience. Karth believed that the committee had been very understanding, but it would not excuse or accept any more mistakes. "We have grown up now." He added that the Subcommittee on Space Science and Applications would "pay particular attention" to Voyager. "If it is authorized and moneys are appropriated by the Congress, I would hope that we will set a different standard by which to gauge ourselves and to which we testify before committees that are responsible for raising the money for the program."[76]

When Congress considered the NASA authorization bill in June 1967, the House and Senate committees both made deep cuts in the agency's requests (table 16). While sustaining the pace of the Apollo program, the House reduced the Voyager budget by $21.5 million and the nuclear rocket

*The Voyager lander proposed in September 1966 for 1977–1979 landings on Mars was quite similar to the Viking lander that would reach the Red Planet in 1976. Similar elements included the tripod landing gear, large direct-link high-gain antenna, smaller relay antenna, and radioisotope thermoelectric generators. The 1966 design already had a boom soil sampler and a television camera, but the scientific experiments would need more definition for a biological mission. The span across the legs of the proposed Voyager lander was nearly twice that of the Viking. Proposed weight was about twice that of Viking. VPE-14 Project Study.*

113

development program by $24 million. An additional $75 million was cut from Apollo Applications, which had been established to provide follow-up activities in the manned program once the first lunar expeditions had been achieved. The Senate denied NASA its entire Voyager request and cut $120 million from Apollo Applications, but authorized the entire amount for NERVA (nuclear engine for rocket vehicle application). Senator Anderson and his colleagues on the Aeronautical and Space Sciences Committee believed that Voyager should be further postponed because the project would use too much of the space science budget. Whereas 21 space science missions were planned for 1967, the number would decrease to 13 in 1970 and to only 2—the Voyagers—in 1973. "It is clear, therefore, that to have a varied mission space flight program in the early 1970's comparable to that now existing in OSSA there would have to be a substantive increase in funding for that Office."[77] Additional dollars would not be forthcoming, and NASA would have to reevaluate its space science activities. In late June, a joint House-Senate Conference Committee worked out a compromise budget that restored $42 million to Voyager for 1968. Excluded from NASA's budget altogether were funds for the proposed Mariner 1971 with the atmospheric probe.[78]

*An automated biological laboratory was developed by Philco Aeronutronic Company to study the possibility of a hard-landing entry probe to make simple assays of the Martian environment. One of several studies for Voyager in the mid-1960s, it grouped science instruments that could be programmed for numerous experiments. None of the projects was flown, but they provided understanding of extraterrestrial biology detectors.*

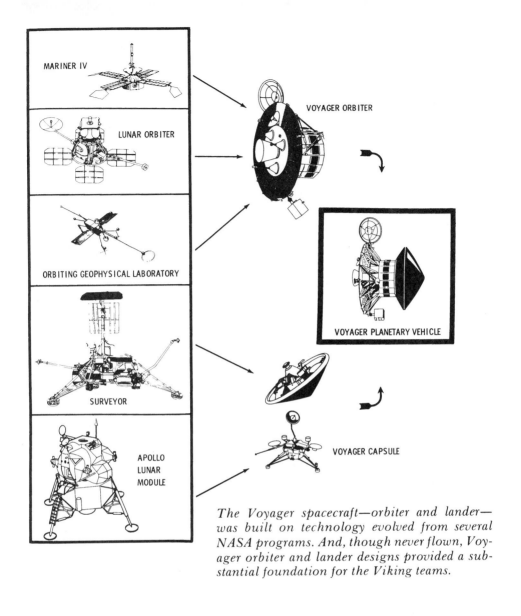

MARINER IV

LUNAR ORBITER

ORBITING GEOPHYSICAL LABORATORY

SURVEYOR

APOLLO LUNAR MODULE

VOYAGER ORBITER

VOYAGER PLANETARY VEHICLE

VOYAGER CAPSULE

*The Voyager spacecraft—orbiter and lander— was built on technology evolved from several NASA programs. And, though never flown, Voyager orbiter and lander designs provided a substantial foundation for the Viking teams.*

Though far from lavish, the funds suggested for Voyager would have been sufficient to begin basic development of the orbiter for a 1973 flight, but this was just the authorization. The appropriation still had to be moved through Congress. Between June and August 1967, while the NASA appropriations were being finalized, riots or violent demonstrations associated with the civil rights movement occurred in 67 American cities. Combined with the unpopular, costly war in Vietnam, the summer of disorder—the third since the burning of Watts in 1965—forced congressional attention to concerns more pressing than sending spacecraft to Mars.[79] At the end of July, as Webb was resolutely refusing to choose between Apollo Applications or Voyager, a Harris survey indicated that the American public no

## Table 15
## Voyager Projected Costs

(in millions)

| Date of Estimate | Missions | Projected Cost |
|---|---|---|
| 8 Mar. 1963 | Four flights with SIB-Centaur | $ 700 |
| 7 Aug. 1964 | Two flights with SIB-Centaur | 450 |
| Dec. 1964 | 1969 test flights, and orbiter-landing capsule mission in both 1971 and 1973 with SIB-Centaur | 946 |
| 14 Dec. 1964 | Four flights with SIB-Centaur | 1250 |
| Project Operating Plan 65-1 | 1969 test flights, and orbiter-landing capsule missions in both 1971 and 1973 with SIB-Centaur (JPL estimate) | 1107 |
| Mar.-Apr. 1965 | Above missions reviewed by Office of Space Science and Applications; kept earlier estimate pending completion of project definition | 946 |
| 10 May 1965 | Above missions | 946 |
| Sept.-Oct. 1965 | 1969 test flights deleted, 1971 landing changed to capsule test, one 1973 mission, and launch vehicle changed to Saturn V (headquarters) | 1000 |
| POP-65-4 | JPL estimate for one 1973 flight and 1971 capsule test | 1300 |
| Dec. 1965 | Landing capsule flights deferred until 1973 and 1975: JPL estimate Hearth estimate | 1578 1200 |
| End of Jan. 1966 | 1973 and 1975 Voyager estimate (Office of Space Science) | 1800 |
| 25 Oct. 1966 | 1973 and 1975 lander missions; cost for spacecraft and lander without launch vehicle | 1429 |
| 18 Apr. 1967 | 1973 and 1975 lander missions with cost of launch vehicles ($400 million); does not include operations costs or $40 million for facilities or $55 million for additions to Deep Space Network | 2200 |

SOURCE: House Committee on Science and Astronautics, *1964 NASA Authorization*, hearings before Subcommittee on Space Sciences and Advanced Research and Technology, 88 1, pt. 3(a), Mar.-May 1963, p. 1621; Donald P. Hearth, "Voyager Cost Estimates," memo for record, 7 Aug. 1964; Robert C. Seamans, Jr., to Donald F. Hornig, 14 Dec. 1964; Hearth to Oran W. Nicks, "FY67 Funding Requirement for Voyager," 10 May 1965; Hearth, "History of Voyager Cost Estimates," memo for record, 15 Feb. 1966; and Hearth, "Estimates of Voyager System Contractors Cost," memo for record, 25 Oct. 1966; Senate Committee on Aeronautical and Space Sciences, *NASA Authorization for Fiscal Year 1968*, hearings, 90 1, pt. 1, Apr. 1967, p. 30.

*Table 16*
*NASA Fiscal 1968 Budget*

(in millions)

| Program | Authorization | | | Appropriation |
|---|---|---|---|---|
| | House | Senate | Confer-ence | |
| Apollo Applications (Skylab) | | | | |
| Requested | $ 454.7 | $ 454.7 | | |
| Approved | 379.7 | 334.7 | $ 347.7 | $ 315.5 |
| Voyager | | | | |
| Requested | 71.5 | 71.5 | | |
| Approved | 50.0 | 0.0 | 42.0 | 0.0 |
| Nuclear rockets program | | | | |
| Requested | 74.0 | 74.0 | | |
| Approved | 50.0 | 74.0 | 73.0 | 46.5 |
| Total NASA budget | | | | |
| Requested | 5100.0 | 5100.0 | | |
| Approved | 479.7 | 4851.0 | 4865.8 | 4588.9 |

SOURCE: NASA, *Astronautics and Aeronautics, 1967: Chronology on Science, Technology, and Policy,* NASA SP-4008 (Washington, 1968), pp. 17–18, 192, 194–95, 237, 320; and NASA, "Chronological History, Fiscal Year 1968 Budget Submission," 8 Nov. 1967.

longer supported large expenditures on space. Detroit Mayor Jerome P. Cavanagh voiced the public's concern: "What will it profit this country if we . . . put our man on the moon by 1970 and at the same time you can't walk down Woodward Avenue in this city without some fear of violence." Cavanagh and others thought "our priorities in this country [were] all out of balance."[80]

Considering the political climate, Voyager still might have survived, but only if NASA were very careful about how it promoted its planetary program. Unfortunately, the Manned Spacecraft Center in Houston chose the first week of August 1967 to send 28 prospective contractors a request for proposals to study a manned mission to Venus and Mars.[81] While not the first such investigation to be suggested, in the summer of 1967 proposing a "Planetary Surface Sample Return Probe Study for Manned Mars/Venus Reconnaissance/Retrieval Missions" for 1975–1982 was a grave mistake.[82] The request infuriated Congressman Karth, who had been fighting an uphill battle to preserve Voyager. He told one reporter that he was "absolutely astounded," especially in view of repeated congressional warnings against "new starts. Very bluntly, a manned mission to Mars or Venus by 1975 or 1977 is now and always has been out of the question—and anyone who persists in this kind of misallocation of resources at this time is going to be stopped."[83] While such advanced study proposals were commonplace among most government agencies, the timing of Houston's request could not have been worse, since previous exercises of this kind sponsored by the Office of Manned Space Flight centers in Houston and Huntsville had been billed as logical extensions of the Voyager missions. This cast Voyager in the role of a "foot in the door" for manned flights to the planets—flights that would cost billions of dollars.

The Manned Spacecraft Center's request for proposals may have been the proverbial last straw, because on 16 August the committee voted down *all* monies for Voyager and the Houston study. On the 22d, the House approved a $4588.9 million budget for NASA, $511 million less than the agency's request. President Johnson did not care to fight the reduction: "Under other circumstances I would have opposed such a cut. However, conditions have greatly changed since I submitted my January budget request." While Johnson went on to say that "these [budget] reductions do not signal a lack of confidence in our space venture," they did signal the end of Voyager.[84]

Despite last minute attempts in October in both the Senate and the House to save Voyager, the program died in the final deliberations of the appropriations conference committee (see table 17 for 1968 budget).[85] After seven years of work, the planetary project had been killed, leaving NASA with no program for the exploration of the solar system. The 1969 Mars Mariner was the last approved flight, since the 1971 Mariner had been cut with Voyager. Much of the responsibility for planning a reduced and revised space science program fell on John E. Naugle, who succeeded Newell as associate administrator for space science and applications. Newell had been appointed in late August to the number three position, NASA associate administrator (see organization chart in appendix G), and Seamans had become deputy administrator on 28 January 1966, having occupied that office in an acting capacity since Hugh Dryden's death the previous December.

When asked what he would do in his new job, Newell responded, "My first assignment will be to develop an orderly, routine planning approach for the agency." The major problem he saw was "defining the major new objectives of the space program." While Newell and his colleagues publicly held out hope for a resurrection of Voyager—"My only hope is that we've sold Voyager and that we're just experiencing a delay because of war and problems on the homefront"—privately they knew that future planetary projects would have to be on a smaller scale, in both physical size and budget.[86] NASA was embarked on a new era—one of ever-tightening budgets and closer congressional scrutiny.

### Table 17
### Final NASA Budget, Fiscal 1968

(in millions)

| Program | Request | Authorization | Appropriation |
|---|---|---|---|
| Apollo | $2546.5 | $2521.5 | $2496.0 |
| Apollo Applications | 454.7 | 347.7 | 315.5 |
| Advanced missions | 8.0 | 2.5 | 0 |
| Physics and astronomy | 147.5 | 145.5 | 130.0 |
| Lunar and planetary | 142.0 | 131.9 | 125.0 |
| Voyager | 71.5 | 42.0 | 0 |
| Bioscience | 44.3 | 41.8 | 40.0 |
| Space applications | 104.2 | 99.5 | 88.0 |
| Launch vehicles | 165.1 | 157.7 | 145.0 |
| Space vehicles | 37.0 | 36.0 | 35.0 |
| Electronics | 40.2 | 39.2 | 35.0 |
| Human factors | 21.0 | 21.0 | 21.0 |
| Basic research | 23.5 | 21.5 | 20.0 |
| Space power | 45.0 | 44.0 | 44.0 |
| Nuclear rockets | 74.0 | 73.0 | 46.5 |
| Chemical propulsion | 38.0 | 41.0 | 35.0 |
| Aeronautics | 66.8 | 66.8 | 65.0 |
| Tracking & data aquisition | 297.7 | 290.0 | 270.0 |
| University program | 20.0 | 20.0 | 10.0 |
| Technology utilization | 5.0 | 5.0 | 4.0 |
| Research & development total | $4352.0 | $4147.6 | $3925.0 |
| Construction | $ 76.7 | $ 69.9 | $ 35.9 |
| Administrative operations | $ 671.3 | $ 648.2 | $ 628.0 |
| TOTAL | $5100.0 | $4865.8 | $4588.9 |

SOURCE: *Space Business Daily*, 27 Oct. 1967.

# 5

# Reorganization and the Creation of Viking

The cancellation of Voyager wiped clean NASA's slate of proposed planetary missions. An unthinkable turn of events, it gave the space agency a unique opportunity to redefine its planetary goals and evaluate the wisdom of earlier projected activities. But, unlike the early 1960s when Voyager was conceived, NASA planners by the end of 1967 had a technological and scientific base on which to build. The nearby planets were not as much a mystery as they had been at the beginning of the decade. And the agency had several proven launch vehicles from which to choose. But more significant, NASA engineers and scientists better understood the technology of spacecraft designed to explore deep space.

An eager group at NASA's Langley Research Center in Virginia was anxious to seek alternative missions to replace the Voyager series, and at the top of their list of possibilities was a Mars landing craft. Having participated on the fringes of the agency's Mars activities for several years, the Langley group created its own new series of proposals, from which the Viking spacecraft evolved. As with many other aspects of NASA's planetary program, Viking's heritage was tied to the many projects—both successful and unsuccessful—that preceded it. At Langley, Viking's roots extended back to 1964, three years before Voyager was canceled.

## LANGLEY ENTERS THE MARS BUSINESS

By early 1964, it was widely recognized within NASA that Mars was the next likely major target for exploration following Apollo's expeditions to the moon. Leonard Roberts, head of the Mathematical Physics Branch in the Dynamics Load Division at the Langley Research Center, became interested in the technological problems associated with vehicles passing through the Martian atmosphere.[1] Langley, by virtue of its extended research into the behavior of airplanes and spacecraft operating in Earth's atmosphere, was generally recognized as the leading NASA center for the study of the aerodynamic and heat-load aspects of the entry design of such vehicles. Pursuing the Langley tradition of researcher-generated study projects, Roberts brought together an informal group of center personnel to examine the possible application of its expertise to the problems associated with landing vehicles on Mars. From that group, he selected William D. Mace, Flight Instrumentation Division; Roger A. Anderson, Structures Research Division; and Edwin C. Kilgore, chief of the Flight Vehicle Systems Division, for a team* that would determine how Langley personnel could best contribute their talents to the investigations of the Red Planet.

Starting from "near zero in knowledge pertaining to . . . interplanetary missions," the Roberts group decided to concentrate on the area in which Langley had talent—vehicle entry aerodynamics. It would work on devising the optimum entry vehicle for landing payloads on Mars. The decision had been influenced by an early look at what other NASA organizations were doing. In Pasadena, the Jet Propulsion Laboratory was the lead "center" for planetary missions. Both Ames Research Center and Goddard Space Flight Center were studying probes that would obtain information about the Martian environment. Langley would examine the specific class of problems related to a vehicle from the time it was released by its transporting craft (orbiter or flyby) until it came to rest on the planet's surface.

After a few weeks of study during which they exchanged telephone calls, cryptic notes, and other informal communications, Roberts and his specialists chose to focus their efforts on the design of a basic, or "baseline," entry vehicle. About two and a half meters in diameter (to fit the Mariner launch shroud), it would weigh 136 kilograms (compatible with Atlas-Centaur capacities). The Langley Mars probe would contain instruments that would make direct measurements of the Martian atmosphere while the vehicle was descending on a parachute deployed from the protective heatshield. About 20 persons in scattered locations at Langley participated in this preliminary planning activity, with the engineering office of the Flight Vehicles and Systems Division becoming the focal point for coordinating all the work. Finding volunteers for the project was no problem, since the Langley people realized that they might be getting in on the "ground floor" of something big. As James McNulty subsequently recorded, during the early period "no sophisticated analyses were made, designs were broad based, and most work was done on scratch paper."[2] Primarily, the Langley team wanted to get a feel for ideas; "a lot of work and concepts were turned out, analyzed, modified, or discarded. . . ." Langley researchers were taking the same kind of initial course that their counterparts at JPL had followed with Mariner B and Voyager.

Two major problems considered by the Roberts group were optimum designs of a heatshield and a descent television experiment. Descent television was considered useful and a "glamorous" idea, but it was scrapped because of weight and the long time lag for transmission and processing of video images. The heatshield also raised the issue of weight allowances.

---

*James F. McNulty and Clarence T. Brown, Jr., were also in the team's early meetings.

Roberts' team looked at heatshields for several different landers—from a simple spherical probe (hard-landing) that would enter the atmosphere at an angle and travel a long tangential path to the surface, to a series of much larger, complex craft (soft-landing). But hard or soft, the landers would need a heatshield to overcome aerodynamic heating and assist in slowing down the craft before touchdown. Writing in the fall of 1964, Roberts noted that during the past decade considerable research had been applied to the design of ICBM and manned entry vehicles for use in Earth's atmosphere, and much of that technology could be adapted for planetary exploration. However, there were some significant differences, "primarily because we face different planetary atmospheres and higher entry velocities."

Although it became obvious that existing heatshield technology would not meet payload weight limitations for the large landers, a solution did appear to exist for the smaller probes. Roger Anderson's Structures Research Division at Langley was working on a new heatshield—a "tension shell" with a peaked cap—in which the payload would be placed below the main ring of the heatshield structure. The membrane, stretched between the payload and the ring, would deflect the entry heat pulse and provide the necessary drag. For the thin Martian atmosphere, this new shield promised to be more efficient than those used for Earth reentry.[3] Concurrently, Langely researchers under William Mace examined the problems posed by sterilizing hardware using intense heat over long periods of time.

In the summer of 1964, Roberts asked the center management to fund a $500 000 industry study of a Mars probe with a tension-shell heatshield. After a vigorous selling job by Roberts, NASA Headquarters allocated the requested funds, half from the Office of Advanced Research and Technology and half from the Office of Space Sciences and Applications. It was December before the request for proposals (RFP) was released, and the six months gave the Virginia team time to define the contractor's tasks.

Preparing a statement of work for the contract proved a challenge. In Langley's first plunge into the interplanetary realm, Roberts and his colleagues discovered it was a difficult task to define on paper exactly what needed to be done. In addition to the probe, NASA Headquarters was urging Langley to examine the lander in more detail. Since the lander had been considered thus far only as it affected the design of the heatshield, this study gave the men at Langley new opportunities. Despite the extra work required, the team was enthusiastic about working on a new lander, since it enlarged the scope and importance of the study. It also gave Langley a chance to enter a domain previously dominated by JPL. A shift away from the Atlas-Centaur launch vehicle to the Saturn IB-Centaur permitted a more realistic examination of larger landing craft. As McNulty said, ". . . it was a new and bigger project—and it was Langley's responsibility."

As it finally evolved, the Langley statement of work for the contractor study contained some familiar ideas and some new ones. While planning in

123

detail for a 1971 probe mission, the contractors would also examine larger, more complex landers for 1973 and 1975. Unlike earlier proposals, Langley's proposal recommended separating the landing probe from the spacecraft before the spacecraft's encounter with Mars. The main part of the craft would subsequently fly by the planet after relaying a short transmission from the probe.[4] Released in December 1964, the request for proposals generated eight responses from industry, which were evaluated in March 1965. A contract was awarded to the Research and Advanced Development Division of AVCO.

This $600 000, seven-month examination was one of three Mars-related studies being funded by NASA in the summer of 1965. First—and foremost—was the Voyager phase IA under the direction of JPL, with Boeing, General Electric, and TRW as contractors. Second, Ames Research Center had contracted with AVCO for a six-month, $300 000 study of a lightweight (11-kilogram), nonsurviving probe. And third was Langley's new contract with AVCO to develop an entry system and survivable lander.[5] The three contracts, two of them managed by Office of Advanced Research and Technology (OART) centers—Langley and Ames—raised many issues that had to be resolved at NASA Headquarters.

Basic to all other concerns was a management problem—how to integrate the Office of Advanced Research and Technology centers into the activities of the Office of Space Sciences and Applications. Langley had no Voyager office as such at this time, but with the increased tempo of Mars activities the Virginia center set up a Planetary Mission Technology Steering Committee, chaired by Leonard Roberts. Through this committee, the center's staff could bring members of Langley into planetary activities without taking them away from their primary responsibilities in their technical divisions. Charles J. Donlan, Langley deputy director, outlined three tasks for the steering committee—guiding the AVCO study, beginning a Langley research program in support of Voyager, and preparing a working agreement defining relations between JPL and Langley.

In the process of overseeing AVCO's work, the steering committee discarded one of its pet ideas, the tension-shell heatshield. The concept had given Langley a foot in the door, but the heatshield had failed to prove out in the wind-tunnel tests. The Apollo and blunt-body heatshields were its equal in performance without some of its structural weaknesses. As one participant noted, "Thus, one of Langley's main selling points—its unique knowledge of tension shell technology—was quietly discarded without notice."[6] Langley's attention shifted to a blunt cone for entry, because it was easier to package than the bigger Apollo heatshield.

In defining the research program, the Langley team demonstrated its bias toward research and technology development rather than the conduct of flight projects. Since the creation of its first facilities shortly before World War I, Langley had been dedicated to applied research. In the NASA era, flight projects were viewed as status symbols, good for public relations and as a source of funding, but the center's managers sought a careful mix of missions and research and strove to keep flight projects subordinate to the research program. At Langley, Voyager-related research in 1965 called for a wind-tunnel test program ($330 000), capsule-heatshield development ($400 000), and parachute development ($865 000). Parachute technology was an important area to be studied, because no parachute then in existence would survive deployment at the extremely high speeds (mach 1.2) needed for a Mars mission.[7]

Defining Langley-JPL working relations was no simple task, because of JPL's unique position in the NASA organization.[8] In July 1965, when the California laboratory was selected as the capsule system manager for Voyager, Homer Newell told JPL Director William Pickering that Langley would act "in a capsule technical support role relating to design, development and testing of the entry system."[9] With management charter in hand, 12 representatives from JPL visited Langley to work out the details of Langley's support, and it was quickly apparent, according to McNulty, that JPL and Langley had some diverse views as to Langley's role. From the Tidewater perspective, it appeared that "JPL was interested in getting Langley out of the 'systems' area which JPL wanted to control and into narrow specific technology tasks (i.e., type of heat shield material) which would support its mission concept." The Langley people, on the other hand, took a broader view. To them, support in the area of entry technology included entry concepts, design, methodology, materials testing, and the like. JPL, in addition, was miffed over the AVCO probe contract with Langley, believing that it might lead to "preferential treatment [of] AVCO in subsequent Voyager capsule procurement."[10] McNulty later wrote that there was much "free discussion but few agreements" between Langley and JPL. Headquarters would have to help define the roles the centers played.[11]

The specialists in Virginia spent the late summer months of 1966 working with the AVCO study and making occasional trips to Voyager capsule advisory group meetings. Like everyone else, the Langley group was surprised at the October shift to the Saturn V launch vehicle. AVCO was redirected to consider the implications of the adoption of the giant booster.[12] More significant, Langley Deputy Director Donlan told the Planetary Mission Technology Steering Committee that the center management wanted to use Voyager as a focus for its research programs, since it was the only major approved NASA activity after Apollo. In addition to seeing Voyager as a source of post-Apollo work, the Langley management could not fail to appreciate the fact that a "real" NASA center might be assigned the Voyager management role instead of the "contractor" laboratory in Pasadena.[13]

AVCO delivered its final report on 1 March 1966, with the following proposed mission highlights:

Experiments

- 3-camera television system

125

- 4 penetrometers to measure surface hardness
- instruments to determine atmospheric composition

Entry capsule
- 4.6-meter-diameter cone
- 925-kg weight

Only technological problem area
- development of parachute for low dynamic pressures.[14]

Delivery of AVCO's results came just a week before the cancellation of the 1969 probe mission and the 1971 Voyager flight. The Langley team embarked on an in-house study of alternative approaches to Voyager landers and landings, giving special attention to out-of-orbit entry versus direct entry from a flyby.

On 2 June 1966, JPL's Centaur-powered *Surveyor 1* became the first American spacecraft to soft-land on the moon. While the landing demonstrated the feasibility of terminal retrorockets, there was some question about the application of other Surveyor mission elements to a Mars flight. Direct entry to the lunar surface was relatively easy, given the detailed knowledge of the moon's motion and the reasonably good views of landing areas from Earth. Mars was a much less well defined target. The absence of any lunar atmosphere also obviated the need for a heatshield and parachute. After the success of the soft-landing rocket system, the Langley team considered using a retropropulsion unit in conjunction with a heatshield and parachute for Mars landers. On 14 August, *Lunar Orbiter 1* orbited the moon, the first American vehicle to do so. Besides mapping the lunar surface in detail for Apollo landing site selection, this Boeing-built, Langley-managed spacecraft demonstrated the center's ability to supervise a major project with a reasonably small staff. Langley also had fewer cost increases and schedule slips with the orbiter project than JPL had with the lander. That fall, successful tests of parachutes similar to those that would be needed for a landing on Mars also spoke for Langley's technical and managerial capabilities.

In August 1966, the results of an in-house study were presented to the Langley Planetary Missions Technology Steering Committee. Reflecting an increasingly complex series of planetary missions for the 1970s, the study made several recommendations regarding Mars landers: employment of a 5.8-meter conical heatshield, the maximum diameter compatible with the Saturn V launch shroud, to provide the fullest aerodynamic braking; development of a standard cone sized to the largest landers so that only one entry vehicle would have to be developed and flight-qualified; and use of the parachute for additional braking after the heatshield had been discarded and before the retrorockets had been fired. This study report, approved by the steering committee, was a rough outline of how Langley planned to land Voyager on Mars.[15]

As a result of Langley's work, Edgar M. Cortright, deputy director of the NASA Headquarters Office of Space Science and Applications, called a meeting for 26 September to discuss that center's role in the Voyager mission.* Earlier that month, a JPL group had described a different approach to landing a spacecraft on Mars. Retrorockets would be actuated at about 6100 meters, continuing to fire through ports in the heatshield until the lander was separated from a protective aeroshell at about 610 meters. Final descent would be slowed by firing the lander engines from 24 to 3 meters. This approach had three major problems: it would be difficult to design ports that would not reduce the effectiveness of the heatshield; the lander and its experiments would have to be protected during separation from the effects of the retrorockets; and, given the unknown density of the Martian atmosphere, the engines would have to have a complicated electronic throttle and carry enough fuel to permit maximum thrust if the density of the atmosphere was at the lower end of the calculated range.

Leonard Roberts described for headquarters Langley's proposed landing techniques, stressing the role of the parachute. The Langley approach had been carefully thought out and analyzed. It was simpler than JPL's approach, more realistic, and practical. Langley's Mars team won their center a major role in the Voyager project—the development of the entry system, or capsule bus as it was called in space engineering jargon.[16]

Langley Research Center personnel took part in three kinds of Voyager activities during 1967. Twenty Flight Vehicle Systems Division engineers under Ed Kilgore worked on design aspects of the capsule bus. Nine engineers under David G. Stone, who had replaced Roberts as the focal figure for Voyager after Roberts had transferred to NASA's Ames Research Center, coordinated all project details. Another 60 research engineers were engaged in developing new technology. Both Stone and Kilgore sat on the NASA-wide Voyager Management Committee, but Stone's job brought him into more frequent contact with the other centers.

### Jim Martin Joins the Mars Team

On 23 June 1967, Langley Director Floyd Thompson announced the appointment of James S. Martin, Jr., as manager of the capsule bus system, thereby forming a project management organization to control all Voyager-related activities at Langley.

Martin had joined the Langley staff in September 1964 after 22 years with the Republic Aviation Corporation. His experiences as assistant chief technical engineer, chief research engineer, and manager of space systems requirements at Republic, as well as his reputation for troubleshooting and no-nonsense management, had been the major reasons Langley Director

---

*Oran Nicks and Hearth represented OSSA; Mac C. Adams, OART; Kilgore and McNulty, Langley; and William H. Pickering and senior Voyager staff members, JPL.

Thompson had recruited him for the Lunar Orbiter assistant project manager job. During the nearly three years he had been on the Orbiter team, Martin had further demonstrated his ability to get contractors to meet the schedule and budgetary requirements of Langley's first major space project. By summer 1967, only one Lunar Orbiter* flight remained, and Martin and his teammates could turn their attention to new projects. The Voyager capsule bus system was their high priority item.[17]

Martin and five engineers set up the Voyager Capsule Bus Manager's Office in June 1967. Plans called for the remaining Lunar Orbiter staff, about 25 more engineers, to join them in September after their last flight. Martin's approach to managing the capsule bus was structured around his people, who would handle project implementation. Ed Kilgore's team would act as consultants and advisers, tutoring Martin's managers. Stone's work on entry systems was controlled by Martin's use of the budget. Dollars would be allocated for only the activities that he thought were germane to the tasks at hand, and all requests for funds had to be justified to the Management Office.[18]

*Cancellation*

Martin had been at his new tasks for only two months when Voyager was denied further funding by Congress. In the wake of this blow, the Langley Planetary Missions Technology Steering Committee convened a "what-do-we-do-next" meeting on 6 September. Eugene C. Draley, assistant director for flight projects, and former supervisor of Lunar Orbiter in the director's office at Langley, told the nearly 50 persons at the meeting something of the background of Voyager's demise. The Office of Space Science and Applications in Washington had been informed by congressional staff members that NASA's budget cuts had been primarily the result of other higher priority programs, not simply disapproval of Voyager. As a result, headquarters requested JPL, Ames, Langley, and Lewis to help define a more modest planetary program. Draley told his audience that Langley's goal was to have a project concept ready for submission by 1 November 1967, and he asked the Planetary Missions Technology Steering Committee to investigate and recommend scientific objectives for such a new project.[19]

Eugene S. Love, chairman of the steering committee, presented a preliminary list of candidate missions. He believed that Mars should continue to be the focus of the agency's interest. "Venus is not nearly so interesting when we consider long term NASA objectives such as ultimately placing men on the surface. In looking at possible unmanned Mars exploration in the 1971-1973 time period at costs much lower than the Voyager concept, a number of approaches are possible." He listed seven of them at the early September meeting:

---

*The first four Lunar Orbiters had returned several hundred detailed photographs of the lunar surface, which would be used in Apollo landing site selection.

a) Direct entry probe, no fly-by spacecraft.

b) Fly-by spacecraft only.

c) Fly-by spacecraft with entry probe.

d) Short period orbiter, no entry probe.

e) Short period orbiter with entry probe.

f) Long period orbiter, no entry probe.

g) Long period orbiter, with entry probe.[20]

All of these alternatives had been considered at one time or another in the course of formulating Mariner and Voyager proposals. In Love's opinion, only the last choice deserved further investigation. "A long period orbiter (a goal covering one complete Martian year) capable of providing color photo mapping of most of the planet's surface over an entire seasonal cycle would provide information of immense and lasting value." The pictures taken during such an orbital mission could be used to compile an atlas that would be of "great value to astronauts in future missions." Scientists would find the images of "inestimable value in assessing past hypotheses and generating new knowledge of the planet." Whereas "color photo mapping of Mars over a seasonal cycle should in itself justify the mission, and should be the primary objective," correlation of the photographs with infrared and radar mapping would yield even greater insights into the nature of the planet.

But orbital photography and scientific measurements, according to Love, were only half the story. "Adequate information on the structure of the Martian atmosphere cannot be obtained from orbit." The addition of a simple entry probe, however, could provide the means for examining the atmosphere and obtaining data essential for refined engineering design of future Martian entry vehicles.

Getting the orbiter and its probe to Mars was still the major problem. Love recommended that "the examination of candidate launch vehicles should be limited to those that are available or will be unquestionably flight proven considerably before the mission time period." He further suggested that the candidate boosters be few.

> The initial study activity should progress as follows: (1) definition of the payload capability for a Mars mission for the candidate launch vehicles, (2) choice of the launch vehicle that gives the best overall capability provided costs are reasonably competitive, (3) definition of the fraction of the payload capability that must go into the orbiter, (4) definition of weight remaining that can be allotted to an entry probe, if any.[21]

At the 6 September 1967 gathering of the steering committee, Chairman Love appointed a subcommittee to recommend a list of scientific objectives for Mars and Venus missions. While the subcommittee deliberated and the committee adjourned for five days, Jim Martin traveled to Pasadena for the sixth meeting of the Voyager Management Committee. Donald P. Hearth told the attendees that the Voyager Interim Project Office would be closed out in early October. To make the best use of the information generated by Voyager, Hearth laid down an orderly plan for terminating existing work and preparing for a new project.[22]

On 11 September, the Langley Planetary Missions Technology Steering Committee met again to discuss the science recommendations. In a fashion reminiscent of earlier JPL reports, the subcommittee emphasized orbiter and probe experiments rather than lander investigations (tables 18 and 19). There was considerable discussion as to the merits of orbiters, probes, and "minimum semihard-landers," and Clifford Nelson requested that a lander not be "locked out" for a 1973 mission. The other attendees agreed, although there was little enthusiasm for sending life-detection experiments to Mars that early. To carry out further study toward a November recommendation to headquarters, Nelson headed a Langley ad hoc study group of 80 engineers divided into 13 working groups.[23]

*Table 18*
*Sample Areas of Scientific Interest*

| | | | |
|---|---|---|---|
| 1. | Orbits | 7. | Atmosphere |
| | | |    Constituents |
| 2. | Rotation | |    Scale height |
| | | |    Density |
| 3. | Size | |    Meteorology |
| |    Mean diameter | |       Clouds, winds, temperature |
| |    Shape | |    Temporal changes |
| 4. | Mass | | |
| |    Mean density | 8. | Surface structure |
| |    Distribution | |    Topography |
| | | |       Relief, morphology |
| 5. | Fields and particles | |    Cartography |
| |    Gravitational | |    Temporal changes |
| |    Magnetic | | |
| |    Electric | 9. | Surface composition, properties |
| |    Trapped radiation | |    Constituents |
| |    Micrometeoroids | |    Temperature |
| | | |    Texture |
| 6. | Ionosphere | |    Radiation |
| |    Existence | |    Albedo and color |
| |    Strength | |    Temporal changes |
| |    Temporal changes | 10. | Internal structure |
| | | |    Constituents |
| | | |    Volcanism |
| | | |    Seismicity |

*Table 19*
*Specific Objectives of an Early Mars Orbiter Probe*

To obtain maximum coverage of the planet's topography with sufficient resolution to identify major geological structures and features, including distinguishing characteristics, or different planetary areas during seasonal changes.

To obtain topographical data over limited areas with sufficient resolution to provide morphological patterns, evidence of vegetation and volcanic activity, and terrain features of geological interest.

To determine the structure, composition, and temporal changes in the atmosphere.

To obtain information on the gravitational and magnetic fields and radiation and micrometeoroid environments.

To obtain information on the extent and nature of clouds.

To observe diurnal and seasonal changes in surface temperature.

## Alternatives for Planetary Investigation

That fall, NASA Headquarters, Langley, and JPL planetary project planners pursued possible alternatives to Voyager for Mars and Venus missions. In Washington, Cortright and Oran Nicks outlined four planetary options for Administrator James E. Webb, Deputy Administrator Robert C. Seamans, and Associate Administrator for Space Science and Applications Homer E. Newell in late September. Nicks later told Jim Martin that the lack of any comments from the managers at headquarters regarding the briefing indicated to him that Webb was still feeling the pressure of the White House's cost-cutting drive.

At a 9 October presentation for Administrator Webb, space science and applications representatives outlined five possible options they believed would help answer the general question: Should NASA plan any flight missions for planetary exploration in the 1970s? As they saw it, the alternatives included (1) providing no funds for fiscal 1968 and 1969; (2) providing the planetary program with a sufficient budget to "maintain technology and pools of scientific, technical and managerial talent to support" subsequent development of planetary missions after Mariner 1969; (3) establishing two 1972 Mariner flights to Venus and two 1973 Mariner flights to Mars; (4) planning for Voyager flight in 1975 if money was made available in fiscal 1970; or (5) initiating the Voyager program in fiscal 1968 or 1969 with a very small budget aimed at producing an orbital flight in 1973 and a lander mission in 1975 (table 20).[24] The space science staff at NASA Headquarters* favored an extension of the Mariner flights (option 3)

---

*Effective 1 October 1967, Newell became associate administrator. In October, John E. Naugle became head of the Office of Space Science and Applications and Cortright became deputy associate administrator for manned space flight. Nicks filled Cortright's old position as deputy associate administrator for space science and applications, and Hearth became director of lunar and planetary programs.

Table 20

*Post-Voyager Proposals for Planetary Exploration Projects*

### Jet Propulsion Laboratory (3 October 1967)

|  |  |
|---|---|
| 1970 | Mariner-Venus Mercury 70, Atlas-Centaur, using Mariner Mars 69 equipment. |
| 1971 | Mariner-Mars 71 orbiter (if funding permits). |
| 1972 | Mariner-Venus 72 flyby, 2 probes, Atlas-Centaur. |
| 1973 | Mariner-Mars 73, orbiter-probe, Titan III (2 flights). |
| 1973 | Mariner-Venus-Mercury 73 flyby (if funding permits). |
| 1974 | Mariner-Jupiter 74, flyby, Titan-Centaur. |
| 1975 | Voyager-Mars 75, orbiter-surface laboratory, 2 on 1 Saturn V. |

### Langley Research Center (5 October 1967)

| | Plan 1 | Plan 2 | Plan 3 | Plan 4 |
|---|---|---|---|---|
| 1971 | Mars orbiter, Titan IIIC. | | | Mars orbiter, Atlas-Centaur. |
| 1972 | Venus orbiter-probe, Titan IIIC (68-kg probe). | | Venus orbiter-probe, Titan IIIC (68-kg probe). | |
| 1973 | Venus orbiter-probe, Titan IIIC (136-kg probe). | Mars orbiter-probe, Titan IIIC (68-kg probe). | Mars orbiter-probe, Titan IIIC (136-kg probe). | Mars orbiter-probe, Titan IIIC (181-kg probe). |
| | (Start in spring 1968 at cost of $893 million, exclusive of launch vehicle.) | (Start in spring 1969 at cost of $339 million, exclusive of launch vehicle.) | (Start in summer 1968 at cost of $566 million, exclusive of launch vehicle.) | (Start in spring 1968 at cost of $378 million, exclusive of launch vehicle.) |

Plan "3-Extended"

1975–  Soft-landed missions to
1977    Mars with 1180-kg landing
        capsule, Titan IIIC-
        Centaur, 14-kg science
        package.

(Start in CY 1971.)

NASA Headquarters
(3 & 10 October 1967)

"Plan 5"

Mariner class spacecraft

1970  Venus-Mercury flyby, Atlas-Centaur, FY 1969 start.

1971  Mars orbiter, Atlas-Centaur, FY 1969 start; JPL using MM '69
      equipment.

1972  Venus orbiter-probe, Titan III, FY 1969 start, Langley.

1973  Mars orbiter-probe, Titan III, FY 1970 start; JPL-developed
      spacecraft, Langley-developed probe.

Voyager class spacecraft

1975  Mars orbiter, lander, Titan III and Saturn V, FY 1971 start.

1975  Mars lander, Titan III, FY 1972 start.

1975  Mars orbiter-probe, Titan III or Saturn V, FY 1972 start.

SOURCE: Donald P. Burcham, "Planetary Extension Program (PEP)—Historical Documents (incl. only pertinent Voyager refs.)," 27 Dec. 1967; and J. R. Hall and J. D. Church, "Schedule and Cost Analysis of Selected Planetary Programs," 5 Oct. 1967.

with plans for work on a mission like Voyager (option 4) to begin in 1970. No budget, or a very small one for 1968 and 1969 (options 1 and 2), would seriously affect the continuation of JPL's work for the space agency. In fact, the first option would have reportedly required "the phase out of JPL after Mariner 69, the loss of the scientific support presently being provided to the planetary program, termination of all contractor efforts and the reassignment of all in-house personnel to other agency programs." Choice number 5 was equally unsatisfactory because the projected costs were too high. But a combination of options 3 and 4 might "provide for continuation of the planetary exploration (without a Voyager commitment) at a reduced level and more effectively use the scientists, engineers, and administrative personnel by focusing their activities at specific missions which incorporate the technologies required for future detailed exploration of the planets."[25]

Combined options 3 and 4 became known as "Plan 5," or the Planetary Extension program. While there were no commitments to specific flights beyond Mariner 69, the managers did have a "wish list" ready if more money became available. Plan 5 was an attempt to keep the planetary team intact by focusing "new technologies (flyby, orbiter, probe and lander) activities toward classes of missions (Venus, Mars, Jupiter and Mercury) and various launch vehicles." This proposal would give the agency a flexibility in choosing future missions, provide a realistic environment for engineers carrying out mission studies, and build a planetary program data bank of mission concepts, technology, and scientific experimental techniques within the limits of current budgets. The agency would use its "supporting research and technology" (SR&T) monies to underwrite technical studies that would permit centers to undertake new projects at some later date without wasting time or talents. Use of SR&T funds would not constitute a new programmatic start, which Congress had banned.[26]

By early November 1967, less than two weeks after Congress had canceled Voyager, Administrator Webb was ready to propose a revised planetary program. His opportunity came during congressional hearings on NASA's proposed operating plan for fiscal 1968. He responded to the inevitable question from Sen. Margaret Chase Smith regarding what the agency planned to do in the field of planetary investigation. The Office of Space Science and Applications was proposing five new Mariner missions (1971–1976), a Voyager-style flight to Mars with two orbiters and two small probes for 1973, and a more ambitious soft-lander expedition for 1975. The 1971 Mariner flight, launched by an Atlas-Centaur, would be a long-term orbiter to make extensive observations of Mars. It would replace the 1971 Mariner proposed earlier by NASA, a flyby craft with a small atmospheric probe. Without the expense of developing that probe, NASA planners expected that the new 1971 Mariner mission would be more economical; they also would use equipment left over from the 1969 Mariner project. The other Mariner flights Webb specifically mentioned to Congress were to Venus in 1972 and 1973 using the Air Force Titan IIIC launch vehicle. The revised Voyager for 1973 had been scaled down so that it could be launched by Titan, as well, rather than by Saturn V, which would cost 10 times as much. However, the 1975 Voyager-style mission was still geared to Saturn.

Webb told the senators that "the conclusion of Mariner V, Lunar Orbiter, Surveyor and deferral of Voyager . . . all occur at the same time—the end of this year." He noted that the decision on the 1969 budget would determine if "these teams, representing an estimated 20,000 to 30,000 man-years of experience, are to be disbanded. Together they have launched 16 spacecraft toward the moon and the planets. It cost over $700 million to do the work represented by their competence." While NASA could use SR&T funds during 1968 "to hold a limited portion of this competence together," Webb stressed that "the President's decision on the 1969 budget and further consultations with this and other committees of Congress will guide our reprogramming action."[27]

Webb's "bold" step toward maintaining NASA's planetary program was influenced by several factors. The principal sources for financing any new planetary efforts were funds that could not be spent on the Apollo Applications Program (AAP). Conceived as a means of exploiting Apollo-developed technology for various manned earth-orbital and extended lunar-based missions, the Apollo Applications Program had also been cut by Congress during the 1968 budget deliberations—from a request of $454.7 million to an appropriation of $315.5 million. Since the number of Apollo applications flights had been sharply reduced and no flights were scheduled before 1970, Webb could argue for more planetary missions without necessarily seeking an overall increase in NASA funds. This proposed alteration of planetary priorities would require overcoming resistance at the White House and the Bureau of the Budget and on Capitol Hill. But Webb believed that space science was a timely and worthwhile cause for which the agency should fight.[28]

As Webb and his headquarters managers prepared for the fiscal 1969 budget process, the centers began to work on plans for executing new planetary missions should the money be made available.[29] JPL was assigned management responsibility for the two Mariner Mars 1971 orbiters, and Langley was directed to manage the Titan Voyager Orbiter 1973 project, which became known as Titan Mars 1973 Orbiter and Lander. On 29 January 1968, President Johnson assured these projects their survival when he said in his budget address to Congress, "We will not abandon the field of planetary exploration." He recommended the "development of a new spacecraft for launch in 1973 to orbit and land on Mars." The new Mars mission would cost "much less than half the Voyager Program included in last year's Budget." Johnson went on: "Although the scientific results of this new mission will be less than that of Voyager it will still provide extremely valuable data and serve as a building block for planetary exploration systems in the future." Although Webb still viewed this new planetary activity as austere, he was glad to see it gain the support of the president.[30]

In a press conference on the budget, John E. Naugle, the new associate administrator for space science and applications, noted that this Mars exploration program would cost about $500 million, rather than the $2400 million for Voyager. Further, "This program of four orbiters and two landers . . . is a minimum program consistent with the need to maintain expenditures at a minimum. Nevertheless, when you compare it to the automated lunar exploration program we have just completed, we think it is an extremely good and sound program." When asked about experiments, Naugle indicated that this topic was still under study. Landed television pictures had a high priority, as did measuring atmospheric pressure and meteorological changes such as wind velocity. Don Hearth predicted a 90-day orbital lifetime for the 1971 orbiters and 180 days for the 1973 craft. But he added, "Bear in mind that *Mariner IV* lasted for three years. So these numbers could be very pessimistic." Hard-landers weighing 360 kilograms were being contemplated for the later mission, which meant that about 10 kilograms of scientific instruments could be landed. This payload was about half the projected instrumented payload for Mariner B in 1961.[31] Though austere, Titan Mars 1973 might actually have the chance to fly (tables 21 and 22).

## Titan Mars 1973

Getting a start on a new series of planetary flights was just a first step on a long road. To get Langley and JPL going, Naugle asked them on 9

### Table 21
### Estimated Costs for Mars Program

(January 1968, in millions)

|  | FY 1968 | FY 1969 | FY 1970 | Total All Years |
|---|---|---|---|---|
| Spacecraft: |  |  |  |  |
| Mariner Mars 69 | $59.2 | $30.0 | $ 5.0 | $125.0 |
| Mariner Mars 71 | — | 18.0 | 40.0 | 86.0 |
| Titan Mars 73 | — | 20.0 | 50.0 | 347.0 |
| Launch Vehicle: |  |  |  |  |
| 1969 (Atlas-Centaurs) | 8.0 | 3.2 | — | 20.0 |
| 1971 (Atlas-Centaurs) | — | 3.4 | 13.0 | 20.0 |
| 1973 (Titan IIIC) | — | — | — | 38.4 |
| Nonrecurring costs for Titan III-Centaur ≈ $30.0 |  |  |  |  |

SOURCE: Donald P. Hearth, 30 Jan. 1968.

*Table 22*
*Mars Program*

(January 1968)

| Year | Mission | | Spacecraft Weight (kg) |
|------|---------|---|------------------------|
| 1964 | *Mariner 4* | Flyby (1) | 260 |
| 1969 | Mariner Mars 69 | Flyby (2) | 385 |
| 1971 | Mariner Mars 71 | Orbiter (2)[a] | 410 (useful[b]) |
| 1973 | Titan Mars 73 | Orbiter (2)[c] (Science instruments | ≈ 455 (useful[b]) 75) |
|  | 2 launches, each with 1 orbiter and lander | Lander (2) (Science instruments on surface | ≈ 365 (total) 14) |
|  | Voyager *(for comparison)* | Orbiter (2) (Science instruments | 1800 (useful[b]) 230) |
|  | 1 Saturn V launch | Lander (2) (Science instruments on surface | 2700 (total) 75) |

Weight Summary

| Year | Mission | | Weight |
|------|---------|---|--------|
| 1971 | Mariner Mars 71 | Useful orbiter | 410 |
|  |  | Propulsion | 455 |
|  |  | Total gross weight at Mars | 865 |
|  |  | Atlas-Centaur capability | 910 |
| 1973 | Titan Mars 73 | Useful orbiter | 455 |
|  |  | Lander | 365 |
|  |  | Propulsion (orbit insertion) | 725 |
|  |  | Total gross weight | 1545 |
|  |  | Titan IIIC capability | c.1130 |
|  |  | Titan-Centaur capability | c.4100 |
|  |  | Titan IIIC—dual burn of spacecraft propulsion | c.2540 |

[a] 1971 orbiter a modification of 1969 flyby.

[b] Spacecraft weight without propellant.

[c] 1973 orbiter same as 1971 except as modified to support lander.

SOURCE: Donald P. Hearth, notes, 30 Jan. 1968.

February 1968 for a study of Titan III–class missions to Mars for 1973. "The objective of this study is to evaluate the baseline mission submitted to the Congress . . . together with all promising alternatives, to permit a mission definition for the 1973 opportunity." Langley's work in fiscal year 1968 was "intended to advance the state of the art of such potential missions and will not be directed at a specific flight project until such a project is authorized by the administrator." The baseline mission included:

1. Two launches in 1973.

2. Launch vehicle to be either a Titan III [D]/Centaur or a Titan III with multiburn spacecraft propulsion for interplanetary injection as well as orbit insertion.

3. Each launch vehicle to carry a Mariner 71 class orbiter and a rough-landing capsule. The capsule may . . . enter the Mars atmosphere [either] directly or from orbit.

4. The 1973 mission is constrained to a total program cost of $385 M[illion], including launch vehicles. This is believed to be consistent with the use of a minimum-modified Mariner 71 orbiter and an 800 pound [360-kilogram] class rough lander. . . .

5. The science objectives should include the following:
   A. Orbiter: Carry payload similar to Mariner 71.
   B. Entry vehicle: Measure atmospheric temperature, pressure, composition, and 3-axis acceleration.
   C. Lander: Transmit limited imagery and measure atmospheric temperature, pressure, wind, soil composition, and subsurface moisture.

The science objectives of a Mars lander mission would have to be tailored to fit physical and budgetary limitations. Naugle asked the people at Langley to consider two alternative missions:

1. Hard-landers, with or without orbiters, direct entry, or out-of-orbit entry.
2. Soft-landers, with or without orbiters, direct entry, or out-of-orbit entry.

Project management was assigned to the Langley Research Center. JPL would provide assistance in such areas as system management of the orbiter or the lander.[32]

The 1973 Mars Mission Project Office under Jim Martin's direction prepared statements of work and awarded study contracts to industry. These studies concentrated on aspects of the "mission-mode" question. General Electric examined the hard-lander possibility; McDonnell Douglas investigated a soft-lander option; and Martin Marietta looked into the virtues of direct versus out-of-orbit entry for the landers. Martin's staff worked with

JPL to ensure the laboratory's support of the orbiter portion of the Mars mission.[33]

## PROBLEMS—MANAGEMENT ASSIGNMENTS AND BUDGETS

During the spring and summer of 1968, Don Hearth at NASA Headquarters and Jim Martin at Langley wrestled with two familiar problems—project management and project budgets. The Jet Propulsion Laboratory management still wanted to control such planetary missions as Titan Mars 73. And the 1968 debates over the fiscal year 1969 budget were threatening the agency's Mars lander goals.

JPL Director Pickering began a high-level management debate in April 1968 with a letter to Charles Donlan, the acting director at Langley.* After cordial comments about the "excellent working relationships" being established between JPL and Langley, Pickering went on to say that his organization agreed with "the previous position taken by LaRC [Langley] representatives relative to Voyager, namely that Project Management and Orbiter System Management should be the responsibility of a single center because the total mission design is so tightly coupled to the Orbiter System functions of acquiring scientific data and transporting an entry-lander to acceptable release conditions." To conform with this management concept, Pickering thought it might be wise to assign "both Project management and Orbiter System management responsibilities to JPL, particularly in follow-up of the Mariner Mars 71 Project." A second alternative would assign project and orbiter management to Langley, with JPL providing "Project-level missions support and Entry-Lander System management." With either approach, Pickering believed his team in Pasadena was the one that should work with Langley in managing the 1973 Mars lander mission.[34]

Eugene Draley, Langley assistant director for flight projects, recorded in a memo for the record that JPL seemed to prefer working on the lander rather than on the orbiter, but Jim Martin's proposed management did not agree with JPL's suggestions. Langley wanted to oversee the project and the development of the lander with JPL supervising the work on the orbiter, which would evolve from the 1971 Mariner orbiter.[35] While sympathetic to the merits of JPL's alternatives, the Langley team wanted to pursue its proposed management scheme for several specific reasons. First, an anticipated tight budget for the 1973 mission required NASA to keep the modifications of the Mariner 71 orbiter to a minimum. Since JPL was responsible for that project, it seemed logical from the standpoint of continuity and cost-effectiveness that the Pasadena facility adapt the 1971 orbiter

---

*Former Langley Director Floyd Thompson had been appointed special assistant to Administrator Webb to evaluate future manned space programs in February 1968. He was scheduled to retire at age 70 in November. Edgar M. Cortright became Langley director on 1 May. Donlan was acting director in the interim.

for the 1973 flight. If Langley were to manage the orbiter, the technological and fiscal risk would increase, since the essential experience and important test equipment were at JPL. Additionally, Langley would have to hire more personnel at an increased cost to the project. Second, the Langley managers believed that their center had entry expertise and other technological experience that would permit them to carry out the lander part of the project more successfully than JPL. Although the California laboratory could claim abilities in this area based on experience with the Surveyor lunar lander, Langley's planners insisted on managing both the overall project and the lander.

Langley's people, having worked hard on planning for a mission to Mars, believed they had won the right to manage the project. Development of the lander was a technological challenge, and they wanted to meet it. According to the planetary experts in Virginia, the lander was important for a host of reasons:

- Landed science remaining pioneering task for Mars exploration.

- Entry science is a new frontier in the Mars exploration program.

- Lander science accorded high priority in 73 mission by [President's Science Advisory Committee] and [Bureau of the Budget] because M71 [Mariner 71] will have accomplished prime orbital science objectives.

- Lander objectives are forcing function in mission design and operations.

- Entry-Lander most challenging technical task of 73 mission.

- 2/3 of variable $ will be spent on lander.

In addition, three other considerations led the Langley people to believe that they should manage the 1973 project. They believed they had a better understanding of experiments that should be carried aboard a Mars lander. Equally important, they argued that Langley needed the management of a major project for the prestige it would bring the center and for developing their management skills.[36]

The management issue was resolved at a May meeting between representatives of Langley and JPL, where after a detailed discussion the laboratory participants agreed to the Langley proposal. In an attempt to improve communications between the two teams, a Mission Design Steering Committee was established, with members from the project management office and from the four major system areas—orbiter, lander, launch vehicle, and tracking and data acquisition. Jim Martin was chairman, with Israel Taback representing the lander system, J. L. Kramer of Lewis acting as launch vehicle delegate, and JPL employees Charles W. Cole and Nicholas A. Renzetti temporarily serving as orbiter and tracking and data acquisition specialists. Walter Jakobowski represented the headquarters Office of Space Science and Applications.[37] Concurrent with the formation of the intercenter design committee, Cortright redesignated Langley's Lunar Orbiter Project Office the Advanced Space Flight Projects Office. The director chose this broad title as "a hedge against the Mars mission getting scrubbed."

As the Mission Design Steering Committee set up working groups to address specific technical topics, renewed budgetary battles were being fought in Washington during the fall of 1968. The Bureau of the Budget cut NASA's initial request by about $1 billion before it went to Congress. Compared to the preceding years, the lunar and planetary proposal was lean, but then so was the total research and development figure—$3.677 billion for fiscal 1969, dropping from budget plans of $3.970 and $4.175 billion for fiscal 1968 and 1967.[38]

*Table 23*
*Lunar and Planetary Exploration Budget Plan, FY 1969*

(in thousands)

| Budget Item | FY 1967 | FY 1968 | FY 1969 |
|---|---|---|---|
| Lunar and Planetary Exploration ................... | $184 150 | $141 500 | $107 300 |
| Supporting research and technology/advanced studies ............... | 22 350 | 19 800 | 30 000 |
| Advanced planetary mission technology ........ | — | 12 000 | 6 700 |
| Data analysis ................................... | — | 600 | 2 600 |
| Surveyor ....................................... | 79 942 | 35 600 | — |
| Lunar orbiter .................................. | 26 000 | 9 500 | — |
| Mariner IV and V ............................. | 13 058 | 3 800 | — |
| Mariner Mars 1969 ............................ | 30 130 | 59 200 | 30 000 |
| Mariner Mars 1971 ............................ | — | — | 18 000 |
| Titan Mars 1973 ............................... | — | — | 20 000 |
| Voyager ....................................... | 12 670 | 1 000 | — |

SOURCE: NASA, "Background Material, NASA FY 1969 Budget Briefing," news release, 29 Jan. 1968.

For whatever consolation it offered, NASA managers and engineers knew that the space agency was not the only organization suffering budget cutbacks. Federally funded science and technology faced bleak times generally. At the beginning of February 1968, the journal *Science* reported, "A scientific community that is already in a state of alarm over a tightening of federal funds in the current fiscal year will find scant cause for rejoicing in the budget that President Johnson presented to the Congress this week." The Johnson administration proposed a five percent increase over fiscal 1968, which would, given inflation and other factors, only keep programs even with the preceding year's levels. The *Science* article concluded that the lesson seemed clear—"there's a long and rocky road between proposing a budget and actually rendering support to the scientist at the bench." NASA's road looked particularly rough, since apparently only two-thirds of the dollars requested for space activities would be appropriated.[39]

On 2 May, the House of Representatives accepted reductions recommended by the Science and Astronautics Committee and made additional cuts before voting 262 to 105 for the FY 1969 space authorization bill. The approved amount, $4 031 423 000, was $1 billion less than NASA had originally proposed to the Bureau of the Budget and about $370 million below the budget submitted to Congress. On 21 May, the Senate Committee on Aeronautical and Space Sciences lopped an additional $27.35 million from NASA's request. The amount finally approved by conference committee in October 1968 was $3.7003 billion.[40]

While waiting for final action on their appropriations bill, NASA officials worked up an interim operations plan based on anticipated reductions. Under the interim plan, work on Apollo, aeronautics, and space applications would proceed at the authorized levels. Activity in other areas would be adjusted, meaning there would be additional personnel cutbacks, with civil service ranks being reduced by 1600 persons and support contractor numbers by at least 2000. Personnel reductions would hit new programs the hardest, since agency leaders believed that Apollo and other ongoing programs could not be pared any further if they were to be executed successfully and on schedule.

Apollo Applications, Titan Mars 73, Saturn launch vehicle development, and the nuclear propulsion program, NERVA, were among the projects most affected by the budget crunch. The Apollo Applications Program would receive about $140 million of the $440 million requested. Only one Saturn IB Workshop would be flown, with an Apollo Telescope Mount. With the exception of the backup launch vehicle and workshop, production on Saturn IB and Saturn V boosters would be terminated. Only 15 giant Saturns would be produced instead of the projected 19. NERVA was once again delayed, with only limited development approved. The plans for a Mars 1973 mission were revised "to conform to sharply reduced funding in FY 1969. The instrumentation to be landed on Mars and the scientific return will be substantially less than in the program presented in the FY 1969 budget."[41] As Don Hearth and his colleagues juggled the various options so that money, limited as it was, could be made available for the 1969, 1971, and 1973 missions, the space agency was mustering outside support for these projects.[42]

## SUPPORT FOR MARS EXPLORATION

Since the winter of 1967, Administrator Webb and others at NASA Headquarters had been generating support for a post-Voyager planetary program from two groups—the Space Science Board of the Academy of Sciences; and the Lunar and Planetary Missions Board, an internal NASA advisory board. The Space Science Board provided high-level endorsement and advocacy for continued planetary exploration, and the Lunar and Planetary Missions Board gave the agency more detailed scrutiny of its planning, especially as it affected the selection of scientific experiments.

From both, NASA managers sought support that would help counter the budget-cutting proclivities of Congress.

### Space Science Board, 1967-1968

Harry Hess, chairman of the Space Science Board, wrote Jim Webb in November 1967 after a briefing on the planetary program by John Naugle: ". . . the Space Science Board met last week and . . . expressed its deep concern over the weakness of the whole NASA science program and the planetary program in particular." Reductions in the NASA budget had led to greater cuts in money for space science, which in turn meant "a loss of some 50 to 75 percent in terms of effective research results." Hess was writing Webb at this particular time because the Space Science Board wanted to have an influence on the agency's planning process. At a time when NASA was cutting back its planetary launches, it was "fairly evident that the Soviets [would] have flights to Mars and Venus at every opportunity as they have had for the last few years." And as the 1967 *Venera 4* mission to Venus had demonstrated, "these are apt to be successes."* The Soviet Union had a "highly successful planetary lander" and, as Hess reminded Webb, "we don't even have one planned in the period to 1975." Unmanned planetary exploration was apparently going to be one of the major USSR space endeavors, and "great discoveries in this area can only be made once. Shall succeeding generations look back on the early 1970's as the great era of Soviet achievement while we did not accept the challenge?"[43]

Hess and his colleagues did not wish to see the U.S. fall behind the Soviet Union. They recommended increased space science activities and a reduction of manned projects like the orbital workshop of the Apollo Applications Program. A planetary science program should take precedence over other NASA activities. These themes were repeated in December 1967, with emphasis on the newly created Mariner and Titan-class Mars spacecraft. While differing in details—the board favored more Venus research—the Space Science Board proposals were basically supportive of NASA's wishes to maintain a planetary exploration program.[44]

The Space Science Board pursued its recommendations with a week-long summer study in June 1968 and published its findings under the title *Planetary Exploration 1968-1975* (see appendix D).[45] While helpful in that they pushed for more planetary missions, the board's proposals were also

---

*Evaluations of *Venera 4* were mixed. Entering the atmosphere of Venus early on the morning of 18 October 1967, the landing capsule touched down in a purported soft landing about two hours later. According to Soviet scientists, the atmosphere as measured by the instruments was almost entirely $CO_2$, with traces of oxygen, water vapor, and no nitrogen. The temperature range was from 40° to 280°C. Atmospheric pressure was 18 times that on Earth. *Venera 4* stopped transmitting data shortly after landing. The Soviet information did not agree with evidence provided by *Mariner 5* or Earth-based radio astronomical measurements. *Venera 4* probably stopped transmitting at an altitude of about 26 kilometers, as the surface pressure is more on the order of 100 times that of Earth's and the temperature at the surface is about 400°C. After a short time, the Soviets stopped claiming that their spacecraft had actually landed on the Venusian surface.

somewhat detrimental, since they did not coincide exactly with the agency's announced goals. In times of extreme congressional scrutiny, Webb and his colleagues at NASA would prefer more closely orchestrated advice. Another source of advice was the Lunar and Planetary Missions Board.

## Lunar and Planetary Missions Board, 1968

To overcome the shortcomings of the President's Science Advisory Committee and the Space Science Board, the Lunar and Planetary Missions Board was established in 1967 to provide NASA with detailed critiques of its proposed missions from a scientist's point of view. But even quasi-internal criticism was sometimes difficult to accept. As the space agency was to learn, scientists tended to be of an independent mind, and their comments often cut more deeply than Webb and his associates would have liked. In fact, this particular group had grown out of a need to resolve conflicts between the space agency and outside scientists.

In January 1966, Webb had invited Norman F. Ramsey, professor of physics at Harvard, to form a panel to investigate NASA's relations with the larger scientific community. The administrator wanted advice on several quite specific issues: evaluation of the Space Science Board's 1965 summer study recommendations on an Automated Biological Laboratories Program, suggestions for a post-Apollo lunar exploration program, and comments on a National Space Astronomy Observatory. Webb was also interested in determining how he might increase scientific participation, confidence, and support for the American space program. As he expressed it to Ramsey, "We in NASA think it is essential that competent scientists at academic institutions participate fully in the next generation of space projects and we believe that we will need new policies and procedures and perhaps new organizational arrangements in order to enable them to participate."[46]

Ramsey's panel responded in August with a series of proposals that would have profoundly altered the organizational structure of the space agency. The scientists were particularly critical of what they saw as NASA's emphasis on engineering at the expense of basic scientific research, citing the "overriding priority of engineering problems associated with launch schedules," which interfered with academic experimenters' control over their payload design. More attention needed to be given to purely scientific concerns: "The time is surely here when we must define maximum success in terms not only of 'getting there' but in terms of scientific accomplishment." Now that the space program had "matured," Ramsey's panel believed that major organizational changes were necessary. Reviving the idea of a general advisory council of scientists to help formulate NASA policy, the group also wanted to reorganize the field centers to give experimenters a greater voice and create a Planetary and Lunar Missions Board that would advise NASA on future Apollo flights and post-Apollo goals.[47]

Jim Webb did not take kindly to most of these recommendations, and at an oral presentation of their suggestions he asked the scientists if they understood the real world of Washington politics. Did they realize that NASA was just a part of a larger governmental, economic, social system and as such could not yield to their demands? NASA's official response, drafted by Homer Newell, was made public about a year later, in June 1967. In a point-by-point critique of the Ramsey report, the agency rejected nearly all of the proposals. A general advisory council was out of the question; certain functions "must clearly . . . remain the responsibility of the Administrator." A permanent advisory body would "blur the lines of authority within the agency." Only the missions board recommendation was accepted, and it was diluted considerably.[48]

Tentatively approved by NASA before the publication of the Ramsey report, the missions board would, in Webb's mind, be a full-time working organization rather than a part-time group of advisers. Each member would be expected to fight for his ideas in a competitive arena instead of pontificating from the cathedral. The term of membership would be limited. By the spring of 1967, the Lunar and Planetary Missions Board, with carefully delineated powers, was in operation. Acting in only an advisory capacity, the board could make proposals to NASA, but the agency reserved the right to reject or accept the advice. The associate administrator for space science and applications, Newell and later Naugle, provided the funds for the board's operations and drew up the questions it was to address itself to. Quite clearly, the administration of NASA did not want the missions board to grow into a general advisory council.

Within this restricted framework, the board had reasonable freedom. NASA granted its members access to internal agency documents, a privilege that the Space Science Board had been denied, and members were permitted to attend major NASA reviews and coordination meetings related to lunar and planetary exploration. Unlike earlier advisory bodies, the Lunar and Planetary Missions Board was asked to evaluate both general and specific objectives. Therefore, it would not only review the "general strategy for manned and unmanned" missions as the President's Advisory Committee and the Space Science Board had done, but also participate "in the formulation of guidelines and specific recommendations for the design of missions and for the scientific payloads to be carried on these missions."[49]

Of the 18 original members* most were familiar faces to NASA's planetary specialists. Twelve were members of the National Academy of Sciences, five were on the Space Science Board, one served on the President's Science Advisory Committee, and four had been on the Ramsey panel. Of the academic scientists, all were full professors, and two were department

*J. W. Findlay, chairman, J. R. Arnold, A. F. Donovan, V. R. Eshleman, T. Gold, C. Goodman, J. S. Hall, H. H. Hess, F. S. Johnson, J. Lederberg, L. Lees, G. J. F. MacDonald, G. C. Pimentel, C. S. Pittendrigh, F. Press, E. M. Shoemaker, J. A. Van Allen, and W. V. Vishniac.

chairmen. Of the nonacademic, two were administrators of research institutes, and the third was vice president of an aerospace corporation. These established professionals were charged with widening NASA's contacts with the scientific community.[50]

Although the missions board never proposed a single comprehensive plan for space exploration, its members did try to bring greater cohesion to NASA's efforts. They wished to avoid a series of disconnected projects; their goal was an orderly exploration of the solar system. They wanted to balance lunar and planetary projects so that one mission would not be pursued or funded at the expense of another. Achieving such goals was at best difficult. As scientists, they favored projects that emphasized science, flexibility in experiment planning, and year-to-year funding of research rather than mission-to-mission budgeting. They also wanted a continuing voice in experiment development, and they fought against one particular attitude prevalent in NASA centers: " Tell us what the experiment is to do, and we will build it, fly it, and deliver the data to the experimenter after it has been collected." As a committee headed by Wolf Vishniac reported in July 1967, "It must be recognized that a proposal of an experiment can no longer remain a one-way street. . . . A continuing dialogue and profound involvement of the scientist with NASA centers is required." According to the scientists, engineers responsible for overseeing instrument development must recognize that they must obtain the scientist's approval at each stage of design, development, and fabrication and his consent for changes.[51] A major recurring theme in the mission board's reports and recommendations was the primacy of purely scientific considerations. The board, in insisting that its recommendations be followed without deviation, failed to acknowledge the realities of the political context in which NASA operated: scientists were but one of many constituents to whom the space agency had to answer.

When President Johnson and Congress dropped their support of the Voyager missions in 1967, the board was, of course, dismayed, but it supported NASA's attempts to pick up the pieces and create a new approach to planetary exploration.[52] Unfortunately, the debate over what would replace Voyager gave way to friction among the mission board members and ultimately between the board and NASA. At the heart of the dispute was Administrator Webb's rejection of the board's alternative planetary program. Dollar, manpower, and facility limitations would just not permit it. Several members of the board, Wolf Vishniac, Gordon J. F. MacDonald, and Lester Lees among them, believed that their leader, John W. Findlay, had yielded to pressure from NASA to water down their recommendations. When the board's ideal, balanced, coherent planetary program clashed with dollar realities, the dream was shattered and the cordial relationship with the space agency was bruised. Many scientists regarded this affair as additional evidence that NASA still maintained its old attitude toward advisory groups—accept only that advice that meets its needs.[53]

146

Although additional conflicts would surely come up in the future, the Lunar and Planetary Missions Board decided to resume normal operations in early 1968. Five working groups were formed—the lunar, Mercury, Venus, Mars, and Jupiter panels. George C. Pimentel, professor of chemistry at the University of California at Berkeley directed the Mars group.* A series of comments was elicited from that group during a familiarization briefing of Titan Mars 73 held at NASA Headquarters on 24 May 1968. All members of the Mars panel agreed that the lander was more important than the orbiter but that too much emphasis was being given to relaying television pictures from the landed craft. The main value of "lander imagery" was to define the landing site, geologically and topographically. Television could tell them what the terrain looked like and how the lander was situated, but it was a supportive activity rather than a prime experiment. The prime experiment, of course, was life detection, but thus far NASA had not included any biological or biochemical experiments in the science requirements for Titan Mars 73. Other lander experiments the panel suggested included mass spectrometry for determining atmospheric composition, x-ray fluorescent examination of soil composition, and determination of subsurface water vapor. The scientists agreed that meteorological experiments should also be examined, and Wolf Vishniac reported that lightweight (one-half-kilogram) life-detection instruments were already available but that they all had the common shortcoming of indadequate sample-gathering capabilities. Of additional concern to the Mars panel, the members considered the question of landing sites (preferably seasonally active ones), the evolution of suitable orbiters, lander lifetime, and the possibility that the Soviet Union would land a spacecraft on Mars in 1973 after sending an atmospheric probe in 1969.[54]

After studying the topic for the entire summer, the Mars panel delivered its report on the scientific objectives for a 1973 Mars mission.[55] Building on the technical studies carried out at Langley and JPL, the panel reaffirmed the importance of a lander for the 1973 flight large enough to carry a meaningful complement of experiments. The group recommended using the Titan IIID–Centaur launch vehicle. Objectives of a lander-oriented mission should include investigation of the Martian atmosphere and surface, especially temperature and moisture variation and distribution patterns and diurnal and seasonal changes in temperature and moisture, since these factors would provide information that would affect the possibility of life on the planet. Although the Mars panel favored including an orbiter in the 1973 mission, a survivable lander was the more important issue. A soft-lander was favored over a hard-lander if the problem of contaminating the landing site by retrorockets could be solved. A soft-lander would permit selection from a wider range of experiments, not just the

---

G. C. Pimentel, chairman, J. S. Hall, W. Vishniac, M. B. McElroy, J. R. Arnold, and L. Lees made up the panel.

choice of the most robust equipment. Foremost among experiments were life-detection devices. "The lander should include an ensemble of complementing experiments relevant to the possible existence of life on Mars, since no single experiment is either completely definitive or unambiguous." Coupled but dissimilar experiments would be one satisfactory approach, such as a mass spectrometer that could detect carbon-containing compounds and a life detector that could search for signs of growing organisms with a carbon base.

In closing their report, the scientists noted that "the current plans of the Langley team are in general harmony with [our] recommendations and they have evolved in a manner evidently responsive to earlier suggestions" by the panel and the missions board. Jim Martin and his Langley team had worked closely with the scientific community and for the time being their effort had paid off with strong support for their plans for the 1973 mission. At the October 1968 meeting of the Lunar and Planetary Missions Board, the Mars panel report was officially approved with only minor alterations. The next big step was defining the mission mode—direct or out-of-orbit entry; hard-lander or soft-lander.[56]

## THE MISSION MODE DECISION

An intensive series of meetings was held at Langley in late October and early November 1968. Part of the mission definition process, the two-week session was under the leadership of Jim Martin. Besides Langley's Titan Mars team, John Naugle, Ed Cortright, William Pickering, Don Hearth, and other senior staff members from headquarters, Langley, and JPL were there. The first week was set aside for contractors, as Hughes, McDonnell, General Electric, Boeing, the Martin Company, and JPL presented their reports and mission recommendations.[57] During the second week's internal agency deliberations, the Mars 73 team summarized the contractor reports and outlined the possible options:

Launch Vehicle (Titan III-C or Titan Centaur)

Support Module (Orbiter or Flyby)

Entry Mode (Direct or Orbital)

Lander (Hard or Soft, 3-Day Life or Extended Life)

Launch Mode if orbiter selected (Combined or Separate)[58]

Viewed dispassionately, it was generally agreed, all the alternatives were technically feasible, but the real question centered on what NASA could afford and realistically recommend to Congress.

Jim Martin's team presented two mission mode alternatives to the NASA managers—(1) a Titan IIIC-powered direct-entry hard-lander with a flyby module, or (2) a Titan-Centaur-boosted orbital-release soft-lander

*Table 24*
*20 Alternative Mission Modes Examined for Viking 73*

| Launch Vehicle | Delivery Mode | Lander | Lifetime | Support Module |
|---|---|---|---|---|
| Titan IIIC | Out-of-Orbit | Soft | Extended | Autonomous |
| Titan IIIC | Direct | Hard | 3-day | Flyby |
| Titan IIIC | Direct | Soft | 3-day | Flyby |
| Titan IIIC | Direct | Soft | Extended | Autonomous |
| Titan IIIC | Direct | Hard | Extended | Flyby |
| Titan IIIC | Direct | Hard | 3-day | Flyby (unfueled Mars 71) |
| Titan IIIC | Direct | Soft | Extended | Flyby |
| Titan-Centaur | Direct | Hard | 3-day | Orbiter |
| Titan-Centaur | Direct | Soft | 3-day | Orbiter |
| Titan-Centaur | Direct | Hard | Extended | Orbiter |
| Titan-Centaur | Out-of-Orbit | Hard | 3-day | Orbiter |
| Titan-Centaur | Out-of-Orbit | Soft | 3-day | Orbiter |
| Titan-Centaur | Direct | Soft | Extended | Orbiter |
| Titan-Centaur | Out-of-Orbit | Hard | Extended | Orbiter |
| Titan-Centaur | Out-of-Orbit | Soft | Extended | Orbiter |
| Separate launches for orbiters and landers with Titan IIIC | Direct | Hard | 3-day | Flyby |
| | Direct | Soft | 3-day | Flyby |
| | Direct | Soft | Extended | Autonomous |
| | Direct | Hard | Extended | Flyby |
| | Direct | Soft | Extended | Flyby |

SOURCE: W. I. Watson, "Viking Project Phase B Report," M73-110-0 [circa Nov. 1968], pp. 7–8.

with extended life and an orbiter with a science package. Given expectations at the start of the meeting, the first option was the mission Martin's people expected to get; the second was the one they really wanted. All of the possible mission configurations were debated in an executive session on 9 November. Don Hearth and Robert S. Kramer discussed the dollar implications of the different missions, and Hearth noted that the out-of-orbit mission, at $39 million for fiscal 1970, would cost $10 million more that first fiscal year than the direct-entry mission.

Cortright spoke on behalf of a soft landing since the hard-lander apparently could not carry enough science for a realistic mission. He noted that the Langley senior staff preferred the Titan IIIC direct mission, as it was

Table 25
Viking Mission Modes
Examined at 8–9 November 1968 Briefing

| Launch Vehicle | Support for Cruise and Relay | Entry Delivery System | | Lander | |
|---|---|---|---|---|---|
| Titan IIIC | Flyby modules New Spinner Stabilized Mars 71 (unfueled) | Direct: | Spinning Stabilized Lifting Nonlifting | Hard: | Limited life, relay only Limited life relay plus direct link Extended life replay plus direct link |
| Titan III-Centaur | Orbiters New Spinner Stabilized Mars 71 Minor modification Major modification for orbital entry | Orbital: | Spinning Stabilized | Soft: Other: | Limited life, relay only Limited life replay plus direct link Extended life, relay plus direct link Autonomous capsules |

If no orbiter above is chosen:

| Orbiter flown for orbital science | Orbiter flown for orbital science and as relay for lander | Separate launch |
|---|---|---|

SOURCE: Langley Research Center, "Titan Mars 73 Mission Mode Briefing," 7-8 Nov. 1968, p. 16.

the most cost-effective and manageable approach and it met scientific needs. With no orbiter to worry about, Langley could concentrate its efforts on the lander. Although a Titan-Centaur orbiter-lander mission would benefit from Mariner technology Cortright did not believe that the smaller lander dispersions—offering more control over the area in which the lander would touch down—promised from such a mission were a significant enough advantage to merit the cost. The addition of an orbiter to the package would not prove a face-saving element should the lander fail, since the lander represented 80 to 90 percent of the project. While the orbiter-lander combination would provide the most scientific information, it was also the most costly and the most complex alternative, both technically and organizationally. Looking at the amount of data that would be returned, Cortright noted that Surveyor had provided over 10 000 photographs, but it had been the first few that had provided the biggest payoff. Since the orbiter-lander approach would cost about $70 million more than the direct entry mode, Cortright believed that the agency should consider the relation between the scientific return and the expenditure. He was not convinced that the extra money would be well spent.

John Naugle's concerns lay in another direction: Which proposal would be the easier to sell to Congress and the new administration? Jim Webb had left NASA in October as a prelude to the end of President Johnson's term, and Thomas O. Paine, Webb's deputy, had assumed the reins of the organization as acting administrator. The significant question was what policy toward space activities would the Nixon administration pursue. With Richard M. Nixon elected to the presidency only four days before the high-level agency meeting, Naugle said that the unknowns of a new administration made it difficult to know what to do, especially in light of the criticisms by some scientists that the planetary program had been too conservative. Still, with all the uncertainties, Naugle favored the more complex mission. He believed that the costs of a lander could be reduced below current predictions and an orbiter with new science would enhance the overall mission. The orbiter had two important functions: orbital photography could be used in landing site selection, and the orbiter could serve as an information relay link, significantly increasing the amount of data returned from the Martian surface. The relay link would permit still further exploitation of the growth potential of the soft-lander for landed experiments. Naugle was willing to try to sell this orbiter-lander option to Paine, to the new president, and to Congress.

After considerable discussion among the NASA representatives, Don Hearth made the following summary of the mode they should recommend to Acting Administrator Paine:

- Soft-lander with extended life and a flyby support module.
- Direct entry.
- Titan IIIC with advantages of Titan-Centaur to be studied.
- Separate launch of Mariner 71 orbiters to be examined by JPL and the Planetary Programs Office.[59]

This proposal met with unanimous agreement, as did the name of the new project—Viking. But on 4 December 1968, NASA announced that Paine and Naugle had selected the more ambitious out-of-orbit option for Project Viking. After listening to the Langley briefing, Naugle believed that an extended-life orbiter with new post-Mariner 1971 experiments was essential to Viking. Looking back, Naugle recalled: "It is a little hard to recapture the mood of the times...but...one of the things that figured in my mind was the fact that we were in competition with the Russians. They had a good strong program of landers, and I . . . felt that we had to establish a good solid scientific mission." If "the Russians landed successfully in '71 or '73, what we landed . . . had to be something that would stand up against what they had done." Acting Administrator Paine for his part was searching for a successful project for which he could assume responsibility, as most people would consider the manned lunar missions to be the work of NASA's second administrator, Jim Webb. In the autumn of 1968 when Paine looked to the future of NASA's program, he believed in the importance of unmanned planetary exploration and enthusiastically endorsed the Viking project in its most advanced form.[60]

NASA chose a soft-lander with a "surface lifetime goal of 90 days" for the Mars project. A Mariner 1971–class orbiter would complement the lander science by providing "wide-area surveillance," which could be correlated with surface data from the landing site. The orbiter would also increase the data returned from the surface by providing a relay link between the lander and Earth. In 1968 NASA decided to employ the Titan IIID–Centaur launch vehicle for planetary missions because of its improved payload capacity. With the Titan IIID–Centaur, the lander and orbiter could be boosted together. Once the two craft reached Mars and went into orbit, the lander would be released. This approach to the mission would permit greater accuracy in landing at a preferred site, lower entry velocities, and more control over entry angles, three vital factors that affected lander survival.[61] The Titan IIID-Centaur would also permit the mission reasonable payload weights:[62]

|  | Titan IIIC | Titan IIID–Centaur |
| --- | --- | --- |
| Total orbital weight | 1136 kg | 3400 kg |
| Lander | 360 | 1000 |
| Scientific experiments | 10 | 30 |

This significantly improved pair of flights—an orbiter and an orbiter-lander, launched about 10 days apart—would cost $415 million, up from $385 million for the smaller, less productive mission discussed during the fiscal 1969 hearings.[63]

After 17 years of promoting, planning, debate, enthusiasm, and despair, NASA could finally get down to the task of designing and building hardware. Although dollars for Viking would always be scarce, this Mars lander would actually journey to the Red Planet. On 6 December 1968, Ed Cortright announced the formation of an interim Viking Project Office at Langley to replace the Advanced Space Project Office (Unmanned):

Effective this date, the following are reassigned to the interim Viking Project Office in the capacities as indicated:

| | | |
| --- | --- | --- |
| Project Manager | : | James S. Martin, Jr. |
| Deputy Project Manager | : | Israel Taback |
| Project Scientist | : | Dr. G. A. Soffen |
| Operations Manager | : | William J. Boyer |
| Engineering Manager | : | Israel Taback |

| | | |
|---|---|---|
| Executive Engineer | : | Angelo Guastaferro |
| Space Vehicle Manager | : | Robert L. Girouard |
| Test Manager | : | William I. Watson |
| Spacecraft Manager | : | Edmund A. Brummer |
| Asst. Spacecraft Managers | : | Royce H. Sproull |
| | | Frank E. Mershon |
| Missions Analysis Manager | : | Norman L. Crabill[64] |

Under the organizational framework set up by Martin and his colleagues, Lewis Research Center would oversee the launch vehicle for Viking, JPL had responsibility for designing and building the orbiter, and Langley would supervise lander and system integration. Following the pattern of Lunar Orbiter, an industrial prime contractor would be selected to develop and build the lander, with Langley personnel members as technical managers. This scheme had been used successfully in numerous other NASA programs, notably the manned spaceflight projects, Mercury, Gemini, and Apollo.

Jim Martin opted for a reasonably simple management structure. Responsibility for the project passed directly from the Office of Space Science and Applications at headquarters through Langley's director to the project manager. All other NASA concerns working on Viking reported to Martin, who clearly established himself as the "boss." Three major tasks would dominate the years before the Viking launch: developing and building the orbiter, developing and building the lander, and selecting and building the scientific experiments. And Martin's team in Virginia would make sure that the necessary work was done on schedule and within the budget.

# 6

# Viking Orbiter and Its Mariner Inheritance

During the closing days of 1968, the engineers at Langley, in consultation with specialists at JPL and NASA Headquarters, completed a Viking spacecraft design. Viking would have two major systems—an orbiter and a lander. While the lander would provide the means for safely delivering the scientific instruments to the surface, house, and provide the necessary power source and communications links for those experiments, the orbiter had a series of equally important functions in the Viking mission. The orbiter would transport the lander to Mars, provide a platform for the Viking imaging system so that proposed landing sites could be surveyed and certified, relay lander science information (pictures and other data in an electronic format) to Earth, and conduct scientific observations in its own right.

Despite early debates among NASA managers, it was only logical that the design and development of the Viking orbiter system be carried out at the Jet Propulsion Laboratory, where the engineering team already had an expertise in the design of planetary spacecraft. After building the Ranger lunar probes and the early Venus and Mars Mariner flyby spacecraft, the California engineers had gone on to build the Mariner Mars 69 flyby craft and were working on the Mariner Mars 71 orbiter when Viking was initiated. The Viking orbiter would borrow heavily from Mariner technology, with such specialized functions as the project demanded being added to the basic chassis.

Early plans for the Viking orbiter called for only a few modifications of the Mariner 71 craft. However, structural changes that permitted mating the lander to the orbiter and enlarging the solar panels led to significant alterations of the basic 1971 orbiter. During the long flight to Mars, the orbiter would have to provide power to the lander, especially during the periodic checkups on the lander's health and during occasional updates of the lander's computerized memory. These additional energy requirements made it necessary to increase significantly the solar panels, from 7.7 square meters to 15.4.

155

The decision to build a large soft-landing craft instead of a small hard-lander led to the requirement for a large orbiter. The orbiter would not only have to transport the lander, it would also have to carry an increased supply of propellant for longer engine firings during Mars orbit insertion, longer than those planned for the 1971 Mariner mission.[1] And an upgraded attitude control system with greater impulse, plus a larger supply of attitude control propellant, would be required to control the combined spacecraft. Table 26 categorizes the Viking orbiter subsystems as compared to Mariner 71, listing subsystems from Mariner requiring only minor changes, subsystems from Mariner requiring extensive modifications, and completely new subsystems designed for Viking.

*Table 26*
*Sources of Viking Orbiter Subsystems*

| Mariner | Mariner Adaptations | New |
|---|---|---|
| Radio | Structure | Computer/command |
| X-band transmitter | Attitude control | Data storage |
| Pyro control | Propulsion | Relay link |
| Omni antenna | Scan platform | High-gain antenna |
| | Temperature control | Science instruments |
| | Packaging | |
| | Data system | |

A brief review of the Mariner 69 and Mariner 71 spacecraft will provide a better understanding of the technological relationships between the Mariner and Viking projects.

## MARINER MARS 69

Born in the winter of 1965, Mariner Mars 69 was supposed to be only a modest improvement over *Mariner 4*. Early plans for a 1969 orbiter and hard-lander mission had been scrapped, and in its place a flyby craft had been substituted that would approach Mars at a distance of about 3200 kilometers, rather than the 13 800-kilometer pass made by *Mariner 4* in 1965.[2] The 1969 spacecraft would also carry more weight (384 kilograms) than earlier Mariners (*Mariner 2*—203 kg, *Mariner 4*—261 kg), because of the performance capability of its Atlas-Centaur launch vehicle. (Detailed information on the Mariner flights is given in appendix C.) Building on Project Ranger and Project Mariner experience, JPL engineers borrowed a number of fundamental mission and systems features for use with Mariner Mars 69. The most important of these was three-axis stabilization (roll, pitch, and yaw), provided by gyroscopes and celestial sensors, switching amplifiers, and cold-gas jets. This attitude control system permitted orientation of the solar panels and thermal shields, which provided temperature control, relative to the sun. The high-gain communications antenna could be aimed toward Earth to improve communications, and the scientific instruments could be directed toward the objects of their study. The attitude control system also permitted the craft to be maneuvered more precisely.[3] Other characteristics of the Mariner spacecraft included an extensive ground command capability and a large number of engineering and scientific telemetry measurements. The ground command capability was used primarily as a backup to the onboard central sequencer, a mini-computer that also reacted to commands from Earth.

Mariner Mars 69 followed the general design pattern of *Mariner 4*. The central body was octagonal with a magnesium framework (127-centimeter diagonal, 46-centimeter depth), with electronic assemblies and onboard propulsion system fitted into the equipment bays on all sides. Four hinged solar panels radiated from the body. On the side of the spacecraft opposite the solar panels was a platform for mounting the television camera, an infrared radiometer, an ultraviolet spectrometer, and an infrared spectrometer. The omnidirectional antenna and the fixed, high-gain, reflector antenna were attached on the side generally oriented toward the sun. Ground stations could communicate with the spacecraft continuously for tracking and the return of scientific data. Images would be stored by an onboard tape recorder for relay to Earth at a reduced play-back rate, since the cameras necessarily acquired imaging data at a rate much higher than the telemetry channel could accommodate.

As they worked on early Mariner and Ranger spacecraft, specialists at JPL had also evolved systems for tracking and controlling spacecraft from Earth, recognizing the requirement for a highly sensitive, steerable antenna (radio telescope) for communication with deep space probes. For continuous long-range coverage, a network of three stations, about equidistant in longitude, was normally sufficient. The first stations were at Goldstone, California; Johannesburg, South Africa; and Woomera, Australia. By the time Mariner 69 was ready to fly, there were eight 26-meter radio antennas and one 64-meter antenna in the Deep Space Network. Signals from the Space Flight Operations Facility at JPL were directed to the spacecraft by the appropriate ground station.[4]

As first established, Mariner Mars 69 had three objectives. The primary goal was to fly spacecraft by Mars to investigate that planet, establishing the basis for future experiments, especially those related to the search for extraterrestrial life. While exploiting existing technology, Mariner 69 engineers also hoped to develop new technology necessary for future missions. A tentatively approved objective to investigate certain aspects of the solar system was dropped from consideration by NASA Headquarters managers in April 1966. Mariner 69 would concentrate its efforts on Mars-related science. Experiment proposals were solicited and received by the Space Science Board, which acted as an advisory body to the NASA Office of Space Science and Applications. As had been proposed several times before, an atmospheric entry probe was suggested, but it was also rejected as before, because it would have significantly increased both the time required to develop the craft and the budget for the project. Scientific payload selection was announced on 26 May 1966.

By mid-1966, the design of the mission and the spacecraft was well under way. Money was the problem faced by N. William Cunningham, program manager at headquarters, and Harris M. Schurmeier, project manager at JPL, and their Mariner 69 team. Successive budget cuts each fiscal year forced the team to defer delivery of certain parts and components, which repeatedly required the engineers to reschedule the assembly and testing of the spacecraft. The budget reductions also forced the deletion of some spare parts and tests and led to several mission design changes. Despite financial constraints, the Mariner project staff was able to expand the scope and effectiveness of the spacecraft. An increase in mission science, for example, affected the planetary encounter phase of the mission. JPL specialists developed an improved telemetry transmission system that would return information at a higher rate than previously possible, increasing the overall volume of scientific return substantially. Since scientists would be using their instruments more frequently, the central control computer and sequencer through which ground controllers talked to the science instruments and manipulated the instrument scan platform would experience greater demand.

As early as September 1966 at the second project quarterly review, it became apparent that the 1969 mission was going to be much more than just a repeat of the *Mariner 4* flight. The instrument scan platform alone had grown in weight from 9 kilograms to 59. Throughout 1967 and 1968, as work progressed on the spacecraft and Earth-based systems, Schurmeier reported to NASA Headquarters that experimenters would be able to take more pictures of the Martian surface with the Mariner 69 equipment than previously anticipated. The accumulated improvements in telecommunications—increased telemetry data rates, expanded communications network, and better computer processing—would lead to a rate of data transmission 2000 times better than anything they had received before.[5] For the scientists associated with the television experiment, this was exciting news. Instead of taking only 8 television pictures during the last day of the spacecraft's approach to Mars, Robert B. Leighton and his colleagues on the television experiment team could gather some 160 images, starting two or three days before encounter with the planet. These approach pictures of the entire planet would bridge the gap between photos taken from Earth and closer images gathered by Mariner 69 craft as they passed by Mars.[6]

Engineers and technicians at JPL assembled components supplied by about a dozen subcontractors into four spacecraft—a proof-test model (PTM), two flight craft (M69-3 and M69-4), and one assembled set of spares (M69-2). While the proof-test model would never fly, it was a very important part of the 1969 project because it had to endure simulated conditions worse than any that were expected during the flight to Mars. The other three units were tested more gently on the vibration table to rehearse the launch and in the thermal-vacuum space-simulation chamber to practice the mission through deep space.

Following several visits to the test bench and much rebuilding and repairing, the craft were pronounced ready for their voyage. While the proof-test model remained behind in Pasadena to continue its service as a test article, the other three craft were sent to the Kennedy Space Center during December 1968 and January 1969. All went well with the preflight checks of Mariner F and Mariner G (preflight designations) until about 10 days before the scheduled launch. On 14 February while the Atlas-Centaur-Mariner F vehicle was standing on the pad undergoing unfueled simulation of launch, the Atlas began to collapse like a punctured tire. Most of the structural strength of the Atlas is provided by the pressure in its fuel tanks. While this balloon-like structure saves a great deal of weight, it means that the pressure must be maintained at a constant level. On this day, a faulty relay switch had opened the main valves, permitting the pressurizing gases to escape. As the Atlas began to sag on its launch tower, two alert ground crewmen sprinted to the scene and shut off manual valves inside the launch vehicle. Pumps restored tank pressure, and the big rocket resumed its original shape. The terrible scar in the thin stainless steel skin of the Atlas made it clear, however, that another launch vehicle would have to be used in its place.

The Centaur and Mariner components were unharmed, and on 18 February KSC personnel moved the Mariner F craft and the Centaur upper stage to the Atlas originally scheduled for Mariner G. Six days later, 24 February, *Mariner 6* began its journey to Mars. After being mated to a new Atlas shipped from San Diego by General Dynamics/Convair, the second Mariner 69 craft was launched on 27 March.[7] As *Mariner 6* and 7 were en route, another group of JPL specialists was at work preparing for the next mission to Mars.

MARINER MARS 71

The battle over NASA's budget during the summer of 1968 had caused the agency's leadership to postpone beginning work on a Mariner Mars 71 project. NASA had begun the year by asking for $4.37 billion for fiscal 1969, or $218 million less than appropriated the preceding year. After the budget cycle was completed, President Lyndon B. Johnson signed an appropriation bill for $3.995 billion on 4 October 1968, the lowest since 1963. This figure, more than half a billion dollars less than the fiscal 1968 budget, sent NASA planners groaning back to their drawing boards.[8]

Despite the tight budget, $69 million was earmarked for the planetary program, to support Mariner Mars 69's flight and preliminary study of Mariner Mars 71 and Viking 73. Two and a half months after the project approval document for the 1971 mission was signed, NASA Headquarters announced on 14 November 1968 that Jet Propulsion Laboratory had been authorized to begin work on the project. Dan Schneiderman was appointed project manager at JPL, and Earl W. Glahn was named program manager at NASA Headquarters.[9]

Mariner Mars 71 was described as part of a continuing program of planetary exploration. Unlike the previous Mariner flights, however, the 1971 mission was designed to orbit the planet with two spacecraft for a minimum of 90 days each. At a December 1968 meeting of the American Institute of Aeronautics and Astronautics, Oran W. Nicks, deputy associate administrator for space science and applications at NASA Headquarters, spoke of the value of orbiter flights and future orbiter-lander missions for the examination of Mars. He noted that *Mariner 4, 6,* and *7* had given "snapshot views of the planet." The two 1971 orbiters would "provide powerful new tools for our survey of dynamic Mars." They were scheduled to "arrive at a time in the Mars cycle when the most striking seasonal changes are evident in the southern hemisphere." A combination of different orbits for the two 1971 craft would provide a complete survey of the entire planet. "The life-times expected from these orbiters will allow observations of the dynamic changes in clouds and surface features over a period of several months."[10] In addition to the improved observations, the two orbiters would meet several other scientific objectives.

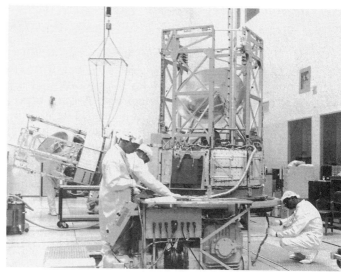

*Mariner F and G spacecraft (below)—to be christened* Mariner 6 *and* 7 *on launch—are tested in preparation for their five-month journeys to Mars to investigate the planet's atmosphere and surface. Solar arrays are not yet installed. At left, an Atlas-Centaur launch vehicle thrusts* Mariner 7 *toward space from Cape Kennedy, Florida, on 27 March 1969, following the* Mariner 6 *launch in February.*

Scientists had four general objectives for the 1971 missions, including the search for "exo-biological activity, or the presence of an environment that could support exo-biological activity." They hoped to gather information that might help answer nagging questions about the origin and evolution of the solar system. A third goal was to collect "basic science data related to the general study of planetary physics, geology, planetology, and cosmology." The specialists were also interested in information that would assist in planning and designing a Viking lander mission on Mars, especially data that would affect landing site selection.

Five specific investigations also demanded the attention of the planetary scientists. The orbiter cameras would provide imagery that could update topographic maps of the planet's surface. The television team, led by Harold Masursky of the U.S. Geological Survey, anticipated photographs of a much higher quality (better resolution) than those taken by the 1964 and 1969 spacecraft. These images, and other orbiter sensors, would also allow the scientists to examine time-variable surface features. Some specialists thought the most obvious of these features—the "Wave of Darkening"—was seasonal. Were the variations the results of moisture, vegetation, or the movement of air-borne dust?[11] The long stay in orbit also would permit study of the composition and distribution of the Martian atmosphere, to gain clues about the planet's weather. A fourth area of study included temperature, composition, and thermal properties of the planet's surface; scientists would be looking for warm spots where life forms might have had a chance to survive. And the Mariner investigators wanted a closer look at the seasonal waxing and waning of the polar caps.[12] Besides studying these five areas, scientists would also be getting information on the internal activity, mass distribution, and shape of the planet.

To meet the objectives, the Mariner Mars 71 mission plan called for two spacecraft to perform separate but complementary missions. Mission A was designed primarily as a 90-day reconnaissance. The orbital path would give the spacecraft instruments a look at a large portion of the planet's surface. Orbiting the planet every 12 hours, the flight path would permit communication with the Goldstone tracking station during a lengthy portion of every alternate orbit. Mission B would study more closely the time-variable features of the Martian atmosphere and surface for at least 90 days, moving in a wide, looping orbit around the planet once every 32.8 hours.[13] Nicks believed that the Mariner 71 orbit missions and the 1973 Viking orbiter-lander flights would be powerful study tools, permitting man to gain at least partial answers to several important questions: "Is there life elsewhere? Has life existed on nearby planets and disappeared for any reason? Can nearby planets be made suitable for life?"[14] But before they could begin to look for answers, the NASA-contractor team had to build the hardware.

Engineers at JPL had a basic philosophy about incorporating changes into each new generation of spacecraft: modifications would be included to

(1) adapt the previous design to unique requirements for the new mission,

(2) overcome difficulties demonstrated in the previous mission, and

(3) incorporate new technology when a major improvement would provide a significant benefit in cost, weight, or reliability.[15]

The Mariner 71 spacecraft designers wanted to carry over as much of the design of the early Mariner spacecraft and ground equipment as possible. As they were quick to point out, the repeated use of experienced personnel, procedures, documentation, and facilities was a benefit to the project during tests, launch, and flight operations. The Mariner 71 spacecraft grew in size, weight, and complexity, however.

*Table 27*
*Mariner 69 and 71 Spacecraft Comparisons*

| Spacecraft Feature | Mariner 69 | Mariner 71 |
|---|---|---|
| Shape | Octagonal magnesium frame | Octagonal magnesium frame |
| Size | 127 cm diagonal; 45.7 cm depth | 138.4 cm diagonal; 45.7 cm depth |
| Solar panels | 112 cm x 90 cm (4); 4.0 sq m | 215 cm x 90 cm (4); 7.7 sq m |
| Launch weight | 412.8 kg | 997.9 kg |

Besides growing much larger than its predecessors, Mariner 71 was also taking on a new major task, orbiting the planet Mars, not just passing by. As a consequence, the *propulsion subsystem* had to be completely redesigned to provide the necessary propulsion capability—a 1600-meter-per-second velocity change—to inject the spacecraft into Mars orbit. The 1971 design incorporated a 1335-newton (300-pound-thrust) engine, instead of the 225-newton (51-pound thrust) engine on Mariner 69. Nearly all the components needed for the 1971 propulsion subsystem (valves, regulators, and the like) had been used on previous spacecraft, but they had not been used in this particular combination. Although the propulsion subsystem was a new design, some inheritance from earlier Mariner systems was realized at the parts level by using flight-proven components.

Mariner 71's *data storage subsystem* was a completely new design, too. This all-digital, reel-to-reel tape-recording unit was, however, derived from earlier development activities at JPL. It incorporated selectable playback speeds of 16, 8, 4, 2, and 1 kilobits* per second, with an eight-track capabil-

---

*Bit is the abbreviation for *binary digit* and stands for the smallest unit of computer-coded information carried by a single digit of binary notation. This form of notation is a system of expressing figures for use in computers that use only two digits, one and zero. A kilobit equals 1000 bits.

ity using two tracks at a time. High-packing density for this electronic information provided a total storage capability of 180 million bits on a 168-meter tape. Data could be recorded at 132 kilobits per second. In this subsystem, there was little or no design-hardware carry-over from previous programs.

Design of the *central computer and sequencer* was altered to increase this onboard system's memory from 128 words to 512 words.* The modification provided the operational flexibility required for orbital operations, permitting repetitive sequences to be carried out. Other changes in the central computer and sequencer led to improved operations between the computer and the sequencer, better checks on stored information, and generally improved control over the spacecraft.

Of the four Mariner 71 onboard science instruments—television, infrared radiometer, ultraviolet spectrometer, and infrared interferometer spectrometer—only one was new to the Mariner series. The *infrared interferometer spectrometer* (IRIS) had been flown on the Nimbus weather satellites. It would provide information on the composition of the Martian atmosphere—measuring water vapor, temperatures at the surface, and the temperature profile of the atmosphere—and would examine the polar caps. Although the instrument was an adaptation of a previous design, many changes had to be made in it so that it worked on Mariner. To Mariner systems engineers, IRIS was a new instrument that they had to incorporate into their spacecraft design.

*Television* was another subsystem that was extensively modified. Installing two cameras on Mariner 71, the engineers could use circuitry, optics, and vidicon components from other systems. But there were difficulties. The Mariner 69 television equipment had developed background noise problems; a considerable amount of processing had had to be done to both analog and digital signals to convert them into usable video images. And the 1969 system had less dynamic range and was not as adaptable as the scientists needed for the orbiter mission. The Mariner 71 team developed an all-digital television system with eight selectable filters in the wide-angle camera, automatic and commandable shutter speeds, and picture sequencing. Another improvement reduced the effects on the optics of long exposure to the harsh space environment. Relying on existing technology minimized development costs and risks and provided the Mariner 71 scientific team a high-performance television system.

Major changes were made in the *attitude control subsystem* to adapt it to the requirements of orbital flight. To accommodate a new autopilot and computer logic changes, the Mariner 71 engineers designed new attitude control electronics and redesigned the inertial reference unit (a device that

---

*A *word* in a computer memory is a binary number containing a specific number of bits and is used as the unit of meaning.

**Mariner Mars 1964**

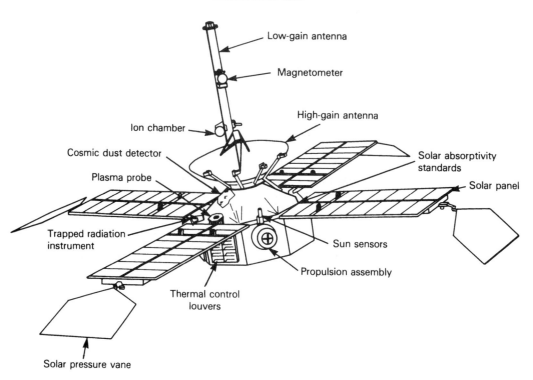

Low-gain antenna

Magnetometer

High-gain antenna

Ion chamber

Cosmic dust detector

Solar absorptivity standards

Plasma probe

Solar panel

Trapped radiation instrument

Sun sensors

Propulsion assembly

Thermal control louvers

Solar pressure vane

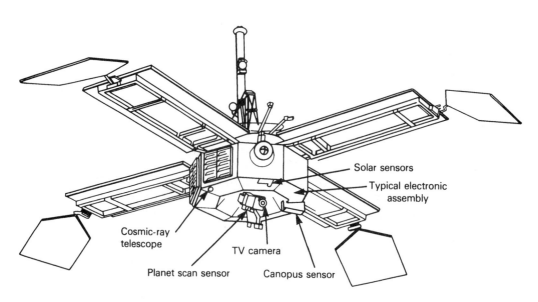

Solar sensors

Typical electronic assembly

Cosmic-ray telescope

TV camera

Planet scan sensor

Canopus sensor

**Mariner Mars 1969**

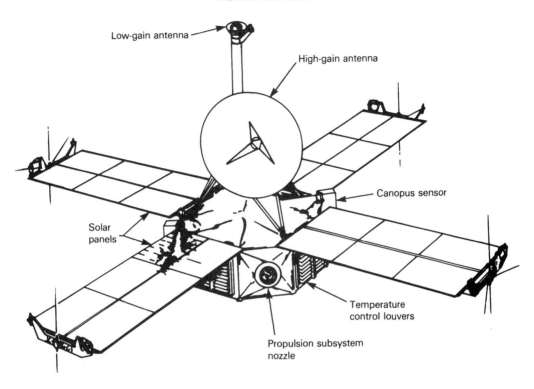

Low-gain antenna

High-gain antenna

Canopus sensor

Solar
panels

Temperature
control louvers

Propulsion subsystem
nozzle

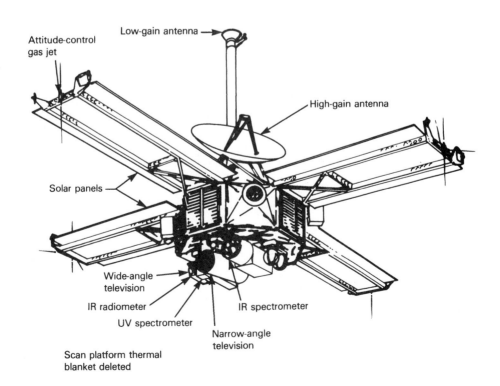

Attitude-control
gas jet

Low-gain antenna

High-gain antenna

Solar panels

Wide-angle
television

IR radiometer

UV spectrometer

IR spectrometer

Narrow-angle
television

Scan platform thermal
blanket deleted

# Mariner Mars 1971

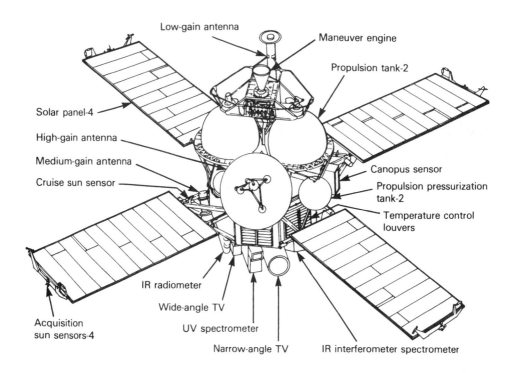

Low-gain antenna
Maneuver engine
Propulsion tank-2
Solar panel-4
High-gain antenna
Medium-gain antenna
Cruise sun sensor
Canopus sensor
Propulsion pressurization tank-2
Temperature control louvers
IR radiometer
Wide-angle TV
UV spectrometer
Acquisition sun sensors-4
Narrow-angle TV
IR interferometer spectrometer

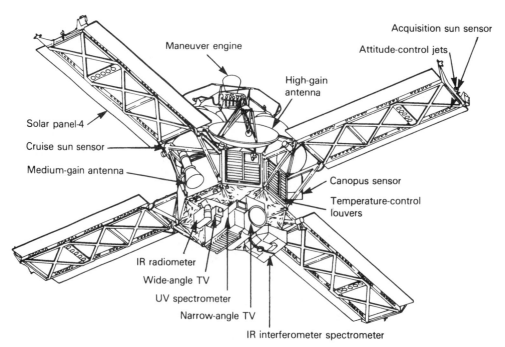

Acquisition sun sensor
Attitude-control jets
Maneuver engine
High-gain antenna
Solar panel-4
Cruise sun sensor
Medium-gain antenna
Canopus sensor
Temperature-control louvers
IR radiometer
Wide-angle TV
UV spectrometer
Narrow-angle TV
IR interferometer spectrometer

Note: Propulsion module and scan platform insulation blankets not shown.

gives continuous indication of position by integration of accelerations from a starting point). They included an acceleration sensor (accelerometer) that would control the firing duration of the propulsion-subsystem rocket engine. To maintain spacecraft attitude stability, gyroscopes were modified from Mariner 69 hardware. Sensors, both solar and star, which help determine the spacecraft's location in space, were considerably altered for the orbital flight. Mariner 71's attitude-control gas-jet system was similar to the 1969 subsystem with only minor modifications.

The *data automation subsystem* was designed to contain a new logic function to accommodate the requirements of the scientific instruments and orbital flight. Integrated circuitry and packaging techniques were directly borrowed from Mariner Venus 67 and the 1969 Mars craft. The *structural subsystem*, or the basic chassis of the spacecraft, was a successful adaptation of the 1969 octagonal frame. Electrical energy requirements were provided by an adapted *power subsystem*, which used new nickel-cadmium batteries and enlarged solar panels like those used in 1969. The *radio subsystem*, which borrowed technology from the Apollo program was altered to eliminate earlier problems. Other systems requiring only minor changes included command, telemetry, antennas, scan platform control, infrared radiometer, and ultraviolet spectrometer. The Mariner 71 final project report notes, "The design changes which were incorporated underwent considerable review and debate prior to approval so that the maximum inheritance could be realized," keeping the total number of changes the engineers had to make in the Mariner hardware to a minimum.[16]

### FIRST PHASE OF VIKING ORBITER PLANNING

Working within this milieu that stressed building on proved technological concepts, the engineers at Langley and JPL also made maximum use of earlier subsystems for the Viking orbiter. First considerations for a design of a Titan-Mars 1973 orbiter mission had begun even before the 1971 Mariner or 1973 Viking flights had been approved. A Titan-Mars orbiter

*Assembly of Mariner 9 at Jet Propulsion Laboratory. The spacecraft's solar panels are spread.*

design team led by Casper F. Mohl was established at JPL in August 1968, with Dalton D. Webb, Jr., as the group's Langley representative.

Casey Mohl was an advanced mission planner at the California lab. He had worked on *Explorer 1* and on several lander capsule studies for Ranger. During the Voyager effort, he had participated in the capsule systems advanced development activities, part of JPL's hard-lander studies. When the laboratory began to work with Langley's Advanced Spacecraft Project Office on the 1973 mission, JPL Director Pickering assigned Mohl and a group of his colleagues to the "pre-project effort," and the men began to study the diameters and weights of possible 1973 orbiters.[17] As they worked, they discovered that every time the Langley people "did something to the lander, it ricocheted back to the orbiter, especially into the [propellant] tank sizing."

Orbiter size was limited by the diameter of the Centaur launch shroud, which was 3.65 meters. Weights considered during the fall of 1968 ranged from 454 to 680 kilograms for the orbiter and 590 to 907 kilograms for the lander. At this early stage in the planning, many suggestions for the mission design were made, including one by JPL engineer Robert A. Neilson that the 1973 flight be made using a 1971 orbiter without scientific instruments or scan platform. Later, of course, such an idea would be unthinkable, but during the mission definition period one of the alternatives called for using the orbiter simply as a bus to deliver the lander to Mars.[18] The two JPL orbiter proposals presented to the Langley Research Center Advanced Space Projects Office on 9 and 30 October did not include any scientific instruments for the orbiting vehicle, as the JPL planners wanted to consider initially only the minimum number of modifications in the 1971 orbiter, just then beginning to take shape on the drawing board.[19]

By mid-November 1968, the JPL advanced planners had gone about as far as they could with the design of an orbiter for 1973 without approval of the project by Congress and the president. But at a 5 December meeting, a very pleased Casper Mohl told the "out-of-orbit" design team that the Titan-Mars 73 project had received the approval of the Bureau of the Budget; they could proceed with the development of an orbiter design while Langley worked on the lander. Although the orbiter science payload would not be defined until the Mariner 69 results were known, John Naugle said that, for planning purposes, the candidate experiment hardware in descending order of priority would include: Mariner 71-style television camera, high-resolution infrared radiometer, infrared interferometer spectrometer, near-infrared mapper, x-ray spectrometer, three-channel ultraviolet photometer, and polarimeter. Projected weights for the orbiter at launch were 1880 to 2130 kilograms, and the lander would weigh between 680 and 920 kilograms, with approximately 70 kilograms allocated for orbital science instruments.[20]

Between mid-November 1968 and mid-February 1969, JPL worked on a "baseline orbiter conceptual design" for the Viking mission, while the project office at Langley concentrated on staffing key management positions. In Pasadena 13–14 February, JPL hosted a review of its conceptual design for the orbiter. The Viking spacecraft (orbiter and lander) was to be launched by a Titan IIID–Improved Centaur, which could lift a combined weight of 3330 kilograms (2513 kilograms for the orbiter and 817 kilograms for the lander). The orbiter and lander would have a minimum life of 90 days after touchdown on Mars. The lander would have communications links directly with Earth stations and through the orbiter, which would serve as a relay satellite.

A key element of the February presentation was the technology that would be borrowed from Mariner 71. For electricity, the Viking orbiter power subsystem was essentially the same as for Mariner 71, providing lander power during transit and early orbital cruise periods. For 50 days of solar occultation during the 1973 mission, the spacecraft would be without the benefit of the sun's energy for one-half to three and one-half hours in each orbit. The increased distance of Mars from the sun during the Viking mission and the revised science instruments also led to some new requirements for the power system. New solar panels were designed, along with a new battery and battery charger. Minor changes were made in the power distribution circuitry, but the core of the entire system was borrowed from Mariner design.[21]

Industry representatives would later write to James S. Martin, Viking project manager at Langley, complaining about JPL's conservative orbiter design. L. I. Mirowitz, director of planetary systems at McDonnell Douglas Astronautics Company in St. Louis, believed that "spacecraft performance could be judiciously improved by considering" some newer components; "for example, the [central computer and sequencer] has a 512 word sequencer weighing [12.5 kilograms], the current state of the art permits use of a lander computer and sequencer that has a 6000 word capacity and weighs [11.3 kilograms]."[22] A. J. Kullas at the Denver Division of Martin Marietta Corporation also believed that weights could be reduced and performance improved by being less conservative than JPL had been in its engineering. In one instance, Kullas suggested that newer kinds of electrical cabling would permit a weight reduction from about 49 kilograms to 39, a saving of 20 percent.[23] While there was no doubt that the JPL baseline orbiter design could be improved, the conservative engineering was not unreasonable in an era of stringent budgets and equally tight schedules. Building on previously proved hardware concepts helped to ensure spacecraft reliability within the budget and on time. The specialists at JPL evaluated alterations to the basic design, and the orbiter did change over time, but conservative engineering prevailed.[24]

*Organizing Orbiter Management*

Early in April 1969, a formal Viking Orbiter Office was set up at JPL to replace the ad hoc arrangements that had existed since the official initiation of the 1973 landing project. Pickering announced the establishment of the management office on the 17th and named Henry W. Norris Viking orbiter manager. Casey Mohl's team went out of business at about the same time, and some of the members of that group joined Norris. A native Californian and graduate of UCLA, Norris had worked in aviation and space activities at General Precision Inc. before joining JPL at the age of 41 in 1963. During the Mariner Mars 69 mission, Norris served as spacecraft systems manager. Kermit S. Watkins, deputy to Norris, came to the Viking project from the JPL Office of Flight Projects, having also been assistant program manager for the Surveyor lunar landers.[25]

Other key personnel members appointed to the orbiter team by Director Pickering included Allen E. Wolfe, spacecraft systems manager, and Conway W. Snyder, Viking orbiter scientist. Wolfe had been spacecraft systems manager for Project Ranger and for the *Mariner 5* Venus mission in 1967. A nuclear physicist by education, Snyder had worked at the California Institute of Technology on Navy rocket research projects during World War II. He joined the JPL physics staff in 1956 and was principal investigator on three space experiments that studied the solar wind, becoming *Mariner 5* project scientist.[26] While Norris, Watkins, Wolfe, and Snyder were essential, highly visible members of the orbiter staff at JPL, they represented only the top of a large pyramid. When the orbiter management held its first weekly staff meeting on 1 April 1969, Norris told the participants that their sessions were not designed to resolve problems, but to discuss them "in sufficient depth to understand and identify items for separate action."[27]

One of the immediate concerns of the project managers was the growing cost of the orbiter as projected in periodic estimates. Early in February, Charles W. Cole, manager of the Advanced Planetary Missions Technology Office at JPL, informed Martin that the hardware for the total orbiter system (two flight craft, spares, and test models) would cost nearly $147 million, while the total amount needed by the California laboratory to get the orbiters ready for flight, with test equipment and facilities, would be $161 million. Cole attributed the high figures to recent increases in hardware requirements, accelerated delivery schedules, and more extensive test procedures. The Viking orbiter would require several major pieces of new hardware (table 28), and the designers at JPL had based their cost projections for this equipment on the master schedule given them by the Viking Project Office. But the people in California did not believe that the schedule was realistic. For example, the JPL engineers were convinced that such an early delivery date for the engineering test model of the orbiter would require a major acceleration of orbiter system and subsystem design plans, which in turn would demand an earlier selection and design of scientific

170

Table 28

Major Test and Flight Hardware to be Developed by JPL for the Viking Orbiter

| Equipment | Purpose or Function of Equipment | Scheduled Delivery Dates | | |
|---|---|---|---|---|
| | | As of 10 Feb. 1969 | As of 13 Mar. 1969 | As of 7 Aug. 1969 |
| Orbiter structural test model (STM) | Also called development test model (DTM). For qualification testing of basic orbiter structure, including vibration, static modal, and separation of orbiter from lander tests. | mid-Feb. 1971 | 15 Sept. 1971 | 15 Aug. 1971 |
| Thermal control test model (TCM) | For thermal qualification of orbiter systems. During tests, TCM to be mated with lander capsule thermal effects simulator to test effects on orbiter of lander heating. Both STM and TCM to be returned to JPL by 1 Aug. 1971 for laboratory testing. | 1 Mar. 1971 | 1 Dec. 1970 | 1 July 1971 |
| Engineering test model (ETM) | To validate physical and functional interfaces between orbiter and lander capsule and between spacecraft and people, procedures, and facilities associated with combined systems tests. To be assembled from early production components for orbiter; flight-qualified parts not necessary. Could be updated after tests for use in Deep Space Network compatibility testing and launch center testing. | 1 Aug. 1971 | 1 Dec. 1971 | 1 Feb. 1972 |
| Proof-test model (PTM) | To demonstrate orbiter design adequacy by performance of qualification tests, including vibration, shock, and thermal/vacuum. Also to be used for propulsion-system-interaction tests. | 1 Feb. 1972 | 15 July 1972 | 1 Aug. 1972 |
| Flight orbiters | Three flight-ready orbiters to be fabricated by JPL, two to be launched, and third to be held as backup before launch and as systems test vehicle during mission. | 1 Aug. 1972<br>1 Sept. 1972<br>1 Oct. 1972 | 15 Oct. 1972<br>15 Nov. 1972<br>15 Dec. 1972 | 1 Jan. 1973<br>1 Feb. 1973<br>1 Mar. 1973 |

SOURCE: "Viking Project Performance and Design Requirements Specification," n.d., encl. to S. R. Schofield, "Minutes of the 17th Viking Orbiter Design Team Meeting Held 20 March 1969," memo, 24 Mar. 1969; Charles W. Cole to James S. Martin, "JPL Resource Requirements for Viking Project," 10 Feb. 1969; Langley Research Center, "Viking Project Orbiter System (VOS) Master Working Schedule," 13 Mar. 1969; and LaRC, "Viking Project Orbiter System (VOS) Master Working Schedule," 7 Apr. 1969.

instruments and related equipment than JPL had planned. These schedule changes would have to be translated into direct dollar increases. But even extra dollars could do only so much toward relieving the problems imposed by the increased tempo. Cole wrote to Martin, "In JPL's opinion, the significant schedule risk . . . is not further reducible by bringing additional money and manpower to bear." What they would need was close coordination among the Viking Project Office at Langley, the lander contractor, and the JPL orbiter team to minimize the risks if they were to build a program that was "suitably balanced and mutually acceptable."[28]

During the spring months of 1969, the orbiter schedules were revised by the project office to give Pasadena teams some more time and the budget a little breathing room. Rising expenditures, however, continued to be a major concern of Viking personnel on both coasts, although evaluating the budget promised to become a more comprehensible, concrete process once the agency selected an industrial contractor to design and build the lander. Only then would they be able to determine a firm figure for the cost of the entire project.[29] In late February 1969, NASA had issued a request for proposals for the lander and, on 29 May, selected Martin Marietta Corporation from the three bidders for the contract. With this choice made (discussed in chapter 7), the Viking project entered a new phase.

Early in June when Jim Martin and his colleagues met with representatives from the new lander contractor and JPL, nine working groups were established. Of these, one of the most important, from the perspective of the budget and scheduling, was the spacecraft interface and integration working group. Formed as the "common ground" for discussion between the Viking Project Office at Langley and the spacecraft builders at JPL and Martin Marietta, this working group allowed the three organizations to exchange information and ideas on spacecraft construction and hardware interface. Donald H. Kindt at JPL was named the Viking orbiter/lander capsule integration engineer. The interface-integration working group met for the first time on 10 and 11 June and, after their sessions, representatives from all three organizations took "action items" home to consider before they met again.[30]

Another aspect of the increased tempo was the further proliferation of committees and working groups. By the end of June 1969, the amount of paperwork reaching Henry Norris's desk at JPL was growing dramatically. All managers in NASA programs, whether government or contractor employees, had to become accustomed to reading thousands of letters, memoranda, telexes, meeting minutes, reports, and other documents in the course of a project. Besides the meetings of the orbiter design team, 28 other conferences had been held by the end of June. The Viking orbiter project staff had held 12 meetings by 2 July, and the Viking orbiter mission design team started a new series of work sessions on 30 June. By the time the orbiter was ready to fly, the personnel of the orbiter design team (and its successor, the orbiter system design team), who oversaw the spacecraft's design and fabrication would meet formally more than 250 times. The mission planners who worked out the flight details for the orbiter—navigation and tracking—met 143 times before the Viking launches.

Although Kermit Watkins noted as early as August 1969 that "we are beginning to become inundated with documentation," all the meetings and paper allowed Norris and his orbiter team to keep abreast of the myriad of details that went into planning and building the spacecraft. At the Viking Project Office in Hampton, Virginia, Jim Martin used similar tools to keep tabs on the progress or lack of progress of the lander. Viking was not brought to fruition by paperwork alone, but the mountain of documents the teams left behind provides some clues to the enormous number of man-hours that went into getting the project off the ground.[31]

During the remainder of 1969, the Viking orbiter personnel worked on a number of key tasks in defining the design of the spacecraft and the nature of its scientific payload. Norris participated in the first meetings of the Viking Project Management Council; Norris, Watkins, and their colleagues worked out the second and third versions of the "Viking mission definition" document; orbiter staff members received a briefing on the preliminary science results of Mariner Mars 69; and the staff took part in the first quarterly review of the whole project. These activities were typical of activities during the next five years.

*Viking Project Management Council*

Jim Martin formed the Viking Project Management Council* in March 1969. Since Viking was the first planetary project in which several NASA centers and contractors would be participating in the design, development, and operation of major spacecraft elements, the project manager believed that a management council would "facilitate common understanding of the overall project objectives and provide a forum where technical and management problems can be freely discussed." At the first meeting, 18-19 August at the Martin Marietta factory outside Denver, each of the systems managers gave a brief status report on his organization's work to the 50 persons attending.

Henry Norris outlined the orbiter design, covering such topics as the relationship between the orbiter and lander during the cruise phase of the trip to Mars, the orbiter's weight budget, and communications equipment for the Viking spacecraft. Noting that orbiter and lander weights were a recurring concern, he told Martin and the other participants at the council meeting that a system of weight bookkeeping must be established between Langley and JPL. By this time, the entire spacecraft was projected to weigh

---

*Membership in the council included J. S. Martin—chairman, W. J. Boyer, H. E. Van Ness, I. Taback, F. W. Bowen—secretary, and E. A. Brummer, Langley; R. H. Gray, Kennedy Space Center; W. Jakobowski, NASA Headquarters; E. R. Jonash, Lewis Research Center; A. J. Kullas, Martin Marietta; and H. W. Norris and N. A. Renzetti, JPL.

3316 kilograms, with the weight of the orbiter at 605 kilograms without propellants. Jim Martin agreed; someone from the Viking Project Office would be assigned to the problem. Norris also reported that procurement had begun for the orbiter components and work was already under way on tasks that would require a long lead-time. The spokesman from JPL noted in summary that additional orbiter personnel at the laboratory would be selected shortly, including some persons that were finished with their Mariner 69 activities.[32]

Once all the systems managers gave their reports, 13 working group chairmen presented information about their work. Norris later told his colleagues at the Jet Propulsion Laboratory that the sessions "proved to be very beneficial in helping to identify and clear the air on a number of interface concerns." In particular, the two days of discussion helped to clarify the roles and responsibilities of individuals and organizations.[33] Equally significant, it gave the managers from scattered geographic locations an opportunity to meet with one another. Face to face, they could take the measure of their colleagues as they worked on problems of mutual interest. This and subsequent meetings of the management council would force the men to work with other human beings, not faceless signatures on memos. The council was just one part of Jim Martin's strategy for forging a team from a group of disparate individuals and organizations.

## VIKING MISSION DEFINITION NO. 2

The Viking project definition document was another element in Jim Martin's attempt to create a viable Mars exploration activity. Revised several times, the document gave project participants a general description of the Viking missions. By August 1969, the document had been updated five times, the latest edition being called "Viking Project Mission Definition No. 2." This 21-page paper was prepared by a group working under A. Thomas Young, the science integration manager, at Langley. Three men had to approve it before it was released 11 August 1969—Gerald Soffen, project scientist; Israel Taback, engineering manager and deputy project manager; and Jim Martin. "Viking Project Definition No. 2" contained a more nearly complete description of the entry and lander science experiments that would be included in the lander capsule and the lander. These experiments had been defined through the work of the Science Steering Group, chaired by Jerry Soffen.[34]

In August 1969, there were eight science instrument teams: orbiter imaging, biology, molecular analysis, meteorology, entry science, radio science, seismology, and ultraviolet photometry. Each of the lander experiments was further described in the "Viking Lander Science Instrument Teams Report," which served as an important reference on the state of instrument design, the scientific rationale for the experiments, and for studies that might lead to ways of increasing the scientific capability of the instruments. The instrument team report and "Viking Project Definition

No. 2" provided the basis for spacecraft design negotiations with Martin Marietta and the starting point for "early Project activity including the initiation of mission, spacecraft and operations design."[35] Although the mission definition was geared toward getting lander hardware design and fabrication started, it also had significant impact on the orbiter design team.

Henry Norris told his people at a 27 August staff meeting that the mission definition had been distributed to all the JPL division representatives. Since this was a controlling document for the project, Norris's team would have to reconcile its "resources," or budget, with its baseline definition of the orbiter. Some differences existed, for example, between the communications requirements as stated in the definition document and as pursued by the JPL engineers. "The main requirement causing a significant impact is that of the orbiter having the capability to communicate with either lander." Norris asked division representatives "to flag any other areas of disagreement."[36]

As Norris and his staff worked on the orbiter design, the mission definition continued to evolve. A number 3 edition would be ready in January 1970 after the final selection of science investigators by NASA Headquarters in December. The number 4 version would be prepared in the early spring of 1971, reflecting any changes that came from the Viking project critical design review. Finally, some time after June 1972, "Viking Project Mission Definition No. 5" would be issued to reflect lessons learned from the Mariner 71 mission. From October 1969 onward, the mission definition documents would be used in conjuction with "project specification" documents to monitor the effort.[37] Meanwhile, the science results from *Mariner 6* and 7 had to be incorporated into the Viking plans.

## MARINER 69 SCIENCE RESULTS

Scientific investigators from the Mariner 69 team presented a series of briefings and press conferences on their findings from the Mars flyby missions. The first major briefing and press conference were held on 11 September 1969, the day the preproposal briefings for prospective Viking science investigators were scheduled in Washington. While less tentative than the results presented at a 7 August press meeting, John Naugle indicated that the September briefings were really only progress reports. The final meeting of the scientists was scheduled for spring 1970, and more detailed accounts of individual experiments would be published in various journals.

Robert Leighton described the results of the television experiment at the September science briefing. "Before the space age, Mars was thought to be like the Earth, polar caps, seasons, . . . rotates in 24 hours, etc." This view of the Red Planet "was largely the legacy of Percival Lowell who popularized the idea of reclamation projects to get the water supposedly from the polar caps down to the equator where the farmers were." Although scientists had rejected the Lowell ideas of an inhabited Mars long before *Mariner 4*, they were not prepared for the stark, lunarlike images acquired during that mission. Pictures from *Mariner 6* and *7*, according to Leighton, showed that Mars was "like Mars," with its own characteristic features, "some of them unknown and unrecognized elsewhere in the solar system."[38]

Leighton noted during the press conference that areas to be photographed by the Mariner 69 missions had been chosen to "cover as many different kinds of classically recognized features on Mars as possible, dark and light areas, oases." *Mariner 6*'s track traversed the equatorial zones and crossed a great many light areas, such as the circular great desert of Hellas, and dark areas, like the region called Hellespontus. *Mariner 7* took a sweep of pictures along a meridian (north to south) that included the south polar cap. The 60-fold increase in the data transmission rate produced for the 1969 spacecraft yielded many more pictures than the scientists had originally hoped.

*Table 29*
*Pictures from Mariner Mars 69*

| Mission | Original Projection | | Pictures Returned | | Total Useful Pictures |
|---------|---------------------|---------------|-------------------|---------------|-----------|
| | Far Encounter | Near Encounter | Far Encounter | Near Encounter | |
| *Mariner 6* | 8 | 25 | 50 | 26 | 428 |
| *Mariner 7* | 8 | 25 | 93 | 33 | 749 |
| Total | 16 | 50 | 143 | 59 | 1177 |

Because of the large number of craters, the television team described Mars as more moonlike than Earthlike. In the *Mariner 6* near-encounter frame 21, which covered a territory of 625 000 square kilometers, there were 156 craters ranging in diameter from 3 to 240 kilometers. There were many hundreds more that were 500 meters across or smaller. The classical area Nix Olympica (18°N, 133°) was identified as a very large, "white-rimmed" crater some 500 kilometers in diameter, with a bright spot in the center. Cratered terrain, the parts of the Martian surface on which craters are the dominant topographic form, were widespread in the southern hemisphere. Although knowledge of cratered terrain in the northern hemisphere was limited, since fewer photographs were available, some cratered areas appeared as far north as 20°. Two kinds of craters were seen in the pictures, large and flat-bottomed and small and bowl-shaped. Flat-bottomed craters were most evident in *Mariner 6* frames 19 and 21, and their diameters ranged from a few kilometers to a few hundred. Shallow, they had a diameter-to-depth ratio of 100:1. The smaller, bowl-shaped craters, best seen in *Mariner*

6 frames 20 and 22, resembled lunar primary impact craters, and some of them had interior slopes steeper than 20 degrees. The flat-bottomed craters were of interest to the Mariner 69 investigators because they were unlike most craters discovered on the moon.

The chaotic terrain was a puzzle. *Mariner 6* frames 6, 8, and 14 illustrated "two types of terrain—a relatively smooth cratered surface that gives way abruptly to irregularly shaped, apparently lower areas of chaotically jumbled ridges." A belt of the latter terrain lay within a band 1000 kilometers wide and 2000 long at about 20° south, between the dark areas Aurorae Sinus and Margaritifer Sinus. Perplexing the scientists because it was nearly craterless, this region of short ridges and depressions was unlike anything on the moon.

Hellas, centered at about 40° south, was the best example of the so-called featureless terrain. At the resolution limit of the 1969 cameras (the cameras could not see objects smaller than 300 meters in diameter), this desert area appeared devoid of craters. Leighton and his colleagues noted: "No area of comparable size and smoothness is known on the moon. It may be that all bright circular 'deserts' of Mars have smooth floors; however, in the present state of our knowledge it is not possible to define any significant geographic relationship for featureless terrain."

Especially bothersome was the fact that pictures taken during the *Mariner 7* traverse showed that the dark area Hellespontus, west of Hellas, was heavily cratered. "The 130- to 350-kilometer-wide transitional zone is also well cratered and appears to slope gently downward to Hellas, interrupted by short, en echelon scarps and ridges." Once the flat floor of Hellas was reached, the craters disappeared. "Craters are observed within the transitional zone but abruptly become obscured within the first 200 kilometers toward the center of Hellas." The possibility of an obscuring haze was rejected because in *Mariner 7* frame 26 "the ridges of the Hellas-Hellespontus boundary are clearly visible, proving that the surface is seen; yet there are virtually no craters within that frame. Thus the absence of well-defined craters appears to be a real effect."[39]

In seeking to explain the relationship of these various kinds of terrain to the light and dark markings noted in telescopic observations, Leighton and his colleagues had a number of thoughts. First, the contrast of light and dark markings on Mars varied with wavelength, as had been known for a long time from telescopic photography. In the violet range of light, "bright" and "dark" areas were essentially indistinguishable since they have approximately the same reflectivity. With increasing wavelength, contrast was enhanced as redder areas became relatively brighter. The distinction between bright and dark areas on the surface was usually more obvious in far-encounter views than in near-encounter views. The clearest structural relationship between a dark and a bright area was that of Hellespontus and Hellas. Chaotic terrain appeared lower in elevation and at the same time more reflective than the adjacent cratered areas. Whether chaotic

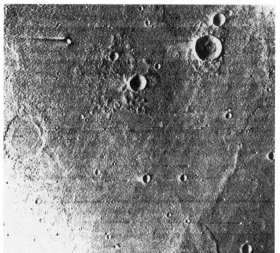

Mariner 6 *took near-encounter photos of Mars on 31 July 1969. Frame 19 (above), 3613 kilometers from the surface, shows flat-bottomed craters a few kilometers to a few hundred wide. High-resolution frames 20 (left) and 22 (below) show smaller, bowl-shaped craters, resembling primary impact craters found on the moon.*

terrain was extensive enough to include previously identified bright areas remained to be determined. Still, some of the areas traditionally thought of as oases were being identified with large, dark-floored craters such as Juventae Fons or with groups of craters such as Oxia Palus. In addition, at least two classical "canals" (Cantabras and Gehon) coincided with the quasi-linear alignment of several dark-floored craters. Other canals, showing up as irregular dark patches, would probably on closer inspection be associated with a variety of physiographic features. Leighton and his colleagues reported another correlation with earlier observations. Some drawings and "maps" of Mars portrayed a circular bright area within the dark region south of Syrtis Major and east of Sabaeus Sinus. In the Mariner 69 pictures, the investigators found a large crater in approximately the same place. The experimenters hoped to devote many hours to a comparison of these new Mariner pictures with earlier maps and photographs in an attempt to identify topographical features.

## Clues to Evolution of Mars

What did the *Mariner 6* and 7 pictures tell scientists about the evolution of the planet's surface? The absence of Earthlike tectonic forms indicated that in recent geologic time the crust of Mars had not been subjected to the kinds of internal pressures that have modified and continue to modify the surface of Earth. Since the larger craters probably had survived from a very early time in the planet's history, the scientists inferred that Mars' interior is, and probably has always been less active than Earth's. The TV experimenters noted that one theory argues that Earth's "dense, aqueous atmosphere may have been formed early, in a singular event associated" with the creation of the planet and its core. Tectonic features, therefore, might be related in origin to the formation of a dense atmosphere, and "their absence on Mars independently suggests that Mars never had an Earthlike atmosphere."

Building their case further for the unearthly nature of Mars, the television specialists commented on the age of the cratered terrains, comparing Martian surface features with similar features on the moon. Both bodies showed heavily cratered and lightly cratered areas, evidently reflecting regional differences in meteoroid bombardment, or response to it, over the life-span of the surfaces. The thin atmosphere on Mars (contrasting with no atmosphere on the moon) possibly had produced recognizable secondary effects in crater form and size distribution. Also, the scientific community generally accepted that the number of craters on the moon could not have been produced in its 4.5 billion years at the estimated present rate of impacts. An early era of high bombardment must have been followed by a long period at a greatly reduced rate. A rate per unit area as much as 25 times that on the moon was estimated for Mars. Since even the most heavily cratered areas seemed to have aged relatively uniformly, "this again suggests an early episodic history rather than a continuous history for cratered Martian terrain, and increases the likelihood that cratered terrain is primordial."

The existence of primitive, undisturbed terrain on Mars would have a number of important ramifications, especially for scientists looking for extraterrestrial life:

> If areas of primordial terrain do exist on Mars, an important conclusion follows: these areas have never been subject to erosion by water. This in turn reduces the likelihood that a dense, Earth-like atmosphere and large, open bodies of water were ever present on the planet, because these would almost surely have produced high rates of planet-wide erosion. On the Earth, no topographic form survives as long as $10^8$ [100 million] years unless it is renewed by uplift or other tectonic activity.[40]

Extrapolating further from this line of reasoning, the scientists found that the Martian environment apparently had not changed much during the life of the planet; thus, there was little possibility of a dense atmosphere or water that could have aided the evolution of primitive life forms.

Norman Horowitz, a biologist at Cal Tech and long-time participant in NASA exobiology studies, thought nothing in the new data encouraged the belief that Mars harbored life. "But the results also don't exlude this possibility." This was essentially what the exobiologists had expected, since Martian life was almost certainly microbial if it existed and would not be easily detected from flyby missions. "We have certainly seen no signs of the noble race of beings that built the canals or launched the satellites of Mars. I'm pretty sure they don't exist." *Mariner 6* and 7 data did strengthen the earlier conclusion that water was extremely scarce on Mars and that was a seriously limiting factor for the search for life. While no clouds, frosts, or fogs had been seen in the new pictures, minute amounts of water vapor had been detected in the atmosphere. "Mars is a cold desert by terrestrial standards. If there is life on Mars, it must be a form of life that can utilize water in the form of water vapor or ice." Horowitz added that it was possible that "extensions of our own terrestrial life, evolutionary adaptations," could live under such conditions. The exobiologist repeated what he had said many times: "The search for life on Mars is not sustained by optimism about the outcome. Anyone who is carrying on this work because he is sure he is going to find life, I think, is making a mistake. The search is sustained by the tremendous importance that a positive result would have, scientifically and philosophically, and until then we are obliged to continue the search." One of the major reasons they were exploring the Red Planet for life was to test their current notions about the origin of life. "We don't want to fall into the logical trap of using these notions to disprove in advance the possibility of life on Mars. We want to get there and make a direct test."[41]

*Effects on Mariner 71 and Viking*

Leighton, during the 11 September 1969 press conference, said that each Mariner spacecraft had "in its turn revealed a new and unexpected, no doubt significant kind of terrain. . . . Now I leave it to you to figure out how many new surprises there are still waiting for us on Mars." While Mars spacecraft evolved from one mission to the next, Leighton believed that he and his colleagues should not "fight the last war" with the Viking space-craft. Instead, they must realize that they were still only in the initial stages of exploring Mars. "Flexibility in design [and] adaptability in execution" were incredibly important.[42]

The distinctive new terrain revealed in the Mariner 69 pictures emphasized the importance of "an exploratory, adaptive strategy in 1971 as opposed to a routine mapping of geographic features." Very early in the first 90-day Mariner 71 mission, all of the planet should be examined with the A-camera, and selected targets should be studied with the higher-resolution B-camera, to correlate the extent and character of cratered, chaotic, and featureless terrains, and any new kinds of terrain, with classical light and dark areas, regional height data, and so on. Leighton and colleagues thought that a second objective should be the search for and exami-nation of areas that indicated the possible presence of local water. The complex structure found in the south polar cap called for close investiga-tion, particularly to separate the more permanent features from those varying daily or seasonally. A look at the north polar cap also promised to be "exceedingly interesting."

"If the effects of the Mariner 6 and 7 results on Mariner '71 are substan-tial, they at least do not require a change of instrumentation, only one of mission strategy. This may not be true of the effects on Viking '73." The Mariner 69 television specialists believed the discovery of so many new, unexpected properties of the Martian surface and atmosphere added a new dimension to selecting the most suitable landing site for Viking. Viking might be even more dependent on the success of Mariner 71 than had been supposed. From the improvement in the image resolution obtained by the 1969 B-cameras, scheduled also for use on Mariner 71, the team thought that an improved system might profitably be included in the Viking orbiter, designed to examine the fine-scale characteristics of terrains even more closely before choosing a landing site.[43]

At its 11 September meeting, the Viking Science Steering Group agreed that a joint meeting of Mariner 69, Mariner 71, and Viking 73 scientists would be useful. Jerry Soffen suggested that such a session would permit a more thorough examination of the *Mariner 6* and 7 information. At the same time, the science strategies for later flights to Mars could be more widely discussed. Plans called for the joint meeting to be held in early 1970 after the final selection of Viking investigators. Generally, Viking interest in the polar regions as a target for primary investigation diminished after hearing the early Mariner 69 reports.[44]

The Viking orbiter science briefing on 12 September concentrated largely on the orbiter imaging system and its role in providing pictures that would help find landing sites. Orbiter science objectives included:

- obtaining information for landing site selection for Viking,

- obtaining repeated coverage of landing sites during the lifetime of landers on the surface,

- obtaining information for selecting landing sites for future missions,

- making scientific investigations using the orbiter radio system, and

- obtaining information for studying the dynamic characteristics of the planet and its atmosphere.

Of the 57 kilograms alloted for orbiter science instruments, more than half (32 kilograms) was set aside for the imaging system. For many months, the specialists would discuss alternative approaches to the design of the camera system, as technical and fiscal issues affected the final design of this important piece of Viking hardware.[45]

## QUARTERLY REVIEW

As another step toward regularizing the management of the Viking project, Jim Martin arranged for the first of a series of project-wide quarterly reviews at the end of the first week of October 1969. Each systems manager was given 90 minutes to summarize progress in his area of responsibility. Henry Norris noted that this process was less detailed than the reports he had given in similar reviews at JPL in the past; instead his presentation was "delivered in tutorial style."[46] What is the orbiter? What is its function? How does it work? What is the progress to date? Are there any problems? If so, do they affect other systems and what steps are being taken to solve the difficulties? Over two days, many, many topics were covered.

The JPL presentations on the orbiter were typical of those given during the quarterly review. Norris opened with a brief overview of the schedule for the orbiter and his projected activities for the next three months. Richard K. Case of the orbiter design team reported on the configuration of the orbiter as it had evolved to date, summarizing telecommunications plans for the orbiter, lander, and Earth stations and briefing the group on steps being taken to integrate scientific experiments. Peter T. Lyman told his colleagues about the orbiter guidance and control propulsion subsystem, a complex subject to master. Lyman, a new member of the orbiter team, was the perfect man to tackle it. After 10 years at the University of California at Berkely, he had worked on Mariner 64 and helped plan hardware for the ill-fated Voyager. During Mariner 69, Lyman had been the project engineer from the Engineering Mechanics Division, overseeing much of the construction of the two successful Mariner craft. G. P. Kautz, in his turn, reviewed the manpower and funding JPL would need to develop the orbiter, closing with a list of the problems it faced.[47]

The quarterly review was followed up by two additional meetings in October. Langley Director Edgar Cortright held a session for the other center directors and key Viking project personnel, and Jim Martin convened a Viking Project Management Council meeting. The consensus was that the project was off and moving at a reasonable pace. Fewer problems seemed to have surfaced than might have been expected at this stage. Harris Schurmeier, the Mariner 69 project manager, noted that Viking was more complex than earlier projects because so many more partners were in the game. With all the different groups involved and with the limited dollars available, he thought the participants needed to establish clearer channels for handling problems.

Jerry Soffen also commented on the need for better communications. Although the quarterly review had been held to secure the participation of the many constituencies in the decision-making and reporting process, many of the scientists had left the meeting before the second day's discussions. Soffen's observation triggered a 45-minute session on how best to integrate the scientists into the project. Nearly everyone agreed that the investigators had to understand the fiscal and technical aspects of Viking so that they could appreciate the relationships of their own activities to the whole enterprise. The scientists would have to learn that their experiments were only a part of a very large undertaking.[48] As the specialists returned to

*Table 30*
*Viking Project Orbiter System:*
*Critical Schedule Activities, 1969*

| Activity | Required Date |
| --- | --- |
| Project spec approved | 1 Dec. 1969 |
| Orbiter investigators identified | 15 Dec. 1969 |
| Concepts approved and first drafts covering orbiter-lander interfaces | 1 Nov. 1969 to 2 Jan. 1970 |
| Orbiter system design concepts and general configuration established to allow subsystem function and design requirements to be prepared | 15 Jan. 1970 |

Critical problems

1. Many activities must start with preliminary data, requirements
2. Schedules must be achieved
3. Little or no recovery time

SOURCE: Martin Marietta Corp., Denver Div., "Viking Project Quarterly Review Held October 7 & 8, 1969 at Langley Research Center; Presentation Material," PM-3700005, Oct. 1969. Since events were to alter the Viking's project's calendar, the systems management offices would be forced to revise their plans many times. This is one early schedule.

*Table 31*
*Viking Project Orbiter System: Baseline*
*Conceptual Design Changes, Expected Weights, 1969*

| Item Changed | Baseline Weight (kg) | Expected Weight (kg) | Cause |
|---|---|---|---|
| Orbiter (less propulsion) | 627 | 606 | 1. Design to "flight loads" analysis<br>2. Use of lightweight solar cells<br>3. Reevaluation of expected subsystem weights |
| Propulsion (inerts and residuals) | 385 | 302 | 1. Substitution of helium for nitrogen as pressurant<br>2. Reduction of required $\Delta V$ = 1575 mps to $\Delta V$ = 1420 mps<br>3. Increase in nozzle expansion ratio from 40:1 to 60:1 |
| Usable propellant | 1420 | 1263 | 4. Use of selected injectors for $I_{sp}$ = 289 sec |
| Lander capsule adapter | 22 | 21 | 1. Design to "flight loads" analysis |
| Lander capsule | 816 | 995 | |
| Spacecraft adapter (includes destruct package and transition adapter) | 149 | 130 | 1. Design to "flight loads" analysis |
| Viking spacecraft launch weight | 3419 | 3317 | |

SOURCE: Martin Marietta Corp., Denver Div., "Viking Project Quarterly Review Held October 7 & 9, 1969 at Langley Research Center: Presentation Material," PM-3700005, Oct. 1969.

their various tasks after the saturating experience of the review at Langley, storms began to gather on the project's horizon.

During the remainder of 1969, one of the questions that nagged NASA managers who were looking for ways to pare the budget was, Is the orbiter essential to the Viking mission? This was an especially difficult question because eliminating the orbiters would obviously save a great amount of money, $100–165 million. For project personnel at headquarters and Langley who thought that the direct- versus out-of-orbit delivery issue had been settled nearly a year before, the revival of this question was disturbing.

On 13 September 1969, NASA's Lunar and Planetary Missions Board, an advisory group, agreed that the orbiters should be preserved, as they would give greater mission flexibility and a higher chance of mission success. When released from orbit, the landers could be expected to touch down in an elliptical area (called a footprint) 180 by 530 kilometers; with a direct entry that footprint would be increased to 500 by 900 kilometers. An orbiter-based mission would use the orbiter cameras to survey potential landing sites, which although not guaranteeing success would permit the Viking team to assess and eliminate obviously hazardous landing regions. But most significant, an orbit relay link would allow two-thirds more information to be sent to Earth than the lander alone could manage. With these considerations, the Lunar and Planetary Missions Board drafted the following resolution:

> A balanced program to develop a deeper understanding of man's neighborhood of the universe should remain a goal of NASA's lunar and planetary program. After examining Mariner 6 and 7 results, the [Lunar and Planetary Missions Board] emphasizes that landing of scientific instruments on Mars in 1973 remains a task of major importance.
>
> The cost of the Viking program now represents a substantial part of the funds at present available to the planetary program. Nevertheless, the [Lunar and Planetary Missions Board] considers the Viking program should go forward as planned.
>
> A Mercury-Venus flyby, the continued exploration of Venus, the introduction of a small planetary orbiter program, and the initiation of a major program to explore the outer planets are all essential to an orderly exploration of the solar system. NASA should develop those programs as required for this exploration.[49]

Although there would be several delays and unexpected twists and turns along the way, this resolution described the basic strategy NASA's planetary programmers would follow during the 1970s. Before it could be implemented, however, Walt Jakobowski and his team in the Viking Program Office at NASA Headquarters had to fight many battles just to preserve the basic Mars orbiter-lander mission. All of their work would be affected by a worsening budget crisis in Washington.

### MONEY PROBLEMS AT NASA

The summer of 1969 was a time for triumph and despair. *Apollo 11* landed on the moon in July, but at almost the same time NASA's budget was cut severely. Despite being an enthusiastic supporter of the Viking project and wanting to pursue an aggressive program of unmanned planetary exploration, Thomas O. Paine, appointed administrator in March, began to preach fiscal restraint to the Viking managers as early as June 1969. He told John Naugle, his associate administrator for space science and applications, that Viking and the other advanced planetary projects would have to be managed wisely because NASA was living in an era of great pressures to reduce the budget. The space agency's expenditures were being subjected to considerable public scrutiny and debate.[50]

Paine's worries were well founded. When the House Committee on Appropriations reported 19 June on the NASA budget request, the projected fiscal 1970 funds were nearly $300 million less than the previous year.

Five days later, the Senate Committee on Aeronautical and Space Sciences recommended a further reduction of $250 million. Late in July, Paine talked with President Richard M. Nixon about the space program as they flew to the Pacific splashdown site of *Apollo 11*. The president said that he personally was very enthusiastic about American space activities, but his administration could not direct large amounts of resources to the space program until the war in Vietnam had been ended. Nixon was reflecting the budget-cutting mood of Congress and the lack of public support for new space initiatives. Reactions to the report of the president's Space Task Group also affirmed the need for a fiscally responsible space program.[51]

To develop goals for the post-Apollo period, President Nixon had appointed a special Space Task Group* in February 1969. Although acknowledging that a new rationale for the American space program had to be sought—competition with the Soviet Union was no longer a realistic justification for NASA's activities—the task group rejected the idea that a manned mission to Mars in the 1980s should be the next great challenge accepted by the United States. The negative responses made on Capitol Hill and in the press to the manned Mars goal reinforced the group's decision. A July 1969 Gallup Poll, for instance, found 39 percent of 1517 persons polled nationally favored attempts to land a man on Mars; 53 percent opposed. Of the 21- to 29-year-olds, 54 percent favored the project and 41 percent opposed, but 60 percent of those over 50 opposed.[52]

As delivered to President Nixon on 15 September, the Space Task Group's report, *The Post-Apollo Space Program: Directions for the Future*, had backed away from an early manned landing on the Red Planet. The focus for the next decades in space was on the development of hardware and systems that would ultimately support a manned mission to Mars at the close of the 20th century. After a presidential briefing on the report, Nixon's press secretary said that the president agreed with the group's rejection of an overly ambitious program aimed at an early landing on another planet but also with its refusal to propose a program that would terminate all manned space activities in the post-Apollo years.[53] Six months were to pass before President Nixon personally reacted to the task group's findings, and by that time Congress, through the appropriation process, had shaped the immediate future for NASA's programs by restricting the agency's budget even further.

As the budget for fiscal 1970 went through successive parings and the public enthusiasm for space projects continued to dwindle, Naugle and his associates at NASA Headquarters grew more and more concerned about the continuing increases in costs for Viking. On 26 August 1969, Naugle wrote Ed Cortright and other top Viking managers to review his "personal

---

*The membership included Vice President Spiro T. Agnew, chairman; Secretary of the Air Force Robert C. Seamans; Administrator Thomas O. Paine; Science Adviser to the President Lee A. Dubridge; and, as advisers, Under Secretary of State for Political Affairs U. Alexis Johnson, Atomic Energy Commission Chairman Glenn T. Seaborg, and Bureau of the Budget Director Robert P. Mayo.

philosophy" on the subject. Naugle told the Langley director that "current indications of an increase over earlier estimates are of concern; particularly in light of the need to minimize Federal expenditures." He was especially worried about "cost overruns which in times of tight budgets, will inevitably result in disruption to the Viking Project or to other projects." While the associate administrator recognized the importance of the Mars mission and while he did not care to "establish arbitrary or unrealistic cost ceilings" that could also jeopardize the success of the effort, he did want everyone in the Mars project to ensure "that Viking [was] tight, efficient, well-engineered, and well-managed." Every effort had to be made to use existing technology "to minimize development risks and associated costs." Naugle recommended a very careful study of the proposed test program to determine if any paring could be done in that area. "While we cannot omit necessary development and tests, neither can we tolerate frills."[54]

But the costs for Viking continued to grow. When first presented to Congress in March 1969, the Viking price tag had read $364.1 million, an unsound estimate. At the time, the design of the spacecraft had not been clearly defined. By August, the expected cost had risen to approximately $606 million, with an additional $50 million for the launch vehicles. In testimony before the Subcommittee on Space Science and Applications of the House Committee on Science and Astronautics in October, Naugle admitted that the total cost of Viking would run about $750 million. Representative Charles A. Mosher of Ohio asked Naugle what he meant when he said that the $750 million "included an allowance for a minimum number of changes." The NASA spokesman responded that past experience with planetary programs indicated that the agency could expect a 15 to 20 percent increase in the cost of a given project. "So, in the case of Viking, we are including in this $750 million estimate about $100 million for mandatory changes or for trouble that we may get into in the project." NASA was using $650 million as its target, but Naugle told the congressmen that "we are only so wise and only so able to foresee into the future."

Representative Thomas N. Downing of Virginia expressed his concern about these projections since they had already grown more than 30 percent in little more than a year. Naugle noted that the figures presented in 1968 were based on a still poorly defined spacecraft. "What we have found . . . is that we underestimated the weight of both the orbiter and the lander." The additional weight could be translated into more man-hours of labor, which in turn could be translated into more dollars. On top of that, the cost of those man-hours had also increased. All the congressmen were disturbed. Joseph E. Karth, the subcommittee chairman, pointed out that his group had to sell these cost escalations on the House floor and it would not be easy. Naugle's statements that everything was being done to keep costs in line were not all that reassuring to Karth, who believed that NASA had "so far failed miserably in that regard." After trying to convince the subcommittee that the agency had "made a substantial effort to accurately determine funding requirements before beginning hardware development," Naugle and his staff renewed their attempts to control the project managers. Since Congress would not suffer another project with a huge cost overrun, Don Hearth and others working for Naugle sought to establish controls over Viking that would prevent sudden and unexpected expenditures by the engineers in the field.[55]

For all their concern and activity, the men at NASA Headquarters could not prevent the budget crisis. When President Nixon signed the fiscal 1970 appropriations bill on 26 November, the total amount—$3.697 billion— was $299 million less than appropriated the previous year. At the same time, the Bureau of the Budget was already beginning to chip away at the dollars the space agency was seeking for 1971. Robert P. Mayo, director of the Bureau of the Budget, found himself in an awkward position; he had promised President Nixon a balanced budget, but finding places where he could reduce expenditures was very difficult. Throughout the fall of 1969, a stiff debate ran between the space agency and the budget people, and some of the meetings Paine, Mayo, and their staffs held were not pleasant.

In light of the Space Task Group's report, Paine reasoned that he could not recommend a budget of less than $4.25 billion for NASA. He told Mayo in a letter: "This is a difficult time. Please do not think me unfeeling toward the many claimants for your scarce budgetary resources." But Paine thought that inefficient agencies were being rewarded with increased budgets while NASA was being penalized. "The people of NASA have produced outstanding results . . . while reducing costs and personnel more than any other area of government. . . . Space offers the President now a highly productive program and his greatest leadership opportunity." Unfortunately, the dollars did not go to the successful.[56]

For Viking, the budget cut was devastating. Before Congress had a chance to consider the budget, Nixon's administration cut $20 million from the amount requested for the Mars lander project for 1971. The picture was unpleasant. With the decline in resources, aggravated by inflation, Administrator Paine had to reduce expenditures.[57]

*Table 32*
*NASA Appropriations, FY 1968–1971*

(in billions)

| Budget Item | FY 1968 | FY 1969 | FY 1970 | FY 1971 |
|---|---|---|---|---|
| Total NASA budget | $4.5889 | $3.9952 | $3.6967 | $3.3126 |
| Lunar & Planetary Programs | .1250 | .0923 | .1388 | .1449 |
| Mariner Mars 71 | — | — | .0454 | .0296 |
| Viking 73 | — | — | .0400 | .0350 |
| Mariner Venus Mercury 73 | — | — | .0030 | .0211 |

Paine was convinced that the only alternative to the delay of Viking was its cancellation. At noon on 31 December 1969, Paine told John Naugle that further analysis of the federal budget for 1971 by the Bureau of the Budget had disclosed a $4-billion problem; NASA had been asked to reduce its request by $225 million. The administrator and his associates considered three ways to cut dollars—delay Viking from 1973 to 1975; cut the Viking orbiter completely and reduce further the Office of Manned Space Flight budget; or eliminate manned flights after the final Skylab flight in 1973. The second and third options would not provide the necessary reduction, and the Bureau of the Budget, with President Nixon's agreement, thought that deferral of Viking was the best step. Naugle spent the rest of that day working out the details of Viking's slip, taking time out to note for the record: "I left at 4:30 pm to welcome the New Year and the new decade in a bleak mood—feeling that two years of careful planning for Viking had been wiped out in four hours by a combination of a budgetary error and the article in the [Washington] Post on Monday, 29 December, by Cohn stating that scientists at the [American Association for the Advancement of Science] Meeting had advocated a reduction in the NASA science program." NASA's space projects were under criticism as part of a general outcry against federal spending that did not contribute to the solution of social problems like pollution and feeding the poor. While scientist Carl Sagan pointed to the Defense Department as the real source of budget misallocations, other "authorities" questioned NASA's current proposals to send manned missions to Mars. Caught in the midst of the antimilitary, antitechnology furor was Viking. During the last hours of 1969, NASA nearly lost another opportunity to land on Mars at all.[58]

After two weeks of scrambling to reorganize the space agency's programs, Tom Paine made a public statement of the changes the 1971 budget would require. Mindful of recent criticisms, he commented:

> We recognize the many important needs and urgent problems we face here on earth. America's space achievements in the 1960's have rightly raised hopes that this country and all mankind can do more to overcome pressing problems of society. The space program should inspire bolder solutions and suggest new approaches. . . . NASA will press forward in 1971 at a reduced level, but in the right direction with the basic ingredients we need for major achievements in the 1970's and beyond.

While NASA diminished its total activities, the agency would "not dissipate the strong teams that sent men to explore the moon and automated spacecraft to observe the planets." Paine listed the following actions as being consistent with the requirements of the 1971 budget:

> 1. We will suspend for an indefinite period production of the Saturn V launch vehicle after the completion of Saturn V 515.
> 2. We will stretch out the Apollo lunar missions to six-month launch intervals, and defer lunar expeditions during the [Apollo Applications

Program] space station flights in 1972 [actually flown in 1973, as Skylab flights.]

3. We will postpone the launch of the Viking/Mars unmanned lander from 1973 to the next Mars opportunity in 1975.

With the closing of the Electronics Research Center in Cambridge, Massachusetts, these actions would reduce the number of persons (including contractors) working on NASA projects from 190 000 at the end of fiscal 1970 to about 140 000 at the end of fiscal year 1971.[59]

Although Viking survived, there was considerable confusion at first over what the modified project would be. Henry Norris and his orbiter teammates officially learned about the change in plans on 12 January 1970.[60] At the Viking Orbiter Staff meeting in Pasadena the next day, Norris explained that they had been asked to examine two alternatives for 1975 Viking missions—the basic 1973 orbiter-lander mission rescheduled for 1975, or a direct-entry lander mission. This renewed debate over what was called "Options A and B" brought a sense of *deja vu* among the working people.[61]

Besides an additional direct dollar cost of about $102.2 million, JPL learned from the program office at headquarters, other problems were associated with deferring Viking to 1975. Steps would have to be taken to bolster morale among the scientists and engineers. The several false starts on Viking's predecessors and the cancellation of Voyager had already discouraged many. As with all complex projects, a strong and highly motivated team was essential for success, and a limited sum of money would have to be made available during fiscal 1970 and 1971 to hold the existing team together and permit some meaningful work on the aspects of the mission that would pose the greatest technical challenges. The balance of the Viking project would be budgeted at 1970 levels, but slipped two years. An additional five percent would be added to compensate for possible inflation.

William J. Schatz of the JPL Propulsion Division pointed out two other problems caused by the delay. A mission in 1975 would require a longer flight time; Mars's position relative to Earth would require a different trajectory. Previously, the mission analysis and design people had used Voyager 1973 work to plan for the 1973 Viking flight. A 1975 launch would require the specialists to start trajectory and flight path analyses from scratch. New calculations would demand more manpower and computer time, both of which cost money. Hardware alterations would also be required. Changes in the materials used for the propulsion systems might be necessary to ensure their reliability, and the use of helium as a pressurant would have to be reevaluated. But beyond these technical considerations was the economic impact of the stretchout. "Of prime importance," said Schatz, was the retention "of a qualified team of engineers at the rocket engine contractor during the stretchout period." The engine manufacturer, Rocketdyne, a division of North American Rockwell, was already laying off personnel, "jeopardizing their ability to support our development program." Other vendors were either closing their doors or dropping assembly lines for certain components because of the general poor condition of the economy. JPL was planning to procure many items it needed for Viking as soon as possible and place them in bonded storage until it was ready to assemble the spacecraft.[62]

During late January and early February, NASA Headquarters, Langley, and JPL personnel continued to evaluate the future course of Viking. After receiving a 28 January briefing by various Viking staff members, John Naugle decided on 10 February that the agency would pursue its original plan to fly an orbiter-lander combination. Positive words of support for the Viking team were put on record by George M. Low, NASA deputy administrator, and Naugle. Both men knew that the real work had just begun, but they appreciated the teamwork displayed during the latest crisis. Low told his colleagues, "Viking holds the highest priority of any project or program in NASA's Planetary Program. Viking holds a high priority among all of NASA's programs."[63]

The Space Science Board of the National Academy of Sciences also underscored the value of continuing with Viking, but the board's endorsement carried some reservations. Philip Handler, president of the Academy, had suggested to NASA Administrator Paine in mid-November 1969 that a Space Science Board review panel be established to evaluate the balance among the scientific disciplines supported by space agency funds. The last such review had been held in July 1966 at a time when National Academy and NASA personnel had assumed that the budget for space activities would continue to increase. Paine accepted Handler's offer, but advised him and his colleagues to weigh carefully the impact of any recommendations to shift money from one project to another. Any recommendations to cancel programs that had already gone through an elaborate approval process within NASA would, in the existing budgetary climate, "almost certainly lead to the curtailment of the on-going [programs] with little chance that additional funds [would] become available for [any] program which the Board feels should be increased."[64]

The Space Science Board team that evaluated NASA's space science activities was known as the Viking Review Panel, reflecting the amount of money being spent on the Mars project and the concern generated by the postponement of the Mars landing. The panel report issued on 24 March 1970 combined praise and concern. NASA was complimented for its work in defining a project that accurately reflected the payload recommendations of the Space Science Board's 1968 study, *Planetary Exploration, 1968–1975*. Cost projections, however, caused some division among the members of the panel. Some believed that the potential return from the Mars mission was so great that $750 million was justified. Others expressed concern that "within the extremely restricted budgetary climate, NASA must set much more limited goals for itself in order to achieve a balanced scientific effort." This latter group feared that Viking's high cost would cause the space agency to lose other "less costly but equally valuable missions."

Some participants in the review were worried about the complexity of the Viking science payload, the most sophisticated payload planned to date, with many new experiments. A two-year delay of the Viking launch might indeed be beneficial. "The additional two years can be devoted to an extensive test of the abilities of the payload, increasing confidence in [it]."

Since it appeared that future budgets for space activities would be low, the Viking Review Panel recommended that "considerably more modest planetary missions" be initiated in the years to come. Single, complicated, expensive projects like Viking were too risky—politically and technologically. Realistically appraising the Viking Review Panel's pronouncement, John Naugle told Paine, "It is, I think, in view of the talk by the scientific community these days, an accurate and as good a statement about Viking as we could expect."[65]

## WORKING TOWARD JULY 1975

Money problems would always haunt the Viking project. The scarcity of dollars especially affected the development of the lander and its science payload and repeatedly tried people's patience and equanimity. Early in 1973, Joseph R. Goudy, the Langley Viking Project Office resident engineer at JPL, commented on budget cuts that led to the dismissal of about 200 employees at the California laboratory on rather short notice:

> These cutbacks have created a different atmosphere and environment, resulting in a change in attitude. Six months ago, when the [Viking Project Office] came in with a new requirement or direction that required additional or premature effort, it was generally accepted with the attitude, "We don't think it's necessary but it's their money; if they want it, we'll do it." Now, with the Orbiter having to take rather severe cuts, this is no longer considered "their" money and the attitude has become much more critical, if not down-right hostile.[66]

Henry Norris, looking for ways to keep his orbiter personnel from reacting too negatively to the repeated budget cuts, tried to convince them—and for the most part he succeeded—that the budget was just one of the many realities that a good engineer or manager had to live with and work around as he tried to do his job.

The tasks assigned to the orbiter teams were laid before them in a five-year schedule, which ended with a pair of mid-summer 1975 launches. The master plan was presented for the first time at the Viking Project Management Council meeting in February 1970, and it reflected the changes brought by the stretchout.

The pace of the work at JPL assumed a rhythm familiar to the people who had worked on other NASA projects. The determining factors, "driv-

*Table 33*
*Viking Orbiter Schedules*

| Event | Proposed before 1 Jan. 1970 | Proposed after 1 Jan. 1970 | Actual Dates |
|---|---|---|---|
| Preliminary design review | May 1970 | Jan. 1972 | 19-20 Oct. 1971 |
| Critical design review | June 1971 | Jan. 1973 | 9-10 July 1973 |
| Start proof-test spacecraft test | Aug. 1972 | March 1974 | Jan. 1974 |
| Qualification test completed | Nov. 1972 | July 1974 | Jan. 1975 |
| Shipment of first flight hardware to KSC | Feb. 1973 | Dec. 1974 | Feb. 1975 |
| Launch | July 1973 | July 1975 | 20 Aug. 1975 9 Sept. 1975 |

SOURCE: Information on the 1970 master plan was taken from Henry Norris, "Viking Orbiter Project Staff Meeting—Minutes of January 13 and 14, 1970," memo, 19 Jan. 1970.

ers" in NASA parlance, for the designers and engineers were master schedules that determined when major hardware components had to be completed so the launch dates could be met. But the realities of designing and building the spacecraft did not always conform to calendar milestones, and the variance led to frequent revisions of the schedules. At every step along the way, the work was formally documented in a large number of Viking project documents. By cross-checking and coordinating these documents, the project manager at Langley could be assured that the orbiter, lander, science payloads, launch vehicles, ground support equipment, flight control facilities, and the tracking system would all function as required when the hardware was brought together and assembled for the launch and flight to Mars. This system of mass documentation, formal reviews, telecons, and informal conversations worked because the people associated with the effort believed in delegated management. Jim Martin's centralized responsibility and authority for Viking was a key factor to the project's success, but equally important was the esprit de corps among the Viking teams at the working level.[67]

The troops at JPL functioned within divisions responsible for specific engineering activities or disciplines. Norris and his orbiter staff allocated funds, prepared plans and schedules, assigned tasks, and received progress reports, but the divisions carried out the actual design and development of the spacecraft and experiment hardware, as well as prepared and operated such facilities as the Deep Space Network and the Space Flight Operations

Facility. Each division chief and his subordinates not only supervised their personnel but also selected the engineers who represented their divisions on the orbiter team.*[68]

The structure of management at JPL did not fit Jim Martin's management scheme. The people at Langley had always worked through a more centralized organization, in which everyone was directly responsible to the project director, and the Viking Project Office was uneasy with the JPL system. Martin knew that the organizational structure of the lab would not likely be changed just for this mission, so he went to Pasadena in the early spring of 1970 to observe firsthand how JPL worked. Specifically, he wanted to know: How had JPL dealt with hardware problems in the past? How did it plan to manage the Viking orbiter in the future? How would it control the flight phase of a mission?[69]

Henry Norris believed that the time Martin spent with division managers and Viking representatives at JPL led him to understand more clearly the lab's approach to project management. Martin was still "not entirely comfortable" with the organization, Norris reported, but at least the project director had been exposed to it and the men who filled the ranks. Likewise, the people at JPL began to appreciate the sources of Martin's concerns and continued to work with the project office to improve and strengthen JPL management control over the teams in Pasadena.[70]

Although they had adopted different approaches, the personnel at Langley and JPL were working toward the same goal. Once the baseline orbiter configuration had been established in February 1969, the next major orbiter goal was the preliminary design review (PDR). This formal review, held on 19-20 October 1971, came at the end of the conceptual phase for the design of the orbiter systems; the specialists were now ready to work on the detailed design of the hardware. Once the basic soundness of all aspects of the orbiter was approved, the teams would head for the next important milestone, the critical design review (CDR). Getting to the PDR had been a major accomplishment, made difficult by the repeated problems with the budget; but the teams at JPL had completed their design work and coordinated their efforts, attending weekly meetings and frequently using the telephone along the way. In fact, more than 60 meetings were held that directly impinged upon the design of the orbiter.

The preliminary design review gave all interested parties a look at the orbiter as JPL planned to build it. Once the conceptual design was complete, work on the design of breadboards, or first working test models, of the basic orbiter subsystems would begin. These designs would be evaluated at subsystem PDRs and, once approved, work on the breadboards would

---

*Divisions and their representatives assisting the Viking orbiter staff at JPL, spring 1970: Quality Assurance and Reliability, G. E. Nichols; Project Engineering, V. R. Galleher; Data Systems, G. F. Squibb; Space Science, M. T. Goldfine; Telecommunications, J. R. Kolden; Guidance and Control, A. E. Cherniack; Engineering Mechanics, W. J. Carley; Astrionics, J. D. Acord; Environmental Sciences Simulation, N. R. Morgan; Propulsion, W. J. Schatz; Mission Analysis, P. K. Eckman; and Technical Information and Documentation, S. B. Hench.

proceed, with their suitability for conversion into flight hardware being confirmed during a series of subsystem critical design reviews. A general CDR for the entire Viking orbiter system would certify the readiness of the orbiter staff to go to the next step—building the flight-ready orbiters.

By October 1971, the orbiter had assumed the basic configuration it would have when launched in 1975. The spacecraft had grown considerably larger than its Mariner Mars 71 predecessor. Most noticeable visually were the larger solar panels and the larger high-gain antenna. But all the internal subsystems were taking on a Viking identity of their own as well. The Mariner inheritance was still there, but instead of directly transferring subsystems from one craft to another, the engineers were borrowing from Mariner experience and know-how. Still, it was this transfer of technological knowledge from Mariner Mars 71 and Mariner Venus 73 that permitted the Viking orbiter personnel to get the craft ready to fly on time with a minimum of problems and money crises.

Jack Van Ness, deputy Viking project manager, recorded in his "Viking Weekly Highlights Report" that the orbiter system preliminary design review was well organized and informative. Only 22 action items remained for solution. "This relatively small number is somewhat indicative of the clarity and thoroughness of the presentations." At the conclusion of the review, the Viking Advisory Review Panel and the Orbiter System Manager's Advisory Panel provided a favorable overall evaluation of the orbiter status. None of the evaluations turned up any critical problems that would give Martin's Viking Project Office cause for concern.[71]

With the PDR behind them, Norris's people began to prepare the detailed designs of the 21 orbiter subsystems. Soliciting requests for proposals from industrial contractors, selecting companies to build the subsystems, and negotiating contracts occupied the months from October 1971 to July 1972. One contract was not let until July 1973. Meanwhile, the various divisions at JPL had begun to work on the subsystems that would be built at the laboratory. Preliminary design reviews for these subsystems began in January 1972 and lasted until late November.

Close on the heels of the PDRs came the subsystem critical design reviews, which spanned January to July 1973. When the subsystem CDRs were completed, a general CDR at JPL 9–10 July 1973 evaluated the entire orbiter system as it had evolved to date. The CDR panel, the Viking Advisory Panel, and the Orbiter System Manager's Advisory Panel all expressed their confidence in JPL's performance and the quality of the teams' work.[72] The technical problems being encountered by the orbiter were the routine kind that appeared during the course of most spacecraft projects—recurring difficulties with poor-quality integrated circuits and an unhappy experience when an early production propulsion tank ruptured because of a metallurgical failure.

During the summer of 1973, only two subsystems caused genuine concern. The infrared thermal mapping (IRTM) subsystem was behind schedule, but by mid-July the Santa Barbara Research Center had the trouble under control, and the subsystem CDR was held that month. The data-storage-subsystem tape recorder's failure to operate at a satisfactory speed put it on the Viking Project Office's "Top Ten Problems" list. In October the "54L" integrated circuits were also added to the list. Overall, however, the orbiter was shaping up as a well-behaved spacecraft, and everyone was pleased. Concern over the orbiter's financial problems was constant, but the project management was confident that Henry Norris's teams were on schedule and doing well. By drawing on Mariner heritage, they had the Viking orbiter under control.[73]

In mid-1973, the orbiter hardware entered the test phase. The first test, called the modal test, was conducted with the orbiter development test model, to determine if the mathematical model used for the engineering load analysis was correct. The modal test ran from late May until the end of July. A week later, General Electric delivered the first computer command subsystem. In late August, the propulsion-system engineering test model was test-fired at the NASA Edwards Test Station in California, while at JPL the flight-data-subsystem breadboard was checked out with other pieces of hardware that were to be linked to it, such as the visual imaging subsystem, the IRTM, and the atmospheric water detector. During the first and second week of September, other tests were run to determine the effect of shock on various orbiter instruments. Joseph Goudy reported to Martin on the 14th that the results from the pyrotechnic shock tests were much better than they had anticipated: "None of the subsystems that were on board for the tests appeared to have suffered any adverse effects. . . ." The sensitive instruments would not be harmed when the spacecraft was explosively separated from the Centaur launch vehicle stage and the lander was explosively separated from the orbiter.[74] In mid-December 1973, JPL completed the vibration stack test of the orbiter and lander development test models. Since this was the first time that orbiter and lander hardware had been mated and tested together, everyone in Pasadena was particularly satisfied when no important questions were raised by the examinations.[75]

With the new year upon them, the orbiter team focused its attention on final assembly of the proof-test orbiter and tests of this first flight-style hardware. These qualification tests would determine the spaceflight worthiness of the orbiter system designs as they had been rendered into hardware. The assembly process took three months as each of the subsystems was checked out and assembled onto the orbiter bus. During April and May, the engineers at JPL conducted the system readiness test, verifying the functioning of all orbiter components. The successful examination of the orbiter hardware prompted Goudy to report to the Viking management at Langley that they were on schedule and that the assembly of the proof-test orbiter had served as a "pathfinder" for the fabrication of the flight orbiters.[76] In the process of building this first craft, officially designated Viking orbiter 1 (VO-1), the spacecraft assembly personnel members at JPL learned some

*Table 34*
*Growth in Capacity of Data Storage Subsystems*

| | Mariner 64 | Mariner 69 | Mariner 71 | Viking 75 |
|---|---|---|---|---|
| Number of tape recorders | 1 | 1 | 1 | 2 |
| Number of tracks | | 4 | 8 | 8 x 2 |
| Recording rate | | 16 200 bits per sec | 132 000 bits per sec | 301 172 bits per sec, tracks 1 through 7; 4 and 16 kilobits per sec, track 8 |
| Playback rate | $8\frac{1}{3}$ bits per sec | 270 bits per sec | 1, 2, 4, 8, or 16 kilobits | 1, 2, 4, 8, or 16 kilobits |
| Storage capacity | 5.4 million bits | 23 million bits | 180 million bits | 640 million bits x 2 |
| Length of tape | 100 meters | 111 meters | 168 meters | 384 meters x 2 |
| Weight | 19 kg | 19 kg | 11 kg | 7.7 kg x 2 |
| Contractor | Lockheed Electronics Co. Inc., Plainfield, N.J. | Lockheed Electronics | Lockheed Electronics | Lockheed Electronics |

NOTE: The data subsystems (reel-to-reel tape recorders) used on the Mariner and Viking spacecraft permitted recording scientific data and subsequently playing it back through the communications subsystem for transmission to Earth. As the number of experiments increased and the amount of data to be stored and played back grew, successive data storage systems became more complex. Each new tape recorder had greater capacity, posing new technological challenges. In Viking, each data subsystem tape recorder weighed 3.3 kg less than the Mariner 71 data subsystem recorder, while having 3.6 times the information storage capacity. That accomplishment took time and caused some real headaches for the Viking managers, but the completed recorders worked very successfully during the missions.

important lessons that would help them build Viking orbiter 2 and 3, the orbiters that would fly to Mars. One problem they encountered was the lack of sufficient work stands, particularly during the installation of the thermal insulating blanket. More stands were ordered, to avoid any bottleneck during the assembly of the flight articles. The proof-test orbiter was moved on 8 May from the Spacecraft Assembly Facility to the Environmental Laboratory, where it would go through the rigors of vibration, electromagnetic interference, pyrotechnic, thermal vacuum, and compatibility tests during the summer of 1974. At the same time, engineers would begin assembling and testing VO-2 and VO-3.[77]

On schedule with satisfactory results, the VO-1 tests were completed in late August. As the JPL team turned its attention to readying VO-3 for early examination, however, unexpected budget problems brought a change in plans.[78] On 27 September, the orbiter staff was forced to order all testing of the third orbiter to cease. The second test team was disbanded; no money was available for testing. VO-3 was put into storage, and the proof-test orbiter (VO-1) was redesignated a flight unit. VO-1 and VO-2 would be the

*The thermal-control model of the Viking orbiter mated to the lander thermal-effects simulator was used in August 1973 to verify the effects solar radiation would have on the spacecraft. The science platform with imaging system and other instruments is attached under the orbiter.*

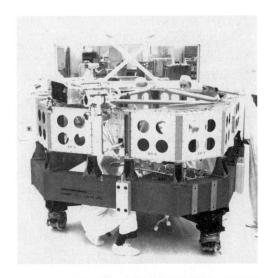

*Building the Viking Orbiter at Jet Propulsion Laboratory in 1974. Men working inside the chassis, right, fabricate the orbiter bus structure. Below right, they attach the propulsion module to the propellant tanks. Below, solar panels are in place on the nearly completed orbiter.*

spacecraft sent to Mars. To ensure the acceptability of the proof-test hardware for flight, a series of meetings were held during the next several weeks.[79] But an orbiter design qualification review scheduled for early October 1974 lost much of its significance, since the change in plans had thrown off JPL's timing. As one participant observed, it was hard for a review panel "to determine if the Orbiter met all of its requirements in spite of all the testing that has been done."[80]

After several more months of work, orbiter VO-1 was verified for flight on 9 January 1975, and the VO-2 tests were completed on the 31st. The orbiters were shipped to the Kennedy Space Center in February, where a series of preflight checks would be made through the spring and summer.[81] The Viking orbiter, remarkably close to early weight predictions (see table 35), was a very carefully tested piece of equipment. For the teams at JPL, the design, development, fabrication, and assembly had, for the most part, gone according to plan, schedule, and budget.

*Table 35*
*Viking Orbiter Specifications, 1969–1975*

| Orbiter Element | Baseline Orbiter Feb. 1969 | PDR Orbiter Oct. 1971 | Flight Orbiter Feb. 1975 |
|---|---|---|---|
| Bus dimensions | | | |
| Long sides | | | 139.7 cm |
| Short sides | | | 50.8 cm |
| Height | 45.7 cm | 45.7 cm | 45.7 cm |
| Distance from launch vehicle attachment points to lander attachment points | | 3.29 m | 3.29 m |
| Distance across extended solar panels, tip to tip | 7.80 m | 9.75 m | 9.75 m |
| Weight with fuel | 2298.6 kg | 2304.3 kg | 2324.7 kg |
| Weight of fuel | 1862 kg | 1404.8 kg | 1422.9 kg |
| Weight of science instruments | 57.6 kg | 65.4 kg | 65.2 kg |
| Visual imaging system | 21.8 kg | 42.05 kg | 40.05 kg |
| Infrared thermal mapper | 13.6 kg | 7.48 kg | 9.30 kg |
| Mars atmospheric water detector | | 15.90 kg | 15.90 kg |

Source: JPL, "Viking Project Orbiter System, Visual Presentation, February 13, 14, 1969" [Feb. 1969]; JPL, "Viking 75 Project Orbiter System PDR, October 19–20, 1971, Presentation Material" [Oct. 1971]; and Martin Marietta Aerospace, Public Relations Dept., *The Viking Mission to Mars* (Denver, 1975), pp. III-25, III-27, III-32, III-33.

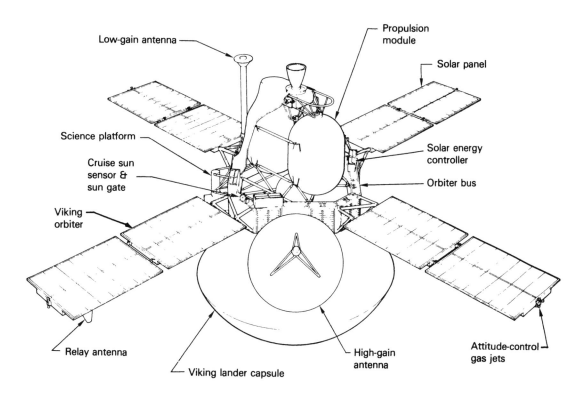

Low-gain antenna

Propulsion module

Solar panel

Science platform

Solar energy controller

Cruise sun sensor & sun gate

Orbiter bus

Viking orbiter

Relay antenna

Viking lander capsule

High-gain antenna

Attitude-control gas jets

*Configuration of the mated Viking orbiter and capsule in cruise mode.*

Carl D. Newby, supervisor of the Spacecraft Development/Mechanical Support Group, oversaw the assembly of the orbiters. It was the biggest spacecraft Newby and his team had built, and because it was so big it was an easy craft on which to work—they had room to move around during the assembly process. Newby pointed out that it requires a special personality to work on space hardware and a special dedication. Fabricators come to view the spacecraft as part of their lives, to care about it. Working in a closed environment, they have to learn to live with one another, as well. Spacecraft builders must be adaptable, very careful, and thoughtful. One false move, one thoughtless motion can destroy an assembly or component worth thousands of dollars or months of time. Damage to a spacecraft usually also requires requalification of the injured components or perhaps requalification of the entire craft. Workers on the Viking orbiters—many had worked on Ranger, most had worked on the Mariners—were very fond of their spacecraft. As Newby repeatedly reminded the specialists at JPL, the orbiter was a "good spacecraft to work on, it was on time and on budget."[82] Building the Viking landers, however, was a completely different story.

# 7

# Viking Lander: Creating the Science Teams

Designing and fabricating the Viking lander was a difficult task. Engineers at JPL could draw on their experiences with Mariner systems as they worked on the orbiter, but the lander team was tackling a new field. The men in California completed the orbiter with relatively few technical difficulties; but the contractors at Martin Marietta in Denver, breaking much new technological ground, encountered many problems. The lander was far more complex than NASA's previous unmanned lander, the lunar Surveyor, and Viking's goals were more ambitious. Viking was twice as heavy as Surveyor; it had two cameras for stereophotography and a complement of very sophisticated scientific instruments, and it was destined to land on a planet far more distant than Earth's own moon. The Viking lander represented a series of clever inventions in answer to specific problems. While this inventiveness can be seen clearly in the creation and fabrication of the biology instrument and the gas chromatograph–mass spectrometer, the NASA-contractor team also developed a host of other new solutions to meet new technological demands.

As with the orbiter, the first priorities of the Viking managers in dealing with the lander were establishing spacecraft specifications, selecting an organization to build it, and forming key teams to do the work— industrial, managerial, and scientific teams. Team-building began months before the official approval of the Viking program when Jim Martin at Langley Research Center selected some of his top people from the Lunar Orbiter team. For deputy project manager Martin selected Israel Taback, spacecraft manager for Lunar Orbiter. Iz, as he was called by his colleagues, had joined the Langley staff in 1942 as a mechanical engineer on graduation from Cooper Union Engineering School. He successively headed the instrument calibration laboratory—a group developing aircraft flight instruments—and the navigation and communications branch at the Langley center. Gerald A. Soffen, Viking project scientist, once noted that while Taback might have looked like a tailor among the engineers and managers, he was the wizard behind the Viking lander. If any one man could be awarded the title "father of the lander," it was Taback.[1]

During the summer and fall of 1968, Taback supervised the progress of the contractors studying various technological approaches for landing on Mars. General Electric was investigating hard-landers; McDonnell Douglas Astronautics was examining soft-landers; the Boeing Company was studying propulsion and landing systems; and Hughes Aircraft was looking into low-cost landers, support modules, and mission reliability. In Denver, the Martin Marietta Corporation was winding up a study of direct versus out-of-orbit entry for the lander.[2] These early studies helped define the shape and size a Mars lander would have for a Titan-launched mission. They also drew attention to subjects that would require special handling. Taback and his associates at Langley worked constantly with the contractors so that their latest ideas for alternative approaches to lander design could be debated and evaluated in NASA circles.

As Taback's people and the contractors worked on general approaches to lander design, Jim Martin took steps to begin definition of the science payload. In August 1968, he established a science instrument working group under the chairmanship of G. Calvin Broome. Broome, who had joined Langley in June 1962, was manager of the photographic subsystem of Lunar Orbiter, overseeing design, fabrication, testing, and operation of the instrument that would photograph the lunar surface. Just 30 years old in the summer of 1968, Cal Broome was given a major responsibility for Viking. His working group, a subdivision of the Mission Design Steering Committee, would oversee all the preliminary planning for the scientific payloads for the orbiter and lander. Essential to its work was an understanding of the interactions among the various lander experiments, especially the interfaces among the surface sampler, biological instrument, and gas chromatograph–mass spectrometer experiments being proposed for the mission.[3]

With the definition of the lander and science hardware taking shape, Jim Martin needed a project scientist. He first took measure of Gerald Soffen during a 1967 briefing, when Soffen, a senior scientist at JPL, described his abbreviated microscope as a possible life detector. The scientist impressed Martin with his technical competence and his enthusiasm for Mars exploration. Jerry Soffen, 42 years old in 1968, was one of the early members of the exobiology community. After receiving his Ph.D. from Princeton University in 1960, he had been a U.S. Public Health Service fellow at the New York University School of Medicine. Shortly after he joined the JPL staff, he took part in devising instruments for detecting life on Mars, in the science planning for Mariner B, and in the development of automated biology laboratories. Before the demise of Voyager, Soffen had been deputy project scientist for that endeavor. With this background, he had the necessary stature in the scientific community that Martin was sure would be needed by the project scientist of a 1973 Mars landing mission.[4]

In August 1968, Edgar M. Cortright, Langley director, asked JPL Director William H. Pickering to assign five JPL staff members to the

Virginia center for six to nine months of temporary duty. Of those requested, four had taken part in capsule systems advanced development activities at JPL. The fifth was Soffen. Pickering and his managers were unhappy about this request. At that time, Langley and JPL were competing over Mars mission proposals, and it did not seem to be in JPL's best interests to send its specialists to help the competition. Pickering told Cortright that if Langley wanted Soffen, then Soffen would have to resign his position and join the civil service staff at Langley. Soffen recalled that he felt like a pawn in a game of planetary chess. Cortright could not promise that the Langley proposal for a 1973 Mars mission would be approved, and if it were not, Soffen could find himself a solitary scientist awash in a sea of engineers in Tidewater Virginia. If he stayed at JPL, he would be able to keep alive his vital contacts with other space scientists, but he might also miss the opportunity to lead the first landed scientific investigations of Mars. Cortright ultimately persuaded Pickering to agree to Soffen's temporary assignment to the Langley Mars 73 planning project, but only after an appeal to John Naugle at NASA Headquarters.[5]

Reflecting on his decision to move from California, Soffen commented that morale and leadership also affected his desire to make the change. In the months immediately following the termination of Voyager, the planners at JPL were in turmoil. At Langley, the situation was different. Cortright and Martin wanted their 1973 project to become a reality, and Martin especially pursued this goal with single-minded zeal. If sheer will and determination could make something happen, then Langley would be the center that landed spacecraft on Mars. Appreciating this aggressive spirit, Soffen forced the issue of his being detailed to Langley by purchasing a house in Hampton, Virginia. In the face of a determined Soffen and a solid front in the NASA management, Pickering had to let Soffen go east.[6]

## A TEAM OF SCIENTISTS

Setting up the science instrument working group and appointing a project scientist* were part of Langley's strategy to gain an early definition of the scientific aspects of the landed mission. Prospective industrial contractors would, in turn, have a reasonably good understanding of the problems in building the lander and incorporating the scientific instruments into it. During the second half of 1968, Jim Martin, Jerry Soffen, and A. Thomas Young began talking to scientists. Tom Young would have a very difficult assignment as science integration manager; he would often be surrounded by the conflicting demands of Martin, project engineers, contractor engineers, and oft-complaining scientists. Another 30-year-old, a mechanical engineer with a second degree in aeronautical engineering, Young was a native Virginian and a graduate of the University of Virginia.

---

*At NASA Headquarters, Soffen's counterpart was Milton A. Mitz, program scientist. On 28 December 1970, Mitz left Viking to join NASA's Grand Tour Project, and Richard S. Young became Viking program scientist.

He had joined Langley in 1961 and managed the mission-definition phase of the Lunar Orbiter project.[7]

Together, Young, Martin, and Soffen went in search of science team members for a 1973 mission. At the outset, it appeared that NASA Headquarters preferred that Langley deal with "inside" scientists; that is, persons already receiving support from the Office of Space Science and Applications. But the managers at Langley wanted to cast their net as widely as possible.[8] Their philosophy was outlined in a document, "Selection Criteria for Team Membership," circulated by Jerry Soffen in early December 1968. It began, "Rarely are scientists assembled in loosely bound organization and asked to perform and make intelligent compromises." As a rule, they act as individuals with considerable control over their own research efforts. For the Mars lander project, a group of scientists would have to work "in concert" to select the best plans for developing instruments that might be used several years later. In addition to projecting the wisest technological approach, the science team would have to handle "engineering problems, financial problems, political pressure, not to speak of scientific unknowns. The quality of brilliance is likely to be in more abundance than wisdom and certainly more than experience." An *"absolute prerequisite"* for membership on the science team was *"complementarity to other members of the team."* The guidelines also noted that usually scientists were identified with a speciality. For this team, however, persons with scientific breadth and an ability to cooperate with others would be more important assets. Strictly discipline-oriented persons would be a liability.

"The most difficult candidates to evaluate are likely to be the new or unknown faces." Some of the newcomers might be "well-meaning-but-not-too-useful" scientists who were attracted to the project because they believed that "the space program might be a nice lark for awhile." Others would not understand that participation in a spaceflight project required a minimum commitment of five years. The burden of ferreting out the good scientists rested with Soffen and his colleagues at Langley and NASA Headquarters. The guidelines cautioned, "An unknown name should not mean that the candidate is relegated to a second rate position." But the NASA managers could not afford to accept an only candidate for a position either, hoping he would "work out." Obviously, "the time for bringing up doubts is during selection not after the choice" was made, when dismissal would be difficult, awkward, and embarrassing.

While "scientists do not like to make decisions any more than other people," someone would have to be the "General" when science and democracy failed to resolve problems. It was, therefore, important for Soffen and his associates to consider which of the scientist candidates would make good leaders. Team leaders certainly had to be good communicators, with their teammates and with other members of the project. One last thing had to be kept in mind during the search: Teams "should not be too large. Five are a democracy, six an assembly, and more than eight lead ultimately to confusion and are often uncontrolled."[9]

The need for three basic lander science teams had been identified by December 1968—imaging, organic analysis, and life detection. The scientists on the imaging team would represent the most mixed set of disciplines, since the goals for that experiment were so broad. Each field of inquiry anticipated some useful information from lander photography. "The biologist has hope of finding something interesting. The geologist expects clues to the surface characteristics. The mineralogist could make some deductions about the surface composition." Cartographers, geographers, and engineers working on landing maneuvers and planning future spacecraft for Mars would all have an interest in the images from the landers' cameras. While most of the specialists who wanted to be included on the imaging team were more interested in the information the system would obtain than the development of the instrument itself, some would have definite suggestions about the technology. To translate these suggestions into specifications that the contractors could use in building hardware, a very talented instrument engineer would also have to be assigned to the team so that Langley's plans for a facsimile camera on the lander could be realized. The name for the camera was borrowed from the technique in telegraphy in which a picture is divided into a grid of small squares. The brightness of each square is converted into an electrical signal, and a sequence of such signals transmitted to a receiving station. The sequence is converted into an equivalent array of light and dark shades, and a "facsimile" of the original picture is produced on a photographic film. In 1968, the facsimile camera for aerospace applications was a relatively new tool, and the imaging science team would have to learn many new lessons in the development of that instrument for Viking.[10]

It was generally agreed that the imaging team leader would need to be familiar with facsimile camera technology, experienced in photo interpretation, and well versed in other major aspects of the mission. He would need a geologist colleague who was a "field scientist familiar with a wide variety of terrain and experienced in interpreting photos." And that geologist would have to be acquainted with the major theories on the formation of Mars. A biologist for this team would be difficult to find according to Soffen. There just was not a large group of "first rate field biologists from which to choose," and of these only a small number were interested in exobiology. Interpreting the images from the standpoint of mineralogy and inorganic chemistry might be done by a geologist, biologist, or related specialist. Analyzing the effects of the braking rockets on the landing zone—called site alteration—might require additional expertise, depending on the mode of terminal descent chosen. Obviously there would be more to Martian imagery than just taking pictures. The photographs would provide many important clues to scientists, and the system would likely be the eyes of the landed spacecraft, relaying important messages to Earth-bound engineers.

For the organic analysis team, five different specialties were required—organic chemistry, gas chromatography, mass spectroscopy, inorganic chemistry, and meteorology. The organic chemist in the group must be a specialist in pyrolysis, since "the central theme of the experiment is the reconstruction of the organic [compounds] from the analysis of the end products of thermal degradation." For pure compounds, this analytical work can be very complex. For mixtures of compounds, the task is exceedingly difficult. "For mixtures in which soil inorganics have been added, the experiment is . . . !!" Since gas chromatography was a science in which the technology was "changing every day," the specialist for this experiment would have to be abreast of those changes. This expertise was especially important because the information provided by gas chromatography would help other specialists understand the makeup of compounds they encountered in other experiments.

The heart of the entire organic investigation was an unusual sensor called a mass spectrometer. This instrument would examine the vapors produced by Martian soil compounds when heated. The vapors would be drawn into the gas chromatograph, which would separate the vapors into their individual components. The components would then be drawn into the mass spectrometer to be ionized (given an electrical charge) and analyzed to identify the constituent components. Profiles for each compound would be converted into digital form and sent to Earth. Results of the organic chemistry analysis would give scientists insights into compounds that might have been produced by any life forms on Mars and identify any organic material that might be present or might be generated at the Martian surface by purely chemical means.[11] The biological experiments were all predicated on the detection of active life processes, but the organic chemistry investigation would determine if any organisms had existed in the past or if the right organic compounds were present for the evolution of life in the future. As a cross-check on the life detectors, the organic chemistry experiment was all-important.

In addition to the analysis of organic compounds, there would also be a need to examine inorganic compounds found at the landing site. Because many of these inorganics are found in volatile form (ammonia, carbon dioxide, carbon monoxide, nitrogen dioxide, nitric oxide, sulfur dioxide, hydrogen sulfide) and appear only as gases in the atmosphere, a scientist would be included on the organic analysis team who was "familiar with such outgassing" and the composition of "juvenile" and secondary planetary atmospheres. A meteorologist could also add to the examination of these atmospheric elements as he studied the dynamics of Martian weather.

Finally, the major instrument planned for the lander was an integrated series of life-detection experiments. By 1968, after several frustrating years of experimenting with sample collectors for Voyager, exobiologists agreed that a Martian biology investigation instrument should have a common source for sample acquisition and analysis if evaluation of the results from the individual elements was to have scientific validity. Because the biology investigation was to be an integrated experiment, Soffen expected several kinds of specialists to be on the biology team. "But more important than the specialties, there should be a good mixture of different attitudes and experiences," since this complex of experiments would undoubtedly be "the most controversial of the payload." For example, the variations on a growth-detection instrument were apparently limitless, so the biology team would have to select the best concepts and then "be willing to defend them as the most reasonable thing to be done." Four kinds of biology expertise were sought for the Viking lander biology team:

> A *microbiologist* is the essential ingredient, one familiar with soil growth conditions and the problems of demonstrating viable organisms from natural soil.
> A *photosynthesis specialist*. Since part of the experiment is likely to be done in the light, searching for the photosynthetic reaction, it is important that someone familiar with these conditions be included.
> A *cellular physiologist-biochemist*. This is usually the same individual as the microbiologist, but in addition it is desirable to find a specialist familiar with intermediary metabolism and the internal biochemistry of organisms. . . .
> One versed strongly in *biological theory*, evolution, genesis, chemical de nova synthesis, genetics. This theoretical job is likely to give the very fabric to the biological goals of the mission. An appropriate person could become the [team leader].

Soffen and his colleagues believed that an engineer with a particularly strong background in developing miniaturized systems would also be an asset to the biology group in the design of the life-detection experiments.[12]

To expedite the development of the lander science instruments, the new Viking Project Office, in concert with the program scientist's staff at NASA Headquarters, organized the science activities into three phases—preparation, implementation, and data analysis. The preparation period would extend from October 1968 to December 1969, culminating in the selection of the Viking scientific investigators for the flight. Implementation would run from December 1969 through the final preparations for launch. The analysis phase would begin with the collection of the first data and end with the shutdown of each of the instruments. Only the lander investigations were identified as requiring a preparation phase, because the Viking managers expected that series of experiments to be more difficult to develop than the orbiter instruments. Orbiter investigators also would be chosen later than lander experimenters.

Associate Administrator for Space Science and Applications John Naugle officially began selecting investigators for the preparation phase 27 September 1968. Although the "solicitation for participation" did not name any specific mission or guarantee the participants in the early phase a place on the flight team, Naugle, program scientist Milton Mitz, and Soffen realized that those chosen in the fall of 1968 to help define the scientific payload for the lander would have an inside track toward selection as investigators for Viking. And everyone—managers and scientists—recognized that the development of an atmospheric probe-lander and the scientific instruments for a Mars lander would "require a long lead time." Considering also the highly integrated payload, the interdisciplinary nature of some of the proposed instruments, and the basic complexity of the lander design, NASA had no choice but to bring scientists into the planning phase at the very earliest point, even if this later made objective selection of the flight team scientists more difficult.[13]

The flight team investigators would be responsible for developing the functional specifications for the instruments and for providing direct guidance in all aspects of instrument design and construction. Including scientists in all stages of experiment definition, design, development, fabrication, testing, and operation was an attempt to preclude a problem that had plagued many of NASA's programs: the conflict between the builders of scientific instruments and the users of the data collected from them. Outside the arena of spaceflight, scientists have traditionally built or at least closely monitored the construction of their own experimental apparatus. Indeed, scientists were often judged by their peers on how well they executed the design of their hardware. With the shift from experiments on the laboratory bench to instruments that had to be integrated into the multiplicity of spacecraft systems, a rift grew between the persons who conceived the experiments and analyzed the results and those who actually built the hardware. An exobiologist might conceptualize an investigation and even build a bench prototype, but any elements of an integrated biology instrument would likely be built by a contractor specializing in the design and fabrication of flight hardware. This new division of labor did not often please the scientists, especially when engineers took an "I know how to do it better than you" stance. To avoid this problem in Viking, Naugle and the other NASA managers wanted the scientists working with the project from the very beginning.[14]

On 11 February 1969, after the headquarters' Space Science and Applications Steering Committee had evaluated the many proposals sent them by potential investigators, Jim Martin sent letters to 38 scientists, inviting them to participate in the preparation phase of project planning. While some familiar names were among the scientists, many were also newcomers to space science. Soffen's objective of incorporating new talent into the teams had been realized. All the invitees accepted, and their first meetings at the Langley Research Center were the inaugural sessions of the Viking science instrument team, 19–20 February, and the Science Steering Group, 21 February.[15] These meetings gave the scientists an overview of the entire project, introducing them to current activities, the project's methods of operation, and the schedule. Scientific objectives were discussed with respect to the existing knowledge of Mars and the investigations planned for Mariner 1969 and Mariner 1970 spacecraft. The scientists were also briefed on their responsibilities and the manner in which the teams and the

Science Steering Group would function. Mission design, engineering facts of life ("engineering constraints"), and hardware design (lander, orbiter, and scientific instruments) were summarized, as well.[16] On 25 February, NASA Headquarters officially announced the selected preliminary Viking science team members.[17] The list was a long one, and the number of teams had grown to eight (see app. D).

During the next six months, each science team planned instrument development. At the February Science Steering Group meeting, Jim Martin had told the team leaders that their science definitions should clearly state the scientific values of the instruments and the definitions "should be so complete that they may be used as a guide in preparing preliminary specifications for spacecraft design." The scientists were responsible for defining their potential hardware needs.[18] Viking planners had initially agreed to include a "science definition" in "Mission Definition No. 2," but that official statement of Viking science objectives promised to be too lengthy.[19] Only the essential data would appear in the mission definition, while the more detailed information would be included in a reference work, "Viking Lander Science Instrument Teams Report." Lander contractors would use both documents as sources of information about the proposed instruments and a guide to scientific rationale as they determined how to increase the scientific capabilities of the lander.[20]

Potential scientific investigators received the "Announcement of Flight Opportunity for Viking 1973" in early August 1969. This package of materials, which included the instrument teams' reports and the mission definition, would guide scientists who wished to work on one of the suggested experiments or who wanted to propose alternative versions of existing experiment proposals or additional experiments.[21] (See app. D for an excerpt from one of the science reports.) Concurrent with the final revisions of the science instrument reports, the Science Steering Group recommended at its July meeting that the weight of Viking lander science instrumentation be targeted at 41 kilograms rather than the original 32 kilograms. The extra weight would permit consideration of a number of important additional goals that had been identified as desirable if a larger payload was possible.[22]

With the completion of three major documents—the "Viking Lander Science Instrument Teams Report," "Viking Mission Definition No. 2," and the "Science Management Plan"—the science instrument team's work was essentially completed. The next step was the reception and evaluation of the science proposals in response to the flight opportunity announcement. More than 300 persons had attended the two day pre-proposal briefing for Viking science. By the 20 October deadline, NASA had received 150 proposals. Since 5 of these were considered dual proposals and 10 presented additional instrument options that had to be studied, the total number of items to be evaluated reached 165. They were divided into nine groups.

*Table 36*
*Viking Science Proposals*

| Lander | | Orbiter | |
|---|---|---|---|
| Experiments | Number of Proposals | Experiments | Number of Proposals |
| Imaging | 14 | Imaging | 17 |
| Molecular analysis | 19 | Proposals for experiments | |
| Active biology | 13 | requiring additional | |
| Meteorology | 11 | instruments | 27 |
| Entry science | 15 | Radio science | 27 |
| Proposals for experiments requiring additional instruments | 22 | | |

As part of the evaluation process, Mike Mitz, program scientist at headquarters, made these proposals available to the four subcommittees of the Space Science and Applications Steering Committee—Planetary Biology, Planetary Atmospheres, Planetology, and Particles and Fields. Each proposal was reviewed by at least one subcommittee. The steering committee recommended 12 experiments and 61 scientists to John Naugle, who concurred on 15 December (see app. D and table 37). Of the 8 lander experiments, 6 had been proposed during the preparation phase of the lander work; 2 were new investigations suggested by outside scientists, and 1 of the major instruments proposed for the lander during the early planning phase, the ultraviolet photometer, would not be flown.[23]

In the course of selecting the scientific experiments for Viking, Jim Martin expressed some reservations to Ed Cortright: "The proposed science payload represents an escalation in science objectives which is likely to lead to cost increases beyond those estimated in our assessment." His concern was especially strong for the experiments not previously examined by science instrument teams. Cost problems could be generated by the entry-science retarding-potential analyzer, the lander-science physical properties investigations, or the magnetic properties experiment. "These additions, when coupled with the problems of using the [gas chromatograph–mass spectrometer] to measure water and adding a gas exchange investigation to the biology instrument, add up to a potential overrun. . . ." Martin was also worried about some of the scientists chosen for the work. He told Cortright that lessons they should have learned over the course of the preceding year were not being implemented. "Specifically, the Biology Team has the same group of men who demonstrated an inability or unwillingness to work together, the [Molecular Analysis] Team has two members only interested

*Table 37*
*Key Dates in Assessment of Viking Science Proposals*

| | |
|---|---|
| 11–12 Sept. 1969 | Pre-proposal briefing for potential experimenters. |
| 20 Oct. 1969 | Proposals due at NASA Headquarters. |
| 23 Oct. 1969 | Copies of proposals due at Langley and JPL; meeting held at Langley to discuss proposals. |
| 3–4 Nov. 1969 | First Space Science and Applications Steering Committee (SSASC) subcommittee meetings to initiate evaluation process, Goddard Space Flight Center. |
| 7–8 Nov. 1969 | Review of science proposals at Langley. |
| 12–14 Nov. 1969 | Second subcommittee meetings, Goddard. |
| 17 Nov. 1969 | Viking Project Office assessments of proposals due at NASA Headquarters. |
| 18–20 Nov. 1969 | Definition of science payload by Headquarters Planetary Program Office. |
| 21 Nov. 1969 | Tentative payload presentation to D. P. Hearth, director, Planetary Program Office. |
| 26 Nov. 1969 | Planetary Program Office recommendations made to SSASC. |
| 3 Dec. 1969 | Recommendations presented to SSASC in writing. |
| 8 Dec. 1969 | Oral presentation to SSASC. |
| 15 Dec. 1969 | Selection of Viking science payload by John E. Naugle, associate administrator for space science and applications, based on SSASC recommendations. |

in water detection who will interfere with achievement of the team's primary objective, and the Entry Team has the same two members who have demonstrated many times an inability to work together."[24]

Martin had good reason to be worried about possible cost escalations. On 3 September, Don Hearth's Planetary Program Office held a Viking science review with Langley personnel, Office of Space Science and Applications program chiefs, and Dr. Henry J. Smith, deputy associate administrator for space science. The objective was to establish weight- and cost-limit goals for Viking science activities. Later decisions about overall Viking costs and flight instruments could be made using these guidelines. Some of the more significant decisions reached at the 3 September review were on reduction of the lander science instruments' total weight, development of backup instruments for the gas chromatograph-mass spectrometer and the biology instrument, and specific dollar limits on science spending.

As a result of the early fall meeting, the science planners reverted to the 32-kilogram limit on science instruments, dropping the 41-kilogram pro-

213

posal made by the Science Steering Group. The major difference between the two weight packages was the addition of a separate mass spectrometer for determining lower atmosphere constituents. Hearth's view was that the additional scientific information they could obtain with that instrument could not be justified when they considered its cost. He believed that the first gas chromatograph–mass spectrometer measurements after touch-down would be sufficient. Weights and costs of the 32-kilogram science payload for the lander were summarized in September 1969 (table 38).

*Table 38*
*Estimates for Lander Payload, September 1969*

| Item | Weight (kg) | Cost (millions) |
|------|------|------|
| Entry science | 4.1 | $ 4.1 |
| Imaging | 5.0 | 6.2 |
| Biology | 5.4 | 11.3 |
| Gas chromatograph–mass spectrometer | 10.4 | 8.5 |
| Meteorology | 2.3 | 3.0 |
| Water | 1.1 | 1.8 |
| Seismometry | .9 | 2.0 |
| Ultraviolet photometry | .5 | 0.7 |
| Total for instruments | 29.7 | 37.6 |
| Integration and test | 2.0 | 5.8 |
| Total | 31.7 | $43.4 |

Cost of the lander instruments was expected to be about $1.36 million per kilogram. The orbiter experiments were projected to cost about $0.56 million per kilogram. Overall costs were broken down as in table 39.

*Table 39*
*Viking Science Cost Projections, September 1969*

| Item | Cost (millions) |
|------|------|
| Lander science | $43.4 |
| Orbiter science | 32.0 |
| Support of science teams | 13.3 |
| JPL support of GCMS development | 5.6 |
| Ames support of biology instrument development | 2.1 |
| Total | $96.4 |

With an additional 10 percent for contingencies, Hearth established a firm ceiling of $107.5 million for the total Viking science package.[25]

Looking at Hearth's estimate in December, Martin believed that they were selecting too many members for the experiment teams. "The total number of team members and participating scientists has increased beyond our budgeted estimates and considerably beyond what the [project office] believes is required to achieve the mission objectives." The budget called for 55 scientists; 61 had been selected. Martin would have been happy with fewer than 40. (By flight time, the number of science team members would grow to 80.) Although Don Hearth's Planetary Science Office had told all the scientists that the payload selection was tentative pending negotiation of a contract for each instrument and an individual contract for each scientist, Martin personally believed that it would be extremely difficult for NASA to drop any scientist or investigation. The "pressure will be on to consider an increase of a few million [dollars] as acceptable; it will come out of our contingency allowance and avoids unpleasantness between [the Office of Space Science and Applications] and the science community."

Martin feared that in a few years when all these reasons for the increased expenditures had been forgotten, he and the Viking Project Office would be held responsible for not properly managing their funds. With only $102 million set aside for total project contingency costs (a small amount compared to other major NASA projects) and the "tight funding environment" that everyone expected to face for several years, it appeared to Martin that "a prudent manager must hold the line against escalation in all areas of the project today." Since he saw considerable cost uncertainty associated with the science instruments, Martin would be especially cautious in this area.[26] Many of his concerns did become problems in the future. There was friction among the members of the biology team, and the costs of the biology instrument and the gas chromatograph–mass spectrometer rose sharply. Most of these difficulties emerged after the January 1970 schedule change from a 1973 to a 1975 launch.

Reservations aside, NASA appeared to be well on its way to organizing a Mars lander mission. In encouraging Joshua Lederberg to work with the biology team, Richard S. Young, chief of exobiology, Office of Space Science and Applications, had written that many details of the biology experiment still needed resolving. Young sought Lederberg's advice on NASA's "method of operation" as much as on "the science involved in these missions." Looking back over the long road since the early 1960s when exobiology was a very new field, Young noted, "The science hasn't changed much since the 'Westex' days [see chapter 3], but we are finally trying to organize in the best way as to achieve some of the 'old' objectives." Young and his colleagues wanted "to make this thing work . . . within the constraints imposed" on them by the administration and Congress.[27] They would need the help of many parties to reach their goal.

## Selecting a Contractor

Selection of a contractor to build the lander and to supervise integration of the lander and orbiter and integration of the spacecraft and launch vehicle paralleled in time the selection of the scientific experiments. On 28 February 1969, Langley Research Center issued a request for proposals on the design and fabrication of the lander and project integration. In addition to the 20 firms directly solicited for this procurement, 12 others requested and were sent copies of the proposal package. Technical and managerial proposals were submitted to NASA by the Boeing Company, McDonnell Douglas Corporation, and Martin Marietta Corporation. All three companies had conducted studies earlier for Jim Martin's Titan Mars 1973 team. In the process, they had developed an enthusiasm for and an expertise in the design of Mars landers.

In April the Source Evaluation Board began with an appraisal of the written proposals and visited the production facilities of each of the three potential builders, where members of the board spoke at length with company representatives. As Administrator Thomas O. Paine noted in his report on the contractor selection process, the board furnished written questions to each firm before its visit. The companies were advised that the questions covered deficiencies and omissions as well as proposal ambiguities and that they were being given an opportunity to support, clarify, correct, or make revisions. After the visits, the board made its final rankings in May 1969.

Martin Marietta received the highest overall final rating; its cost proposal was between those of the other two bidders. The Denver-based division's technical proposal was well organized, according to the judges on the board; its strong points were "outstanding mission analysis and plans for maximum science return, the communications system, the terminal descent radar analysis, a common deorbit and descent engine, and landing gear design." Weak points included "the power system design and uncertain subsonic stability of the aerodynamic configuration." NASA specialists believed these to be "readily correctable" problems, and Martin Marietta suggested that the inflatable-balloon decelerator (ballute) and parachute combination, which had been proposed for slowing and stabilizing the lander once it was separated from its aeroshell, be replaced by a more conventional parachute.

Boeing received the second highest overall ranking and offered the lowest cost. Boeing's proposal contained "a well-conceived mechanical design, a redundant and flexible communications system, and an excellent plan for launch and flight operations." Proposal weaknesses centered on a method suggested for dealing with the scientific instruments and the investigators, the power system design, and deorbit propulsion. The latter two areas would require "major proposal revisions," according to the source board. Boeing had planned to join forces with General Electric and Hughes Aircraft Company—GE as the subcontractor for entry, power, data handling, and attitude control systems; Hughes as the subcontractor for terminal landing subsystems, terminal guidance and control, terminal propulsion, and landing gear. While the combination of these three companies offered much "specialized experience" and while the Boeing-GE-Hughes team plan was well organized, NASA officials thought there were "potential management and operational problems" in this arrangement.[28]

McDonnell Douglas, with the highest cost estimate, was ranked third. Technical weaknesses outweighed the strengths of its proposal. And the potential strength of its management team was outweighed by its decentralized facilities, which were not as well suited for Viking as those at Martin Marietta or Boeing.

Following the Source Evaluation Board presentations, Paine met with a few key NASA employees to obtain their views on the board's findings. Administrator Paine, Associate Administrator Homer E. Newell, and NASA General Counsel Paul G. Dembling subsequently met and agreed to award the contract to Martin Marietta.[29] Paine explained that his choice for the lander contractor was influenced by the fact that the firm had "applicable company experience, technical capability and the most outstanding facilities . . . which are specially tailored to Viking requirements." Martin Marietta's participation in early Voyager activities and its decision to maintain a team effort with more than 100 persons during the 1967–1969 period had "established a strong and highly motivated" group from the top management down through the working personnel.[30]

On 29 May 1969, Paine announced that NASA planned to award a cost-plus-incentive-fee/award-fee contract for $280 million.[31] The lander system as proposed by the contractor was technically evaluated by the engineers at Langley to identify changes that should be made before the formal contract negotiations between NASA and Martin Marietta began. These alterations were documented in a "shopping list" of 18 items over which Langley and the new contractor negotiated. With the changes, the contract figure totaled $299.1 million in the contract approved by Paine 20 October. Martin Marietta's fee was targeted at $14.52 million, but the incentive provision permitted the company to earn more money if the contract was concluded at less than the projected cost of $299.1 million and it penalized the company for any cost overruns. For every dollar above the target, Martin Marietta would lose 15 cents from the fee, while any cost savings would bring an additional reward of 15 cents per dollar.[32]

The statement of work that accompanied the contract for "Viking lander system and project integration" was kept as general as practical so that the number of changes in the contract could be kept to a minimum. Other large NASA projects like Gemini and Apollo had produced thousands of contract modifications. David B. Ahearn in the Langley Procurement Division sought from the beginning to produce a Viking contract that would ensure that the work was done properly but with a minimum of paperwork. During the life of the contract, the number of alterations made in that document numbered about 300.[33]

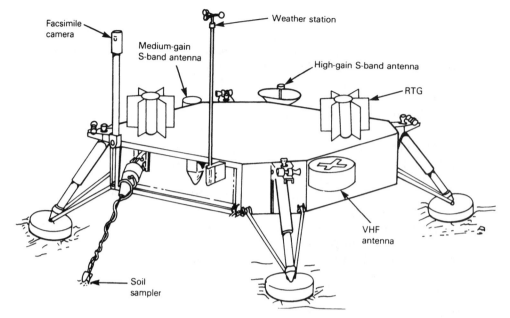

Facsimile camera

Medium-gain S-band antenna

Weather station

High-gain S-band antenna

RTG

Soil sampler

VHF antenna

*The Viking lander design went through a number of versions in 1968 and 1969. Above, one of the four-legged configurations presented at the Viking science instrument team's meeting 19–20 February 1968 was to be powered by radioisotope generator and battery. One not shown arrayed solar cells on the lander's flat top to provide power. Although RTGs posed heat problems, the Viking Project Office preferred them. Below, the three-legged September 1969 design added a second camera for stereophotography and moved the meteorology instrument to the high-gain antenna mast.*

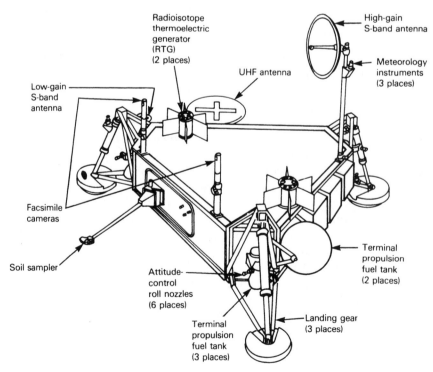

Radioisotope thermoelectric generator (RTG) (2 places)

High-gain S-band antenna

Low-gain S-band antenna

UHF antenna

Meteorology instruments (3 places)

Facsimile cameras

Soil sampler

Attitude-control roll nozzles (6 places)

Terminal propulsion fuel tank (2 places)

Landing gear (3 places)

Terminal propulsion fuel tank (3 places)

Very early in the contract, a major modification, made necessary by the two-year launch-date slip, was negotiated between NASA and Martin Marietta. On 13 January 1970 following the administrator's unexpected announcement of the change in plans for Viking, the Langley Research Center Contracting Office notified the contractor to stop all work authorized under the contract. That week meetings at JPL, Martin Marietta, and Langley began reprogramming for the new game plan. Martin Marietta studied two possible alternatives for a 1975 launch (table 40).[34]

*Table 40*
*Alternatives for 1975 Viking Launch*

| Option A:<br>Viking 1973 Mission<br>Slipped to 1975 Opportunity | Option B:<br>Direct Entry<br>Lander Mission in 1975 |
| --- | --- |
| Orbiter–lander–Titan III–Centaur | Lander–relay module–Titan III–Centaur |
| 1973 management and contractor team | 1973 lander contractor to supply relay module |
| 1973 science and scientists | 1973 science and scientists |
| Type II trajectory | Type II trajectory |
| Use added time to minimize technical risk, optimize hardware use, minimize schedule risk, and minimize cost. | |
| FY 1969, 1970, 1971 funds held to $87.5 million. | |
| First priority in study | Second priority in study |

By mid-February, the Viking Project Office authorized Martin Marietta to proceed with the first option and lifted the stop-work order. Through the end of fiscal 1971 (30 June), only $87.427 million would be made available for the project, so Martin Marietta would not be able to hire as many persons as planned. Nor would it be able to increase employment levels as rapidly as it had hoped under the 1973 schedule. JPL also had to make changes in its manpower projections. Although Martin Marietta would employ a smaller total number during the life of the lander contract, those who did work on Viking would be employed for a longer time. As a consequence, the total cost of the lander grew by another $44 million (see also graphs in appendix C.)[35]

The immediately apparent increase caused by the shift from a 1973 to a 1975 launch was $141 million. While other factors would drive Viking costs

*Table 41*
*Viking Cost Increases Because of Launch Delay*

(in millions, as of June 1970)

| Component | Viking 1973 | Viking 1975 (as of June 1970) |
|---|---|---|
| Lander | $313 | $360 |
| Orbiter | 202 | 257 |
| Other | 94 | 133 |
| Total | $609 | $750 |

even higher, the economics of delaying the project two years to meet the political pressures on the fiscal 1971 budget were expensive for NASA and American taxpayers.

## SCIENTISTS, INSTRUMENTS, AND SUBCONTRACTORS

The Viking project stretchout also affected management of the scientific experiments for the Mars mission. Originally, the Viking Project Office had planned to negotiate contracts with the scientists and select instrument subcontractors during the first weeks of 1970, and most of the science teams had met in early January to review their plans. With the switch to a 1975 mission, that schedule had to be reevaluated and the activities reprogrammed. On 13 January the science teams, except those working on the biology instrument and the lander imaging system, were told to terminate their Viking activities.[36]

Jerry Soffen advised all of the scientists in late January that the Viking Project Office's main goal was to make the transition to a revised schedule as smooth as possible, while protecting against any unnecessary cost increases or further schedule delays. "During this transition period," Soffen hoped that the scientists would "not lose sight of the Viking objectives," and he reminded them that "scientific research has never been an easy way of life. We expect to find favorable aspects of this Viking deferment in the form of improvements in the investigations and the better use of Mariner 71 results."[37] The Viking Project Office worked out a procedure for keeping the science team leaders in the instrument definition process during the transition without having to include them in formal contract negotiations. After selection of a subcontractor to negotiate to build a science instrument and before negotiations began, a technical review would be held. Martin Marietta, the Viking Project Office, the science team, and the subcontractor (or "vendor") would thoroughly review the procurement drawing, especially where changes in specifications were required. The science team leader could participate in discussions leading to prenegotiation specification. Then, during negotiations, any additional changes would be coordinated with the team leader through the Viking office.[38]

For the scientists as a group, the next big gathering scheduled was the Viking science review in mid-April 1970. By that time, Martin Marietta had chosen Itek Corporation's Optical Systems Division to develop and build the lander camera system and was evaluating biology instrument proposals from Bendix Aerospace Systems Division and TRW Defense and Space Systems Group. JPL was in the process of evaluating a breadboard model of the gas chromatograph–mass spectrometer, and Martin Marietta's planning for the construction of the upper-atmospheric mass spectrometer breadboard was under way.[39]

For three days, 13–15 April, 42 scientists (about two-thirds of the total team membership) met with representatives from the project office and lander contractor. After receiving reports from the Viking managers the first morning, each team leader presented a 10- to 20-minute summary report on the status of his experiment that afternoon. On the 14th, a series of concurrent team meetings gave the scientists time to talk with their teammates and discuss matters of common interest with other teams. Later that day, a number of special science meetings took up investigative considerations affecting more than one team, such as site alteration, organic contamination, landing site characteristics, atmosphere. The final day of the gathering was given over to a session of the Science Steering Group. The scientists found all the meetings educational but agreed that the smaller "think" groups they had participated in the second day were particularly stimulating. Viking's schedule may have been stretched out, but nearly everyone agreed that much work would still have to be done by all to meet the 1975 launch date.[40]

The pace of work was moderately slow at first because of the limited money available, but in retrospect that may have been fortunate, because many technological problems lay ahead. Three scientific instruments—the ones given first priority for the dollars available—were particular problems: the gas chromatograph–mass spectrometer, the biology instrument, and the lander imaging system.* While the story of these instruments is a tale of amazing accomplishment, the facts also indicate that if Viking had flown in 1973 it probably would have been launched without the gas chromatograph–mass spectrometer and the biology instrument. Without those experiments, Viking would have been a vastly different mission. Those instruments were ready to fly in 1975, and the story of their design and fabrication deserves to be told. For the men and women who worked the extra hours, sweated out the successive problems, and reveled in personal

---

*Thomas A. Mutch has described the history of the lander cameras in *The Martian Landscape*, NASA SP-425 (Washington, 1978), pp. 3-31.

221

satisfaction when the experiments actually worked on the surface of Mars, it was "their" lander, "their" experiment, and "their" triumph.

### Gas Chromatograph–Mass Spectrometer (GCMS)

Development of a GCMS prototype had initially been assigned to the Jet Propulsion Laboratory by Langley in August 1968. This responsibility remained with JPL when the Viking project was officially established. Before selecting a contractor to build the flight hardware, the California lab had the task of developing, fabricating, and testing a lightweight portable breadboard of the GCMS that could be used to carry out surface organic analysis by pyrolysis. Gas chromatography and mass spectrometry in the laboratory were one thing; shrinking the equipment to a size that could be placed on a spacecraft was another.[41] Requirements for such an instrument were not easy to meet for a laboratory model; restrictions put on the design to qualify it for spaceflight made it extremely difficult.

Pulverized Martian soil would be placed in the instrument and heated to temperatures up to 500°C. The gases given off would be carried into a gas column, a long tube packed with coated glass beads that would selectively delay the passage of gases according to their adsorptive qualities. The column would then be heated progressively to 200°C at a rate of 8.3°C per minute. Each level of temperature would release different organic molecules, separated into narrow family groupings. A palladium separator unit, porous only to hydrogen, would filter out that gas, leaving only the vaporized organic compounds, which would be drawn into the mass spectrometer to be ionized. The stream of ions would be focused in the electrostatic and magnetic sectors of the device. When the stream of focused ions struck the electron multiplier tube, generating electrical impulses, that activity would be amplified and recorded, producing a profile of each compound. Finally, the profiles would be converted into digital signals that could be transmitted to Earth.[42]

Although the GCMS was a complex piece of equipment, no one predicted the difficulties that JPL encountered in its development. At first, dollars and failure to agree on priority for the instrument's development were causes for delay. But by the summer of 1970, serious engineering and managerial problems were plaguing GCMS development.[43]

In September 1970, Cal Broome told Jim Martin that the GCMS, nominally under the purview of Henry Norris's Viking Orbiter Office, was a stepchild not getting proper supervision because of the decentralized management structure at the lab.[44] A five-day GCMS engineering model review, held 25–30 January 1971, was a disaster. Jack Van Ness told Langley Director Cortright that between 200 and 300 "request for action" forms resulted from the review; he anticipated that 100 to 150 of those items would be assigned to JPL for its attention. "It is expected that the major output of the review will be a critical reassessment of the requirements imposed upon the instrument and its subsystems, with an eye towards reductions in instrument complexity."[45] Two weeks later, Van Ness reported that JPL had taken steps to strengthen its managerial control. John J. Paulson, head of the GCMS project office, would henceforth report directly to Robert J. Parks, assistant laboratory director for flight projects. This shift put the GCMS on the same management plane as Mariner Mars 71 and Viking Orbiter. The Viking Project Office hoped this visibility would help solve some of the stepchild's troubles.[46]

Jim Martin was not pleased. At a Science Steering Group meeting 2–3 March 1971, he indicated that funding increases, technical problems, and schedule slips had caused him and his colleagues considerable concern about the future of the GCMS. Although the recent management change at JPL was encouraging, the instrument's progress would be watched closely during the next few months. If progress was not satisfactory, Martin would have to consider an alternate or less ambitious design.[47] The project manager's attitude toward the GCMS difficulties was not enhanced by his unhappiness over the science subsystem preliminary design review at Martin Marietta on 1–2 March. The part of the PDR covering the science experiments integration laboratory (SEIL), to be built in Denver, was particularly unsatisfactory. Martin told the lander contractor that the SEIL PDR would be repeated and that no funds would be spent on equipment for that instrument until a satisfactory review had been held.[48] (The SEIL was canceled in July 1971; instruments tests would be performed on the system test bed lander at Martin Marietta.)

On 18 March, the GCMS engineering breadboard was operated for the first time as a completely automated soil-organic-analysis instrument. Several problems of the kind usually associated with first tries were encountered, but everyone in the Viking Project Office interested in the development of the GCMS considered it a major step forward.[49] Meanwhile, an ad hoc GCMS requirements review panel, established by Martin after the unsuccessful engineering model review in January, met to discuss possible ways of simplifying the design.* Preliminary results of the ad hoc panel's study were presented at the June 1971 Science Steering Group meeting. Martin noted several discouraging facts at this session: by this date the start of GCMS science testing had slipped by six months (from early 1971 to October 1971); after four years of work the breadboard was just ready; and the GCMS was now getting too heavy. Originally projected to weigh about 9.5 kilograms, the GCMS was weighing in at about 14.5 kilograms. The ad hoc panel presented five GCMS design variants with weight projections between 11 and 14 kilograms, but they requested and were given more time to study the science impact of these alternatives.[50]

---

*Panel members included Chairman H. B. Edwards, K. Biemann, T. Owen, R. S. Young, J. J. Paulson, and G. C. Broome.

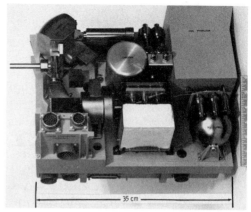

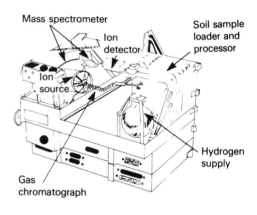

Mass spectrometer

Ion detector

Soil sample loader and processor

Ion source

Hydrogen supply

Gas chromatograph

*The development model, top left, of the gas chromatograph–mass spectrometer was the first step toward spacecraft hardware. After a breadboard model, completed in October 1971 to perfect functioning of the instrument, designers worked on weight, size, and modifications to integrate it into the lander. The mockup, top right, is 35 centimeters wide. Finally, the flight GCMS is tested and prepared for its long journey through space to investigate Mars.*

As the reconsideration of the GCMS continued, the Viking Project Office sponsored the first "Viking science symposium,"structured to provide extended discussions of the chemical and biological premises on which two of the project's major investigations—biology and the molecular analysis experiment—were based. While much of the material presented was old information to seasoned Mars hands, for many of the attendees it was the first time they had been exposed to these scientific assumptions underlying the Martian search for life. In addition, several new interpretations of old phenomena or refined Mars data were presented for discussion. Alan Binder of the Illinois Institute of Technology's Research Institute suggested an alternative explanation for the so-called "wave of darkening." The most common reason given for this phenomenon had been an increase in atmospheric humidity as water sublimed from one polar cap and moved toward the other. New observations indicated that the wave, which progressed at a speed of 30 kilometers a day, might actually be a wave of brightening. Earth-based photometric measurements had compared dark areas to bright areas on the assumption that it was the bright areas that were unchanging. If the bright areas were getting brighter, then water or vegetation was not needed to explain the change. Instead, the explanation might be some simple mechanism, a dust storm, for example. Some microbe hunters who saw this as one more strike against the possibility of Martian water might not have been pleased, but the reasoning was more consistent with other investigations that indicated limited water on the Red Planet.[51]

Toby Owen of the State University of New York at Stony Brook and Michael McElroy of Harvard reported that *Mariner 6* and *7* had provided new clues about the composition of the planet's atmosphere. It was 95 percent carbon dioxide. Nitrogen probably existed in quantities less than 4 percent, and perhaps as little as 0.5 percent. Traces of carbon monoxide, molecular oxygen, ozone, and water vapor were likely. While these were not very encouraging comments for those who wanted to find life on Mars, Carl Sagan repeated his oft-given summary that the only way to make such a determination was to go there and check out the planet. Such an examination might not end all speculation, but it would certainly give them better data. To make that trip worth the effort, the GCMS and the biology instrument would have to work.

The problems encountered with the gas chromatograph–mass spectrometer were not made any better by renewed money problems. A special meeting held 19 September to discuss the budget led to some very bitter reactions by several scientists. Martin told those investigators that they would have to reduce their projected costs by a further $17 million to $22 million. Before the next discussion of the science budget reduction in early October, Jerry Soffen received some amazing letters in response to his comments about scientific priorities. There was a decided lander-versus-orbiter outlook among the scientists, and a dichotomy between the build-the-experiment-hardware-yourself group and the more theoretically oriented investigators.

Harold P. Klein, biology team leader, was among the first to write. He concluded that it was more important to get results from the lander than from the orbiter. "I say this for a number of reasons: by 1975, we will have had several missions to the planets—with flybys and orbiters, but no lander mission; we have learned a great deal about Mars from the Mariner series and there is no doubt that these have shaped our views of the planet, and that Mariner 9 should add immeasurably to this store of information." But there had never been a direct measurement made from or of the surface of Mars.

"What I am emphasizing is something which scientists recognize as first order science—i.e., it is generally easier to refine your techniques, and repeat your experiments with more sophisticated equipment than to start investigating in unknown territory." But Klein noted that "it is much more exciting to try something completely new and different—to do something first." He would be willing to sacrifice the orbiter imaging system rather than subtract anything from the landed group of experiments.

> On the lander, we are proposing a number of investigations—and while these will all be "first time" investigations, and therefore of great potential interest, it is obvious that some are concerned with answering really colossal questions and others are not. It is no surprise—at least to me— that there is a direct relationship between the magnitude of the scientific question being asked, and the complexity, uncertainty and, therefore, the expense involved in the equipment concerned with each investigation.

Klein would prune the orbiter science to only that needed to support the lander. While dropping the large imaging payload, he would maintain the atmospheric water detector and the infrared thermal mapping device. He hoped that no lander experiments would have to be eliminated, but if deletions were necessary the big experiments—the GCMS, the biology instrument, and lander imaging—must be preserved.[52]

Don L. Anderson, seismology team leader, was equally strong in his opinions. "First of all, I feel that Viking was poorly conceived from the beginning, and this, of course, was headquarters' fault." With that shot across the NASA bow, he continued:

> The way science was selected was ill-conceived, and headquarters was repeatedly warned that one does not decide what needs to be done and then try to find someone to do it. In the past, the scientists designed the experiments and, by and large, the instrument. The Viking scientists have little experimental experience and virtually no equipment experience. They were chosen because they expressed an interest in an area—not because of any demonstrated wisdom on the important problems of Mars or of the solar system. As a group they cannot provide you guidance in scientific policy matters of priority. As individuals they are ineffective, because of the system, in riding herd on their own experiments, particularly the costs.

Translated, the exobiologists might be asking the "colossal" questions, but it was Anderson and his colleagues who were doing experiments with which they had first-hand experience. They could create hardware and deliver it at a reasonable cost and on time. Anderson accepted, to a degree, that "one can argue that the first mission to Mars should have biological emphasis," but the realities were "that the biological and organic experiments were not ready when the payload was selected, are not ready now, and probably will not be ready in 1975." Anderson admitted that physical measurements, such as seismology, were relatively easy, but that complex experiments like the GCMS and the biology investigation were more difficult than anything NASA had ever flown. One could argue parenthetically that the molecular and biological investigations were closer to real laboratory science than anything ever done before in space. These experiments required more than data gathering; they demanded elaborate manipulations of sample materials in miniature laboratories. As he noted, such biological investigations as these were "not even routine measurements on the earth." They were "not ready to fly a biological mission to Mars. Even if the instruments are ready the chances are high that they will not work on Mars, and if they do, will give ambiguous results." This team leader represented one camp of scientists who wanted to make "straightforward" measurements; Klein and his associates preferred to pioneer a new "first order" science in space. There were strong arguments for both points of view, which did not make Soffen's or Martin's tasks any easier. The Viking Project Office managers had their hands full—with complicated and troublesome hardware, independent and troublesome scientists. A firm discipline would have to be applied to both.[53]

The issues raised in the September–October 1971 Science Steering Group meetings would not be resolved immediately. But the discussions led to several changes, as the minutes recorded:

1. Reduction of science team support—By deleting certain efforts of the scientists, holding fewer meetings, and supplying less assistance. . . . This will save $3 M[illion].

2. Reduction of the Molecular Analysis Investigation—Current technical problems with the GC/MS have resulted in substantial cost increase over the original estimate. Most team leaders agree to the importance of the investigation but feel that there should be a cost ceiling. By reducing the requirements and simplifying the instrument, it should be possible to assure technical feasibility and to bring the costs down to a level consistent with the present project plans ($35 M). This involves a reduction of the number of samples analyzed, deletion of direct [mass spectrometer] analysis and [deletion of a detector portion of the gas chromatograph]. The cost saving is $3.0 M.

3. Relaxation of the Biology Instrument Requirements—Two major requirements involving temperature control and waste management, and several minor ones, can be relaxed at considerable savings. . . . The total cost reduction of $2.0 M has been agreed upon.

4. Limitation of Viking Orbiter Science Mission Planning. . . . The saving is $1.0 M.

5. Reduction of Meteorology Investigation . . . to result in a "weather station" type experiment. . . . The saving would be $1.6 M.

6. Limitation of the Physical Properties Investigation to Current Baseline. . . . [The saving would be $0.15 M.]

7. Use of fixed masts for the Viking Lander Cameras. . . . The cost saving is $0.3 M.

8. End Mission B at the beginning of conjunction. . . . The savings are essentially in operations: $0.5M.[54]

These changes totaled up to a possible saving of $11.5 million. Decisions that were postponed at that meeting included eliminating photometric calibration of the orbiter camera ($1.6 million) and deleting the X-band radio ($1.1 million), the image-motion-compensation device for the orbiter camera ($0.4 million), the retarding-potential analyzer from the entry science experiment ($2.3 million), and deleting either the infrared thermal mapper ($3.3 million) or one of the biology experiments ($1.9 million). (Deletion of the orbiter imaging system was also seriously considered at this time. That proposal is described in chapter 9.)

Between October 1971 and March 1972, there were numerous conversations among Viking Project Office personnel members, JPL authorities, and the contractor, Litton Industries, about the fate of the GCMS. Jim Martin was not very happy with JPL's management of this activity, and he told the lab on several occasions that he wanted JPL to monitor the contract the way Martin Marietta was monitoring its science subcontracts. He did not want JPL trying to build the GCMS; that was Litton's responsibility. As early as October 1971, Martin was considering finding another organization to handle the GCMS contract, and the project office awarded Bendix Aerospace a contract to study the feasibility of using an organic analysis mass spectrometer (OAMS) in place of the GCMS. Similar in the information that it produced, the OAMS did not use a gas chromatograph. To demonstrate his concern, Jim Martin added the GCMS to the "Top Ten Problems" list on 26 October. "Specifically the problem is the systems design and program redefinition of a simplified GCMS." Shortly thereafter, Klaus Biemann and his colleagues of the molecular analysis science team requested that Alfred O. C. Nier, the entry science team leader, be added to their group because of his background in mass spectrometry.[55]

The addition of Nier to the GCMS activity was another blow to JPL. He had written to Jerry Soffen in September 1971: "While I regard a *properly* devised and managed GCMS experiment as one of the most important things we could do on Mars, the history of this endeavor leaves so much to be desired I really wonder whether it has not disqualified itself already." Nier thought that JPL's record in this area was "dismal." Nier also shared Don Anderson's complaint about the GCMS scientists' lack of experience in inventing and building instruments. He believed that it was "most unfortunate that in NASA's selection of the team some regard was not given to this factor in view of JPL's weakness in this very difficult area." By these statements, Nier did not mean to detract from the caliber of the individuals on the GCMS team, but he felt that it was necessary to underscore the nature of the problem facing the project managers.[56]

Continued troubles with development scheduling for the gas chromatograph–mass spectrometer and the lack of confidence among the scientists
in JPL's ability to manage the instrument's development and fabrication
led Martin to transfer the management of the GCMS instrument contract
from JPL to his Viking Project Office at Langley. As a preparatory measure, he announced that effective 29 February 1972 Cal Broome, lander
science instruments manager, would report directly to the Viking project
manager. This shift was one more step to tighten control over the lander
science payload and give those experiments the visibility that they seemed
to require. Further—as a consequence of Klaus Biemann's presentation on
the GCMS and the OAMS made at the February Science Steering Group
meeting, in which Biemann had noted that each instrument had advantages
and disadvantages that could not be directly compared—Martin decided in
favor of continuing the development of a simplified version of the GCMS.
His action was prompted primarily by the cost projections, which indicated
that it would be cheaper, by about $7.5 million, to retain the GCMS and
transfer management of the instrument to Langley. NASA Headquarters
approved this recommendation on 10 March, and Martin appointed Joseph
C. Moorman as the GCMS manager and J. B. Lovell as the Viking Project
Office resident engineer at Litton Industries. Although the development
and fabrication of the instrument was still far from ensured, at least a more
responsive management-contractor structure had been established to deal
with the problems that would emerge later.[57]

### Viking Biology Instrument

Nearly everyone associated with the Viking project realized the Viking
biology instrument was going to be a technical and scientific challenge, but
no one was able to predict just how much time, energy, and dollars would
be required by this complex scientific package. Devising a biology instrument that held three experiments in a container less than 0.027 cubic meter
in volume and weighing about 15.5 kilograms was more of a chore than
even the most pessimistic persons had believed. Certainly the TRW Systems
Group personnel who won the Viking biology instrument subcontract in
competition with Bendix Aerospace Systems Division did not expect its
original estimated cost of the completed flight instruments and test articles
to soar from $13.7 million to more than $59 million.[58] A box about the size
of a gallon milk carton, the instrument contained some 40 000 parts, half of
them transistors. In addition to tiny ovens to heat the samples were
ampules containing nutrients, which were to be broken on command;
bottled radioactive gases; geiger counters; some 50 valves; and a xenon lamp
to duplicate the light of the sun. It was a complicated and sophisticated
miniature laboratory.

The Viking biology instrument was originally conceived as essentially
the integration of four individual life-detection schemes. According to

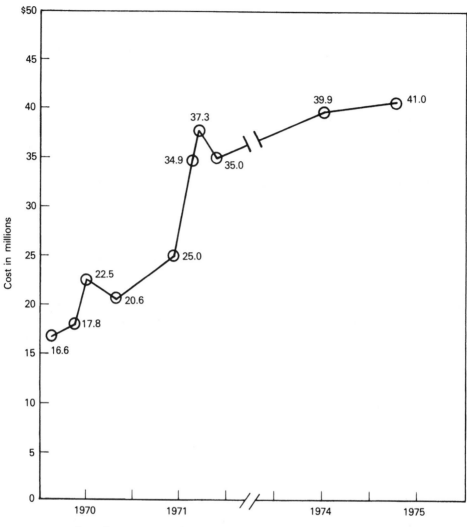

*Gas chromatograph–mass spectrometer cost projections.*

Loyal G. Goff, Viking Program Office, NASA Headquarters, "the transition from these early hardware models to an integrated, automated, and miniaturized flight unit capable of surviving all of the environmental conditions of sterilization, launch, cruise, and landing was a horrendous undertaking." These environmental requirements, with the performance specifications, demanded considerable examination and testing of the materials used in the biology instrument. The initial design concepts for the experiment were developed by Ball Brothers Research Corporation, Boulder, Colorado, and the Applied Technology Division of TRW Defense and Space Systems Group, Redondo Beach, California, under contracts managed by NASA's Ames Research Center.[59]

On 3 September 1970, when the TRW team was given the go-ahead by Martin Marietta, four direct biological tests had been selected for the instrument that could examine the Martian soil for traces of living organisms through the measurement of some aspect of the metabolic process. Three of the procedures could in principle detect "resting" metabolism, although all would be more reliable if growing organisms were present. The first experiment, originally called carbon assimilation but later known as pyrolytic release, would be performed with a minimum addition of external substances (that is, only radioactive carbon dioxide [$^{14}CO_2$], radioactive carbon monoxide [$^{14}CO$], and water vapor) to the samples. Experiment two, originally known as Gulliver and subsequently called labeled release, was to add extremely diluted solutions of labeled (carbon 14) organic matter to the Martian soil samples under conditions that barely moistened the samples. Experiment three, called the gas-exchange experiment, provided for adding greater amounts of organic materials and water to the samples. Because it was rich in nutrients, Jerry Soffen and others referred to this as the "chicken soup" experiment. The fourth experiment (subsequently eliminated) was the light-scattering experiment, or Wolf Trap as it was better known. Requiring the growth of organisms in the sample, this investigation provided the least Marslike environment because it would suspend the sample in an aqueous solution. But if microorganisms did grow, they would turn the liquid cloudy, and the light sensor would detect the change. Together, the four experiments represented a range from very dry to saturated solutions, and experimenters hoped they would provide a check on each other while giving Martian microbes a choice of environments in which to grow.[60]

The first year of work leading up to the preliminary design review was spent making a breadboard model for each of the experiments. The PDR, originally scheduled for July 1971, was postponed three months so that a number of changes could be made in the biology instrument design. In October, TRW submitted new "estimated cost at completion" figures to Martin Marietta; the cost had risen to $20.2 million. TRW had greatly underestimated the complexity of the task, which accounted for about half of the $6.5-million jump. The rest was due to modifications in the experiment definition.

The 4–6 October preliminary design review in Redondo Beach, California, disclosed a number of problem areas in the design and management of the Viking biology instrument. Rodney A. Mills, Walter Jakobowski's deputy, feared that Martin Marietta and TRW could both be blamed for poor management.[61] Of particular concern were the complexity of the waste management system, which would store the water and organic materials after they had been tested; the complicated nature of the sampling system; the increasing instrument weight, which would lead to higher costs; and the numerous elements that, should they fail, would render the whole instrument useless. On 1 July 1971, Jim Martin issued Viking project directive no. 6: "It is project policy that *no single malfunction shall cause the loss of data return from more than one scientific investigation.*"[62] Each of the biology experiments was considered to be one scientific investigation under this philosophy, and there were numerous "single point failures" that could terminate the data return from the instrument. At the October PDR, no single experiment stood out as a particular problem, but Martin, Broome, and their colleagues were worried about the overall complexity of the TRW design.[63]

During November and December 1971, TRW and Ames Research Center personnel under Harold Klein worked to simplify the biology instrument. Deleted from the design were the Martian gas pump, the onboard carbon dioxide gas system, one control chamber each for the gas-exchange and light-scattering experiments, and related valves, plumbing, and wiring. But it became apparent at a biology instrument review in late December that more drastic changes would have to be made. During the final days of January 1972, Martin concluded that one of the experiments would have to be eliminated to reduce the volume, weight, complexity, and cost of the package. Walt Jakobowski and Richard Young from NASA Headquarters met with representatives from the Viking Project Office, Martin Marietta, and TRW on 24–25 January to discuss ways to remedy the problems, especially cost, which had escalated to $33 million.[64]

That meeting was not a satisfactory one from Jakobowski and Martin's point of view. TRW was not able to suggest any acceptable engineering cost reduction items without removing two or more experiments. Additionally, all of TRW's cost reduction proposals had high-risk factors for scheduling, testing, or both. Martin Marietta personnel who had reviewed TRW's schedule and manpower figures were also unable to offer any alternatives. To find solutions to their problems, Martin formed an ad hoc panel for the examination of imposed and derived requirements on the Viking biology instrument under the chairmanship of Howard B. Edwards of Langley's Instrument Research Division. While that panel met to determine which, if any, of the scientific and engineering requirements could be relaxed or eliminated to reduce cost, weight, size, and complexity of the overall instrument, Klein, Joshua Lederberg, and Alex Rich, biology team members who were not affiliated with any particular experiment, met to discuss priorities for deleting one of the experiments.

Dropping an experiment was a painful experience for the men who made the recommendation and those who implemented it. By 13 March, NASA Headquarters had decided that the light-scattering experiment, the investigation based on the least Marslike premise, should be terminated. The men in Washington cited possible difficulties in interpreting results and a potential for further cost growth as reasons for their action. It was John Naugle's unhappy responsibility to tell Wolf Vishniac that his Wolf Trap would not be included in the Martian biology instrument. Noting that "this was one of the more difficult decisions" that he had had to make since joining NASA, Naugle told Vishniac that they had to "simplify the biology experiment—its history of growth in cost and complexity had

232

forced this position." In deciding how to reduce costs, the managers at NASA had tried to consider both scientific and engineering factors:

> On the science side, we are assured that the deletion of the light scattering experiment, while undesirable, is the least damaging in terms of data lost. I won't go into detail here since you have talked at length with Drs. Lederberg and Rich on this subject. On the engineering side, it seems that the light scattering experiment might be considered one of the least complex in terms of number of parts and detail of design, but is one of the more difficult to actually build into a problem free device.

Following advice from all members of the biology team, Naugle stressed the desire that Vishniac continue to participate as a member of that group.[65]

Although the biology team seldom acted as a cohesive group, the decision to eliminate the light-scattering experiment did draw members together temporarily. As a group, they aired their dissatisfaction with the decision, the manner in which it was made, and the limited likelihood that it would reduce significantly the cost of the biology instrument. At a biology team meeting in March, Dick Young and Jerry Soffen were on the hot seat as they once again explained the need for cost reductions in an era of tight budgets. Klein, the team leader, wrote to Naugle on behalf of the whole group:

> Naturally, the Team is not very happy that the scope of the biological experiments was reduced. . . . This science reduction is all the more difficult to accept because it is not at all clear just what factors dictated this decision. Recent discussions with TRW . . . leave little doubt that no savings in weight or in volume will follow from the elimination of the light scattering experiment. . . . Whether, at this late date, any cost savings will accrue from the deletion is also problematical.

While stopping short of mutiny—and still promising to work hard—Klein said that the team wanted a better explanation of why Wolf Trap was dropped.[66]

Understandably, Wolf Vishniac was not happy with the decision. He criticized Lederberg and Rich for not being familiar with the development status of his experiment: "I am shocked to find that a judgment on the value of an experiment was based upon such complete ignorance on the present state of the instrument. . . ." Much of the discussion regarding Wolf Trap concerned "matters which have long ago been settled and solved." Some of the data the NASA managers had used in their decision-making process had been gathered by the Ames Research Center. Vishniac was told by persons at Ames that they had sent headquarters "some old reports which we had lying around." When the scientist asked why "old" material was used, he was given some surprising news: "It doesn't really matter, we have long ago decided that light scattering is to be eliminated." The more Vishniac investigated the elimination of his experiment, the more he was displeased.

He believed that there had been some anticipatory preparation for dropping Wolf Trap. And according to Vishniac, Lederberg and Rich were not really suited for or capable of making an informed decision. "Their aloofness from the team, their ignorance of the mechanical details and the apparent predisposition of Ames to leave out the light scattering experiment makes me question the value of their recommendation."[67]

In a compassionate review of the decision and the process by which it had been made, Naugle tried to allay Vishniac's frustration and anger. The associate administrator pointed out that something had to give, as the budget could not be increased. They had been forced to review and revise all of the Viking experiments on the orbiter and lander. If Lederberg and Rich had not participated in the examination of the biology instrument, someone entirely unfamiliar with the instrument and the search for life on Mars would have.

> We recognized that we were asking them to undertake a very difficult and personally distasteful job of reviewing four experiments which had originally been very carefully selected and had just recently been certified as complementary and an excellent payload for Viking, and recommending which of the four could be removed with the least impact on the overall biology experiment. They reluctantly agreed.
>
> In the guidelines we gave then we said the decision should be primarily made on the basis of the scientific merits of the experiments since there was no substantial engineering factor to use to select the experiments to be deleted. . . .
>
> Dr. Lederberg and Dr. Rich's recommendations were clear—that all four experiments should fly, but if one must be dropped, it should be the light scattering experiment. They also make it clear that although the experiment should be dropped, the experimenter (Dr. Vishniac) should not!

Naugle thought that the deletion would "contribute" in a very real way to the solution of their Viking payload problem. "I am assured that we will save at least two or three pounds [0.9–1.4 kilograms] by this action. This will be applied directly to the weight deficit already incurred by the biology package." Additionally, space would be saved for other biology requirements, at a saving of at least $2.3 million.[68] In the short run, the projected cost of the biology instrument did drop, but by the fall of 1973 the cost estimates would escalate wildly, leading to another major review of the biology package.

Wolf Vishniac faced other disappointments in the loss of his Mars experiment. While he continued to participate constructively in the biology team's work, he no longer had any NASA funds to support his research projects and personnel. Vishniac soon discovered that he would have to pay a high price for having gambled on spaceflight experiments. He had been the first person to receive exobiological research support from the agency,

234

but now that the money was gone he discovered a hostility on the part of many scientists directed toward those who had accepted "space dollars." In spring 1973, Vishniac wrote to Soffen telling him that he could not attend a particular meeting. "I will do whatever is essential in the Viking Program but I simply must place my priorities on my university work. The consequences of my change in status in the Viking Team have been far-reaching as you know, not to say disastrous." He was finding it difficult to obtain support for laboratory research because of his work with the space agency. The National Institutes of Health had refused a grant application; "I was told unofficially that it received a low priority because I was 'NASAing' around." The National Science Foundation had decided not to renew a grant for Vishniac, partly because of his association with NASA. The exobiologist told Soffen that "it is essential that I recapture some sort of standing in the academic world and I must therefore limit my participation in Viking to essentials only."[69]

In 1973, Vishniac was still pursuing his research into the origins of life and the possibility of life on other worlds when he fell 150 meters to his death in Antarctica's Asgard Mountains. Searching for life in the dry valleys of that bitter cold and windswept region, Vishniac was attempting to prove that life forms could adapt to extremely hostile environments. Early in 1972, he had found microorganisms growing in what had previously been thought to be sterile dry valleys. This discovery by Vishniac and his graduate student assistant Stanley E. Mainzer, using a version of the Wolf Trap light-scattering instrument, was a bit of good news for the believers in life on other planets but a contradiction of the findings of Norman Horowitz and his colleagues Roy E. Cameron and Jerry S. Hubbard, who in five years of research had yet to detect any life forms in that barren land.

The dry valleys of South Victoria Land, Antarctica, with a few other ice-free areas on the perimeter of that continent, formed what was generally agreed to be the most extreme cold-desert region on Earth. The area was also the closest terrestrial analogy to the Martian environment. These valleys, which covered several thousand square kilometers, were cut off from the flow of glaciers out of the interior of the continent by the Transantarctic Mountains. Although the valleys were ice-free, their mean annual temperature was –20°C to –25°C, with atmospheric temperatures rising to just the 0°C mark at the height of the summer season. Liquid precipitation and water vapor were almost nonexistent, and the limited snowfall usually sublimed to the vapor phase without ever turning to liquid. It was in this region that Horowitz's colleagues discovered what was believed to be the only truly sterile soil on the face of Earth. From their research in the dry valleys, Horowitz and his associates concluded:

> These results have important implications for the Mars biological program. First, it is evident that the fear that terrestrial microorganisms carried to Mars could multiply and contaminate the planet is unfounded.

*Scientists attend a Viking planning meeting at Langley Research Center in 1973. Left to right are Dr. William H. Michael, Jr., leader of the radio science team; Dr. Wolf Vishniac, assistant biology team leader; and Dr. Richard S. Young, Viking program scientist from NASA Headquarters in Washington.*

The Antarctic desert is far more hospitable to terrestrial life than is Mars, particularly in regard to the abundance of water. In other respects, too—such as the ultraviolet flux at the surface—Mars is decidedly more hostile than the Antarctic.

Second, Martian life, if any, must have evolved special means for obtaining and retaining water. . . . This has been known for some time. What is new in these findings is that even under severe selective pressure microbial life in the Antarctic has been unable to discover a comparable mechanism. To some this may suggest that life on Mars is an impossibility. In view of the very different histories of Mars and the dry valleys . . . we believe that such a conclusion is not justified.

Finally, the Antarctic has provided us with a natural environment as much like Mars as we are likely to find on Earth. In this environment, the capacity of life as we know it to adapt and survive is pushed to the limit. The concentration of living things around the sources of water in the dry valleys and their rapid thinning out in the most arid locales may be useful as a model of the distribution of the life we may, if we are lucky, find on Mars.[70]

But in 1972, Vishniac detected microorganisms with Wolf Trap in exactly those regions that Horowitz had declared sterile. Life had found ways to survive in the inhospitable, Marslike dry valleys. In December 1973, Vishniac went back to Antarctica to learn more about these hardy microbes. He wanted to know where they obtained their life-sustaining water and nourishment. Alone on a steep slope in the dry valleys, Vishniac slipped, fell, and died.[71] Vishniac and his Wolf Trap life detector had been successful on Earth, but he would not see Viking go to Mars, and his instrument would not be applied to searching for elusive Martian microbes. A man who had done much to give exobiology legitimacy as a field of research was gone. The loss of Vishniac from the biology team was repeatedly felt in later years. He had been an arbiter and a man of good cheer. As the biology instrument continued to increase in cost and to raise more and more technological hurdles to be overcome, a man with his talents and humor was sorely needed.

During the first half of 1973, work progressed on the design development units for the biology instrument and the gas chromatograph–mass spectrometer. Science tests for the biology instrument had begun in mid-December 1972, with biology team members participating in the trials of their experiments. GCMS testing began in early May. After the first round of testing, the Viking managers held a critical design review on 23-25 May for the biology instrument, and even though they discovered no major problems with the package, Martin Marietta and the Viking Project Office were less than pleased with the review. The GCMS critical design review in mid-July disclosed only three major concerns, which was encouraging news considering the problems that piece of hardware had caused earlier.

Unhappily, new trouble with the management of the biology instrument surfaced in mid-July. At a meeting held at TRW, Jim Martin learned that completion of the design development unit had slipped by three weeks and the projected delivery date of the proof test capsule unit was behind by five weeks. The problem, Martin found, was failure to plan ahead; TRW lacked the skilled manpower to assemble and check out these crucial units. As the July session went on, the discussion of the biology instrument came "unglued," according to Martin; he feared that the work at TRW was "out of control" with no credible schedule or cost plan.[72] By that autumn, the situation was even bleaker. On 15 October, Ed Cortright wrote to George Solomon, vice president and general manager at TRW. Cortright had been monitoring TRW's handling of the biology instrument problems with the intent of reporting to Hans Mark, director of Ames Research Center. His report was to give the center better data for judging prospective contractors—of which TRW was one—of experiment hardware for the Pioneer spacecraft scheduled to visit Venus. Cortright's report to Ames would not be favorable. He thought that TRW, Martin Marietta, and NASA had underestimated the complexity of the biology instrument task: "The original TRW proposed cost was grossly underestimated with the result that the current estimate at completion is $30.9 million, which is $18.4 million or 147 percent over the original estimate." Of that amount, $12.4 million was TRW's overrun; $6 million had been spent on redefining the experiments.

Cortright told Solomon that the TRW management had placed too much emphasis on the company's previous performance and had been reluctant to face the fact that the biology instrument was getting into serious trouble. "You are currently beset with a rash of technical problems

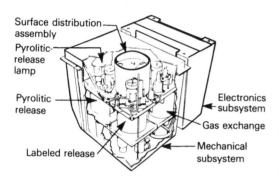

Surface distribution assembly
Pyrolitic-release lamp
Pyrolitic release
Labeled release
Electronics subsystem
Gas exchange
Mechanical subsystem

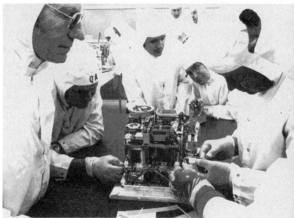

*The development model of the Viking biology instrument's mechanical subsystem, top left, conveys some of the external complexity of the experiment. The mockup, top right, minus the essential electrical and plumbing connections exposes the hardware to view. At lower left, a diagram shows the biology instrument after deletion of the light-scattering experiment. At lower right, in final stages in 1975, the automated equivalent of a well-equipped biological laboratory makes up a package of less than 0.03 cubic meter to land on Mars.*

which further threaten schedule and cost. It is clear that if the job were on schedule, there would be more time to adequately cope with the necessary fixes." Impressed with the steps Solomon had taken to strengthen management of the biology package, Cortright nevertheless believed that "heroic action" would be necessary to ensure "a successful experiment on the surface of Mars."[73] Two weeks later, after the schedule had slipped even further and the biology instrument had been put on Martin's Top Ten

Problems list, Cortright again wrote the general manager about the "potentially catastrophic" situation and sent a similar letter to Richard D. De Lauer, TRW Space Systems executive vice president. To De Lauer, Cortright bluntly said, "It is imperative that you bring to bear on these problems the most talented individuals you can find within your Company, and elsewhere, and quickly weld them into a problem solving team to get this job done. I know you have taken steps in this direction and I cannot fault individuals who are currently working the problems. However, I must believe that you have not yet applied your maximum effort, for which there is no longer any substitute."[74]

The problems at TRW were twofold. The engineering tasks imposed by the experiments were very difficult, and TRW's management of the project was poor. At very low temperatures, valves and seals failed, and other hardware difficulties surfaced as the initial pieces of equipment were tested. But most serious was the absence of a strong, driving manager at the California firm overseeing the work. In November 1973, production of the flight units was essentially stopped while the biology instrument was redesigned. But design quality and workmanship problems persisted, causing test failures and schedule difficulties. To meet the launch date, TRW was required to conduct design-development concurrently with qualification testing and fabrication of the flight units. By the first of February 1974, several independent analyses of the situation at TRW pointed to the possibility that the final flight units of the biology instrument would not be ready until July 1975. That would be very close to the scheduled launch dates (August and September) and too late for adequate preflight science testing.

Cal Broome, who had been appointed NASA biology instrument manager in December 1973, in a private note to Jim Martin on 7 February 1974, stated that his own view of the situation at the subcontractor's was that the "engineering organization, and, to a lesser extent, the manufacturing organization [at TRW], are running out of control." Furthermore, "The TRW engineering 'culture' simply cannot accept scheduling and discipline in connection with engineering problems." Broome was also worried that others would not share his opinion of TRW's failings and simply view his pessimistic outlook as a case of Broome having panicked again; but Hatch Wroton, the Martin Marietta resident engineer at TRW, and Dave Rogers, the JPL resident at TRW, had independently assessed the biology instrument's status and agreed with Broome's bleak prognosis.[75]

During the remaining months before the Viking launches, time lost in the schedule would be made up, only to be lost again when some new difficulty appeared. In July 1974, Martin had Walter O. Lowrie, lander manager at Martin Marietta, and Henry Norris, orbiter manager at JPL, study contingency plans for flights without the biology instrument and single flights of the Viking spacecraft in 1975, 1977, and 1979. Days later, progress on the instrument at TRW looked more promising, but by the end of the year, when the performance verification tests of the completed instruments were being conducted in Redondo Beach, new doubts about meeting the schedule plagued the Viking managers.[76]

The seesaw between failure and progress finally stopped in the early spring of 1975. On 7 March, Martin wrote to the three men who had seen the biology instrument through some of its most difficult moments—Eugene M. Noneman, TRW; Hatch Wroton, Martin Marietta; and Roy J. Duckett, Viking Project Office: "I was pleased today to be advised that Viking biology instrument S/N 106 is in its shipping box ready for delivery. I believe that you and your team members have achieved a very significant and important milestone. While there is still much work ahead of us, having a flightworthy biology instrument ready to ship to the Cape is a gratifying accomplishment." Martin extended his personal congratulations to every member of the team.[77] On 28 May, Cal Broome could at last recommend to Jim Martin that the GCMS and the biology instrument be removed from the Top Ten Problems list. Those had been the final items on the list of troubles. The hardware units were finally ready for shipment.

*Table 42*
*Viking Biology Instrument Schedule, 1971–1975*

| Milestone | Original Contract Delivery Date | Actual Delivery Date | Delay in Months |
|---|---|---|---|
| Preliminary design review | July 1971 | Oct. 1971 | 3 |
| Critical design review | Aug. 1971 | May 1973[a] | 9 |
| Design-development testing complete (S/N 001) | July 1971 | Dec. 1973[b] | 17 |
| Qualification unit delivery/ qualification testing complete (S/N 102) | Sept. 1973 | Mar. 1975[c] | 18 |
| Proof-test capsule unit delivery (S/N 103) | June 1973 | Nov. 1974 | 17 |
| Flight unit-1 delivery S/N 105 on Viking lander capsule #1 | Jan. 1974 | Mar. 1975 | 14 |
| Flight unit-2 delivery S/N 106 on Viking lander capsule #2 | Apr. 1974 | Mar. 1975 | 11 |
| Flight unit-3 delivery (S/N 104) | July 1974 | Apr. 1975 | 9 |
| Spare flight unit | Added Dec. 1973[d] | Deleted Oct. 1974[e] | — |

[a]Martin Marietta contended that a realistic CDR was not completed until Mar. 1974.
[b]Design development testing was completed on a nondeliverable unit; one of the deliverable units was canceled; the other deliverable unit's mechanical subassembly was simulated in system test bed testing.
[c]Qualification testing was different from original plans and not as comprehensive.
[d]This unit, not included in the original contract, was added in Dec. 1973.
[e]Unit deleted Oct. 1974 when requirement for spare lander was eliminated.

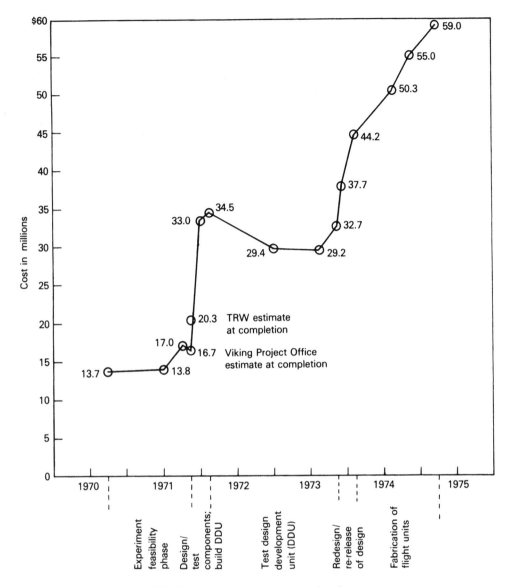

*Biology instrument cost projections.*

The cost in individual time and effort on these two items had been high; the dollar costs were equally great. By launch, the GCMS bill read $41 million, and the biology instrument had cost $59 million.[78]

There was, of course, more to the Viking lander science package than the gas chromatograph–mass spectrometer and the biology instrument. Each of the other instruments went through a similar series of problems met and problems solved. The GCMS and the biology instrument were unique because of the magnitude of the difficulties and the expense. With time, all problems with the instruments were resolved, and interaction among the scientists improved. Still, each team remained a collection of individuals, and among the teams only a loose confederation existed. Before the missions were flown, a stronger discipline would have to be forged. Operation of the orbiters and landers would be a complex task, and each sequential operation would have to be carefully planned and precisely executed. Jerry Soffen, Jim Martin, and Tom Young had many difficult tasks ahead of them, and one was establishing tighter control over the Viking scientists without stifling their inquisitiveness—exercising discipline so as to get maximum science return, but not in such a manner as to eliminate flexibility when scientific targets of opportunity appeared.

As the Viking science teams and their instruments matured, Jim Martin faced other technical problems with the lander, each of which had to be solved before the spacecraft could fly. Complexity and technological challenges abounded. Building Martian landing craft was genuinely hard work.

# 8

# Viking Lander: Building a Complex Spacecraft

The Viking lander represented a careful melding of the demands imposed by the scientific mission and the high degree of reliability required of the spacecraft subsystems. Weight and volume considerations affected the size of each subsystem. After the Voyager program with plans for an 11 500-kilogram spacecraft was abandoned in 1967, a follow-on study concluded that a spacecraft weighing 3700 kilograms could be transported to Mars by a Titan-Centaur–class launch vehicle. The lander and its flight capsule would account for more than a third of this weight (1195 kilograms). At the start of the mission, the orbiter and lander would be housed in a 4.3-meter shroud atop the Titan-Centaur. The landed spacecraft would be 3 meters at its widest point and 2 meters tall from the footpads to the tip of the large disk S-band high-gain antenna. While weight and volume limitations helped to shape the Viking lander, data about Martian atmospheric pressure obtained during the Mariner 69 mission were also influential.

Mariner 69's occultation experiment indicated that the atmospheric pressure at the surface of Mars ranged from 4 to 20 millibars, rather than 80 millibars as estimated earlier. This information had a definite impact on the aerodynamic shape of the Mars entry vehicle being designed, since weight and diameter would influence the craft's braking ability. Langley engineers had determined that aerodynamic braking was the only practical method for slowing down a lander as large as Viking for a soft touchdown. The entry vehicle would have a diameter of 3.5 meters, an acceptable ballistic coefficient that would help ensure Viking's safe landing on Mars.

Since electrical power requirements were thought of in terms of the weight that the power apparatus would add to the spacecraft, the design engineers sought creative means for getting maximum results from a minimum amount of power. Low-power integrated circuits were used extensively both to conserve energy and to keep the package small. In addition, power switching techniques were devised to reduce energy requirements. As John D. Goodlette, deputy project director at Martin Marietta, noted, the design rule was "turn off unneeded consumers."[1] When power had to be used, the equipment was designed with multiple power levels, or states, so that only the minimum power required to achieve the immediate function would be consumed.

Once separated from the orbiter with its 700-watt solar panels, only 70 watts of radioisotope-thermoelectric-generated power would support the long mission on the surface. Because of this limitation on landed power, the radio transmitters could be used only sparingly, a factor that in turn controlled the amount of data that could be sent to Earth.

The Viking lander was a highly automated spacecraft for a number of reasons. Since there was only a 20-minute one-way communications opportunity between Earth and Mars during the landings, control of the lander from Earth from separation to touchdown was not practical. The entire function of navigation—from obtaining an inertial reference to locating a local surface reference—had to be accomplished by the onboard computer. After landing, the spacecraft would be out of direct communication with Earth for about half of each Martian day. And because of electrical power limits, the communications between lander and mission control in California would amount to only a short time each day. The lander, therefore, had to be capable of carrying out its mission unattended by Earth. Mission specialists could send the lander new assignments or modify preprogrammed ones, but for the most part the craft was on its own as it did its day-to-day work.

## LANDER MISSION PROFILE

Jim Martin and his colleagues hoped the lander mission would follow the ideal schedule: Final prelaunch activities begin 56 days before launch with the terminal sterilization of the entire lander system within its bioshield. The craft must survive a 40-hour sterilization cycle, during which temperatures will reach a maximum of 112°C. During this preparation period, the lander is functionally passive except for its two mass-spectrometer ion pumps. Following a checkout, the propellants, pressurants, and flight software are loaded, and the lander is mated with the orbiter. After the first prelaunch checkout, initiated by the orbiter under local control of the guidance computer, the spacecraft is encapsulated, transported to the launch pad, and mated with the launch vehicle, followed by the second and final prelaunch checkout. All major communication with the lander before separation is accomplished through the orbiter communications link.

During the launch and boost phases, only the power and pyro controllers, the data acquisition system, and the tape recorder are active. After the spacecraft separates from the launch vehicle, the orbiter commands the lander computer to separate the bioshield cap and begins the lander cruise state. During cruise, the lander is largely passive. Only the data acquisition system, ion pumps, and thermostatically controlled heaters on propulsion equipment, the biology instrument, and the inertial reference unit are powered. The heaters prevent the freezing of propellants and biology nutrients. Heat also controls viscosity of the gyro flotation fluids. The primary housekeeping chore during the cruise phase is monitoring the thermal balance and the equipment when it is powered.

The tape recorder is activated about every 15 days to ensure its later performance. An update to the computer requires the activation of the computer and the command detectors and decoders. The portion of the computer memory used during prelaunch checkout procedures is modified during the cruise so that it can perform other operations during the mission. The gas chromatograph–mass spectrometer requires a venting-and-bakeout sequence to rid the analyzer section of absorbed gases. For bakeout, with its high peak-power demand, the lander batteries are first conditioned and charged using orbiter power; the computer, detectors, and decoders are powered up; and a six-hour bakeout sequence is commanded from Earth, followed by a week-long cooldown period to reestablish the proper thermal equilibrium. About five such cycles in two groups are required, each accompanied by mass-spectrometer readings, which are analyzed to determine the performance and health of the instrument. After each activity, the lander is powered back to cruise state and, after the final bakeout of the gas chromatograph–mass spectrometer, a cruise check is made and the batteries discharged. About 52 days before reaching Mars, the final conditioning and charge cycle is undertaken for the lander batteries.

Before the lander separates from the orbiter, a four-and-one-half-hour checkout verifies the lander systems' health. A group of orbit commands precedes this last check, during which local control is assumed by the lander computer and power is transferred from the orbiter to the lander. At checkout completion, the computer memory is read out, the batteries are recharged on internal power, and the computer reverts to standby. After cruise checkout, power is transferred back to the orbiter, which assumes control. The next events prepare the lander for its release. For eight hours, the radioisotope thermoelectric generators recharge the lander's batteries.

Twelve hours before separation—318 days into the mission—an orbiter commander turns on lander command detectors and decoders, placing the lander under the control of its own computer. Mission control commands update descent information and carry out checkout decisions made by the operations team. The commands are directed to the lander via its S-band receivers. A memory readout follows update, and the lander assumes a standby mode. This sequence is repeated three and one-half hours before separation. About two and one-half hours before separation, direct orbiter command starts the separation sequence. Final preparations begin with warming up the inertial reference unit to its operating temperature. At 37 minutes before separation, a final "go" is uplinked from Earth and received by the lander 15 minutes before separation. At this point, valve-drive amplifiers, pyrotechnic controllers, entry thermal control, and relay communications link are activated. A final check verifies that the inertial reference unit has transferred to the entry condition and that all systems are go. If these checks fail, the lander is powered down and transferred to the update mode. If the checks pass, the telemetry system is

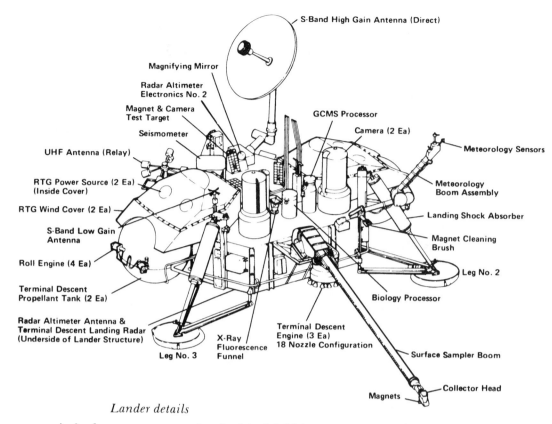

S-Band High Gain Antenna (Direct)

Magnifying Mirror

Radar Altimeter
Electronics No. 2

Magnet & Camera
Test Target

Seismometer

UHF Antenna (Relay)

RTG Power Source (2 Ea)
(Inside Cover)

RTG Wind Cover (2 Ea)

S-Band Low Gain
Antenna

Roll Engine (4 Ea)

Terminal Descent
Propellant Tank (2 Ea)

Radar Altimeter Antenna &
Terminal Descent Landing Radar
(Underside of Lander Structure)

Leg No. 3

X-Ray
Fluorescence
Funnel

Terminal Descent
Engine (3 Ea)
18 Nozzle Configuration

GCMS Processor

Camera (2 Ea)

Meteorology Sensors

Meteorology
Boom Assembly

Landing Shock Absorber

Magnet Cleaning
Brush

Leg No. 2

Biology Processor

Surface Sampler Boom

Collector Head

Magnets

*Lander details*

switched to an entry mode, the bioshield base connectors between orbiter and lander are separated, and the lander-orbiter separation pyrotechnic devices are fired.

Immediately after separation, attitude control–deorbit propulsion is readied by opening the isolation valves. After inertial reference unit calibration, attitude control is initiated by orienting for the deorbit burn. The burn is delayed until the lander capsule is far enough away from the orbiter that the orbiter's solar panels will not be damaged or contaminated. The pitch-yaw engines supply the deorbit impulse with a 23-minute burn. The control system ensures that the lander is in the proper position for the entry science experiments to function. The retarding-potential analyzer and the upper-atmosphere mass spectrometer collect data during the three-hour descent.

*Entry and Landing*

After orienting the lander in preparation for entry into the Martian atmosphere, the control system turns on the radar altimeter, which assumes the high-altitude search mode. On sensing 0.05 g with the longitudinal accelerometer, the attitude control system is adjusted, and the computer begins radar-altimeter data processing. Aerodynamic forces quickly trim the entry vehicle to about a –11° angle of attack, corresponding to the lander's

246

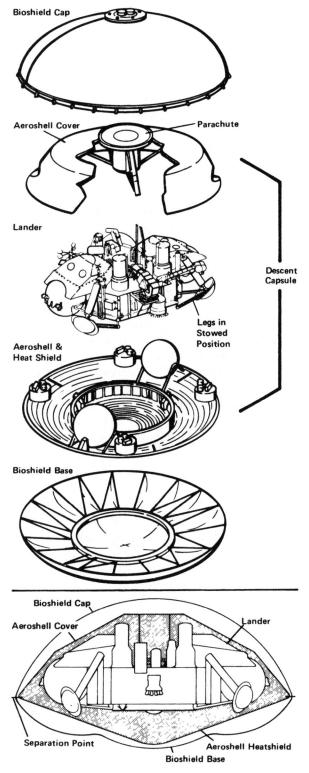

*Landing capsule system*

offset center of gravity. Instruments collect additional entry science data for pressure and temperature during the remainder of the deceleration period.

At 5.5 kilometers above Mars, the computer begins parachute deployment based on radar range to the surface. Terminal-propulsion valve-drive amplifiers power up, the aeroshell separates from the lander, and the terminal-roll-propulsion isolation valves open within about seven seconds after parachute deployment. Radar-altimeter changes occur with separation of the aeroshell, and a lander body-mounted antenna switches into use. The four-beam doppler terminal-descent and landing radar is also activated to sense velocity relative to the surface. The lander's legs are deployed from their stowed position.

At about 1.5 kilometers above the surface, the computer initiates another radar-altimeter mode change and shortly thereafter opens the terminal-propulsion isolation valves. The parachute–base cover assembly separates from the lander, and the lander descends toward the surface under three-axis attitude control. The control system and engines halt the horizontal velocity acquired while on the parachute by tilting the entire lander upwind. At the same time, residual vertical velocity is stopped. On sensing 610 meters to the surface, the radar altimeter switches to low-altitude mode; the low-altitude mode for the terminal-descent and landing radar begins at 100 meters. At about 50 meters, vertical navigation continues inertially, ignoring radar-altimeter data. At 17 meters, the terminal engine-shutdown switches are armed, and a constant velocity descent is initiated to maintain a speed of 1.5 meters per second until landing-leg touchdown. Velocity steering continues, using the terminal-descent and landing radar. On sensing closure of the terminal-engine-shutdown switches, the computer commands shutdown of the terminal propulsion system by closing a pyro-activated isolation valve, backed up by a software timer.

## Landed Operations

The landed mission begins with several housekeeping chores, which include shutting down all descent guidance and control equipment except the computer and the inertial reference unit; the latter operates five more minutes to establish the local vertical altitude and the direction of north. This information is used to compute the direction of Earth so the high-gain antenna can be accurately pointed the following day. Protective devices are armed but not yet activated, the telemetry is set to the highest relay data rate mode of 16 kilobits per second, and the first real-time imaging sequence is begun. A multiple readout of about 25 percent of the computer's memory follows.

After deploying the high-gain antenna and the meteorology boom, opening the camera dust-removal valve, and opening the cover to the biology-processor and distribution assembly, all mission pyrotechnic events are completed. A second real-time imaging sequence begins and continues until the orbiter disappears over the horizon. The relay link fades out about 10 to 12 minutes after landing and, at 15 minutes after, the transmitter is shut off. The meteorology instrument and the seismometry instruments are turned on, and the high-gain antenna is stowed to its normal rest position. Finally, the adaptive mission is begun by activating the mission sequence of events.

Before the Viking landers had the opportunity to perform this complex series of events on Mars, managers, scientists, and engineers faced a multitude of problems on Earth.

## SCIENCE DATA RETURN

The goal of obtaining the greatest amount of scientific information possible from the Martian surface was the major influence on the design and structure of the lander. During 1975 and 1976, Mars and Earth would be at their maximum separation distance, about 380 million kilometers. Since the distance would vary during the mission and since the length of relay opportunities would also vary, several data transmission links were built into the lander equipment for direct communications with Earth (1000, 500, and 250 bits per second were available at a single transmitter output of 20 watts). A second communications link, UHF through the orbiter, was functionally redundant with the direct link. The orbiter relay had three transmitter power levels (1, 10, and 30 watts) and two data rates (4 and 16 kilobits). Since available communication time was severely limited by the power available, typical communication periods would be about 1 hour for the direct link transmitters and 20 minutes for the relay link transmitters. With these link times, data rates, and power output, the rate of scientific data returned to Earth would be about 1 million bits per day for the direct link and 20 million for the relay link. Since the relay link was the more efficient from an energy standpoint, the mission planners would use the orbiter link for the majority of the mission's activities.

Several electronic tricks could be played with the data transmitted (telemetered) to Earth. Because of the short transmission times, "housekeeping" engineering data would be telemetered in real time. Much of the scientific data would be sent on a delayed schedule, having been stored on the tape recorder. Bits of immediate data and delayed data could be electronically interleaved. Although this combination of information cut in half the amount of data that could be returned, it did guarantee the return of important scientific and engineering data during the crucial communication periods. Furthermore, each instrument was constructed to convert its scientific information into a digital code. The imaging system would produce large amounts of digital information, but the biology instrument and the gas chromatograph–mass spectrometer would send much lower volumes of data. With the exception of the imaging system, the lander instruments could automatically communicate with the guidance, control, and sequencing computer when their storage capacities were full. At that time, the data would be dumped into bulk storage. Imaging-data storage or direct transmission, however, had to be preplanned because of the very large amounts of digital information.

Considerable technical sophistication was required to execute the scientific experiments, digitize the information collected, store the data, manipulate it, and transmit it to Earth on cue. This technological complexity and sophistication had a direct dollar equation: developing such a complicated machine in a small package against a specific deadline required a large budget. The world in which NASA operated, however, was full of budget restrictions.

The stringent post-Apollo fiscal scene forced the space agency's managers to work hard and be tough with their personnel and their contractors. Legislators who favored tighter federal budgets argued that such activity was a natural part of NASA's job, but a decade earlier many of these same senators and representatives had willingly appropriated extra dollars when

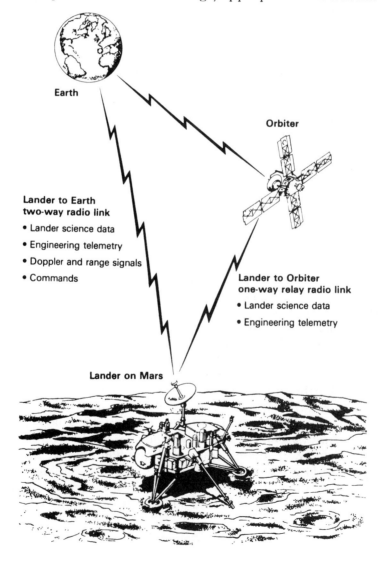

**Earth**

**Orbiter**

**Lander to Earth
two-way radio link**

• Lander science data

• Engineering telemetry

• Doppler and range signals

• Commands

**Lander to Orbiter
one-way relay radio link**

• Lander science data

• Engineering telemetry

**Lander on Mars**

Apollo managers needed them to solve the problems associated with winning the race to the moon. Post-space race hardware was also expensive, and the Mars landing was a complicated project. The Viking managers were committed to accomplishing their mission in a scientifically valid manner and within a reasonable budget, but more dollars—and the spirit of the Apollo era—would have made it easier. Ingenuity and good management would have to substitute for extra appropriations.

Management's warnings about costs began to sound like a broken record to many of the scientists in the Mars venture; but, like it or not, scientists had to think about money as much as about science. In the fall of 1973, the total project cost had been estimated at $830 million. During the spring and summer of 1974, that figure grew substantially, and despite additional parings the estimated cost at completion reached $930 million by the fall of 1974. That amount, however, did not include the extra dollars the biology instrument ($7 million) and the gas chromatograph–mass spectrometer ($4 million) would demand from fall 1974 to spring 1975. These two instruments long occupied prominent places on Jim Martin's Top Ten Problems list.

*Table 43*
*Cost History of Viking Lander and Selected Subsystems*
(in millions)

| Date | Estimated Cost at Completion | | | | | Total Lander Actual Cost |
|------|---------|------|------------------|------|-----------------|-------------------|
|      | Biology | GCMS | Lander Camera | GCSC | Total Lander |  |
| June 1970 |  | 17.8 |  |  | 360 | 19 |
| Sept. 1970 | 13.7 | 20.6 | 9.8 | 3.4 |  |  |
| Aug. 1971 | 17.0 | 25.0 | 12.9 |  | 401 | 62 |
| Feb. 1972 | 34.5 | 35.0 | 17.4 |  | 381 | 107 |
| July 1972 | 32.3 | 35.0 | 18.1 | 10.2 | 420 | 149 |
| Apr. 1973 | 29.2 | 35.4 | 22.9 | 10.2 | 430 | 286 |
| Mar. 1974 | 44.2 | 38.7 | 23.1 | 24.1 | 512 | 411 |
| July 1974 | 50.3 | 39.9 | 27.4 | 24.7 | 543 | 451 |
| Sept. 1974 | 55.0 |  | 23.5 | 28.1 | 559 | 473 |
| Mar. 1975 | 59.0 | 41.0 | 27.5 |  | 545 | 545 |
| June 1976 | 59.5[a] | 41.2[a] | 27.3[a] | 28.1 |  | 553.2[a] |

[a]Actual cost incurred.  GCMS = gas chromatograph-mass spectrometer.
GCSC = guidance, control, and sequencing computer.

## TOP TEN PROBLEMS

Martin began the Viking Top Ten Problems list in the spring of 1970 to give visibility to problems that could possibly affect the launch dates. Viking project directive no. 7, issued 4 October 1971, codified the concept:

"It is the policy of the Viking Project Office that major problems will be clearly identified and immediately receive special management attention by the establishment of a Top Ten problems list." To qualify for this dubious distinction, the problem had to be one that seriously affected "the successful attainment of established scientific and/or technical requirements, and/or the meeting of critical project milestones, and/or the compliance with project fiscal constraints." Anyone associated with the Viking project could identify a potential priority problem by defining the exact nature of the difficulty and forming a plan and schedule for solving it. When Martin made an addition to his list, a person in the appropriate organization was charged with solving the problem, and someone in the Viking Project Office monitored his progress. Weekly status reports were datafaxed from the field to Langley. At Martin Marietta, William G. Purdy, vice president and general manager of the Denver Division—through Albert J. Kullas and later Walter Lowrie, his project directors—sent weekly status bulletins on the lander's top problems, since that system seemed to have the greatest number of difficult components and subsystems. In the spring of 1972, Martin told Cortright he hoped the supervisors of employees who had one of their tasks assigned to the top 10 list would not be penalized. Martin, not wanting a stigma attached to identification of a problem, was concerned that at Martin Marietta assignment of a problem might "automatically be considered as a mark of poor performance" when promotions or raises were given. Generally, the nature of the crucial problems was so complex that punishing one individual would not solve the problem. As with the gas chromatograph–mass spectrometer and the biology instrument, the novelty of the technological task was often the source of the trouble.[2] Some problems seemed to stay on the manager's worry list forever. Others made repeat performances.

At times, Martin found it necessary to bring a particular problem to the attention of a specific subcontractor. Depending on the clout needed behind the message, Martin would sign the letter or enlist the aid of Langley Director Cortright or Martin Marietta Vice President Purdy. In extreme cases, the letter would be sent out over the signature of the NASA administrator. Early in 1973, the Viking Project Office identified six subsystems that required Administrator James C. Fletcher's personal touch: inertial reference unit, subcontractor Hamilton Standard; terminal-descent and landing radar and radar altimeter, Teledyne-Ryan; guidance, control, and sequencing computer, Honeywell; lander camera, Itek Corp.; upper-atmosphere mass spectrometer, Bendix Aerospace; gas chromatograph-mass spectrometer, Litton Systems.[3]

Fletcher wrote the president of each company asking for his personal pledge of support for Viking and seeking his fullest cooperation in resolving the problem. The administrator usually asked them to come to Washington to discuss the issues further. By setting off an alarm in the front office, NASA managers from Fletcher and his deputy, George Low, to Jim

## Table 44
## Top Ten Problems

| Item | Added to List | Deleted from List |
|---|---|---|
| GCMS progress and schedule | 4 May 1970[a] | July 1971 |
| Lander gear design | 4 May 1970[a] | |
| Site-alteration program schedule | 4 May 1970[a] | |
| Solder joints, failure mode under sterilization | 4 May 1970[a] | Dec. 1971 |
| Post-Mars-orbit-insertion orbit determination convergence | 4 May 1970[a] | May 1971 |
| Lander weight growth | 4 May 1970[a] | |
| Orbiter weight growth | 4 May 1970[a] | |
| Site alteration | 4 May 1970[a] | Aug. 1971 |
| Wet tantalum capacitor failure under sterilization temperatures | Jan. 1971 | Feb. 1971 |
| Completion of data requirement list/data requirements description | Jan. 1971 | Feb. 1971 |
| Lander gear design | Feb. 1971 | Aug. 1971 |
| Lander weight contingency | Mar. 1971 | Oct. 1971 |
| Orbiter weight contingency | Mar. 1971 | Aug. 1971 |
| Lander materials | Aug. 1971 | Oct. 1972 |
| Lander processes | Aug. 1971 | 1 June 1972 |
| Lander parts program | 27 Aug. 1971 | 1 June 1972 |
| GCMS configuration and schedule | Oct. 1971 | 2 Feb. 1972 |
| Balloon-launched decelerator test | 2 Feb. 1972 | July 1972 |
| Radar-altimeter design-development schedule | Feb. 1972 | 24 July 1973 |
| Lander entry weight | Mar. 1972 | 26 Apr. 1973 |
| Proof-test-capsule schedule | 31 Mar. 1972 | 26 May 1972 |
| Guidance, control, and sequencing computer development-test schedule | 21 July 1972 | 15 Jan. 1975 |
| Aeroshell radar-altimeter-antenna engineering release | 22 Aug. 1972 | 5 Jan. 1973 |
| GCMS development-test schedule | 1 Sept. 1972 | 6 June 1975 |
| Viking lander camera development schedule | 1 Sept. 1972 | 2 Apr. 1974 |
| Titan-Centaur-shroud qualification program | Sept. 1972 | |
| Upper-atmosphere mass spectrometer development schedule | 12 Jan. 1973 | 31 Oct. 1974 |
| Surface-sampler boom motor | 8 Feb. 1973 | 10 Sept. 1974 |
| Proof-test-capsule component delivery | 26 Feb. 1973 | 19 Feb. 1974 |
| Flight-team facility space | 11 Apr. 1973 | 1 Nov. 1973 |
| Inertial reference unit | 12 Apr. 1973 | 3 Oct. 1973 |
| Seismometer instrument | 22 Aug. 1973 | 20 Dec. 1973 |
| Viking-orbiter-system data-storage-subsystem data recovery | July 1973 | 20 Dec. 1973 |
| 54L microcircuit particle contamination | Oct. 1973 | 4 Nov. 1974 |
| Proof-test-capsule schedule recovery | 19 Apr. 1974 | 21 Oct. 1974 |
| Lander-test-sequence development | 19 Apr. 1974 | 12 Dec. 1974 |
| Building 264 construction of facilities funding | 27 June 1974 | 29 Aug. 1974 |
| Guidance, control, and sequencing computer flight software | 2 July 1974 | 15 Jan. 1975 |
| GCMS processing-and-distribution-assembly shuttle block design | 29 Oct. 1974 | 6 June 1975 |
| Biology instrument | 26 Oct. 1973 | 6 June 1975 |

[a]Although the Top Ten Problems concept was not officially recognized until October 1971, the system was used before that date. In Jim Martin to Henry Norris, "Viking Top Ten Problems," 4 May 1970, these items were listed.

Martin hoped to impress on subcontractors their obligations to the Mars project.[4] At times, the NASA administrator had to be extremely blunt. Once his letter resulted in a meeting with John W. Anderson of Honeywell, Inc., about the guidance, control, and sequencing computer. This worrisome piece of hardware, the key to the lander's performance, controlled and arranged the sequence of all lander functions from separation from the orbiter through completion of the mission. Without this brain and central nervous system, the lander would be worthless. Schedule delays and cost increases in developing the guidance, control, and sequencing computer were in large part the result of the requirements established for this piece of equipment—energy efficiency, small size, reliability, heat resistance, longevity. Each lander had a guidance, control, and sequencing computer, made up of two completely redundant computers with 18 432-word plated-wire memories. The overall computer was 0.03 cubic meter (26.7 x 27.3 x 40.6 centimeters) and weighed 114.6 kilograms. Its most advanced feature was the two-mil plated-wire memory, making small size and low power consumption possible. Either of the twin computers within the guidance, control, and sequencing computer could operate the lander, but only one would be used during descent. The computer would have to work flawlessly if the landing was to succeed. Once the lander was on the surface, either computer could control the craft.

As prime contractor for the lander, Martin Marietta had responsibility for the important computers. In May 1971, the firm asked for proposals from 11 firms to subcontract for the guidance, control, and sequencing computer and received 5 responses. After an unusually complicated contracting process, Honeywell, Inc., was selected as the builder, largely because of its plans to use the two-mil plated-wire memory. Work began at Honeywell's Aerospace Division in Saint Petersburg, Florida, in November 1971. Honeywell also had a contract with Martin Marietta for the development of the lander's data-storage memory, a digital-data-storage device used in conjunction with the lander's tape recorder. The data-storage memory would have the same plate-wire memory units. Combined projected costs for the guidance, control, and sequencing computer and the data-storage memory in 1971 was $6.1 million, with a ceiling cost of $6.8 million.

A preliminary design review for the computer was held on schedule in April 1972. At this review, plans called for development testing to be completed by December 1972, following which Martin Marietta and NASA personnel would hold the critical design review. Because of difficulties in fabricating the sense-digit transformers, the plated wire, the memory tunnels, and memory planes, the critical design review was rescheduled for March 1973. As problems continued with component deliveries and memory fabrication, the date for the review was slipped several times. Finally held in August 1973, the critical design review indicated that the design was acceptable in theory, but more development tests were required. Because Martin Marietta needed early delivery of the computers to keep lander fabrication and testing on schedule, Honeywell had to proceed with a parallel program in which the development units and the flight-style computers were built at the same time.

Throughout 1973, Honeywell had difficulties with the plated-wire memories. Engineers could not produce sufficient quantities of the plated wire with the proper magnetic characteristics, and they had problems in fabricating the matlike tunnel structures in which the wire was manually inserted. The magnetic keepers applied to the exterior of the sandwiched memory planes also became troublesome for Honeywell. The subcontractor faced another setback when faulty plated-through holes were found in many of the printed circuit boards, which had been purchased from a commercial supplier. These 18- x 23-centimeter Honeywell-designed circuit boards had up to 16 layers of circuits and 3000 plated-through holes for making electrical connections. A great many of the original circuit boards had to be scrapped and reordered from another supplier.[5]

The various problems with the guidance, control, and sequencing computer led NASA Administrator Fletcher to ask Honeywell Vice President Anderson in for a serious talk. Anderson had previously met with Fletcher on 15 February 1973, and subsequently Deputy Administrator Low told Fletcher that some of the computer problems had apparently been solved as a result. But, still unhappy, the administrator wrote again: "During our meeting I was . . . disturbed by the inference in one of your remarks, that Honeywell is unable to put forth its best efforts on this job because of the type of contract. . . . I hope I was mistaken in my impression that this is so, and I trust that Honeywell will fully live up to all of its obligations." Only Fletcher could talk this firmly to corporation executives. Jim Martin, for all his crustiness, was not in such an authoritative position.[6]

Anderson responded in the positive manner Fletcher was seeking: "In spite of my comments of philosophical concern, I had hoped to have left you with the conviction that Honeywell was applying the best of its resources in a prudent and expeditious fashion. I believe it would be agreed by both the people from NASA and the Martin Company that we are going to find solutions to problems."[7] However—despite all the efforts of the agency, Martin Marietta, and Honeywell—the guidance, control, and sequencing computer, which first made the top 10 list on 21 July 1972, was not removed from the chart until 15 January 1975.

By late 1973, Honeywell had exceeded the $6.8-million ceiling by nearly $3.5 million. Working under a fixed-price contract, the contractor had no profit incentive to improve the situation. Martin Marietta took several steps, at NASA's urging, to improve the Honeywell operation. The contract was changed from a fixed-price to a cost-plus-incentive-fee contract, with Honeywell accepting a $3.5-million loss. The project was rescheduled, and its cost reestimated to $24 million. Honeywell doubled the number of employees assigned to the computer, as special teams worked to solve specific problems and expedite production. Alternatives to the two-mil plated-wire memory were also examined. While the engineers in Florida attacked these problems, Martin Marietta began a backup program to develop an alternative memory system.[8]

Also in late 1973, Martin Marietta established a resident management office with a staff of 20 at Honeywell's Saint Petersburg factory. This team, and NASA personnel from Langley, assisted Honeywell's managers with scheduling, managerial, and engineering tasks. Everyone was clearly concerned about the fate of the computer. Langley Director Ed Cortright told Deputy Administrator Low on 30 October 1973 that a meeting with top computer industry experts indicated that Honeywell's problems were not unique. "It appears that the major difficulty is one of schedule time available to reliably produce and test computers needed to support the building of flight" landers. Money might buy more people, but neither people nor dollars could purchase more time.[9]

In January 1974, the Viking Project Office team decided to change all the flight-model computer memories to a new two-mil plated wire known as "coupled-film wire" because it had a second magnetic layer. It was easier to produce, had higher electrical output, was less subject to mechanical damage, and was affected less by temperature changes. Honeywell became more optimistic about meeting schedules. The first flight-model computer was delivered to Martin Marietta in April 1974, nine months late according to the original schedule. Although this proof-test-capsule model had a great many deficiencies, it did permit Martin Marietta to go ahead with its lander tests. Unhappily, delivery schedules continued to slip during 1974 as Honeywell faced more technical difficulties. Faulty components were uncovered. One lot of transistors was rejected. More unsatisfactory printed circuit boards came to light.[10]

Continuous monitoring of the subcontractor's troubles was rewarded, however, in late 1974 when the computers were finally ready for delivery. On 15 January, Jim Martin received the following message from Walt Lowrie at Martin Marietta:

> Oh ye of little faith—We gave birth to the last computer today. I don't know how you feel on the subject but it would appear to me that this top ten has now died of old age.
> Seriously—although the path has been extremely tortuous I really feel we now have an excellent computer on Viking.[11]

Martin removed the lander computer from his list of major problems. Thus it went, step by step—problems identified and then solved. At this stage in Viking's existence, there was very little glamor, just long hours, hard work, and an occasional antacid.

## TESTING THE LANDER

Another phase of the lander's evolution was the multiplicity of tests to which the components, subassemblies, and assemblies were subjected.

The guidance, control, and sequencing computer was the Viking lander's brain. At right, magnetic wires as fine as human hair are inserted into the computer at Honeywell Aerospace, Saint Petersburg, Florida. In testing below, the HDC-402 computer, part of the lander's computer, looks like pages of a book. At bottom, Jim Martin (second from right) on 10 January 1975 congratulates Barton Geer (left), director of system engineering and operations at Langley; R. Wigley, Honeywell's Viking program manager; and F. X. Carey, Martin Marietta resident manager at Honeywell. GCSC flight article 2 and the qualification unit are in the foreground.

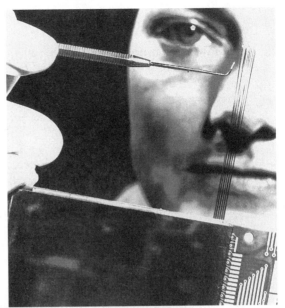

Again, this was not terribly exciting work, but it was essential to producing spacecraft that could be relied on to function far from Earth.

As with the Viking orbiter, a number of simulators were developed to verify analytical predictions of lander system performance, to investigate the effects of the thermal and dynamic environment on the craft, and to permit tests of subsystems, such as the scientific experiments. The major Viking simulators included:

*Lander (structural) dynamic-test model (LDTM).* A flight-style structure with partial flight-style or equivalent propulsion lines and tanks. Mass (weight) simulators were used for nonstructural hardware. The LDTM was used for structural vibration, acoustic noise, separation tests, and pyrotechnic shock evaluation.

*Lander (structural static-test model (LSTM).* A flight-style structure used for qualification of the primary structure under steady-state and low-frequency loads.

*Orbiter thermal effects simulator (OTES).* A simulator used to study the orbiter's thermal and shadowing effects during the lander-development thermal environmental tests.

*Proof-test capsule (PTC).* A complete Viking lander capsule assembly assembled from flight-style hardware, used for system-level qualification.

*Structural landing test model (SLTM).* A ⅜, geometrically scaled model of the lander, dropped at various velocities and attitudes to determine landing stability boundaries. The ⅜ scale was chosen because the Martian gravity was ⅜ that of Earth's.

*Thermal-effects test model (TETM).* A full-scale model incorporating developmental thermal control systems and flight cabling test harness. Flight equipment thermal effects were simulated by special equipment. The TETM was used to verify the system developed for controlling the temperature of the lander.

*Electrical thermoelectric generators (ETGs).* Generators used in testing in place of the radioisotope thermoelectric generators (RTGs). ETGs had electrical heating elements that simulated the electrical and thermal characteristics without the hazard of nuclear radiation.[12]

There were three broad categories of tests: system development, qualification, and flight acceptance. Development tests determined the levels of performance that components and subassemblies would have to meet to be acceptable. They also provided early identification of design deficiencies. These trials used primarily the dynamic-test model, the orbiter thermal-effects simulator, and the thermal-effects test model. Qualification tests used hardware attached to the proof-test capsule and the static-test model. During the "qual tests," hardware was subjected to stresses and environmental conditions that exceeded any expected during the mission. Environmental tests included heat compatibility, acoustic noise, launch sinewave vibration, landing shock (drop test), pyrotechnic shock, solar vacuum, and Mars-surface simulation. These and additional tests were performed at the component and subsystem level. The flight acceptance tests were performed on flight hardware before qualification testing. Only the thermal sterilization and solar vacuum tests were made with assembled flight landers.

### Environmental Tests

The proof-test capsule, encapsulated in its bioshield, was subjected to the heat compatibility test to verify that the system could withstand heat sterilization. During this test, the chamber atmosphere consisted of dry nitrogen containing about three percent oxygen and other gases. The capsule was subjected to 50 hours of 121°C heat, and flight landers were exposed to 112°C for 40 hours. Components were subjected to five 40-hour cycles and three 54-hour cycles at 121°C.

The vibrations of liftoff were computed by analysis of data from earlier Titan-Centaur flights and the February 1974 proof flight of a Titan IIIE–Centaur D-1T launch vehicle (this flight and preparation of the Viking launch vehicle are discussed in appendix E). Despite the necessary destruction of the Centaur stage on this flight after its main engine failed to start, some information was gained to help define the ground-based simulations (launch sinewave vibration tests) of the low-frequency vibrations encountered during launch, stage separation, and spacecraft separation. Through combined analysis, flight-derived data, and simulations, the engineers were able to determine if the lander components could withstand the predicted vibrations.[13] The acoustic noise test simulated the effects of the sounds of powered flight. Levels of the individual components were determined by earlier tests using the lander dynamic-test model and proof-test capsule.[14]

Random vibration tests were applied only at the component level, to screen out faulty workmanship and design defects. Laboratory simulations of the levels of vibration encountered during actual flight proved not to produce satisfactory data. Borrowing from procedures devised during the Apollo program, the vibration levels were raised to a level that would screen out bad components but not damage good ones. Component vibration levels were the same for both qualification and flight acceptance testing, but the latter was shorter so that multiple tests could be run without exceeding the qualification test levels. In the pyrotechnic shock tests run at the system level, a series of pyrotechnic devices was fired to simulate the effects of actual mission events and at the same time demonstrate the actual performance of the pyrotechnically actuated mechanisms. Components were subjected to vibrations similar to those expected with the Viking pyrotechnic devices and to contained explosions that replicated the impact of explosions and gas pressure buildups on specific assemblies.[15]

Solar vacuum tests, held in a nearly complete vacuum in Martin Marietta's test chamber (4.5 meters in diameter and 20 meters high), simulated the worst predictions for thermal heating and cooling during the flight to Mars. Both the effects of heating and cooling and the performance of the lander's thermal control system were evaluated. Each mission phase was completed twice for the qualification tests of the proof-test capsule and once for the flight acceptance tests of the flight landers.[16]

Deorbit-entry-landing thermal simulation tests, conducted on a component level, duplicated the effects of entering the Martian atmosphere—pressure increase, entry heating, and the post-landing cooldown. Components were placed in the vacuum test chamber at 1/760 of an Earth atmosphere, heated to a temperature of 149°C, and held there for 530 seconds. Chamber pressure was then raised to 5/760 of an Earth atmosphere with cooled nitrogen gas, to provide an atmospheric temperature of -101°C. In this manner, the lander's passage through the Martian atmosphere with the attendant heating and cooling was duplicated. The change of 250°C represented the wide range of temperatures that the lander would be exposed to on Mars. Such extremes were part of the reason the engineering of the lander had been such a complicated task. For all components, the most critical period would be the 15 to 20 minutes after landing, since by that time all equipment would be operating and the entry heat buildup would not have had time to dissipate.

In the landing shock tests, the proof-test capsule, with landing gear extended, was dropped from a height necessary to achieve a velocity of 3.36 meters per second on impact. Each drop produced the worst possible dynamic loads on a different landing leg and footpad. In addition to these drop tests, the shock stresses generated by the opening of the parachute were evaluated analytically and then measured during the balloon drop tests (balloon-launched decelerator tests) at the NASA White Sands Test Facility in New Mexico in the summer of 1972. They were carried out successfully despite postponements caused by uncooperative weather. As a consequence of these tests, new techniques were developed to unfurl the parachute progressively, minimizing the deployment shocks to the lander.[17]

During the Mars-surface simulation tests, the lander configuration of the proof-test capsule was subjected to thermal conditions worse than those expected on the surface of Mars. By subjecting the lander to different conditions and varying the vehicles' internal electrical power, three basic tests were performed—hot extreme, cold extreme, and the predicted norm.[18] In consultation with the Science Steering Group, the test engineers chose argon for the chamber atmosphere during the cold extreme, because preliminary data from the Soviet Mars probes had indicated that as much as 30 percent of the planet's atmosphere might be composed of this rare gas.* Since argon promotes electrical corona and arcing in electronic components, the test teams were to determine whether there would be any adverse effects on lander subassemblies if the concentration of argon was that high.

## Science End-to-End Test

One of the most significant activities during the lander testing cycle was the science end-to-end test (SEET), conducted during the Martian-

---

*Subsequent Viking data indicated that the argon content in the Martian atmosphere was only about 1.5 percent.

*Table 45*
*Mars Surface Thermal Simulation*

| Parameter | Hot Extreme | Cold Extreme | Nominal |
|---|---|---|---|
| Shroud temperature | -129°C ± 5°C | -151°C to -81°C | -112°C ± 5°C |
| Chamber pressure and atmosphere | 2 ± 0.1 mb, $CO_2$ | 35 ± 2 mb, Argon | 4 ± 0.1 mb, $CO_2$ |
| Solar radiation | 1078 ± 47 watts/m² | 0 watts/m² | 539 ± 31.5 watts/m² |
| Solar duration | 12.33 hrs. | None | 12.33 hrs. |
| Vehicle power | 1395 watt-hrs | 1345 watt-hrs | 1371 watt-hrs |
| Ground simulator | -46°C to +24°C | Uncontrolled | Nominal |
| Thermal coating | Degraded | Original | Original |
| ETG thermal output | Maximum (680 ± 2 watts) | Minimum (630 ± 2 watts) | Nominal (673 ± 2 watts) |
| Test duration (PTC) | 3 days | 4 days | 3 days before hot extreme; 3 days before cold extreme |

surface simulations at Martin Marietta. The two major SEET objectives were "to verify the adequacy of the implementation of the scientific investigations from sampler collection to interpretation of resulting data by the scientists" and to "familiarize the Viking scientists and other flight operations personnel with *total operation of the investigations* and their respective characteristics." In the course of carrying out these basic objectives, any hardware or procedural problems were to be resolved, to avoid similar difficulties during the actual mission.[19]

Getting the science end-to-end test started took some effort. It was postponed several times because of problems with the motor used to load samples into the oven heating assembly of the gas chromatograph–mass spectrometer. When the pumpdown of the vacuum test chamber began on 17 September 1974, the proof-test capsule lander used in the operation had a GCMS simulator aboard instead of the actual test unit. SEET was also run without the biology instrument. Despite the absence of these two major experiments, the test was useful.

The lander systems were examined rigorously. During the thermal vacuum chamber operations, a Martin Marietta computer facility sent commands via cable to the guidance, control, and sequencing computers. The plated-wire memory, once a leading top 10 problem, performed very well in the simulated Martian atmosphere. In addition, JPL processed data recorded on computer tapes from lander subsystems much as data would be during the real mission. Tests of the ultrahigh-frequency (UHF) radio link

*Viking simulators went through intensive environment tests to ensure the final spacecraft would function far from Earth. Above left, Viking program technician Alonzo McCann adjusts a cable on the proof-test capsule-decelerator assembly, as lander and capsule are prepared for January 1974 heat-verification tests. One of the lander cameras is to the right of center. Above right, a technician watches as an acoustic shroud is lowered over the proof-test capsule before acoustic tests in mid-June 1974. At lower right, the proof-test capsule is lifted out of the vacuum chamber at Martin Marietta, Denver, in October 1974 after a month-long series of rigorous tests to qualify it for operations on Mars.*

for data transmission and the lander tape recorder also indicated that those systems were ready for flight.

Other subsystems were given a thorough examination: the surface sampler, the lander's imaging system, the weather sensors, the x-ray fluorescence spectrometer, the seismometer, and the biology sample processor.

The multipurpose surface sampler (boom-and-scoop assembly) successfully delivered soil samples to the x-ray spectrometer and biology processor unit and to the GCMS position. The only significant problem occurred when the sampler arm snagged on the holder of a brush used to clean a magnet on the magnetic properties experiment. This problem was cleared up by minor hardware modifications and a new mission rule that prohibited cleaning the magnets until all of the biology samples had been taken. The lander facsimile cameras made nearly 100 images, including pictures of trenching exercises with the backhoe and of particles adhering to the magnets. The meteorology instrument performed well in Marslike conditions that could not be duplicated in a standard wind tunnel. Although the biology instrument was not on board, the processor containing the screens and cavities for the measurement and separation of the materials scooped up by the surface sampler was tested and proved satisfactory.[20] During the seven days (18–23 September) that it took to simulate five days of experiments on the surface of Mars, many important lessons were learned as procedural and hardware "glitches" were encountered and overcome, and much needed experience was gained with the meteorology, seismology, camera, x-ray fluorescence spectrometer, and magnetic properties experiments.[21]

Priestley Toulmin, team leader for the inorganic chemical investigation (x-ray fluorescence spectrometer) had been uncertain about the merits of SEET as it was planned, however. Toulmin's experiment, a late addition to the lander science payload, would determine the nature of inorganic compounds (minerals) in the Martian soil. As early as 1968, the Space Science Board had suggested it in recommendations to NASA for planetary explorations. But the priority given inorganic analysis was much lower than that assigned the search for biologically derived compounds—although, with the exception of this experiment, the original payload for Viking had followed the board's suggestions closely. Information gathered from the lunar samples returned by Apollo astronauts and early *Mariner 9* results suggested the need to reconsider the utility of inorganic analysis. Mariner 71's findings were particularly evocative because they indicated that Mars was geologically younger and more active than had been expected. As a result, in the fall of 1971 the space science community lobbied the NASA management, especially John Naugle, associate administrator for space science and applications, to include an inorganic experiment on the lander. Of two possible investigations, the one designed by Martin Marietta and the team led by Pete Toulmin was selected. (The other instrument, designed by a team led by Anthony L. Turkevich at the Enrico Fermi Institute of the University of Chicago, had been under development for a longer time, but the XRFS was expected to cost less, be lighter, and require less space and power.)[22]

As time for SEET approached, Toulmin was concerned about the manner in which it would be conducted. Both he and Klaus Biemann, team leader for the molecular organic analysis (the GCMS), had insisted strongly on the inclusion of "blind" samples in the analyses to be done by their instruments. These materials, unknown to the teams, would be identified by the results of the experiments, to simulate the interpretative work of the actual mission. In addition to making certain that this aspect of the SEET experience was carried out, Toulmin told Jerry Soffen in early September 1974 that he was concerned about the validity of the trials since the x-ray fluorescence spectrometer to be used in the test was different from the actual flight article. The test version had several shortcomings that had already been corrected in the flight units. A final reservation centered on the seeming inflexibility of the test plans.[23]

By the end of the science end-to-end test, however, Toulmin believed in its worth. He had previously discussed with Jerry Soffen "some reservations and qualifications the Inorganic Chemical Investigations Team felt were applicable to that program." In most instances, Toulmin believed that "the events proved us correct in our concerns regarding the state of the hardware, the software, and ourselves" and they had predicted several of the breakdowns that occurred. But in one major respect Toulmin felt he and his colleagues had misjudged the testing program: "I . . . grossly underestimated the tremendous value of the experience for those who participated in it. We learned things about the operation of the instrument and its relations with the rest of the lander, and about the recognition, diagnosis, and correction of problems and malfunctions that we would never have learned by any other method." Although the actual mission would differ greatly from the simulations, "it was an invaluable introduction to a whole new world." In his report to Martin, Toulmin singled out "for special mention the three unflappable controllers of the SEET data room: Henry von Struve, Frank Hitz, and Ron Frank."[24]

Phase B of the science end-to-end test was less satisfactory. Begun on 7 October with the reworked gas chromatograph–mass spectrometer, it had to be terminated on the 10th when additional problems were encountered with that instrument. These difficulties led to a special test of the GCMS in conjunction with the biology instrument's performance verification test in February 1975. Despite some additional functional difficulties, Klaus Biemann was able to identify from the GCMS data tapes the five compounds in the blind samples.

Whereas the mass spectrometer went through the end-to-end functional and operational exercise, the biology instrument did not. The biology instruments were delivered too late for proper testing. By the time the hardware became available, limited time, money, and manpower argued against the thorough test. To questions about the adequacy of the functional testing of the hardware on the proof-test capsule lander in Martin Marietta's thermal vacuum chamber and the biological operation of the experiments, Cal Broome told Martin on 30 June 1975, less than two months before liftoff, "The current planning assumes that the testing already accomplished is adequate, i.e., the combination of Biology [performance verification] at the lander level (instrument 103) and soil biology at the instrument level (instrument 102 and 103) is adequate to provide assurance of proper operation on Mars." He added, "There is no question that this program does not provide ultimate verification, i.e., operation of a flight instrument in a lander with real flight sequences and verification of proper results," but said, "Our position has been that risk of the current approach is acceptable." Broome was responding to a NASA Office of Space Science inquiry about the possibility of conducting a biology end-to-end test after the Viking spacecraft had been launched.[25]

Four major factors influenced the scope of the biology instrument acceptance test program. One, the introduction of soil or experiment nutrient into an instrument would render it unusable for flight. Cleaning the instrument was impossible without destructive disassembly. Thus, the functions of the flight instruments (S/N 104, 105, and 106) had to be tested only by simulating their operations on Mars. Soil testing was necessarily limited to components and units not reserved for flight use. Two, the complexity of the instruments, the multiplicity of their functions, and the operational pace (one minute between commands) meant that complete functional tests would be extremely time-consuming. The minimum time required for an entire end-to-end electrical and pneumatic checkout of a biology instrument was one month on a round-the-clock schedule. Only abbreviated functional tests could be performed. Three, given the long turnaround time required to repair and retest instruments if a component failed, all components and subassemblies had to be tested before assembly in the integrated instrument, where accessibility was a problem. And four, a substantial number of design changes were incorporated into the flight units after the manufacture and test of the qualification unit (S/N 102), requiring additional qual tests. Functional tests were then carried out to ensure that the flight instruments had not been harmed by the qualification test stress levels.

Each flight version of the biology instrument was subjected to a sequence of acceptance tests: operational system checkout, vibration test, functional verification, thermal verification, sterilization, and operational system checkout. The operational checks were computer-controlled, testing the electrical functioning of the instrument. Mechanical and structural quality was verified through vibration tests, while the functional verification tests were complete validations of all instrument systems. Computer-controlled electrical and pneumatic sequences assessed individually the functioning of each critical component or subassembly. The thermal verification tests were performed with the biology instrument in a Marslike atmosphere of carbon dioxide through a temperature range of $-18°$ to $30°C$. Instruments were sterilized in a biologically filtered nitrogen atmosphere at

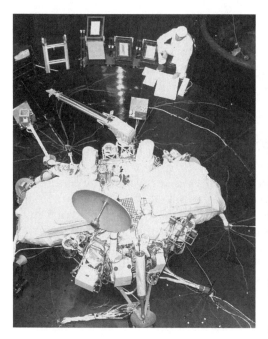

*Science end-to-end tests sought to verify complete performance of the Viking scientific instruments and familiarize scientists and flight operators with the total operation of Mars investigations. Above, a technician prepares the proof-test-capsule lander for the environment and SEET tests. At right, sample boxes are positioned for testing the lander's surface-sampler assembly.*

120°C for 54 hours. The total acceptance test spanned three to five months, depending on problems encountered during the process.[26]

Although this was a busy test schedule, no flight-model biology instrument had been tested as part of the total lander system, and in the fall of 1975 Harold P. Klein, leader of the biology team, and his colleagues argued for such a test. Langley and headquarters personnel resisted any lengthy additional testing. Such an examination could not take place before January 1976 and would interrupt a number of schedules. In late September, the Viking Project Office proposed a committee led by Gary Bowman, biology instrument team engineer, to take an in-depth look at the biology instrument test data from a lander systems point of view. From the team review, areas of specific concern could be identified and a decision about additional tests made.[27]

Klein responded on behalf of his teammates in November after Bowman's group and the biology team had looked at the testing issue again.

Ideally, the biology team would have liked to install the flight-model S/N 104 biology instrument in the proof-test capsule at Denver, to make biological examinations of soil samples, but the S/N 104 unit had to be kept sterile until after the mission, when additional tests might be necessary.

What the biology team could do was install the proof-test-capsule unit (S/N 103) on the proof-test-capsule lander to observe real data being processed from the biology instrument detectors through the lander system. The biology instrument simulator was not similar enough to the flight hardware to provide a meaningful test of the lander–biology instrument interface, but the test could simulate the sequence of biology instrument operations from soil collection through processing, analysis, and data return. Not only would experimenters have a chance to see if the instrument would function as planned, but they could watch their hardware in action, in preparation for the days when the instruments would be operated on Mars.[28]

Jim Martin and his staff on 25 November 1975 decided at least part of the tests the biology team wanted could be carried out during the flight-operations-software verification tests scheduled for the proof-test-capsule lander in February 1976. Only the tests that would not require extra funds could be done. Martin told Klein: "We have neither the dollars to extend the test nor the people to analyze the data." Other aspects of the biologists' plans for testing were likewise impossible:

> . . . your request for lander/biology tests with transmitters/antennae in real operational modes is also difficult to accommodate. As you know, this test would require use of an anechoic chamber (very expensive) or moving the entire lander to an outdoor location to avoid RF reflections (also expensive). We made a fundamental decision in 1973/1974 that the lander [electromagnetic compatibility] test program had to proceed without a real biology instrument because such an instrument did not exist until much too late. Instead, we have relied upon the positive results of a rigorous EMC test on the instrument at TRW. In today's dollar limited environment, the dollars to plan, set up, and conduct another radiated EMC test for biology are prohibitive. We must rely on analysis and instrument level test experience.[29]

While not enthusiastic about any additional biology testing, Martin informed Noel Hinners at NASA Headquarters that the "potential return from [the partial testing he had agreed to] is sufficient to incorporate it into our plans." He believed that the project management had "done everything reasonable to satisfy the concerns of the Biology Team as to the adequacy of the pre-landing test proram." Martin wanted to turn to other more important issues: "Following the test, we must and will devote the full biology flight team resources to preparation for landed operations, . . . including training contingency analysis and preparation of pre-canned sequences to be ready for the multiplicity of possible required reactions to data from Mars."[30]

## REORGANIZATIONS AND ADDITIONAL CUTBACKS

During the remaining year and a half before the Viking launches, a number of changes were made in the top management structure at NASA. The first of these was announced by Administrator Fletcher on 5 March 1974. Rocco A. Petrone, director of the Marshall Space Flight Center, was appointed NASA associate administrator, the number three position at headquarters, replacing Homer Newell who retired in late 1973. John E. Naugle, named Petrone's deputy, continued to act as associate administrator for space science until Noel W. Hinners, director of lunar programs in the Office of Space Science,* was selected in June to fill the space science slot. When Petrone left NASA for a job in industry in April 1975, Naugle assumed his duties on an acting basis until 23 November, when he was appointed to that position.

Fletcher in March 1974 also announced a headquarters reorganization, with two primary objectives. First, he sought to consolidate under one senior official, the associate administrator, the planning and direction of all NASA's research and development programs. And second, by creating a new position—associate administrator for center operations, to whom the center directors would report—the administrator funneled the responsibility for the field centers to one office. George Low, deputy administrator, temporarily took on this new task until Edwin C. Kilgore was appointed in May 1974.

Fletcher stressed that the changes were necessary in this era of consolidation, an era of tightening budgets and reducing manpower levels.

> As we approach the time when the Space Shuttle becomes operational, there needs to be a mechanism for the orderly phaseover from conventional launch vehicles to the shuttle; at the same time we need to take an innovative and coordinated approach in planning and developing all of our future payloads—manned and unmanned, science, applications, and technology. Our aim is to achieve this consolidation of all Aeronautics and Space Activities through the office of the Associate Administrator.

NASA's administrator believed that the future of the agency's activities depended entirely upon the strength "of NASA's most important resource—the 25,000 people located primarily at our field centers." This figure was down from a peak of nearly 36 000 in fiscal year 1967.[31]

Petrone and Hinners had the unenviable task of keeping Viking project costs from escalating further. When Petrone assumed his responsibili-

---

*In December 1971, a reorganization set up an Office of Space Science and an Office of Applications, replacing the Office of Space Science and Applications.

ties as associate administrator in March 1974, the projected completion cost of Viking had risen to $927.5 million, and nearly all of the cost problem was associated with the lander—the biology instrument, the gas chromatograph–mass spectrometer, and the guidance, control, and sequencing computer were among the leading troublemakers. As table 46 illustrates, the price of the orbiter was repeatedly pared to help pay for the lander. Money for support activities was held relatively constant. Actual costs for the orbiter and support activities were below the June–July 1970 estimates, but the lander was costing nearly $200 million more than it was projected to in 1970.[32]

*Table 46*
*Viking Cost Projections, 1974*

(in millions)

| Date | Lander | Orbiter | Support | Total Estimated Cost at Completion (Estimated Total + APA[a]) | Cumulative Total |
|------|--------|---------|---------|-----------------------------------|------------------|
| July 1970 baseline | $359.8 | $256.0 | $134.2 | $750.0 + $80.0 = $830.0 | $ 51.0 |
| Dec. 1970 | 359.8 | 256.0 | 134.2 | 750.0 + 80.0 = 830.0 | 54.5 |
| June 1971 | 358.0 | 256.3 | 135.7 | 750.0 + 80.0 = 830.0 | 81.8 |
| Jan. 1972 | 384.6 | 256.7 | 143.7 | 785.0 + 44.7 = 829.7 | 150.6 |
| June 1972 | 414.4 | 252.3 | 134.5 | 801.2 + 28.2 = 829.4 | 223.8 |
| Dec. 1972 | 426.1 | 251.3 | 132.2 | 809.6 + 19.8 = 829.4 | 366.6 |
| June 1973 | 436.2 | 247.5 | 143.0 | 826.7 + 11.3 = 838.0 | 466.5 |
| Dec. 1973 | 456.7 | 241.0 | 140.3 | 838.0 + 0.0 = 838.0 | 595.2 |
| Mar. 1974 | 511.9 | 242.4 | 140.2 | 894.5 + 33.0 = 927.5 | 646.7 |
| Apr. 1974 | 518.2 | 242.8 | 140.2 | 901.2 + 18.8 = 920.0 | 667.9 |
| Dec. 1974 | 545.2 | 242.1 | 139.1 | 926.4 + 3.6 = 930.0 | 805.2 |
| July 1975 | 548.7 | 243.0 | 138.0 | 926.2 + 3.5 = 929.7 | 855.2 |
| July 1976 | 558.2 | 243.0 | 134.1 | 935.3 + 0.3 = 935.6 | 898.9 |
| Jan. 1977 actual costs | 558.2 | 240.5 | 115.8 | 972.4[b] | 914.5 |

[a]Allowance for program adjustment (APA), or reserve funds.
[b]Estimate through end of prime mission.

In October 1974, Petrone and Hinners tightened the purse strings considerably. Viking budget ceilings were established for fiscal 1975 and 1976, and deviation from these amounts required Petrone's personal approval. Before any increase in the budget would be permitted, Petrone wanted to see documented evidence of steps taken to squeeze the dollars from elsewhere in the Viking budget. The reserve funds (allowance for program adjustments) were directly controlled by Petrone. Hinner's staff provided Petrone with weekly status reports on project costs and manpower levels for Martin Marietta, JPL, TRW, and Honeywell throughout the winter of 1974.[33]

Two important management changes also took place at the centers during the summer of 1975. At Langley in September, Ed Cortright, after 27 years of government service, retired and entered private industry and also served as president of the American Institute of Aeronautics and Astronautics. He was replaced by Donald P. Hearth, who since leaving the Lunar and Planetary Programs Office at NASA Headquarters in 1970 had been deputy director of the Goddard Space Flight Center in Greenbelt, Maryland. On the West Coast at the Jet Propulsion Laboratory, Bruce Murray had been appointed in April to succeed William Pickering, who was retiring after having led the laboratory since 1954. Hearth and Murray were old Mars program men. Occasionally they had disagreed over budget, manpower, and managerial issues during Mariner and the early years of Viking, but they would cooperate on the team that would launch, fly, and land the Viking spacecraft. Present from nearly the beginning of the search for life on Mars, Hearth and Murray would see the fruition of years of work from the inner NASA circle.[34]

In September 1974 when the second flight orbiter was canceled and the proof-test orbiter converted to a flight article, the third lander, the backup, was also terminated. By this move, Petrone and Hinners hoped to save an additional $9 million. As the project moved closer and closer to the billion-dollar mark, members of Congress had told NASA that no further reprogramming of funds, like shifting $40 million of the fiscal 1974 budget to Viking, would be allowed. In an across-the-board cost-reduction exercise, Jim Martin's project office searched for ways to save dollars to cover the expense of such items as the biology instrument and the GCMS.[35]

Three landers had been planned originally to ensure that at least two would be ready for launch. Had one of the prime landers suffered a last minute problem that required a violation of sterilization procedures and then reassembly and resterilization, the backup could have been used. With this third lander gone, only parts would be available for substitution should either flight lander have preflight troubles. The need for a backup orbiter had never been as critical as for the lander, since the orbiters did not have to go through the subassembly and completed assembly rigors of sterilization. Resterilization of either lander would have required precious time during the 65-day Viking launch window. The process at the Cape would require about 5 days, although only about 48 hours would actually be spent in the oven at microbe-killing temperatures.

If the first lander should fail at the time of launch, the second lander could replace the first with a minimum of lost time. If difficulties occurred during the second launch, however, it could take up to 27 days to remove the lander from its sterile capsule, disassemble it, find the malfunction, repair it, reassemble the lander, and then resterilize it. Under such a contingency, Martin and his people believed that they could carry out the work and still launch the second craft in time; it would be tight, but if the lander was repairable they thought they could get it on its way.[36]

270

## PREPARING FOR LAUNCH

The first Viking flight hardware arrived at the Kennedy Space Center (KSC) during November and December 1974. This material included the Titan IIIE core vehicle (liquid-fueled rocket stage), the solid-fueled rocket motor components (strap-on booster stages), and the Centaur upper stage.* All of the elements were as close to flight configuration as practical when delivered, so that the major tasks remaining were only assembly and testing. The Centaur standard shroud, the "nose cone" that protected the orbiter and lander during ascent through Earth's atmosphere, was delivered ready for the addition of such bolt-on items as electrical harnesses, instrumentation, and insulation. Upon delivery, launch vehicle B, which would be used for the first mission, was prepared for the mating tests scheduled for April 1975.

Viking lander capsule 1 arrived at the Cape on 4 January 1975, and engineers made a detailed inspection and subjected the capsule and lander to a series of verification tests, which included compatibility checks between the S-band radios and the Deep Space Network. Last minute modifications followed, based on the test information, after which the radioisotope thermoelectric generators were installed and the lander system was finally built up for mating tests. Meanwhile, the first Viking orbiter arrived on 11 February and was put through the same rigorous verification tests.

Up to this point, the flight lander and orbiter had never been physically or electrically in direct contact, having been assembled over 1600 kilometers

---

*The Titan IIIE core vehicle was shipped by C-5A aircraft from Denver, where it had been manufactured by Martin Marietta. The Centaur stage, built by General Dynamics Convair Division, was also flown to Florida on a C-5A from the factory in San Diego. United Technologies Chemical Systems Division shipped the solid rocket motors from Sunnyvale, California, by rail.

*Work progresses on the Viking lander 1 (foreground) and 2 at the Martin Marietta plant in Denver in the fall of 1974.*

apart. Viking orbiter 1 and Viking lander capsule 1 were mated for the first time on 8 March. More than two weeks of interface and system testing indicated that they would work together satisfactorily. The next hurdle was encapsulating the orbiter-lander assembly inside the Centaur shroud on 27 March. The specialists in Florida would then run some additional tests before the whole unit was moved to launch complex 41 where the Titan IIIE stood assembled. After the assembly had been hoisted and mated to the launch vehicle on 31 March, another series of tests were carried out on this 48.5-meter-high stack of hardware. A flight events demonstration, Viking orbiter precount, Viking lander prelaunch, and terminal countdown—all were completed successfully.

After mating tests, the orbiter and lander were removed from the launch vehicle and returned to the assembly facility for flight compatibility tests. The Viking flight team monitored these examinations from the Viking mission control and computing center in building 230. The Deep Space Network provided communications for telemetry and spacecraft commands. Concurrently, the second orbiter and lander were going through the checkout process so successfully that it became feasible to use either of the two craft for the first launch. This additional capability gave Jim Martin and his people a dose of extra confidence.

As work on the hardware moved along according to schedule, the men who would control and command the craft during the flight were also simulating mission activities. Members of the orbiter performance and analysis group participated in seven separate tests during April. For each activity through launch, the group had at least one test exercise that would prepare them for the real thing. The flight path analysis group simulated a midcourse maneuver exercise on 14 April, and the results were so successful that a repeat exercise was canceled.[37]

May was an equally active month at Kennedy, with some occasional troubles. Grounded circuitry delayed for two days the important plugs-out test (during which the spacecraft was on internal power) of Viking lander capsule 1, and some communications problems between ground data system and the Deep Space Network required additional tests. Orbiter performance and analysis group personnel experienced some difficulties with a computer program and had to reschedule orbiter simulations. Still, build-up and checkout of both Viking spacecraft were proceeding according to the latest schedules. All flight equipment, except for the gas chromatograph–mass spectrometer, had been installed on the first lander. Viking orbiter 1 was undergoing the system readiness test at the end of May, while installation of the high-gain antenna was begun on Viking orbiter 2.[38]

A lightning bolt that struck the Explosive Safe Area Building caused momentary excitement. Electrical charges from the strike induced currents that damaged two pressure transducers on the orbiter propulsion module S/N-005. After a quick review, the Viking managers decided not to fly this unit. Instead, S/N-006, being readied for the second launch, was assigned to the first spacecraft. Once again, the modular approach to building space-

272

*Viking orbiter 1, top left, is mated to Viking lander 1 at Kennedy Space Center on 8 March 1975. Above, technicians lower the launch shroud over the spacecraft on 27 March. At left, the shrouded orbiter and lander move toward 31 March mating with the Titan IIIE launch vehicle, for more tests.*

craft had paid off. To be able to substitute assemblies when required was clearly advantageous. Caution was a major element in preparing for a successful mission. Orbiter propulsion module S/N-005, its propellants unloaded, was refurbished as a spare. The previously designated backup was upgraded to flight unit status and assembled to the second orbiter. Buildup and checkout continued into June, interrupted now and then by thunderstorms and lightning alerts. To protect personnel and hardware, safety regulations at KSC stipulated that all activities had to be halted when a lightning alert was declared.[39]

A major milestone many people had worried about was passed when the first lander capsule (VLC-2) was successfully sterilized. Much of the trouble with the design, development, and testing of the lander subsystems had centered on building components that could withstand the high temperatures required to kill all terrestrial organisms. Eliminating microbes without degrading or destroying the hardware had been one of the major challenges of the project. Viking lander capsule 2 was placed in the sterilization chamber at Kennedy on 15 June. For more than 43 hours, the craft and its capsule were subjected to temperatures up to 116.2°C as heated nitrogen gas swirled around the hardware. The poststerilization short test verified that all subsystems were functioning properly. A number of minor glitches arose, but none proved to be a major concern.

Once the Viking management was assured of the first craft's good health, the second, VLC-1, was moved into the sterilization chamber for almost 50 hours. While lander 1 was readied for propellant loading, lander 2 and orbiter 2 were mated for a last time, officially becoming the Viking A spacecraft. By mid-July, the long process of designing, building, assembling, testing, and flight preparation was drawing to a close. The Viking A spacecraft was mated to its Titan launch vehicle on 28 July at launch complex 41. The 3500-kilogram spacecraft was ready to go to Mars. Preparations for Viking spacecraft B were proceeding for the second launch, while emphasis on personnel training increased during the last two months before the first liftoff.

System-level flight operations test and training continued with a series of verification tests. Verification test 3 on 12 June checked out the portion of the mission that included the launch of spacecraft B while spacecraft A was in its cruise phase. All the verification tests up to this point had been classified "short-loop"; their data—commands and the like—had been generated inside the Spaceflight Operations Facility at JPL. Beginning with verification test 4, data were exchanged between JPL and the tracking stations in Goldstone, California, and in Spain, test 4 verifying the design and execution of the spacecraft B midcourse maneuver. Verification test 1B was still more elaborate, and the loop was even longer. Simulating the launch portion of the Viking A mission, computers at the Kennedy Space Center generated data for the Viking Mission Operations Facility at JPL. Deep Space Station 42 at Tidbinbilla, Australia, also participated in this test, since it would be responsible for first communication with the spacecraft after launch. The launch phase of this simulation was normal, but trainers threw in a malfunction—an early cutoff of the Centaur engine—to test the reactions of the flight team. The team had to plan and execute an early emergency maneuver with the orbiter propulsion system to place the spacecraft on the proper trajectory to Mars. While no one really expected the Centaur upper stage to give any problems (it had been performing well for nearly a decade), the trainers wanted the flight team to prove its readiness for any contingency.

With these tests completed, the flight team was certified by the successful operational readiness test on 6 August.

*Table 47*
*Viking Demonstration and Training Tests*

| Date | Test | Nature and Results |
|------|------|--------------------|
| 2 July | DT-2 | Processing uplink commands to lander through orbiter for cruise checkout. Data processing went well, but flight team needed more training. |
| 13 July | DT-3 | Fifty-hour cruise operation test culminating mock midcourse maneuver. Working around the clock, flight team met several problems. Successful test. |
| 25–26 July | DT-1 | Three-part exercise. Part 1 covered spacecraft powerup through launch to 6 hours into mission. Part 2 covered midcourse maneuver. Part 3, conducted at request of Deep Space Network personnel, covered lander memory-readout sequence. All 3 parts successful. |
| 10 July | TT-1 | Simulation of midcourse maneuver with simulated emergencies. Not successful. |
| 28 July | TT-1 rerun | Successful retest of TT-1. |

SOURCE: R. D. Rinehart and H. Wright, "Daily KSC Status (FAX)," memos dated 23, 24, 25, and 26 June 1975; and VPO, "Mission Operations Status Bulletin," no. 7, 23 June 1975, and no. 8, 8 July 1975.

During the last week before liftoff, final preparations were made:

| | |
|------|------|
| 29 July | Orbiter precountdown checkout and lander cruise-mode monitoring tests completed. |
| 30–31 July | Lander computer prelaunch checkout. |
| 1 August | Composite electrical readiness test completed. |
| 2 August | Super Zip installed on Viking A shroud. (Super Zip is a linear explosive charge used to separate the clamshell halves of the shroud after launch.) |
| 3 August | Pyrotechnic ordnance devices installed on Viking A. |
| 6–7 August | Propellants loaded into Titan IIIE launch vehicle. |

Although a faulty valve and a battery discharge problem would delay the beginning of the journey to Mars by nine days, Viking was otherwise ready. Many had labored mightily to get the project to this point, and the adventure was about to begin. A great amount of work lay ahead of the Viking teams, however, before the landers could touch down on that distant, alien planet. One of the most important tasks, preparation for which had paralleled hardware development, was the selection and certification of scientifically valid but technologically safe landing sites on Mars. Before examination of the Martian environment could begin—and even while the Viking spacecraft headed out through space—many hours would be spent looking for safe havens for the two landers.[41]

# 9

# Safe Havens: Selecting Landing
# Sites for Viking

Since the basic goal of Viking was to conduct scientific experiments on the surface of Mars, the selection of landing sites was recognized early as a topic of major importance. Once the decision was made in November–December 1968 to make a soft landing from Mars orbit, the project engineers and scientists began a long colloquy in which they weighed the demands for lander safety (a crashed lander equaled no science) against the desire to land at locations most attractive scientifically. At first, the discussions were necessarily general in tone; the scientific knowledge (in terms of both physical data and visual images) was still very limited. *Mariner 4,* flying by the planet in July 1965, had yielded new information and the first extraterrestrial images, revealing a heavily cratered, moonlike surface. From *Mariner 4*'s perspective, the planet appeared to have eroded very little. Some scientists concluded that this meant there had not been much wind or water activity on the surface. Other scientists pointed out that *Mariner 4* had sampled only 1 percent of the Martian surface; they wanted to see the other 99 percent, and they wanted to see it more closely.

The Viking Project Science Steering Group began to consider the interplay between landing sites and Viking lander science during its first meeting in February 1969. *Mariner 4* had raised as many questions as it had answered, and data from Mariner 69 (*Mariner 6* and *7*), soon to be launched, would not be available until next year. Donald G. Rea, deputy director of planetary programs in the Office of Space Science at NASA Headquarters, during this first Science Steering Group meeting raised the landing site question when he asked for thoughts on how best to use the orbiter in support of the landed science program. Thomas Mutch, a geologist, began the discussion. The lander imaging team he headed had not considered landing site selection, since members thought orbital images were of little value in the site selection process. They assumed that orbital photographs would not be able to pick up geological features smaller than a football stadium (i.e., resolutions in the 100- to 1000-meter range). Ground-based scientists could not possibly see the lander or smaller scale hazards that could affect its safety, and Mutch's team did not believe that orbital pictures would help them pick either a good science site or a good landing spot.[1]

Wolf Vishniac of the biology team disagreed. Orbiter imaging could provide a valuable means for differentiating between places of low and high biological potential. He also believed that a possible strategy for selecting a landing site might be to set one craft down in a dark area and the other in a light area. The difference between Mutch's evaluation and that expressed by Vishniac was in itself illuminating. Mutch was thinking in terms of the small-scale features (measured in centimeters) that the lander would be able to see. Vishniac was basing his comments on the large-scale light and dark features observed through Earth-based telescopes. Between these two scales lay an unknown range of Martian topographical features that would mean the difference betweeen a safe landing and a crash.

The second steering group meeting, at Stanford University a month later, heard additional possibilities for using orbiter imaging in selecting a landing zone. On the large scale, Seymour Hess, the Viking meteorologist, expressed the hope that the orbiter could find a large, flat area on which the lander could be placed, so his weather station would function more effectively. He preferred a place with no surface "relief for 10's to 100's of kilometers." Surely the orbiter images could spot such a tableland. But Klaus Biemann, a chemist from MIT, noted that in the search for life forms, as well as in the molecular analysis his team would make, it was preferable that the first lander sit down in a warm, wet, low site. His ideal site demanded the fewest degrees below freezing, the highest traces of water in whatever form it might be found, and the highest atmospheric pressure (i.e., the lowest elevation) possible; life would most likely survive under those conditions.

In addition to the imaging system, the water-vapor mapping and thermal mapping experiments being planned would give the Viking team clues to the best sites while the lander was still attached to the orbiter, but the exact role of the orbiter would become clear only with time. Defining the mission occupied the Science Steering Group for the remainder of 1969 and most of 1970.[2] By August 1970, Jim Martin believed "that the definition of Viking landing site characteristics, the definition of data and data analysis needed to support the selection of sites, and the integration of engineering . . . capabilities and constraints" should be more coordinated.[3] A. Thomas Young, Viking Program Office science integration manager, led a landing site working group,* which met for the first time as a body at MIT on 2 September 1970. Martin opened the proceedings, indicating that "the actual Viking landing sites would be selected through this group."

C. Howard Robins, Jr., deputy mission analysis and design manager, reminded the group that the Viking system requirements were not being developed for a single ideal mission. Instead, his teams were planning for a

---

*Other members of the working group were C. H. Robins and G. A. Soffen, Langley Research Center; W. A. Baum, Lowell Observatory; A. Binder, Science Applications Institute; G. A. Briggs and C. B. Farmer, JPL; H. Kieffer, University of California at Los Angeles; J. Lederberg, Stanford University; H. Masursky and H. J. Moore, U.S. Geological Survey; and C. Sagan, Cornell University.

broad spectrum of missions based on the desire to set the lander down anywhere in the latitude band 30° north to 30° south. The hypothetical landing sites being used to develop the "preliminary reference mission" had not been selected for their scientific merit. They had been chosen simply to give the analysis and design specialists something to work with in creating spacecraft design requirements. Finally, he reported that his office would develop the "operational mission design," which would guide the conduct of the real missions, by working hand in hand with the landing site working group.

The working group members began to discuss the desirable features and characteristics of Viking landing sites, with Tom Young suggesting that initially they ignore any potential system or mission constraints. Carl Sagan led off the brainstorming session by considering the problem in terms of three primary areas of investigation—biology, geology, and meteorology. Comments on biology centered on the availability of water, atmospheric and surface temperatures, and ultraviolet radiation. Each of these three variables could affect the possibility of finding life forms.

The meteorologists wished to observe four related phenomena over a period of time—seasonal darkening, the daily night-day cycle, long-term meteorological variations, and the annual polar-cap regression process. They also hoped the lander could be in a position to observe dust devils, ground fog, and ice clouds. William Baum of the Lowell Observatory's Planetary Research Center presented a status report on Earth-based motion studies of clouds on Mars. Cloud patterns were being mapped under the International Planetary Patrol Programs hourly each day, and recent daily photographs had shown significant changes, but he could not say how these alterations might be correlated with seasonal or other patterns.

The first working group meeting closed with a discussion of the relationship between the Mariner 71 mission (*Mariner 9*, launched 8 May 1971) and Viking. Dan Schneiderman, Mariner 71 project manager, hoped Viking personnel members would participate in that mission as observers during the first 100 days and thereafter as users of the orbital cameras to look for potential Viking landing sites. Martin assured the working group members that they would have an opportunity in October to discuss topics of common interest between Mariner 71 and Viking.[4]

## FINANCIAL PROBLEMS THREATEN ORBITAL IMAGING SYSTEM

August to October 1970 was a busy time for the Viking project managers and the landing site working group. General discussions quickly gave way to deliberations over specific problems. One of those specifics was the orbiter visual-imaging subsystem, which had been identified as a candidate for elimination or modification to reduce costs substantially. The project stretch-out required paring costs, and Jim Martin and his colleagues sought ways to do so while still saving the orbiter and other key elements of the proposed mission.[5]

The Science Steering Group had identified three alternative approaches to orbital imaging that would save dollars—the Viking camera system already proposed; a slight variation of that system in which the image motion-compensation device was eliminated at an estimated $1-million saving; or a modified Mariner 71 imaging system (using improved optics), at a possible saving of $8 million.[6] At the July 1970 Science Steering Group meeting, Viking project scientist Jerry Soffen had told his colleagues that the cost reduction exercise in progress made it necessary for them to decide which investigations or parts of investigations were the most important scientifically. Each science leader had to defend the costs and merits of his team's experiment and recommend ways to conserve money. When Mike Carr—orbiter imaging team leader and an astrogeologist from the U.S. Geological Survey, Menlo Park, California—had defended the orbiter television camera system, he had argued that the costs were as low as they could be. When asked if the Mariner 71 camera system could be used on Viking as well, he had said emphatically, no.

Carr's orbiter imaging team reported in October that the orbital imaging from Viking would substantially enhance the scientific value of all the other experiments.[7] The imaging system would improve the probability of a safe landing, help define the environment in which the lander experiments would be performed, and permit comparisons of the landing site with other regions on Mars. The team was convinced that the proposed Viking camera system would yield superior pictures. "A modified 1971 camera would provide only minimal support for the Viking mission and would add only little to our knowledge of the planet. The Viking camera system outperforms the [Mariner] 71 camera in . . . very fundamental ways." Mariner 71's camera was a slow-rate vidicon unit, requiring a cycle time of 42 seconds to capture a single image. Viking's fast vidicon worked in a tenth of that time. To get overlapping coverage with the Mariner 71 A-camera, it would have to look at a larger area, losing detail in its resolving power. Mariner 71's B-cameras had a resolution comparable to the Viking system, but with a slow vidicon system it could not produce contiguous frames of coverage and would leave gaps between pictures. Viking's cameras would yield high-resolution and overlapping images, so the Viking team could get the photographic images they needed of the entire landing area in a single pass.

The fast vidicon camera system put other demands on the team, however. On the orbiter, the camera would require a fast, reliable tape recorder to store all the electronic bits into which the images had been coded. The telemetry system and ground-based recorders must be capable of handling the data flow, and the image-reconstruction and processing computers and related equipment would have to process that data as quickly as it was received. But Carr believed that this elaborate complex of machines and men was essential to Viking's success. "The Viking camera will always outperform the [Mariner] system by delivering more resolution per area covered, by allowing greater flexibility in choice of filters and lighting conditions and making more effective use of a lower periapsis."*

These performance differences were important to site certification. "With only two landers judicious choice of landing sites is essential to ensure that they will result in maximum scientific return." According to Carr and his colleagues, orbital imaging would be the key to site selection by providing:

(1) Numerical terrain data (crater statistics, slope frequency distributions, etc.) such that the landability of different sites can be compared and assessed.
(2) Distribution frequencies of features such as craters, ridges, block fields, that are potentially detrimental (or advantageous) to lander experiments.
(3) Absolute and relative elevation measurements as a supplement and check to radar and [infrared] data.
(4) Information on the geologic nature of the potential landing sites.
(5) Information on seasonally variable clouds, condensations, and surface albedo differences both locally and regionally around potential sites.[8]

The orbital imaging team was sure that the difference in results from the Mariner 71 and the Viking systems would be striking. Mariner 71 would be unable to portray objects smaller than 1 kilometer in diameter, while resolution with the Viking system, judged to be about 45 meters, was "close to the limit from which data can be extrapolated to the scale of the lander [2–3 meters]." The orbiter imaging specialists contended that using a modified Mariner 71 system would render the imaging "virtually worthless for obtaining terrain statistics and the distribution of specific features at the scale of the lander or making useful elevation measurements." To make their point, they used 80-meter- and 1-kilometer-resolution photographs of the *Apollo 14* landing site on the moon to illustrate how sensitive geological and topographical analyses were to this change. Most telling was the team's comment that the state of Martian imagery after Mariner 71 would be "roughly comparable to that of the Moon before any spaceflight program."

Besides searching for landing sites, the experts hoped the orbiter imaging system would return data on the activity of the Martian atmosphere, provide a much better understanding of the geological processes, and perhaps even yield clues to the existence or nonexistence of life. And there was the future to look to, they suggested. "The Viking landers will not be the last spacecraft to land on Mars. Others will surely follow and sites will have to be selected. Our whole lunar experience has been that the prime

---

*Periapsis is the point in an elliptical orbit at which a spacecraft or satellite is closest to any body it is orbiting. Its opposite, or highest point, is the apoapsis. Specifically for Earth orbits, the terms are perigee and apogee; for the moon, perilune and apolune; and for the sun, perihelion and aphelion.

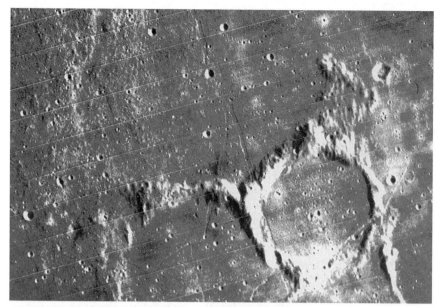

*Orbiter imaging team leader Michael Carr used* Apollo 14 *photos to explain the difference in image resolution between the cameras of Mariner Mars 71 and Viking orbiters. Resolution of about 80 meters for the top photo of the Apollo 14 landing site is slightly worse than the effective ground resolution of the Viking baseline camera. Chief justification for choosing the site was the presence of the Imbrium Basin ejecta, indicated by rough terrain in the west part of the photo. In the bottom photo, at a resolution of about 1 kilometer (comparable to that of the Mariner Mars 71 camera), the area looks bland and uninteresting and the ejecta is not detectable. Details of the terrain are inadequate for assessing landing conditions and topographic and geologic content of the area.*

consideration in selecting any landing site is the availability of imagery."
No judgment could be made about the relative merits of different sites for
engineering or scientific purposes without adequate images. "In the past, a
[lunar] site without imagery has been rejected immediately. There is little
reason to believe that for Mars the decision making process is going to be
significantly different." It was "imperative to collect as much imagery as
possible to provide a decision making base for future missions."

Finally, the imaging team turned to political considerations.

> One of Viking's characteristics is its high-risk, high-gain mode of
> focusing on a search for life. Negative results on all the biologic experi-
> ments is not unlikely; the seismometer may never see a quake. To run a
> billion dollar mission and obtain largely negative results would be
> embarrassing politically for the project as well as for NASA as an agency.
> Whether negative results reflect the lack of life, or the wrong kinds of
> experiments or the wrong landing locations might be difficult to see. . . .
>
> Thus, the high-resolution imaging system may be considered as the
> "meat and potatoes" low-risk but guaranteed-significant-gain experi-
> ment in the mission.

It was excellent insurance against critics who might say that Viking had
been too narrowly focused. The orbiter imaging team urged that the
Mariner 71 camera system be dropped from further consideration.[9] The
landing site working group recommended to the Science Steering Group
that the Viking system be retained, and the steering group and NASA
Headquarters concurred.[10]

A year later, money problems recurred. On 19 September 1971, the
Science Steering Group met in a special session where the science team
leaders got the bad news. Despite all efforts to reduce costs in management
and engineering phases, Jerry Soffen had to tell his colleagues they must
reduce the overall science costs by $17 million to $22 million. Several
methods were mentioned, but each team quickly put in writing reasons why
its own experiment should be exempted from the reductions.

The 6–7 October meeting of the steering group at the California
Institute of Technology concentrated solely on money matters. Three
options for reducing costs were discussed at length. The first called for
deleting some routine activities—holding fewer meetings, and the like;
perhaps as much as $3 million could be saved here. By simplifying the gas
chromatograph–mass spectrometer and the biology instrument, another $5
million or so might be cut. Reducing science activities on board the orbiter
could save another million. Other parings and deletions brought the total
potential savings to just over $22 million. The second option called for
eliminating the gas chromatograph–mass spectrometer, which had a predicted $35-million price tag, but the Science Steering Group preferred not
to act on this item until it had a better feel for the technical feasibility of
building the instrument.

Option 3 was the removal of the imaging system from the orbiter. As Hal Masursky recalled the scene, Soffen said, "We have a 17–22-million-dollar problem and the Orbiter Imaging System costs 25 million. Any suggestions?" Most of the steering group members were reluctant to recommend removing those cameras until they saw the Mariner 71 photographs. They would make that recommendation only if the Mariner images showed a bland, uninteresting surface. Mike Carr and Hal Masursky believed the imaging system was necessary for site certification regardless of what data Mariner 71 produced. C. Barney Farmer, team leader for the Mars atmospheric water-vapor detection experiment, expressed his concern about the whole idea of using these meetings to effect cost reductions. He went on record indicating his reluctance to recommend the removal of any full investigation. The group postponed a decision on the third option until January–February 1972. Money had been, was, and would continue to be a problem. Still, it was only one part of the problem of searching for a landing site.[11]

### PREPARING FOR SITE SELECTION

Besides considering the imaging system and discussing desired landing site characteristics at its October 1970 meeting, the landing site working group also considered what it could gain from Mariner 71. Dan Schneiderman introduced the group to the Mariner project, and Edwin Pounder reviewed mission operations plans for both the prime 90-day mission and the extended mission (for the remainder of the first year in Mars orbit). Pounder went on to outline problems and promises of the project, one of the promises being data that would assist the Viking team in landing site selection. Patrick J. Rygh and Robert H. Steinbacher briefed the working group on mission operations and participation by scientists.

In turn, Hal Masursky and Carl Sagan told the Mariner specialists what the Viking team hoped to learn from Mariner 71. What they wanted was not in the written mission plans but was rather, How do we learn as we go along and then modify our plans accordingly? In NASA shorthand, this tactic was called the adaptive mode—acquiring data from a spacecraft and quickly using it to modify the mission. The Viking team was certain it would need this skill, and it would require discipline, planning, and timely responsiveness to succeed. In the plans for Mariner 71, data processing was not scheduled to catch up with acquisition for a year, and Masursky feared that unless adequately supported, the complete process could take 5 to 10 years, which was obviously too slow to be of value to Viking. Years of work had to be compressed into weeks. On occasion, time for data processing would have to be whittled down to days and even hours.[12]

At its next meeting, 2–3 December 1970, the landing site working group made its initial recommendation for landing sites, so that Howard Robins' mission planning staff could proceed with its work. These proposed sites had been chosen after only four months. Carl Sagan, who had been urging that the site selection process be completely documented, prepared a convenient summary of the thinking—as he saw it—that went into the choices. "The following is a preliminary attempt to integrate coherently a range of ideas which have been suggested on the Viking landing site question, to point out inadequacies in the existing data, and to serve as guide for future discussion." He noted that the "present cycle of discussion on landing site selection is to aid development of the Viking Project Reference Mission #1," a theoretical model that would be used in planning mission operations and designing the spacecraft. Since in some respects this was a training exercise, there was no commitment to the specific landing sites they had selected.

In considering landing sites for the two Vikings, some factors would be certain to change. But those that would likely remain unaltered fell into two categories, engineering and scientific. Under the engineering heading, the 30° south to 30° north latitude range for landing sites was dictated by the angle at which the spacecraft would have to enter the Martian atmosphere to obtain optimum aerodynamic deceleration and proper thermal conditions. Second, nearly all of the working group members agreed that the lander should sit down where atmospheric pressures were the highest. As on Earth, high pressure corresponds with lower elevation, but whereas sea level pressure on Earth averages about 1013 millibars, surface pressures on Mars are 100 times lower. Pressure at the lowest elevation was believed to be close to 10 millibars and at the top of mountains less than 1 millibar, but the uncertainty in these values was 20 or 30 percent at the time. The Viking scientists hoped that Mariner photographs and ground-based radar studies would give them more exact information on atmospheric pressure relative to topographical features. A third engineering concern was the effect that Martian surface winds would have on the spacecraft. The Mars engineering model with which the team was working predicted winds of less than 90 meters per second, but Sagan noted that newer calculations indicated the possibility of winds up to 140 to 200 meters per second.

> If such winds are encountered during landing maneuvers, the survivability of the spacecraft is very much in question; and such winds, even after a safe landing, might provide various engineering embarrassments. It will shortly be possible to predict which times and places are to be avoided. . . . Such considerations obviously require further theoretical study and (with Mariner Mars '71) observational study. But they do indicate how new parameters, not previously considered, can severely impact landing site choices. Such considerations imply that any landing site selected at the present time should not be too firmly imbedded in the Project's thinking.[13]

Other technical factors affecting the choice of a landing spot included the time of day on Mars at touchdown, the size of the landing target, and a pair strategy calling for one very safe (but perhaps less interesting) site and one of greater scientific potential. Depending in part on progress made in developing the lander tape recorder, Sagan thought that it might be desirable to land in the late afternoon to ensure that some lander images of the planet would be transmitted to the orbiter before it passed out of view of the lander, giving the team at the Jet Propulsion Lab maximum assurance of obtaining at least some initial pictures of the surface. They had to face the possibility that the lander could die while the orbiter continued on its way around Mars; it would be 24.6 hours before the orbiter passed over the lander a second time. Should a late afternoon touchdown be called for, those areas with dense cloud development at that time of day would have to be excluded. Turning to the target, or landing ellipse, Sagan indicated that it was currently 400 by 840 kilometers, which would eliminate areas appreciably smaller than this zone. The pair strategy had been devised for reasons of "survivability." One landing site would be selected with "safety considerations weighed very highly"; if the first mission failed on entry, the team would want to have a preselected, extremely safe site for the second lander. "It is therefore necessary to consider some sites almost exclusively on engineering grounds." Sagan hoped planners could "back off from this requirement a little bit and seek out safe contingency sites with at least acceptable science." Alan Binder had made this same point earlier but somewhat more bluntly: "The engineering criteria must reign since it hardly need be mentioned that a crashed lander is not very useful even if it did crash in the most interesting part of the planet."[14] Sagan wrote, "Before any Viking lander is committed to a given site, there must be reasonably extensive Mariner Mars '71 type data, including but not restricted to imagery." He thought that selection of alternative candidate sites should be based on Mariner 71 data, and certification of the various candidates should be based on Viking data, which would be of higher resolution.

Sagan's report then turned to the working group consensus on science criteria for the landing sites. Many members believed it would be useful to pair the first two landing sites in such a manner that each one would be a control for the measurements made at its companion location. A reason for varying from this plan would be positive results from the biology experiments on the first lander; then the Viking team might wish to land the second craft as near the first one as possible to determine if the results could be duplicated. The best guess at the time was that Martian life, "or at least that subset of Martian life which the Viking biology package is likely to detect," would be found where there was water near the surface. But there was still considerable debate about the nature and amount of water that might be found. Low atmospheric pressures and temperatures always below 0°C did not augur well for the presence of liquid water. Still, Sagan and others believed that it was possible to have life-sustaining water present in other forms.

The uncratered terrain observed in the *Mariner 4* photographs was of possible interest. Sagan hypothesized that such terrain must have been recently (in geological terms) reworked. "Whatever the cause of the reworking, but particularly if it is due to tectonic activity, such locales are much more likely *a priori* to have had recent outgassing events and therefore to be of both geological and biological interest." Taking into consideration all these factors, Sagan listed his six favorite landing spots, but several of his colleagues came up with other suggestions of their own.[15]

After considerable freewheeling debate of the kind that characterized many of the working group's meetings, the group recommended three sites for each lander. It wanted to find water, and it wanted to land one craft in the north and one in the south. The mission planners indicated that it would be best to land the first Viking in the northern latitudes, or during the Martian summer. Immediately following the working group sessions, the mission analysis and design team subjected the six candidate sites to a preliminary examination, and its first quick look revealed no apparent difficulties. On 7 December, Jim Martin directed Martin Marietta to proceed with the design of the two Viking missions using Toth-Nepenthes (15°N, 275°)* for the touchdown area of the first lander and Hellas (30°S, 300°) for the second craft.[16]

Early in February, Dan Schneiderman and Jim Martin signed a "Memorandum of Agreement for Viking Participation in Mariner '71 Operations." Two areas were identified for direct Viking participation—mission operations and scientific data analysis. Viking personnel would work as part of the Mariner team. The Viking data analysis group would be housed in the Science Team Analysis Facility at JPL, and a Viking representative would act as an observer at the Mariner science recommendation team meetings, watching the interplay between the science advisers and the mission operations personnel.[17]

The Viking landing site working group did not meet again until April 1971. Meanwhile, the mission planners and the Martin Marietta Corporation evolved the "Mission Design Requirements Objectives and Constraints Document," which outlined for the first time in detail how the two missions would be conducted from launch through operation of the science experiments on Mars. Members of the landing site team and the Science Steering Group met in joint session on the afternoon of 21 April to discuss that document and mission planning in general, but earlier that day the landing site team had considered at length its participation in the Mariner 71 operations.

Tom Young opened the morning session, noting that Robert A. Schmitz would serve as manager of the Viking–Mariner Mars 1971 participation group. His duties included overseeing the Viking data analysis team, which would examine areas related to proposed Viking landing areas. This team would be drawn from two groups of scientists, those who would be working as part of the Mariner 71 operations team—Geoffrey

---

*Longitude on Mars is always determined in a westerly direction, 0–360°. For more on Martian place names, see T. L. Macdonald, "The Origins of Martian Nomenclature," *Icarus* 15 (1971): 233–40.

Briggs, Michael Carr, Hugh Kieffer, Conway Leovy, Hal Masursky, and Carl Sagan—and part-time participants from the Viking team.* Schmitz also was to act as the Viking observer on the Mariner 71 science recommendation team, which would give him a much broader understanding of the entire Mariner project.

Hal Masursky raised two problem issues in data management for Mariner 71, computer data processing and preparing Mars maps. The flow of data from the Mariner spacecraft would be so rapid that only one-fourth to one-third of the information could be processed in real time or near real time by the Mariner 71 system. At that rate, Masursky predicted it would take 18 months to get a complete set of reduced data records, a serious lag for Viking planners who wanted to use this information to land their spacecraft. And to prepare maps from Mariner 71 photography, stereo plotters and computers for analytical cartography, as well as more experienced cartographers, must be brought in. The photogeologist noted that these problems would be discussed with the JPL Mariner people later in the month. But at Carl Sagan's request, these issues were raised that afternoon at a joint session with the Science Steering Group. The advisory body agreed that modest expenditures of Viking funds would be justified if supporting Mariner 71 data processing would contribute to the success of Viking. Masursky would prepare a letter to Jim Martin that clearly defined items that needed support and justifications for using Viking funds.[18]

## MARINER 9'S MISSION

Mariner 71 did not get off to an auspicious start, as *Mariner 8*'s launch from Kennedy Space Center on 8 May 1971 ended in failure. Anomalies began to appear in the Centaur stage main engine after ignition. It shut down early, and the Centaur stage and spacecraft fell into the ocean. An investigation team determined the cause of the failure and worked out corrective actions before the 30 May launch of the second Mariner 71 craft.

At 6:35 p.m. EDT, *Mariner 9* began its 398-million-kilometer direct-ascent trajectory toward Mars. Weighing 1000 kilograms at liftoff, the spacecraft carried six scientific experiments: infrared radiometer, to measure surface temperatures; ultraviolet spectrometer, to investigate the composition and structure of the atmosphere; infrared interferometer spectrometer, to measure surface and atmospheric radiation; S-band radio occultation experiment, to study the pressure and structure of the atmosphere; gravity field investigations; and the high- and low-resolution television imaging system, to map the surface of the planet. After a journey of 167 days, *Mariner 9* went into Mars orbit on 13 November 1971, becoming the first spacecraft to orbit another planet. Orbital parameters were close to those planned, and the spacecraft circled Mars twice a day (11.98 hours per

---

*C. Snyder, T. Mutch, D. Anderson, W. Baum, A. Binder, B. Farmer, R. Hutton, J. Lederberg, H. Moore, T. Owen, R. Scott, J. Shaw, and R. Shorthill.

revolution) at an inclination of 65°. Technicians referred to *Mariner 9*'s path as 17/35—after 17 Martian days and 35 revolutions of the spacecraft the ground track would begin to repeat itself, giving the specialists the same images under essentially the same solar illumination. The Mariner planners had chosen a periapsis altitude of about 1250 kilometers to ensure some overlap when two consecutive, wide-angle A-frame images were recorded looking directly downward at the surface. Gaps between images acquired before or after periapsis could be filled in on a subsequent cycle 17 days later.[19]

The NASA team sent *Mariner 9* to the Red Planet at a time when the southern polar cap was shrinking and the southern hemisphere was undergoing its seasonal darkening, and the spacecraft instruments were designed to observe these phenomena. But Mars gave the Mariner scientists more than they had bargained for. On 22 September 1971, as the spacecraft made its way to its destination, ground-based astronomers noticed a brilliant, whitish cloud, which in a few hours covered the whole Noachis region of Mars. What they saw was the beginning of the greatest, most widespread Martian dust storm ever recorded.*[20]

The progress of the storm was amazing. It spread from an initial streaklike core, some 2400 kilometers in length. On 24 September, the dust cloud began to expand more rapidly to the west, blanketing a large area from the east edge of Hellas (a proposed Viking landing site), west across Noachis in three days, a distance two-thirds of the way around the planet. To the north, Syrtis Major was beginning to disappear beneath the haze. On 28 September, a new cloud developed in Eos, a region later found to be part of the canyon lands of Mars. Peter Boyce, of the Lowell Observatory in Flagstaff, Arizona, reported that his observations taken in the blue-light spectrum had shown a reduction in contrast for several prominent features days before the dust cloud was visible to astronomers. This indicated that Martian dust had been drawn up into the atmosphere some time before the actual cloud could be seen. By the end of the first week in October, clouds or storms had engulfed nearly the entire planet. A zone about 12 000 kilometers long had been obscured in only 16 days. Prospects were dim for a successful mapping of the planet when *Mariner 9* reached Mars on 13 November. At Mariner mission control, there were some worried people, and the Viking team worried along with them.

On 8 November, the first pictures of Mars came back from the spacecraft. While these were essentially calibration shots designed to check out the television system, they were large enough to give a reasonably good view

---

*C. Capen of the Lowell Observatory theorized in February 1971 that such a storm was possible. Since 1892, astronomers have observed substantial dust storms each time an Earth-Mars opposition coincided with Mars' closest approach to the sun—1892, 1909, 1924-25, 1939, 1956. Because of the eccentricity of its orbit, the radiation received by Mars at perihelion is more than 20 percent stronger than usual. This increase substantially raises atmospheric and surface temperatures, and the resultant instabilities give rise to swirling columns of air that lift dust and debris into the Martian sky.

of the planet. But the dust was all-pervasive; no detail could be discerned. One scientist, in a bit of gallows humor, suggested that they must have visited Venus by mistake, since that planet is perennially blanketed by clouds. His remark was not well received. With the loss of *Mariner 8*, the Mariner 71 project planners had completely reworked the missions they had scheduled for the two spacecraft. *Mariner 8* was to have mapped the planet while *Mariner 9* looked at the variable features of Mars, and both of these tasks were of great interest to Viking planners. The redesign of two missions into one had been accomplished while *Mariner 9* traveled toward the Red Planet.

Mariner personnel members began a series of preorbital sequences to gather science data on 10 November. Originally they had hoped that these long-distance photographs of the whole planetary disk would provide them a global view of the surface. These images would have helped fill the gap between the low-resolution views obtained by *Mariner 4, 6,* and *7* and the higher resolution closeups they were hoping to take with *Mariner 9*. The first preorbital science picture revealed a nearly blank disk with a faintly bright southern polar area and several small dark spots. The intensity of the storm "shook everybody up," according to Hal Masursky, "because we could in effect see nothing." The key to their elaborate mission plan was a series of photographs that would be used in developing a control net for photomapping. That work was supposed to be done during the first 20 days after the spacecraft went into orbit, but they couldn't see a thing! The revised plan was dumped, and the Mariner operations team searched for items to photograph while waiting for the storm to subside.

Working with classical maps of Mars and more recently acquired radar data, the *Mariner 9* television crew was able to demonstrate that one of the dark spots they could see in the science picture coincided with Nix Olympica (Snows of Olympus). That mysterious feature, often seen topped with bright clouds or frost deposits, was known from radar measurements to be one of the highest areas of the planet. Nix Olympica, towering through the dust clouds, was revealed as a very high mountain, the first Martian surface feature other than the polar cap to be identified by *Mariner 9*. Computer enhancement of the 14 November images revealed volcanic craters in the summits of four mountains protruding through the pall of dust. This unexpected information led to the discovery that Nix Olympica and the three nearby dark mountains were actually enormous volcanoes, which would dwarf any found on Earth. But only these large features were visible. Other mapping sequences of orbital images produced a series of nearly featureless frames. Unhappily for the Viking team, adaptive photography brought pictures of things that did not aid its search for a landing site, like images of the Martian moons, Phobos and Deimos.

By 17 November, craters in certain regions began to appear in the television images as light-colored, circular patches. In similar fashion, an irregular, bright streak appeared running along the "canal" Coprates, through Aurorae Sinus into Eos, the region of chaotic terrain identified by *Mariner 6* and 7. Radar measurements had shown a depression several kilometers deep in this region. Indeed, the evidence, as incredible as it sounded, had indicated the presence of a huge canyon some 3000 kilometers long and varying in width from 100 to 200 kilometers. Beneath all that dust was a world of amazing topography. The Mariner and Viking science teams anxiously awaited their first clear view of that scene. In late November and early December, the dust storm seemed to be subsiding, but a couple of weeks later that trend slowed to a standstill. Worried scientists were relieved when the clearing process began again during the last days of the year.[21]

While the dust storm had a significant impact on the *Mariner 9* mission, its persistence through the month of November had a devastating effect on two Soviet probes launched on 19 and 29 May. Each of these craft weighed 4650 kilograms (nearly eight times the weight of *Mariner 9*) and consisted of an orbiter and a lander. The lander, containing a sterilized scientific package, was designed to enter the Martian atmosphere protected by a conical heatshield. Once the shield was discarded, the scientific instrument unit would descend on a parachute, and at about 20 to 30 meters above the surface the lander would be slowed further by a braking rocket. Those were the Soviet plans. On 27 November, just before *Mars 2* entered orbit, the lander was ejected from the spacecraft to begin a 4½-hour journey to the surface. But something went wrong, and the lander crashed into the Martian surface at 44.2°S, 313.2°. Five days later, *Mars 3* approached the planet and released its scientific cargo. After the descent, the craft landed safely at 45°S, 168°, and relayed a television signal to its parent craft in orbit. Success was short-lived, as the signal stopped after only 20 seconds. Soviet space scientists concluded that both failures were due to the storm raging on the surface. Unable to decipher the electronically coded television data, the Soviets could not determine what the surface looked like. Not only did the Soviet landers fail, but the dust storm outlasted the lifetimes of the imaging systems on both orbiters. Complementary data would have been useful for both the *Mariner 9* and Viking teams, but the planet would not cooperate. Viking was likely to be the first craft to take pictures on the Martian surface, but only if it landed safely. And for many NASA planners, that was still an open question.[22]

When the Viking Science Steering Group met at JPL in December 1971, one of its primary concerns was to learn what *Mariner 9* could tell it that would affect Viking. Although the men participated in a weekly Mariner science evaluation team meeting designed to summarize the most recent scientific findings, they did not learn anything positive. The severe dust storm had foiled their efforts. Hal Masursky and his colleagues concluded that the Martian atmosphere might never completely clear, especially in the low areas, during the Mariner mission. If *Mariner 9* did not acquire the reconnaissance data they required, Viking would have to perform the task, which made the instruments on the Viking orbiter even more important. The Viking Project Office would have to keep "its options open" and give thought to several different models of the Martian surface, to be prepared for whatever Viking might encounter.[23]

Clouds were clearing over Mars the third week of February 1972, however. During orbits 139 to 178, one mapping cycle of interest to Viking had been completed, covering the region from 25° south to 20° north. A second mapping cycle was in progress, and a third later that month would cover a Viking area yet to be determined. The coverage was reported to be very good.[24]

Mariner scientists devoted the February session of the Science Steering Group to reports summarizing their recent data and comments on the implications for Viking. Most of what they had to say had already been made public during an early February press briefing held at NASA Headquarters. Bradford Smith, deputy team leader for the television experiment, had told reporters that the Martian atmosphere had begun to clear slowly in December, with more rapid progress during the first week of January. Pictures now available of the Martian surface led the science team to conclude that the planet was a far more dusty place than they previously had thought. But at that same press conference, Hal Masursky had some positive words about the dust storm. The first 30 to 40 days of the mission had given the scientists an opportunity to study the dynamics of the Martian atmosphere. "It will be 15 years . . . before such a large dust storm can be seen" again. The storm, however, forced the mission planners to devise a reconnaissance scheme for looking at the planet from a higher altitude and photographing any clear areas with the high-resolution camera. Once the clearing trend started, the Mariner team began a new series of mapping sequences that were at least as complex as the original mission plan.

The mapping process revealed a fantastic planet, strewn with features that caught scientists' immediate attention. Huge volcanoes with attendant lava flows were found in the Tharsis region. And features that had been observed previously—such as three dark areas called North Spot, Middle Spot, and South Spot—were now clearly volcanoes. The caldera, formed by the collapse of the cone, of North Spot was 32 kilometers across, while the width of South Spot's crater was 120 kilometers. But these volcanoes were all dwarfed by Nix Olympica, which was renamed Olympus Mons. To the east of Tharsis, the Mariner team found a high plateau, much of which was 8 kilometers above the surface, that evidenced complex fault zones. Some areas had been uplifted; others had been depressed; in places large blocks had been tilted. "We think this indicates a very dynamic substratum under the Mars crust," Masursky noted. He showed the press some slides of the great chasm, which was some 4000 kilometers long and hundreds of kilometers wide at points. Looking at this complex of valleys and tributaries so recently obscured by dust, Masursky commented, "We are hard put to find a mechanism other than water to form this kind of complex, erosional channel. If it were not Mars, and if water weren't so hard to come by there, we would think that these were water channels." This thought, pregnant with many possibilities, would require considerable analysis.

Masursky told the Science Steering Group that the scientific community was changing its thinking about Mars. After the 1969 flyby missions (*Mariner 6* and *7*), scientists still tended to believe, from the 165 low-resolution photographs taken from a distance of 3400 kilometers, that Mars was a dead primordial planet. But the *Mariner 9* photographs illustrated a very different kind of place. The crispness of the edges on the volcanic piles and the absence of cratering seemed to indicate that these volcanoes were, geologically speaking, young. Just how young was uncertain. The fault zones showed that the crust had been broken many, many times. Mars evidently was a dynamic, geochemically evolved planet and not just a static accumulation of cosmic debris as some experts had theorized after the Mariner 1969 flights. With the realization that Mars was an active planet in geological terms, the search for possible life forms became more exciting.[25]

Next at the February meeting, Al Binder described some of the work the Viking data analysis team was doing. A preliminary contour elevation map of the zone of interest to Viking had been compiled from 1967, 1969, and 1971 Earth-based radar observational data, which had been combined with *Mariner 9* S-band occultation findings. To help determine the topography of Mars, the S-band experiment correlated the effects of temperature and pressure differences on radio signals through the thin atmosphere. Such maps would give clues as to which regions deserved a closer look and more detailed mapping later in the summer of 1972.[26]

Jim Martin opened the second day of the Science Steering Group meetings on 17 February with a summary of the cost status of the project, particularly of the experiments. What followed could only be called a tough session. Each team leader explained what was being done in his project area to cut costs and under close cross examination defended his budget against future cuts. Everyone felt the pressure, so Mike Carr was not shy about arguing strongly for his orbital cameras.[27] Prefacing Carr's presentation, Conway W. Snyder, Viking orbiter scientist, described eight possible camera choices for Viking:

| *Alternative Choices* | *Savings (in millions)* |
| --- | --- |
| Delete cameras altogether | $17.80 |
| Use Mariner TV cameras | 3.30 |
| Use augmented Mariner TV cameras | 3.15 |
| Mariner engineering | 5.30 |
| Viking imaging system without image motion compensation | 0.40 |
| Viking imaging system without photometric calibration | 1.30 |
| Viking imaging system without image intensifier | 0.70 |
| Delete above 2 items | 2.00 |

Carr proposed that the photometric calibration and the image intensifier be dropped. This modified imaging system would permit double coverage but at one-half the resolution of the originally proposed system. The

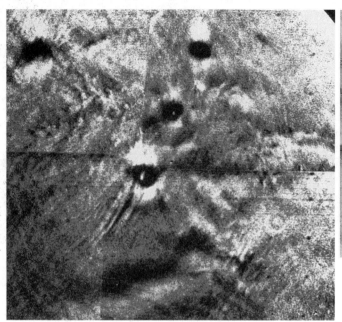

In a mosaic (above left) of photos taken by Mariner 9 just before going into orbit of Mars in November 1971, computer processing reveals subtle details and swirls of dust. There is no suggestion that the dust storm is dispersing. Arsia Silva, most southerly of the three dark volcanic peaks, is slightly below the equator and 200 km in diameter. Streaks are probably wind-driven clouds. Bright patches near the dark spots are artifacts of processing. Olympus Mons (above right), gigantic volcanic mountain photographed by Mariner 9 in January 1972 as the dust storm subsided, is 500 km across at the base, with cliffs dropping off from the mountain flanks to a surrounding great plain. The main crater at the summit—a complex, multiple volcano vent—is 65 km across. Mons Olympia is more than twice as broad as the most massive volcanic pile on Earth. The meandering "river" in the photo below is the most convincing evidence found that a fluid once flowed on the surface of Mars. The channel, Vallis Nirgal, some 575 km long and 5 to 6 km wide, resembles a giant version of a water-cut arroyo, or gulley, on Earth. Mariner infrared spectral data, as well as Earth-based instruments, showed very little water on Mars, however. The Martian valleys also resemble sinuous rilles on Earth's moon believed to be associated with lava flows, but no lunar rilles display the branching tributaries seen here. The channel was first seen on 19 January 1972.

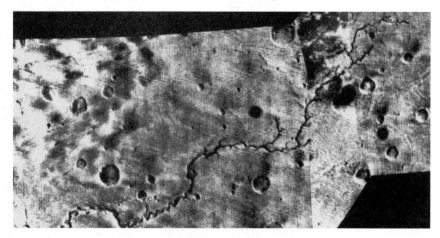

Mariner 9's wide-angle TV camera on 12 January 1972 photographed the vast chasm at right, with branching canyons eroding the plateau. These features in Tithonius Lacus, 480 km south of the equator, represent a landform evolution apparently unique to Mars. The resemblance to treelike tributaries of a stream is probably superficial, for many of the "tributary" canyons are closed depressions. Subsidence along lines of weakness in the crust and possibly deflation by winds have sculptured the pattern. The photo, taken from 1977 km away, covers 376 by 480 km. The mosaic of two photos below, taken of Tithonius Lacus region from 1722 km, covers an area 644 km across and shows a section of Valles Marineris. Pressure measurements by Mariner's ultraviolet spectrometer registered a canyon depth of 6 km (the Grand Canyon in Arizona is 1.6 km deep). The dotted line is the UVS instrument's scan path. The profile line below shows measurements converted to relative surface elevations. The photo on the following page shows the full length of the canyon system.

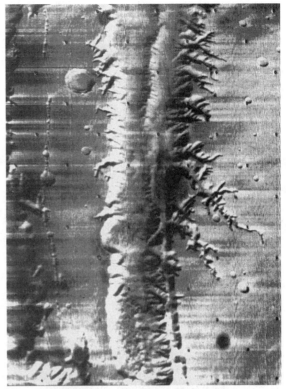

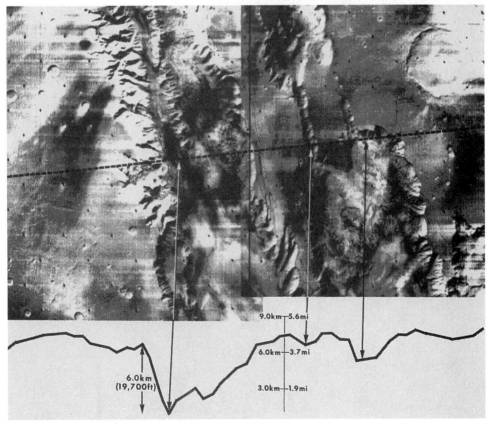

9.0km—5.6mi

6.0km—3.7mi

6.0km
(19,700ft)

3.0km—1.9mi

*Panoramic view of the equatorial region on Mars was made from pictures taken by Mariner 9 from late January to mid-March 1972. Several hundred frames were scaled to size for the composite, which extends from 10° longitude at the right edge to about 140° at left. The photo map stretches more than one-third the way around Mars and covers about 28 million sq km, about one-fifth the planet's surface. The equator bisects the mosaic horizontally. At left, the complex of newly discovered giant volcanic mountains includes Olympus Mons, the largest. At least nine huge volcanoes have been pinpointed in Mariner 9 photos. Through the center runs the enormous canyon system Valles Marineris, 4000 km long, some 200 km wide at points, and nearly 6 km deep. (Portions are shown in the previous photos.) On Earth, the canyon system would extend from Los Angeles to New York.*

modified Viking imaging system would also permit all data to be put onto one tape recorder. The reduced resolution (about the same as the Mariner B-frame high-resolution images) was acceptable to the orbiter imaging team since the more important requirement for contiguous images could still be met. Snyder had pointed out that contiguous or overlapping photos could be obtained with the modified *Mariner 9* cameras, but that the process of acquiring such photos on multiple passes would be long and inefficient. The orbiter imaging people took the position that the Mariner systems were not very suitable for Viking site certification; they wanted the modified Viking imaging system. They would, of course, have preferred the original system but were willing to give up parts of the initial concept to help pare the budget.

An executive session of the science group was held the next day. Once again, each team leader explained how he might save money, and NASA Associate Administrator for Space Science Naugle presented his perspective on the budget problem. After a few brief words of praise and the good news that Viking had passed a major hurdle—its fiscal 1973 budget had been established—Naugle stated that the best program operating policy always called for setting a cost ceiling and adhering to it. He did not intend to give Viking financial relief because such a deviation from policy could have long-term disruptive effects on other aspects of the agency's program. True, there were funds being held in reserve, but Naugle stressed that they were a hedge against possible problems during the hardware development phase. Noting that the cost of the science payload had risen from $110 million to $160 million, the associate administrator made it clear that it was now necessary to make hard decisions to avoid more forced cost reductions in the future. While final decisions were not due until 1 March, Naugle gave his preliminary thoughts about cuts: he favored the proposed $2-million modification of the imaging system (Snyder's last alternative).[28]

## CANDIDATE SITES

With money problems temporarily set aside, the landing site working group turned once again to site selection. The "Viking '75 Project Landing Site Selection Plan," distributed the second week of February 1972, spelled out the entire process the Viking teams would follow in finding sites. The plan carefully delineated responsibility distributed among the groups within the Viking organization.[29]

At the top of the pyramid, John Naugle's Office of Space Science at NASA Headquarters would have overall responsibility for reviewing the project's proposed landing areas and approving final selections. Jim Martin's Viking Project Office at Langley would oversee the six groups whose activities influenced the selection process. Martin Marietta Corporation's Denver Division, in its role as mission planning coordinator, would have to keep track of all flight and engineering considerations that might influence or be influenced by the landing spots ultimately chosen. Jet Propulsion Laboratory, supervising the design of the orbiter, would ensure that the craft could actually perform the tasks required of it. The United States Geological Survey was charged with making a series of Mars maps (regional, area, topographic, and geologic) to support the site selection process and with analyzing the terrain in the territory mapped.[30] The landing site working group, which established science criteria for landing areas, applied those criteria to candidate spots and recommended the best sites to the Science Steering Group. And the Science Steering Group, after reviewing recommendations, formulated its own site selections for Martin's project office—a simple format for a complex task.

Twenty-five members of the landing site working group met for their fifth meeting, at JPL on 25 April 1972, to discuss a wide variety of topics. James D. Porter of Martin Marietta, Viking mission analysis and design program engineer, brought the working group up to date on the engineering constraints that impinged on the site selection process. One was obvious: north or south of 25° latitude, the spacecraft in orbit would not receive adequate solar radiation on its solar panels to keep its batteries charged. Without that power, the orbiter could not relay messages to Earth. Other problems concerned the surface the lander encountered: slopes it touched down on had to be less than 19°, free of rocks and other hazards greater than 22 centimeters in height. Porter was also worried about winds during descent. A landing area that had winds greater than 70 meters per second was automatically eliminated. Porter's presentation was a status report, and he would be keeping the landing site working group informed as new restrictions were discovered.

As the day's discussions progressed, a lively debate developed over the nature of the processes that had shaped the Martian terrain. Areology, the scientific study of the planet Mars, was still less than a precise enterprise. Tim Mutch, in considering the terrain map (1:25 000 000 scale) that the Viking data analysis team had developed, questioned how the working group could extrapolate terrain information from such a map to determine topographical features as small as 22 centimeters. Several men present believed that rock sizes in the centimeter range could be determined from ground-based radar, since it would supposedly provide information on Martian features that small. Combining radar data with high-resolution images similar to the Mariner B-frame pictures had worked well in selecting landing sites on the moon. Others suggested that the radar-photo analysis approach would not be as simple on Mars; the varying kinds of terrain created by different processes would make interpretation of radar data more difficult. At this meeting, the rift between believers in radar and believers in photography first appeared. That division would widen and characterize many of the discussions held, right up to the time of the Viking landings.[31]

After additional consideration of physical characteristics for landing sites, Howard Robins turned the meeting's attention toward the 35 sites that had been proposed for *Mariner 9* photographic coverage. *Mariner 9* had taken 6876 photos covering 85 percent of the planet. At the time of the meeting, the spacecraft was powered down and would remain so until June, because its position relative to Mars and the sun no longer gave its solar cells adequate exposure. *Apollo 16* was a second factor leading to the suspension of the Mariner mission; during mid-April the Goldstone, California, 64-meter deep space antenna was being used to return Apollo's color television pictures. On 4 June 1972, *Mariner 9* would begin its "extended mission"—to complete the mapping of the planet and take landing site photographs for Viking.

After a rather lengthy discussion, the landing site working group recommended that all 35 sites be photographed.[32] With a dual purpose in mind, I. George Recant, Viking science data manager, decided it would be useful to rank the 35 sites. The information would be valuable "in the inevitable trade-offs which have to be made in the negotiations with the Mariner Project in targeting areas for photography." And he thought that the evaluation exercise would identify "many of the considerations which may be required by the [working group] in the landing site selection process."[33] The next two working group meetings, previously scheduled for June and July, were slipped to August and September, at which time the group would have to determine six 30° by 45° regions that would be topographically mapped by Hal Masursky's Branch of Astrogeological Studies of the U.S. Geological Survey at Flagstaff, Arizona.[34]

During May, George Recant, Tim Mutch, Bob Schmitz, and Travis Slocumb evaluated the 35 areas according to engineering safety and scientific interest, with safety considerations outweighing science by more than five times. After much juggling, which Recant noted was subjective in many ways because "no quantitative methods were used in evaluating most of the criteria," they came up with a "relative rating" of the candidates.[35] Schmitz took the target preferences and worked out a photography schedule with the Mariner team, and on 6 June he advised Martin that three narrow-angle, closeup B-frames and one wide-angle A-frame coverage would be attempted for each target. The B-frames would be large enough to cover an entire landing ellipse. He noted further that sites with a relative score between 90 and 75 would be covered first, 74 to 60 second, and below 60 last. Finally, 24 of the 32 sites—3 sites were dropped from consideration—would be photographed during the first nine weeks of work that summer.[36]

Typical of the complexities brought on by continuous evaluation of data was the proposal to add 4 more targets to the list of 32. On 9 June, Hal Masursky, Al Binder, and James Gliozzi, representing the Viking data analysis team, wrote a memo to Bob Schmitz. The 4 additional sites "are in areas which have become accessible on the basis of Binder's recent revision of the Mars Topography map and updated Viking Lander capability." Masursky and his colleagues pointed out that "these sites are typical of some of the most striking geomorphologic features of the Martian surface which have not been considered in previous targeting exercises." They also presented alternate choices for landing sites should engineering constraints continue to change. If these areas were not established as candidates soon, they feared that these particular kinds of terrain would never be considered if no B-frame pictures were taken, even though upgraded lander capability might warrant selection of such spots. Looking back at their experience with Project Apollo, the data analysis experts realized that no pictures of a location meant its immediate rejection from consideration as a landing site. They hoped to forestall that kind of decision.[37]

It quickly became clear that the Viking planners might be asking for too many photographs. The Mariner team had to divide the attitude control gases aboard the spacecraft between Viking's requests and Mariner's experiments. One investigation in particular, the relativity experiment scheduled for September during solar conjunction, would require a major expenditure of control gases. Since early spacecraft maneuvers had consumed more propellant than anticipated, the total number of Viking target sites to be photographed had been reduced to 24. As of early July, 15 areas had been photographed—once with the wide-angle A-frame camera and three times with the high-resolution B-frame camera. Tim Mutch complained about this cut to Naugle, who while sympathetic could only note that although the Mariner pictures would be an important factor in the Viking landing site selection "Viking Orbiter capability for reconnaissance and site certification can also be used as needed."[38]

*A mosaic (at right) of photos taken by* Mariner 9's *high-resolution camera B of a Mars feature about 130 km long by 64 reveals dunelike ridges in what shows as a dark patch in a large crater in the photo (at left) taken by wide-angle camera A. Highest resolution of camera B at lowest planned point in orbit could reveal features as small as 60 m and cover an area 16.4 by 20.8 km, while a camera A frame covered 164 by 108 km with a resolution of 800 m.*

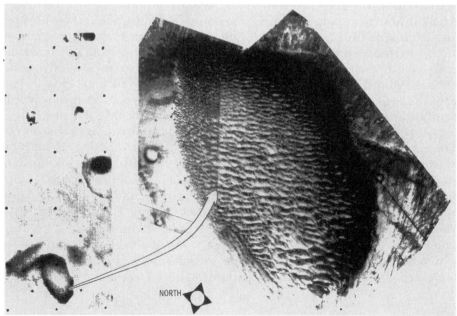

On 19 July, Tom Young distributed copies of the 19 Viking target photographs taken by *Mariner 9* to date, in an attempt to accelerate the site selection process so that operational mission design could begin as early as possible. On 13 July, he told the landing site working group that he hoped it would choose regions of primary interest during its 4–5 August session. At a meeting to be held late in September or early in October, Young expected the working group to identify and debate candidate landing sites. Everyone would have his say, but in two months the group would pick primary and backup sites for each lander. A review by the Science Steering Group and the Office of Space Science would follow immediately. It was a tight, busy schedule, but Young believed it was necessary to make the best use of the project's resources and give the scientists time to participate in the mission design process.[39]

When Young met with the landing site working group at Langley Research Center in August 1972, he summarized the many preliminary steps already taken to finding landing areas for *Viking 1* and *2* on Mars. All this had been necessary and useful training for the actual selection process. "I want to be sure we understand the seriousness of the actions we'll be taking. Consider[able] design effort will be expended on designing the mission starting in December and changes will be costly and have schedule impacts," Young emphasized. Therefore, "the sites we are selecting will be *the* landing sites unless we learn something significant from future analysis of our data or [from] a Soviet landing in 1973." He meant that no site changes would be made for minor reasons; they could react only to important findings or new safety considerations. "I want us to select the best sites in December that our collective wisdom will permit."[40]

The debate that followed Young's statements demonstrated just how divergent opinions were among the 33 specialists present. Jim Porter, who kept the minutes of the meeting, noted that during a discussion of the Mars atmosphere each investigator appeared "to have his own technique for determining atmospheres" and total correlation was not achieved. There were similar debates over radar analysis, the fate of the Soviet Mars landers, and other topics. Hal Masursky gave the group additional cause for concern when he pointed out that the visual impressions of Mars had been constantly changing from the beginning of the mapping mission. For example, features were just now becoming visible on the floor of the region called Hellas as the dust in the atmosphere dissipated. He expected his whole outlook on landing sites to alter by February when the skies would be clearer and orbital photographs more revealing. Jerry Soffen ranked the experiments proposed for Viking, giving the search for life the highest priority, which meant that water or evidence of water in the past would give a region good marks as a landing spot. And there were other considerations:

- The geoscience investigation should be made in areas of heterogeneous and differentiated characters.

- The meteorology investigation should be in locally smooth areas.
- Entry science preferably should have one mission in the northern hemisphere.
- The sites should be selected so that orbiter science has favorable viewing conditions.

Regardless of their interests, the specialists all had to work within an established "landing site strategy": The Viking sites would be selected using Mariner and Earth-based data, with a primary and a backup site for each lander. The preselected Mission A primary site would be examined by the Viking orbiter's science instruments before the first landing, to make sure there were no surface changes or atmospheric hazards. If the site was certified, the lander would be committed to it; if not, the backup site would be the next choice. If results from the Mission A orbiter and lander supported the preselected Mission B sites, certification would be made in the same way as for Mission A, but spacecraft B could be retargeted to a new area if data indicated the need. A new site would be certified by orbiter instruments, although certification would be more complicated because the site would not have been studied intensively beforehand.

On 5 August, "with maps, overlays, theories and opinions abounding," each interest group was given an hour to indicate its primary choices for landing regions. Two stood out—15°N, and 0° to 10°S longitude. *Viking 1* would be targeted for the north; *Viking 2* would be sent south. Before the end of September, the eight regions chosen would be examined in detail.[41]

Before the working group adjourned that day, it placed a conference telephone call to Joshua Lederberg of the biology team. Professor of genetics at the Stanford University School of Medicine, Nobel Laureate, and long-term supporter of Mars exploration, Lederberg carried considerable influence. He restated the biologists' desire to land at low, wet places, preferably near river basin deltas, but he raised another possibility. Why couldn't Viking land far north, 65° or higher, touching down where the polar cap had recently retreated? He had originally expressed the desire to go north in a handwritten memo to Howie Robins in June, believing the zone between 30°S and 30°N to be too restrictive. "I am about to leave the U.S. for about a week; but on my return will prepare a statement of dismay (for the record). *Biology* is assertedly a prime goal of Viking. The [Mariner] 71 *data* surprised us by indicating that Mars' $H_2O$ is principally poleward. Yet here is the box we are in for Viking '75: to be choosing 'optimal' landing sites *within* the least promising zone."[42] Since many members of the working group responded favorably to Lederberg's proposal to go farther north, Tom Young had his mission planners look into the engineering constraints. Jim Martin subsequently authorized Martin Marietta and JPL to make a limited study to determine what would be involved in landing between 65° and 80°N, but Young advised the landing site working group they would "approach this subject with caution and much reservation."[43]

302

The 28 September 1972 landing site working group session at Langley was well attended. After Soffen introduced new member Noel Hinners of the Apollo Lunar Explorations Office at NASA Headquarters and Jim Porter briefed the group on engineering constraints, Young addressed the key topic: Should they try to land a craft near the north polar region of Mars? His answer was no. It could be done, but the risks to the landing craft did not seem to justify the potential gains. Equally significant, not only would it cost between $2 million and $20 million more depending on hardware changes, it would also slow the project's schedule. Young believed the combined increase in money and time was bad news, but he was ready with a detailed reply to the scientists' grumblings. First, landing in the north polar region was technically feasible, an important consideration for future missions. And second, the Viking Project Office understood the high scientific interest in the far north. But third, studies indicated high-risk or high-cost schedule impact because additional communications equipment would have to be developed to ensure adequate links between the orbiter and the lander. With a fixed launch date, the risk was just too high. Because of a November 1976 solar conjunction—Mars would be out of view from Earth—a delay in launching the spacecraft would cut into the prime cycle of science data gathering. All factors considered, Viking 75 would not try to land closer to the northern polar region. John Naugle also restated the budget limitations, firmly reminding the scientists that no additional money would be made available. Information gathered by *Mariner 9* regarding the polar regions would have to be filed away for use on some future mission to Mars. Considerably more discussion ensued about sites in the 30°S to 30°N latitude range, as each member of the working group had the opportunity to indicate his preferences. At the conclusion of the meeting, they recommended 10 sites.

*Table 48*
*Candidate Landing Sites Selected August 1972*

| Mission A | | | Mission B | | |
|---|---|---|---|---|---|
| No. | Latitude | Longitude | No. | Latitude | Longitude |
| 1 | 20°N | 158° | 7 | 2°S | 148° |
| 2* | 20°N | 77° | 8 | 2°S | 186° |
| 3 | 19.5°N | 34° | 9 | 9°S | 144° |
| 4 | 12°N | 158° | 10 | 9°S | 181° |
| 5 | 12°N | 77° | | | |
| 6 | 12°N | 267° | | | |

*Denotes lower priority.

The group's next step was to review these candidates and select one primary and one backup site for each mission at the next meeting, scheduled for 4 December 1972 at Orlando, Florida.[44] But it was not that simple. On 19 October, Tom Young telexed 20 members of the landing site working group; Hal Masursky and William Baum of the Planetary Research Center, Lowell Observatory, had recommended changes in the 10 sites.

*Table 49*
*Changes in Candidate Landing Sites, October 1972*

| Site No. | 28 September Locations | | Masursky Recommendations | |
|---|---|---|---|---|
| | Latitude | Longitude | Latitude | Longitude |
| 1 | 20.0°N | 158° | 21°N | 157° |
| 2 | 20.0°N | 77° | 19°N | 65° |
| 3 | 19.5°N | 34° | No Change | |
| 4 | 12.0°N | 158° | 8°N | 163° |
| 5 | 12.0°N | 77° | 10°N | 80° |
| 6 | 12.0°N | 267° | 10°N | 269° |
| 7 | 2.0°S | 148° | No Change | |
| 8 | 2.0°S | 186° | No Change | |
| 9 | 9.0°S | 144° | 9°S | 141° |
| 10 | 9.0°S | 181° | No Change | |

The alterations had been proposed so the Viking team could get maximum Mariner B-frame high-resolution pictures. Tom Young polled the working group by telephone on the 20th—15 had no objections; Barney Farmer, Richard Goldstein, Jim Porter, and Toby Owen had specific comments; and Al Binder could not be reached. Changes like these would become part of the routine process of landing a spacecraft on the surface of another planet, and this was just the beginning.[45]

*Polar Option Revisited*

By the time the landing site working group next met, in December 1972, *Mariner 9* had completed its mission and Joshua Lederberg had thumped on the desk of NASA Administrator James C. Fletcher. On 27 October when *Mariner 9* used the last of its attitude control propellant, a command was sent from JPL's Mariner Mission Control that shut down the spacecraft's transmitters. Despite initial setbacks, the mission had mapped the entire planet, permitting Viking scientists to gather far more images of candidate landing zones than they had originally anticipated. The infrared and ultraviolet instruments aboard Mariner had also observed large portions of the planet. As the data were being analyzed in November, Lederberg met with Fletcher and Naugle to express the scientists' concern that the

polar region was not being given a fair chance because the "engineers"—
Lederberg's shorthand for the project management—had not done their
homework earlier and examined the polar regions for possible landing
zones. The upshot of his arguments was the decision by Fletcher, in consultation with George M. Low, his deputy, and Naugle, to hold in abeyance
any final action on a polar region landing until they had heard from all the
science team leaders.[46]

Armed with the latest in a growing series of maps based on *Mariner 9*
data, the landing site team met 4–5 December at Martin Marietta's Orlando
facility to pick the primary and secondary landing sites for *Viking 1* and *2*.
When Tom Young opened the session, he admitted that the question of a
polar landing site had not yet been resolved, but the working group would
go ahead with the original task of naming the landing sites in the equatorial band. In turn, John Naugle commented briefly on the strategy for a
polar landing mission: "Send Mission A to the equatorial zone and target
Mission B for the polar regions. Then if A succeeds, allow B to continue to
the polar landing site, but if A fails, retarget B to an equatorial site." The
Viking Project Office would provide a work plan and cost estimate to
NASA Headquarters by 15 January 1973. Young responded that the ground
rules would be kept open but no hardware changes would be made yet. If a
decision was made to go to the pole, a fifth and sixth site should be selected;
that is, a primary and secondary site in the polar region. NASA intended to
hold a press conference in late December to announce the landing targets,
and Young wanted any decisions reached by the working group before then
withheld until the briefing.

The December 1972 announcement was not made; it was 7 May before
any decision was made public. Between the December 1972 and the April
1973 sessions of the landing site working group, there was a great deal of
argument, debate, or spinning of wheels—depending on one's perspective.
Unanimity over where to land was difficult to achieve. During the 4
December dialogue, which lasted some 12 hours, only one site was selected.
The group agreed that the primary site for the first lander would be at
Chryse, 19.5°N longitude, 34° latitude. On the fifth, after another lengthy
discussion, site 10 from the list, Apollinares, 9°S, 181°, was picked as the
prime target for the second lander; site 9, Memnonia, 9°S, 144°, would be the
backup. A secondary target for the first lander was not selected, because of
concern over the strength of the surface at site 1 (21°N, 157°) and because site
2 (19°N, 65°) had undesirable elevation characteristics.[47]

A backup target for *Viking 1* and the question of going further north
continued to be nagging problems into the early months of 1973. An ad hoc
group* for identifying north polar region sites for review by the working
group met 14 December at Stanford University. Five sites were proposed.

---

*N. L. Crabill, J. A. Cutts, C. B. Farmer, J. Lederberg, H. Masursky, L. Soderblom, G. A. Soffen,
and A. T. Young.

305

*Table 50*
*Polar Landing Sites Proposed December 1972*

| Site No. | Latitude | Longitude |
|----------|----------|-----------|
| 12 | 73°N | 350° |
| 13 | 74°N | 225° |
| 14 | 63°N | 0° |
| 15 | 63°N | 85° |
| 16 | 63°N | 160° |

During the next several weeks, each of the science team leaders indicated his group's thoughts on the polar landing, as Jerry Soffen had requested during a December Science Steering Group meeting.[48]

Mike Carr was one of the first scientists to express his opinion: "In general the Orbiter Imaging Team has conflicting responses." On one hand, team members were enthusiastic about a polar landing coupled with a successful equatorial touchdown, because of the potential benefits to lander science. But on the other, they were apprehensive about the impact of such a landing on the orbiter imaging experiment. If the orbiter was used to support a lander in the polar regions, the craft could not be employed to photograph other areas of the planet as planned, because the orbiter's path would have to be altered considerably to accommodate a polar site. Carr said, "We are unable at this time to adequately, confidently assess some of the implications of a polar landing because of inadequate study of the problem."

Carr was also concerned that landing one spacecraft so far north would curtail the two "walks" around the planet, during which the orbiter would photograph Martian features at higher resolution. Four years would have passed since *Mariner 9* did the same; to lose this comparative photography would reduce measurably the understanding of the processes at work shaping the planet's features. If there were great pressure to go north, "the disadvantages could be tolerated if the Lander were to go to a site that exhibits uniquely polar phenomena *ie* one that is at least as far north as 75°. We would be very reluctant to accept these substantial disadvantages for a site at 65°N, which is not likely to be significantly different geologically from an equatorial site."[49]

Physical properties investigator Richard W. Shorthill said that he had discussed the "north polar site" with his team, and it had not favored the proposal. From the start, this group had considered safety to be the prime requirement for a Viking landing site. "Considering the present state of knowledge we cannot support a North polar landing." There was no radar coverage of the polar regions—"*no* information on surface roughness at the scale of the spacecraft, *no* information on the mechanical properties of the surface materials." Mariner imagery of the polar regions had been either ambiguous or too obscured by dust for a reliable evaluation. He went on:

We believe that soils with excessive amounts of water or ice are incompatible with the Viking lander surface sampler system as well as the GCMS [gas chromatograph mass spectrometer] experiment. The behavior of soil with intersticial ice in a Martian environment [is] poorly understood. One could visualize a sublimation process that yields a porous under dense surface material more than 40 cm thick. On the other hand the surface could be wind swept yielding a rock like surface composed of soil and ice.

Furthermore, the safe landing of the first spacecraft at the equator would not ensure that vehicle's longevity on the surface. The team believed that the second lander should also be sent to a relatively safe place. Shorthill also looked at the question of money. He was strongly opposed to cutting back on science funding to provide changes in the lander and orbiter so that they might operate in the polar region. "Any new funds NASA might make available for design changes required for polar operations could better be used to increase mission success in areas where previous cutbacks have reduced the chances of success"—areas such as testing, mission planning, team activities, and continued assessment of the surface properties of the equatorial landing sites.

After first evaluating the polar proposal on its scientific merits, Seymour L. Hess of the meteorology team had subsequently developed reservations. He found it "incredible that a project with such severe financial problems has accepted the addition of a thirteenth experiment and now seems to be about to swallow an additional major cost for the polar option." He also believed that the polar site proposal would be bad for the entire project. "One of the major sources of our troubles is that NASA has been extremely ambitious in the total amount of science it is scheduling in comparison to its fiscal resources. To add this new ambition is, in my opinion, fiscal recklessness."[50]

Harold P. Klein, biology team leader, had another point of view. Klein told Soffen that the biology team had once again reviewed the polar versus equatorial site question at its 11 December meeting and regarded the presence of water as "the most critical parameter in the search for life on Mars." After listening to all the facts and opinions, the team believed that liquid water would be less likely on the equatorial regions than in areas closer to the poles. "We are not, of course, assured that the polar regions will afford opportunities for the production of water under or near the ice caps, but we feel that these regions afford a significantly better prospect for this than the more equatorial zones." Therefore, Klein reported that his team strongly supported the proposition that at least one landing be made in the polar region.[51]

Tim Mutch made a personal response. He had written one letter to Soffen that was a "dispassionate, scientific-engineering analysis," which came to a slightly negative to neutral conclusion on the polar landing proposal. After thinking about what he had written, Mutch concluded that he had probably missed the real issue: "The point is that Viking is an exciting journey of exploration. The fact that it survives NASA's budget cuts is partly attributable to its wide appeal. Scientific skepticism not withstanding, laymen are intrigued by this bold search for life on another planet." Looking at Viking in this framework, Mutch had asked himself what all the talk about a polar landing really meant. "We are maintaining that we should keep open the possibility—only the possibility—of going to a northern latitude after the first lander has been on the surface almost two weeks and has been working perfectly for that period." A second successful landing did not guarantee a doubling of the scientific knowledge gained about the planet. Indeed, exploring near the poles might yield less information than gained at the equator. The Lederberg scenario was only a "long shot," which may or may not be worthwhile. But to Mutch, the single, most important aspect of the mission was exploration. If the lander is a success, then Viking as a project will be a success. Keeping the polar option open permitted them to continue playing the role of bonafide explorers. "The public will appreciate this (and ultimately we're responsible to those taxpayers who foot the bill). In essence we've identified the two most disparate areas on the planet and we're considering going to both. It's not that different from any polar journey. You equip yourself as best you can. You set some intermediate reasonable goals, and if all goes well you make a dash for the pole."

Although the polar option would increase the cost by at least $2 million to $3 million, Mutch felt that it should be preserved. He thought that Administrator Fletcher should award the additional money. "Failing that, it does not seem unreasonable to absorb it within existing Viking budgets—even though there will be associated pain."[52]

Not all of Mutch's colleagues agreed with him, and Young and Soffen continued to receive letters concerning the matter as late as the day of the next landing site working group session, 8 February. Robert Hargraves of the magnetic properties team seemed to favor trying a polar landing. "If 'A' is successful, the prospects of a 77 or 79 Viking Mission are dim, and the engineering risk is not horrendous, I'd say let's try." Hugh Kieffer, representing the infrared thermal mapping team, said that it "moderately opposed" a north polar landing site considering only the infrared thermal mapping experiment. But when members looked at the pole from the standpoint of lander science, they noted that it was a proper objective for Viking; however, it would be difficult to validate such a site for safety. Kieffer, therefore, was "hesitantly in favor of a polar site." C. Barney Farmer, leader of the Mars atmospheric water detector experiment team, was also somewhat ambivalent in his analysis. A polar landing would not be good for his group's experiment, but from the overall science strategy it seemed to be the thing to do. He favored "a polar B-site strategy." The molecular analysis team, led by Klaus Biemann, had met on 10 January at JPL, deciding unanimously in favor of trying a polar landing. It had not considered the funding and risk problems, assuming that the final decision-makers at the Viking Project Office and NASA Headquarters would take them into account.[53]

When the landing site team met at Langley in February, it began to tackle its two problems—a backup site for the first lander and a possible site in the polar latitudes. All aspects of the polar problem were reviewed again, as Carl Sagan outlined the positive and the negative. Sagan had come to the side of the those who favored safety. In a 12 January letter to Young, he had said, "When I total up the pros and cons I find that the scientific advantage of a polar landing site, while real, is far outweighed by the risks." He believed the successful landing of the second spacecraft in the polar region was actually much less than 50-50. When he equated that to losing $200 million—the total projected cost of a future Mariner-Jupiter-Saturn probe—he considered the risk unjustified. But Sagan had other worries on his mind. He was still concerned about the possibility of the first lander's disappearing in quicksand at one of the equatorial sites and favored further study of the meaning of the radar data, expressed in terms of dielectric constants, so the surface-bearing properties of Martian soil could be better evaluated. Generally, he believed too much stress was being placed on visual images at the 100-meter scale and not enough on radar, which could indicate surface irregularities at the 10-centimeter scale. He pointed out:

> It is perfectly possible for a candidate landing site to be smooth at 100 meters and rough at 10 cm, or *vice versa*. Cases of both sorts of anticorrelation are common enough on Earth. It has been alleged that at least in studies of the Moon there is an excellent connection between roughness at 10 cm and roughness at 100 meters. A detailed statistical study of such correlation should be prepared and subjected to critical scrutiny. The [U.S. Geological Survey] seems to be the obvious organization to prepare such a study. . . . However even if such correlations exist for the Moon, it is by no means clear that they exist for Mars. A similar study should be performed for the Earth. This can readily be done by cross-correlating Apollo and Gemini photography of the Earth with radar studies. . . . Until such a connection is clearly shown for the Earth—and I have grave doubts that such a strong correlation exists—we would be foolhardy to attach very much weight to the 100 meter appearance of candidate landing sites on Mars. Unfortunately visual appearance has been given high weight in [working group] deliberations.[54]

Arguments and debate over, the majority of the working group favored northern sites 12 and 13 as primary and backup targets for a polar landing. Tom Young forwarded the group's recommendations to the Science Steering Group that

1. The Mission "A" landing sites with all factors considered be:

|  |  | Latitude | Longitude |
|---|---|---|---|
| Primary | Site  3 | 19.5°N | 34.0°W |
| Backup | Site 11 | 20.0°N | 252.0°W |

The landing site working group on 8 February 1973 debates the choice of targets on Mars for Viking landers. In a voting process on one proposal are (above, left to right) Richard Young, Harold Klein, Henry Moore, William Baum, Noel Hinners, Harold Masursky, and Walter Jakobowski. At left, Norman Crabill and Masursky discuss pros and cons. From left below, Carl Sagan, Tobias Owen, Terry Gamber of Martin Marietta, and Barney Farmer consider arguments for a proposed site.

2. The Mission "B" landing sites, if NASA decides that the sites will be in the North Polar Region, be:

| | | | |
|---|---|---|---|
| Primary | Site 12 | 73.0°N | 350.0°W |
| Backup | Site 13 | 73.5°N | 221.5°W |

3. The Mission "B" landing sites, if the sites are in the equatorial region, be:

| | | | |
|---|---|---|---|
| Primary | Site 10 | 9.0°S | 181.0°W |
| Backup | Site 9 | 9.0°S | 144.0°W |

While approving the A sites for the first lander, the Science Steering Group could not come to an agreement over the second lander's destination. Several members of the group still wanted additional information regarding which areas had the highest probability of containing water in liquid form. In a joint memo to the steering group, Soffen and Young noted that any additional delays would have "a significant impact on the Viking mission design schedule and other Viking planning"; completion of the recommendation by 1 April was extremely important.[55]

As the delays mounted, the Viking management grew restive. Some unknown person suggested that when ultimately chosen the Mission B site should be called "Crisis Continuum," but at higher levels that sense of levity was not shared. On 20 February, John Naugle reported to Administrator Fletcher that the polar latitude site issue had still not been resolved. In reviewing the problem, Naugle went over the "presence of water" issue that was dividing the scientists. "It appears that the regions most recently studied by the Viking Landing Site Working Group may not be good sites from the point of view of availability of *liquid* water because of low temperatures, even though large amounts of water ice are known to exist." Furthermore, *Mariner 9* data being analyzed suggested that the optimum

*Viking landing site working group discussions continue on 8 February 1973, with (left to right) Michael Carr, Gerald Soffen and Thomas Young at the table, Robert Hargraves, Burt Lightner, Arlen Carter, and Priestley Toulmin (back to the camera).*

sites for water availability might be around 55° north. Naugle added that this region had not yet been studied in detail by the landing site specialists. The possibility of water and the geographical differences between the polar and equatorial zones were the two reasons some of the scientists favored landing in the northern latitudes. Naugle said that the Office of Space Science saw sufficient justification, in recommendations from a majority of the landing site working group to plan for a polar landing. "If we do decide to do a polar mission, we have a majority decision . . . to recommend a site at 73°N to you." Liquid water was the nagging question. Where were they most likely to find it? Naugle promised the administrator a recommendation by early April. On the following day, 22 February, Fletcher returned a copy of Naugle's memo with a terse, handwritten message in the margin:[56]

> John N—
> I have two questions.
> (1) Does Lederberg (& his committee) agree that the chances of life are best at 73°?
> (2) Does liquid water have to exist *now* or could it have existed once, for life "signatures" to be detected?
> From my own point of view, the main reason to consider polar landings was to increase the probability of finding life, not to study vastly different geological regions.
>
> JCF

The biologists would commit themselves only to the statement that the chances of finding life "are highest wherever liquid water is present for at least transient periods." As for Fletcher's second question, Naugle reported:

> For an active biota to exist, liquid water must exist at least transiently. Biological "signatures" (e.g., organic molecules) can exist if liquid water was ever present and life existed in the distant (millions of years) past. The difficulty here will be in determining whether the organic molecules detected are the product of biological processes or nonbiological (or prebiological) processes. Viking will detect organic matter, but may not be able to clearly distinguish between biological and nonbiological types. We have newly developed techniques available in the laboratory now . . . but such sophisticated analyses will have to wait for post-Viking or return sample missions.

A special meeting had been held 28 February to consider the question of possible locations for liquid water on the Red Planet. Naugle summarized the results of the "water hole tiger team's" session for Fletcher. No one could say positively that there was any locale on Mars where liquid water could be found. "This does not preclude the possibility for liquid water; it simply means that based on what we know now about the surface of Mars, we cannot determine where liquid water may exist even in transient form." Therefore, it was not realistic to select any site using only the liquid water argument. In the absence of a firm consensus among the biologists, Naugle had to report that he was not much closer to a recommendation than before.

He told Fletcher that he would have the necessary information following the next meeting of the landing site working group.[57]

*Destinations Determined*

The big day was 2 April, and Tom Young wasted no time in laying down the ground rules for this important meeting of the landing site working group. Selection of Viking targets "should be based upon the best knowledge that we have today." In choosing a backup for the second lander, Young wanted the group to assume that the first one had made a successful touchdown. Finally, site selection should be based only on scientific, cost, schedule, and risk considerations—not on policy constraints.[58]

While Naugle and Fletcher were puzzling over a polar landing site, the Viking scientists were changing their minds about the north. The more they talked about liquid water at higher latitudes, the more they thought about the low temperatures they would find there. Could they expect living organisms to survive during the transient appearance of liquid water as that compound passed from the solid to the vapor form or vice versa? Biologist Wolf Vishniac had looked into that question during February but had not turned up any evidence to support the belief that bacteria could grow or survive at temperatures much below -12°C.[59] Studies made with the *Mariner 9* infrared interferometer spectrometer had disclosed surface temperatures ranging from -123°C at the north polar region to +2°C near the equator.[60] The search for unfrozen, active life forms in the northern latitudes on Mars seemed unrealistic.

On 2 April, Lederberg conceded that 73°N no longer appeared to be a rational goal. Now the biologists were seeking a region where condensation might be anticipated that reached a temperature as high as -13°C—they wanted to land between 40° and 55°N. Long hours of discussion followed, during which the working group voted not once but several times on where to send the Viking spacecraft. Site 3 (19.5°N, 34°) was selected as the primary target for the first lander, and site 11 (20°N, 252°) was chosen as the backup. The group narrowed the second mission to two candidates, 16 and 17, but remained undecided over which should be the primary target. The Science Steering Group subsequently made that decision, recommending number 16 (44.3°N, 10°) as the Mission B primary site and number 17 (44.2°N, 110°) as the backup.[61]

Looking back on the ordeal of choosing sites for Viking, Tom Young used the word "traumatic" to describe the process. "We really thought that we were embarking on a reasonably simple task. . . ." But it had been very difficult to focus all the engineering and scientific issues on each specific site; "everytime we thought we [had] it, we would find another problem. . . ." One of the complicating factors had been the continuous stream of new knowledge about Mars. Their immediate need for information had forced them to take a quicker and harder look at the recent *Mariner 9* and Earth-based data than they would have under normal conditions. And each new piece of data changed the Martian picture as the specialists tried to select their targets.[62]

Hal Masursky, who had worked with Lunar Orbiter, Surveyor, Apollo, and earlier Mariner missions, thought the lengthy debate over landing sites had been not only useful but essential. However, he thought that the biologists and organic chemists had not thought through the landing site question; they had had to be educated about the nature of the planet and the spacecraft's capabilities. According to Masursky, some of the scientists developed many of their ideas as the debates went along, and they were forced to analyze quickly new facts at the table. The photogeologist recalled that while Tom Young and Jim Martin had kept pressing the working group for timely decisions, the managers had obviously understood the need for extended debate and had never tried to stifle interchange of ideas. Young and Martin, despite all kinds of external pressure, had managed to protect the scientific integrity of the landing site working group.[63]

Results of the landing site search were made public on 7 May 1973. John Naugle announced that a valley near the mouth of the six-kilometer-deep "Martian Grand Canyon" was the target for the first lander. Known as Chryse, the region had been named for the classical land of gold or saffron of which the Greeks had written. If all went well, Naugle told the assembled press corps, the first Viking would be set down on or about the Fourth of July 1976. The backup to the Chryse site was Tritonis Lacus, Lake of Triton, named for the legendary river in Tunisia visited by Jason and the Argonauts. The second Viking was targeted for Cydonia, named for a town in Crete, with Alba, the White Region, as backup. Soffen told the press that NASA hoped *Viking 1* would be heading for a very safe but interesting target. The scientists had decided early that the first site should be sought in the northern hemisphere (because it would be Martian summer there), at the lowest elevation possible (higher atmospheric pressure and better chance of water in some form), on the flattest, least obstructed region they could find (for landing safety and weather observations). But the second mission had been a different story. The biologists wanted water, and after much debate and study they hoped to find it in the 40° to 55° north latitudes. Their Mission B sites were just above 44°.

But what about these specific sites? What were they really like? Hal Masursky spoke to this point. From the *Mariner 9* photographs, he could demonstrate that about 50 percent of the planet was pockmarked with large craters not unlike the southern highlands of the moon. "We think this is the ancient crust of Mars that was differentiated very early . . . and continued to be bombarded by cosmic debris. . . ." Large basins on the planet recorded that epoch. The largest basin, Hellas, was nearly twice the size of the Mare Imbrium, a giant lunar crater 676 kilometers in diameter. To the north, the planet appeared to be smoother and younger, and scientifically more interesting. Chryse was at the point where a number of "stream" channels appeared to empty onto a plain (Chryse Planitia). Essentially featureless in the Mariner A-frame photographs, there was reason to believe that the area was covered by fine wind-borne materials. The photogeologists believed that there was a likelihood of finding fossil water* in this region, since highland materials would have been deposited here during earlier epochs of water-caused erosion. According to Masursky, "It looks like . . . the Mars environment has been different enough so that there was surface flowage of enormous amounts of water . . . into this great northern basin, and our landing site is at the mouth of that great channel system." The B site, Cydonia, combined the greatest chance of atmospheric water with a low, smooth plain. Masursky thought that this was an optimal landing target. Whereas the first area saw the drainage of the highlands and would likely provide soil samples representing transported highlands materials, the second site, near a large volcanic complex, was covered with basalt flows partly blanketed by wind-blown debris. "We think this combination of sites gives us the best possibility of fossil and present water, and our best samples to test the evolution of the planet."[64]

Selection of the Viking landing sites based on the data available in 1973 was only a first step toward ensuring safe havens for the spacecraft on Mars, and the Viking scientists recognized that additional data should be obtained from Earth-based radar observations and Viking orbital photography. Still, the work of the landing site working group provided the foundation for subsequent debate about the safety of the two Viking landers. The second phase of the landing site story focuses on the certification of the chosen sites, a process that started in May 1973 and was still going on hours before the landers were released for descent to the Red Planet. As is so often true of first steps, site selection was the initiation to a more bewildering process— landing site certification.

---

*Stream channels are equivalent to fossils in that they are evidence that water once existed.

# 10

# Site Certification—and Landing

With site selection behind them, the landing site working group members faced two more tasks—completing their individual commitments to science teams and certifying the Martian landing areas they had chosen. They were reasonably hopeful that the targets were safe ones, but they could be certain only after they had examined additional Earth-based radar studies and the orbital pictures *Viking 1* would send back from Mars. Careful certification of each Viking site would have to be carried out before the landers descended to the Martian surface, but no one expected certification to pose real difficulties now that they were over the hurdle of finding suitable targets. These expectations would be dashed in June 1975.

## PLANNING SITE CERTIFICATION

### *Certification Team*

In August 1973, Jim Martin selected Hal Masursky to lead the landing site certification team, with Norman L. Crabill from Langley as his deputy. This group, which functioned as an operational organization rather than a planning body, included members from orbiter imaging, infrared thermal mapping, Mars atmospheric-water detection, and mission planning and analysis teams, as well as radio astronomers.[1] Together they designed a strategy for landing site certification, which R. C. Blanchard of the Viking Project Office presented at the February 1974 meeting of the Science Steering Group. Blanchard broke the certification process down into four periods: 1. Pre-Mars orbit insertion (MOI) for *Viking 1*. 2. Post-MOI and prelanding for *Viking 1*. 3. Postlanding for *Viking 1*; pre-MOI for *Viking 2*. 4. Post-MOI and prelanding for *Viking 2*. Blanchard noted that before the first Viking spacecraft orbited Mars, new sources of data that might possibly affect the landing sites could include Earth-based radar studies of the planet, Soviet missions flown before June 1976, and scientific observations made by Viking as it approached Mars. Analyzing all new information would help them make a "go/no-go" decision concerning the desirability of landing at the prime site latitude and, they hoped, would contribute to "A-1" site (first choice for first lander) certification.

*Viking 1* would make extensive observations of the prime site, with special emphasis on the low-altitude photographs obtained during the close approach (periapsis). In addition, two or three picture pentads (groups of five photos) would be taken on each revolution to permit comparison of images taken at different exposures (due to the elevation angle of the sun). The A-1 site would also be studied by the orbiter water-vapor detector and infrared thermal-mapping instrument to determine if the scientists' preconceived notions about the target were valid. *Viking 1* would also observe the second lander's primary target (B-1) from low altitude with two picture swaths and one high-altitude pentad. Should the A-1 site be found acceptable (certified), then the lander would be targeted for that site. If it was not acceptable, then the backup site (A-2) would be examined. Once a landing area was chosen, orbiter trim maneuvers would fix the spacecraft's periapsis near that site.

During the third period, postlanding for *Viking 1* and preorbit insertion for *Viking 2*, information sources available to Earth control would include B-1 site data from the first orbiter, entry and landed science data from the first lander, evaluation of the first site certification procedure, and approach observations made by *Viking 2*. The team would then make its commitment to the B mission target. Once the second craft was in orbit, the men would confirm a B-1 site using additional data from the second orbiter and the further assessment of *Viking 1* science results. Blanchard assured the Science Steering Group that the A-1 and B-1 targets chosen by the landing site working group definitely would be used, unless compelling arguments materialized to require a change. Further, the scientists were reminded that the certification team would continue to be influenced strongly by considerations of safety during the first landing, but hoped that during the second landing it could look for a more scientifically interesting site even if less safe than the first.[2]

The first new data the Viking team received came from the Soviet missions.

## Soviet Attempts to Investigate Mars

Much to the dismay of everyone working on Viking, the four flights the Soviets sent to Mars in 1973 raised as many issues as they settled. *Mars 4* and 5 were launched on 22 and 25 July, followed by *Mars 6* and 7 on 5 and 9 August. *Mars 4* came within 2100 kilometers of the Red Planet on 10 February 1974 but failed to go into orbit when the braking engine did not fire. On 12 February, *Mars 5* went into orbit. As no effort was made to detach landers, Western observers assumed that these two Soviet craft were designed to operate as orbiting radio links between landers aboard *Mars 6* and 7 and tracking stations on Earth. *Mars 7* approached its target on 9 March, but the descent module missed the planet by 1300 kilometers when some onboard system malfunctioned. On 12 March, the remaining vehicle separated from its carrier ship, which then went into orbit around the sun.

*Mars 6* descended directly to the surface and provided telemetry for 120 seconds before it crashed.[3]

Soviet scientists reporting on the descent and crash-landing of *Mars 6* calculated that it landed at 23°54' south latitude and 19°25' longitude in the region called Mare Erythraeum. The landing site was "situated in the central part of an extensive lowland region," part of the global zone of depression extending for several thousand kilometers north and south of the Martian equator. Most of the landing zone (about 75 percent) was heavily cratered. Part of this terrain analysis was based on *Mariner 9* data, but the characteristics of the actual landing zone were determined by the radar-altimeter readings obtained during the parachute descent of the Soviet craft. Additionally, *Mars 6* instruments indicated "several times" more water vapor in the atmosphere than previously estimated, news over which Viking scientists were cautiously optimistic, since it enhanced the possibility of discovering some kind of life forms. *Mars 5* photographs provided additional data on the planet's surface features, and while most of the Soviet findings correlated with previous knowledge and predictions there was one major anomaly.[4]

One of the experiments carried on the *Mars 6* lander was a mass spectrometer designed to determine the gaseous composition of the Red Planet's atmosphere. Although the recorded mass spectrum data were not recovered, engineering data on the operation of the vacuum pump appeared to indicate unexpected quantities of noncondensable gases. Soviet scientists interpreted the data as an indication that the atmosphere might contain as much as 15 to 30 percent argon (contrasting with 1 percent in Earth's atmosphere). The Americans had been operating on the assumption that the thin Martian atmosphere contained less than 3 percent argon. A concentration approaching 15 to 30 percent would force some rethinking about Mars and about Klaus Biemann's mass spectrometer experiment. It would mean that the Martian atmosphere had been much denser in the past than the specialists had believed. That would have made the existence of liquid water possible, but it posed a question—what had happened to those atmospheric gases? That was the puzzler. A great concentration of argon would also require some changes in the use of the gas chromatograph–mass spectrometer, since inert gases like argon tended to impede its operation. Obviously, the Soviet Mars missions had not answered many of the U.S. questions, but they had added another element of excitement to the first Viking landing. Everyone would watch closely the results of the entry science team's experiment to see just how much argon it detected as the lander made its way to the surface.[5]

## THE SIGNIFICANCE OF RADAR

As Klaus Biemann puzzled over argon in the Martian environment, others on the Viking team were tussling with an equally troublesome issue, radar. As a tool to study planetary surfaces at great distances, radar seemed to have immense potential. A signal of a known strength could be transmitted from one of the large radio astronomy antennas on Earth—Arecibo in Puerto Rico, Goldstone in California, or Haystack in Massachusetts—to the moon or Mars. The returned signal could then be compared with known signal characteristics and a judgment made about soil composition, the dimensions of slopes and rocks, and other characteristics of a specific area. Radar promised to give information on a scale of a few centimeters, where orbital imaging would tell the site certification team only about features that were larger than a football field. Radar thus promised to be a powerful tool for certifying landing sites, except that not everyone believed in its promise, making it a controversial issue. Furthermore, this technique could examine only a restricted range of latitudes on Mars.

While the Viking Project Office had been planning all along to use radar as an aid to landing site certification, Carl Sagan, once again acting as a catalyst, forced the issue early in 1973.[6] On 3 February, Sagan wrote to Jim Martin, and beneath the hyperbole of his prose Martin found some specific steps that could be taken to rectify what Sagan saw "as serious shortcomings in the landing site selection procedures." What worried him most was the interpretation being placed on some of the radar signals received from Mars. Some of Sagan's colleagues saw visually smooth areas as sand or dune fields, but he hypothesized that the low reflectivity of the radar was not due to the scattering effect of sand grains and surface ripples but to the absorption of the signal by a deep layer of dust. "At a recent landing site working group meeting we were all entertained to see a Viking lander sinking up to its eyebrows. . . . While a similar suggestion that lunar landing spacecraft would sink into surface dust has proved erroneous, it by

*The 305-meter-diameter radar dish antenna at the Arecibo Observatory of the National Astronomy and Ionosphere Center nestles in the Puerto Rico hills at left below. At right below, the 64-meter dish antenna of the Deep Space Network's Goldstone, California, tracking station faces toward space. Arecibo's reflector surface consists of 38 778 aluminum panels, each about 1 by 2 meters, attached to a network of steel cables. The radar feed mechanism, mounted on a 600-ton triangular platform, is suspended by cables above the dish.*

no means follows that quicksand is not a hazard for Mars." He reminded the project manager that the Soviets had suggested that quicksand might have been the cause for *Mars 3*'s failure.

As a consequence, Sagan made some "explicit recommendations." First, he believed more serious theoretical work was needed to understand better the meaning of returned radar data. Second, Earth-looking radar on satellites and aircraft could bounce signals off terrain thought to be analogous to that on Mars, and as a data base was established scientists could compare radar returns from unknown Martian surface areas with known Earth terrains. Third, Sagan thought that major support should be given to Arecibo, Haystack, and Goldstone Observatories so they could examine Mars in detail during the 1973 and 1975–1976 oppositions. He noted that the Arecibo staff was resurfacing its 300-meter radar dish and would be installing a new transmitter. Once these renovations were completed, the observatory would "have a very impressive Mars mapping capability, which should be exploited to the fullest."

Turning to visual imaging, Sagan repeated his concern that smooth surfaces at the 100-meter scale might be rough at 10 centimeters. Had lunar surface data been analyzed to determine if there was any relationship between roughness at the two scales? Hal Masursky's people might look into this matter. And similar correlation of Earth photos should also be studied. He seriously doubted that one could make judgments about the nature of the surface or the scale of the lander from any photographs the orbiter was likely to produce. Sagan believed that radar, properly understood and interpreted, was likely to be more useful in site certification than all the photographs that would be taken.[7]

Sagan's concerns were important ones. Jim Martin and Tom Young considered his recommendations, and on 23 March 1973 Martin wrote to Edgar M. Cortright, director of the Langley Research Center. Martin planned to take three actions as a consequence of Sagan's letter. Arecibo, Goldstone, and Haystack radar facilities would make nearly simultaneous observations of the same areas on Mars during 1973. Since the latitude base that could be studied was limited to 10° to 20° south, none of the candidate sites could be examined, but the information would be valuable because it would contribute to the specialists' understanding of radar's potential in such investigations. The Arecibo team also agreed to make studies in the 1975–1976 period and prepare a quick analysis of its data in the weeks before the scheduled landings.

The second action taken by the Viking Project Office was to set up a radar study team, which would undertake to eliminate some of the ambiguity in interpreting radar data. On 1 March, Tom Young and Jerry Soffen met with Von R. Eshleman and G. Leonard Tyler of Stanford University's Center for Radar Astronomy, where they had been engaged in an active program of analyzing and interpreting lunar radar studies. Tyler agreed to lead the team that would work toward improving interpretation of Mars radar information. Martin told Cortright, "As you are aware, some of the areas with low radar reflectivity are candidate landing sites. We must better understand the meaning of the low radar reflectivity to assure that the current sites are acceptable or guide the selection of proper alternatives." Tyler had his work cut out for him, and Martin arranged for a retreat at which a small group could consider thoroughly the implications of radar studies for Viking.[8]

Tyler presented the results of his study to the landing site working group meeting at Langley on 4 November 1974. Basing his conclusions on data obtained from all three radar facilities, Tyler noted that correlation between radar features and Project Mariner imagery was poor. His study group had learned a great deal: the Martian surface was very heterogeneous on the large scale; Mars tended to have greater variation in surface reflectivity than Earth or the moon; Mars appeared smoother than the moon to the radar; the 100-meter resolution of the orbiter camera system seemed likely to give appropriate information for extrapolating down to the scale of the lander; and data for the 15° to 20° south band of the planet could not be applied to latitudes in the north without variation. Jim Porter, keeping minutes for this meeting, reported that both Tyler and his colleague Gordon Pettengill "laced their presentations strongly with tutorial material which greatly enhanced the ability of the group to understand and correctly interpret their findings."

After listening to Tyler, the landing site working group was unanimous in the opinion that the A and B sites were still the best targets. Although the four targets A-1, A-2, B-1, and B-2 were still believed to be in the correct order of precedence (the Chryse site, A-1, receiving a strong vote of confidence), the team became less enthusiastic in its endorsement of the B sites. They also raised some questions about the C sites that had been located recently at 9° south. The need for new sites had been raised in early 1974 when some of the working group members began to get nervous about what the orbiter's cameras might find. Should the prime and backup sites prove unsatisfactory or if operational difficulties should develop with the spacecraft that would require the selection of some other safe landing spot, they wanted a pair of "super safe" sites where radar, photographic, and topographic information indicated that the spacecraft would have the best chance of landing undamaged. A special subcommittee* had been established to look into possible C sites and make recommendations as early as possible.[9]

The work of the C site subcommittee took longer than the working group anticipated. After meeting in December 1974, the group met again on 6 February 1975 at the Jet Propulsion Laboratory to recommend the study of three latitude bands (8.5°S, 4°S, and 4°–6°N) that would be visible to either the Goldstone or Arecibo radars during August to November 1975. The radar specialists would observe each of these regions as it became

---

*Subcommittee members included Chairman H. Masursky, N. L. Crabill, J. D. Porter, L. Kingsland, G. L. Tyler, T. Owen, H. Moore, G. A. Soffen, and G. A. Briggs.

accessible and recommend sites based on combined radar and visual criteria to the Landing Site Staff, the new name of the certification team, in September 1975. They would repeat the process in November after the 4° north coverage. From these observations, the Landing Site Staff would develop a final recommendation in April for Tom Young, who had become mission director. A detailed alternate mission design (for the C sites) would be developed between December 1975 and May 1976 by Viking flight team members at JPL.

A general feeling among the subcommittee members was that the second mission should be targeted for one of the C sites, since the available radar data indicated that some regions on Mars were very unsafe for landers. The B sites were so far north that radar coverage would never be possible. Norm Crabill wrote in the minutes of the 6 February 1975 meeting that apparently radar data could be used to reject sites, but it was doubtful that it was sufficient to confirm a site. On the other hand, Sagan and some of his colleagues did not want to rely on photos alone. Despite all their earlier work, the landing site specialists were still nervous about their efforts to find suitable landing points for Viking.[10] Putting aside nagging uneasiness, the Science Steering Group and the Landing Site Staff met in a joint session at Langley to consider the recommended process for selecting the C sites. After more discussion of radar as a tool, further explanations of this complex business by Len Tyler, and additional considerations of the argon problem, the joint group approved the proposed plan for C site selection.[11]

## EVOLVING A CERTIFICATION PROCESS

Along with radar and C site problems, site certification remained an open issue. In late May 1975, the Viking Project Office released one of the major products of the Landing Site Staff, a draft of the "Site A-1 Certification Procedure." This document described how the landing specialists would establish the acceptability of A-1 for landing and how they would

*During the 24 February 1975 landing site working group meeting, Len Tyler explains his complex radar studies of the Martian surface to (left to right) B. G. Lee, William Michael, Thomas Mutch, Don Anderson, Richard Shorthill, Gary Price, and Robert Hargraves.*

recommend a target point that maximized the probability of a safe touchdown. Their recommendations would be based primarily on their analyses of low-altitude photographs taken during the first 10 orbiter revolutions. Key to the certification process would be the stereophotographic swaths of the A-1 site taken during the fourth and sixth revolutions—called P4 and P6 photos, as the revolutions were numbered from Viking's periapsides. After playback on video recorders, reconstruction, image processing, and enhancement, these photo frames, in the form of stereo pairs, would be analyzed at the U.S. Geological Survey's Flagstaff facility, where ground elevation, slope angles, and surface roughness would be estimated. This information would be used with photo mosaics made from the orbiter frames to produce geologic maps of the proposed landing site region. Earth-based radar and telescopic observations, oblique and high-altitude photos, as well as Mars atmospheric-water detector and infrared thermal-mapping coverage of the landing site region, would provide supportive information about the nature of the surface. From this data base, landing site specialists would prepare a safety assessment report of the target zone.

During the site certification process, the Landing Site Staff would provide a series of recommendations. Just before the craft was inserted into orbit of Mars, the team would decide either to execute the normal insertion maneuver and proceed with data acquisition or to modify the maneuver if a dust storm or some other anomaly were detected during approach. After playback, processing, and inspection of the imaging system frames from a pass over the A-1 site, it might be necessary to adjust the timing of the orbit or the pointing of the camera platform to obtain optimum coverage of the site on subsequent passes. Four such data-acquisition-adjustment opportunities were planned that would affect the camera sequences at or near P3, P4, P6, and P10. The Viking team would then have to answer the crucial question: would it land the craft or reject the site the team had selected? Recommendations would be made at three points before lander separation from the orbiter. A preliminary commitment to A-1 would be made seven days before separation (at about P9), based on a preliminary assessment of available data. A firm commitment to land would be made three days later, and a precise target point would be established. A final commitment to land, made just before separation, would be determined after examining photos taken during the previous five days to confirm the absence of dust storms and high winds.[12]

In the time that remained before the spacecraft reached Mars, the Landing Site Staff continued extensive preparations for completing site certification as scheduled during the critical period between orbital insertion and lander release. In June and July, a functional test checked the ground-based hardware that would process photos from the orbiter and make the photomosaics and maps. The weakest link in the several-hundred-million-kilometer chain from Mars to the photo analysis labs in northern Arizona seemed to be the 850 kilometers the photographs traveled across the western U.S. Continental Trailways bus express, a leased army aircraft, and datafax were used to strengthen the connection between Los Angeles and Flagstaff. The team did parts of the test a second time, verifying the readiness of the processing equipment and the personnel.[13]

During the last months of 1975 and early 1976, the staff gave considerable attention to timing. Since so much depended on timely certification, scheduling became a paramount concern. The landing site specialists, working closely with the mission design team and the orbiter performance and analysis group, were ready by early February 1976 to test the timeline in what they called the "SAMPD-1" test, an exercise developed by B. Gentry Lee's Science Analysis and Mission Planning Directorate.[14]

## FLIGHT TO MARS

While the Viking team calculated, planned, and debated where to land, the two spacecraft and their launch vehicles were delivered to the Kennedy Space Center. After completion of the prelaunch checkout, the countdown for *Viking 1* began on 11 August 1975. At 115 minutes before the planned launch command, a thrust-vector-control valve—essential to launch-vehicle directional control—failed to respond properly when tested, and the countdown was halted while the valve was examined. Technicians found that a slight leak of propellant had caused corrosion. The valve probably would have worked, but the project management was not willing to take chances with a $500-million payload. The launch was rescheduled for 14 August.

Before the faulty valve could be replaced, another problem was discovered on the 13th. A check of the orbiter's batteries showed they were producing only 9 volts instead of the required 37, having been discharged by a rotary switch that had been turned on inadvertently after the first postponement. Even though the problem was quickly traced and the managers were convinced that it was the result of a failure outside the spacecraft, the batteries still required replacement, a process that would require much time. The entire spacecraft had to be removed from the Titan-Centaur launch vehicle and replaced by the second Viking. Jim Martin and Tom Young had been prepared for such a contingency—the second spacecraft had been tested and was also prepared for launch. This dual readiness for liftoff prevented a costly delay.

Countdown was resumed, and the launch was completed without further incident. *Viking 1* was on its way at 5:22 p.m. EDT, 20 August 1975.[15] The shroud was jettisoned, the spacecraft separated from the launch vehicle, and the solar panels deployed. The star Canopus was acquired by the star tracker on the first try. *Viking 1* was off to a good start.

Repairs were quickly made to pad 41, and the second launch vehicle was readied. Batteries replaced and tested, the first spacecraft was mated to the Titan-Centaur. But new troubles were discovered in the orbiter's S-band radio system during precountdown checkout. When the difficulty could not be solved on the pad, the spacecraft was removed from the launch vehicle for a second time. After replacing part of the S-band hardware, the Viking flight team was ready to try again.

The second launch was a cliff-hanger. The countdown was going smoothly as storm clouds began to gather near the Cape. Seven minutes before the scheduled liftoff, meteorologists at the launch site said that if Viking were not launched within ten minutes the flight would have to be scrubbed because of cloud cover, high winds, and possible lightning. *Viking 2* left its pad at 2:39 p.m. EDT 9 September, just three minutes before the order would have been given to cancel. About five minutes after the Titan and its cargo disappeared into the clouds, an intense rainstorm began and lasted for more than an hour. Eight minutes into the flight, all telemetry from *Viking 2* was lost. Six minutes later, the stream of electronic data returned, and the craft went flawlessly on its way to Mars. Jim Martin may generally have discounted luck in the course of Project Viking, but on 9 September 1975 *Viking 2* was a lucky spacecraft.[16]

The Vikings had begun a journey half way around the sun. For the next 10 months, the landers would be kept in hibernation, with just enough activity to allow the flight team to monitor key systems. When the flight contollers tried to charge the second lander's batteries en route to Mars, the battery charger did not respond to the command. After several days of detailed analyses and tests on the "test lander" at Martin Marietta's Denver factory, the specialists concluded that something inside the battery charger had failed, and they used the backup charger to bring all batteries up to full charge. During November, a complete system checkout indicated that both landers were in excellent condition. Throughout the remainder of the

*The Titan III–Centaur launch vehicle thrusts* Viking I *upward on 20 August 1975 on its 800-million-kilometer journey from Cape Canaveral, Florida, to a 1976 landing on Mars. The spacecraft was to go into orbit of the planet in mid-1976 and, after verification of the landing site, the lander would separate from the orbiter and descend through thin atmosphere to land gently with its scientific instruments and cameras.*

cruise, lander and orbiter science instruments were prepared and calibrated for Mars operations.[17]

Jim Martin and Tom Young noted in a postmission journal article that on paper the mission operations strategy appeared sound, but the "complexity of the mission made us duly cautious." NASA had never had to operate four spacecraft (two orbiters and two landers) at one time, and the Viking managers had sought to guarantee success by extensively testing the hardware and exhaustively training the flight team. Ground system tests verified the readiness of computer programs and all the interrelated equipment scattered across the United States. Compatibility tests between the ground system and the spacecraft led to many software modifications to facilitate command signals. A comprehensive simulation system trained the flight team and checked out the readiness of the entire system, while intentionally introduced emergencies tested the ability of men and hardware to adapt to unforeseen circumstances. Martin said time and time again that he did not believe in luck. In this highly complex business, one should rely only on hard work and brains. The Viking teams tested and retested their systems, and the results often meant personnel reassignment, schedule changes, and the modification of operational concepts.[18]

Of all the tests conducted during the first half of 1976, the most important was the "A-1 Site Certification Timeline Validation." SAMPD-1, Science Analysis and Mission Planning Directorate test 1, was designed to evaluate the *Viking 1* site certification decision process. Participants had agreed beforehand that this would not be a true simulation, since the data from the test would not be run through the computers. Some of the processing equipment was still not ready and, without better information about the landing sites, to simulate photos of those areas would have been difficult. SAMPD-1 would be an intensive review of exactly what data would be available at each step and how that information would be produced and distributed. From this drill, the Landing Site Staff hoped to identify any necessary procedural changes.

Conducted in early February, SAMPD-1 was judged by different parties a success—or a failure. As mission director, Tom Young was satisfied because the exercise had allowed the flight team to evaluate the certification process and discover its weaknesses. But Gentry Lee, science analysis and mission planning director, looked back on the SAMPD-1 operation as a disaster. Flight team members had repeatedly arrived at certification meetings without knowing why they were there or else had attended them because they had had nothing else to do. After the test, Lee took steps to alleviate the confusion. He asked Norm Crabill, deputy chairman of the Landing Site Staff, to prepare a schedule for all regular meetings of the staff to be held during the actual certification process. Crabill was also called on to devise a procedure that would let all the participants know when Landing Site Staff decisions would cause changes in the flight team's plans. Updating documents, plans, and schedules was a major enterprise, matched only by the need to keep everyone working from the same revised materials.

*Table 51*
*Major Training Tests for Planetary Operations*

| Test | Date | Purpose |
|---|---|---|
| Uplink development exercise (demonstration test 4) | 2–15 Dec. 1975 | To design primary mission for *Viking 1* for 12 days following touchdown. Also to train for SAMPD and prepare for demonstration tests. |
| Science Analysis and Mission Planning Directorate (SAMPD) test 1 | 8–12 Feb. 1976 | To evaluate site certification process. |
| Continuation of demonstration test 4 | 22 Feb.–2 Mar. 1976 | Simulated events of *Viking 1* mission from 52 hrs before separation to 8 days after touchdown, to demonstrate capability to perform all necessary sequences and respond to data gathered. |
| Demonstration test 5 | canceled | Demonstration test 4 success obviated test 5. |
| Demonstration test 6 | 31 Mar.–4 Apr. 1976 | Simulated events on orbiter 1 from 24 hrs before Mars orbit insertion to 4 days after insertion, to test downlink and uplink processes. |
| Demonstration test 7 | 7–10 Apr. 1976 | Simulated lander and orbiter operations from day 11, to test activities of active science mission following first sampling of Mars's soil. |
| Demonstration test 4R | 18–22 Apr. 1976 | Detailed simulation of mission from 30 hrs before separation to shortly after touchdown, to retest sequences for separation, entry, and landing. |
| Training test 5 | 26–29 Apr. 1976 | To test landed sequence for 8th Mars day and separation activities, with introduced anomalies. |
| Training test 3 | 2–4 May 1976 | Simulation to test preseparation and separation activities, with introduced anomalies. |
| Training test 4 | 10–11 May 1976 | To train for Mars orbit insertion, with introduced anomalies. |
| Operational readiness test | 2–3 June 1976 | Final dress rehearsal for MOI of *Viking 1*. |

In all, 40 "action items" resulted from SAMPD-1, all requiring resolution before *Viking 1* reached Mars, but once those actions were taken the actual certification process would proceed more smoothly.[19] Subsequent demonstration and training tests were more successful, with each exercise pointing the way toward readiness for the active science part of the mission. On 2 and 3 June, about two weeks before *Viking 1* was to enter orbit of Mars, the last full-dress rehearsal was held without a hitch. The Viking flight team was finished with simulations. It was time for the real thing.

VIKING 1 AT MARS

"Planetary operations" began 40 days before orbital insertion, a date chosen arbitrarily. As the first spacecraft approached Mars, the pace quickened on Earth. Much lay directly ahead—final instrument calibrations, optical navigation, course corrections, approach science observations, Mars orbit insertion, spacecraft navigation, landing site certification, entry, landing, and finally initiation of the landed science experiments.

During preparation for a planned final *Viking 1* course correction maneuver 9 June 1976, 10 days before orbital insertion, the orbiter's telemetry revealed a problem. Helium gas was slowly leaking through the gas regulator that pressurized the orbiter's propulsion system. As Tom Young later described it, a ladder series of pyrotechnically operated valves opened and closed the line from a large helium bottle to the gas regulator, and that regulator was in turn connected to the fuel and oxidizer tanks. The regulator was leaking at a rate that would pose a serious problem. The gas did not leak overboard; it leaked into the fuel and oxidizer tanks, and the pressure could rise so high that not only would the engine stop functioning it would explode.[20]

Should the Viking controllers run the engines in an extra course correction maneuver, or should they fire the last remaining pyrotechnic shutoff valve in the line between the helium tank and the regulator? Another midcourse correction would use up the extra pressure without closing off the gas line, and if the pressure continued to rise after the maneuver it would not be excessive at the time of the final orbital insertion maneuver. But, alternatively, the flight team could close down the pressure line and open it again just before insertion. Jim Martin did not favor the second option because a valve failure would abort the mission. That was a risk he would not take. Martin held "a fairly hairy meeting" at JPL that day, at which he and Young favored another midcourse correction, while nearly all the other members of the team wanted to close the valve. Even John Goodlette, the project's chief engineer, preferred closing the valve. But when Martin telephoned Administrator Fletcher and John Naugle at NASA Headquarters, it was with the news that he was overriding his advisers' suggestion. He was going to make another course correction maneuver on 10 June.[21]

After that burn was executed, the leak continued, but a second engine burn on 15 June reduced the pressure in the tanks to an acceptable level. After orbital insertion, the line between the helium tank and the faulty regulator was closed and the remaining helium posed no further threat. These two maneuvers slowed the spacecraft down, delaying insertion by 6.2 hours. Additional maneuvers could have held the arrival time constant, but the men in Pasadena preferred not to waste the spacecraft's propellant.

Orbital insertion of *Viking 1* required a long engine burn—38 minutes of thrust, which consumed 1063 kilograms of propellant and was more than twice the time of the engine burn required by *Mariner 9* to enter Mars orbit. Viking had to be slowed from its approach speed of 14 400 kilometers per hour to 10 400 kilometers per hour for insertion into orbit. To bring the spacecraft to the proper point at its first periapsis, the mission flight path analysts placed it in a long, looping 42.6-hour revolution of the planet, reaching first periapsis at the time originally scheduled for the second. Previously computed timelines could be maintained with only a minimum of modification.

Great precision characterized Viking's navigation throughout the mission. After orbit insertion, the orbital period was only 12 minutes shorter than planned, even though the mission could have accepted a much larger error at that stage. And periapsis was only 3 kilometers above the predicted 1511. Other parameters were equally precise. A 21 June trim of the initial orbit adjusted the period to 24 hours 39 minutes 36 seconds, by lowering the apoapsis of the orbit from 50 300 to 32 800 kilometers without changing the periapsis. This placed *Viking 1* in the desired orbit, bringing it over the landing site in Chryse once each Martian day. Because of the 42.6-hour first revolution, for scheduling purposes there never was a "first orbit." The P1 calibration photos were lost, and the first photographs of the Chryse region were not received until the third revolution.[22]

### Crisis over Chryse

On the evening of 22 June 1976, the Landing Site Staff was holding its fifth meeting in what was to stretch into a series of 48 sessions before both Viking spacecraft were on the surface. During their early discussions, the scientists had concentrated on the readiness of men and machines to certify the landing regions.[23] In the midst of another theoretical session on the problem of extrapolating downward from the scale of the images produced by the orbital camera system to the size of the lander, reality intruded. At 6:09 p.m. PDT, the first picture of the landing site appeared on the overhead television monitor in the meeting room. Gentry Lee later told the press, "You would have believed that all the people in that room were ten years old because we all got up and forty of us ran over to the scope and watched it come in line by line." Mars as viewed by *Viking 1* did not look like the planet photographed by *Mariner 9*. Their landing site, chosen after years of debate, lay on the floor of what looked like a deeply incised river bed.

Surprise, shock, and amazement only began to describe the specialists' reactions to this first picture.[24]

Mike Carr recalled his feelings when the orbiter imaging team members began to look at the P3 data in detail. "We were just astounded— both a mixture of elation and shock. . . ." They were elated at the quality and detail of the pictures but shocked at what they saw. All their data-processing schedules had been based on a preconceived notion of what Mars should look like, and this was not it. The night of 23 June stretched into morning as building 264, which housed the Viking scientists working at JPL, became a beehive of activity. The orbiter imaging team was busy arranging photographs into mosaics, counting craters, and evaluating the geological nature of the region. All that they saw—the etched surfaces, the multitude of craters and islands in the channels (all at the 100-meter scale)—told them that the A-1 site was not a suitable place to land.[25]

The Mars of Viking was strikingly different from the Mars of Mariner for two reasons. First, the Viking cameras permitted the imaging team to see far more detail. And second, they could discriminate ground features more readily because the Martian atmosphere was much clearer. Hal Masursky remarked that large lava flows in the Viking photographs were totally invisible on Mariner images. "There was enough fuzzy in the air so all that stuff just vanished into gently rolling topography. We can see the sharp edges of little tiny lobate lava flows standing on one another." From studying the Mariner findings, the photogeologists had come to believe there were very few small craters on Mars; now they found fields of them. Masursky recalled, "Jim Cutts wanted us to . . . count all these thousands of craters. . . . That's interesting, but it wasn't necessary for site certification. You can take off your socks and count all the craters you need" to know that it was a dangerous place to land.[26] Masursky and his colleagues now understood that the dust had never really settled during the *Mariner 9* mission. Instead of a blurred surface, they now saw a fantastic array of geological detail. Mars was at once an intriguing and forbidding planet.

There were other problems, too. At the Landing Site Staff meeting on the 23d, Gentry Lee said that he was nervous about the analysis effort. Great attention had been given to planning the gathering of data, but the analysis was diffuse. Carr and Masursky shared his concern. As the data continued to pour in, it was obvious that more discipline was needed in evaluating the hazards (craters, depressions, knobs, and islands) and mapping the geological structure of the landing area. Meanwhile, new computer programs had to be written and additional consoles rounded up and plugged into the computer at the California Institute of Technology. A series of task groups was established to take on the work, and a group of JPL summer interns (engineering undergraduates) was put to work counting craters and other hazards. Carr reported that there was a period of floundering, but the landing site team soon got reorganized and back on the track. From that point, despite the long hours, the team worked more efficiently.[27]

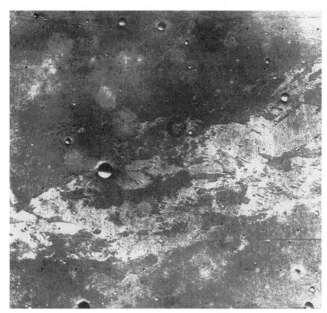

*The first closeup of the Chryse region on Mars—the A-1 candidate landing site photographed 22 June 1975 from* Viking 1 *orbit—changed the Viking schedule. A channel floor with depressed areas and irregular edges, as well as the many craters, did not make an inviting area for the lander. The center of the photo is at about 18°N latitude, 34° longitude. Other photos (opposite) followed.*

At a 24 June Viking press briefing, Lee explained what was going on behind the scenes. Between 300 and 400 persons participated in the site certification process. When the pictures came down from Mars, JPL, the Astrogeology Center at Flagstaff, and several other organizations went to work. Every night, a Landing Site Staff meeting was held, divided into two portions—operational and analytical. Were the photos, mosaics, maps, and the like acceptable and on time? What did it all mean? To find a safe place large enough for a landing ellipse, the team would need more photo coverage, possibly to the northeast or northwest of the prime A-1 site.[28] Apparently the spacecraft could go either direction without upsetting the timetable for a Fourth of July landing.[29] The next 28 hours were just the beginning of a very busy, tension-filled period.

At noon on 25 June, the press heard from Lee and Carr. The Viking Project Office had decided to move the P6 photo coverage 60 kilometers farther to the northeast than previously planned, to avoid the southwestern part of the original landing ellipse where the so-called etched terrain, or scablands, were. Just before midnight on the 24th, the latest P4 photos had come in to fill the gaps in their mosaic. Lee, Carr, Masursky, and nine other members of the imaging team had sat there for more than 30 minutes sliding ellipses around on the mosaic in an effort to find an area where it might be safe to land. So far, there was no safe haven.[30] To study these images, Mike Carr had three groups working for him. Bill Baum led the analysis of atmospheric phenomena. Ronald Greeley was in charge of geological mapping. Jim Cutts and Win Farrell were making quantitative analyses of landing site hazards, and Henry Moore oversaw the mapping of these craters, knobs, and hummocks.

Two landing site meetings were held on the 25th. The discussions centered on one key question, "Do we continue at A-1, or do we prepare to go to A-2?" Masursky summed up the situation:

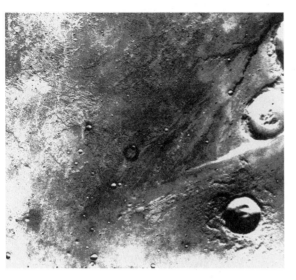

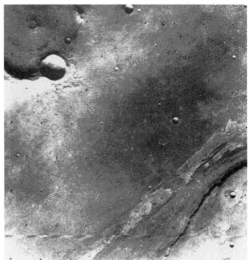

More photos of the Chryse region on 22–23
June 1976 told the unhappy story. The A-1
landing site was not safe. Above left, a camera
on orbiting Viking 1 photographed an "island"
in a rough channel complex with eroded rims.
A closer view above right shows a channel and
craters. At right, "islands" with etched layers
of rock are in the channel of Ares, largest channel in Chryse, in a pair of high-resolution photos. Meteorite-impact craters pepper the surface. Below, a mosaic of 12 photos is the center
third of a strip taken by Viking 1 on its second
low pass over Mars on 23 June. A-1 lies toward
its center.

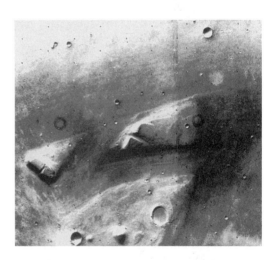

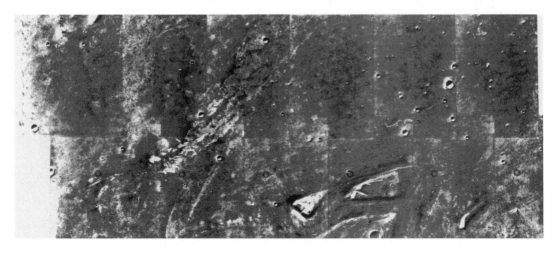

The P3 reconnaissance coverage was successful in that it provided some room to maneuver the ellipse when the original location turned out to be unacceptable, and we have some overlap on the last Arecibo radar coverage at 17–18°N.

The resolution of the Viking 1 pictures is several orders of magnitude better than Marine 71 at Chryse. We can now see and identify objects as small as 130 meters, so we have a powerful tool for looking at surface texture. . . . Mariner 71 had 2 passes over Chryse—one at very high sun angle, the second at a low sun angle. . . . During the second pass, the planet was much closer and more surface textural details could be seen. The planet appears much more clearly *now* than even during the last of the Mariner 71 mission. The channels (scablands or etched terrain) we are now seeing were the vaguest kind of markings in the '71 pictures; on Viking 1 pictures we can also see layering and ejecta blankets, so we are in a much better position to evaluate site hazards and adjust the ellipse location. We are moving the ellipse around to avoid islands, craters, ejecta blankets and etched terrain. The current location is about 19.35°N and 32.5°W [the so-called A-1 South site] centered in cratered lunar mare type terrain unit. Many successful landings have been made on the moon in that kind of unit.[31]

At the end of the staff meeting, a straw vote indicated that 20 members of the group favored staying with A-1, while 24 wanted to move on to A-2. Jim Martin did not vote, but he indicated that he would take all the views into account before the decided which course to follow.

Martin was not long in making his decision known. A landing site meeting that lasted most of the day on 26 June suspended the regular order of business to let the group concentrate on two options outlined by Gentry Lee. The spacecraft could be moved immediately to the northwest for a possible landing in the region called Chryse Planitia. Photographs of that area would be compared later with Arecibo radar coverage scheduled for 4 and 5 July. Or they could reject Chryse altogether and go directly to A-2. Martin explained that he had decided not to land at A-1 (19.5°N, 34°) or the alternate A-1 (19.35°N, 32.5°) on 4 July because project specialists did not understand the processes that had formed some of the visible topographical features. Without a clear understanding of the geology at the 100-meter scale, predicting what the surface would be like at the scale of the lander would have been nearly impossible. Now that the decision had been made to give up the attempt to land on the Fourth of July, a new strategy could be established.

The team's major concern was that so little time was left for determining a course of action for *Viking 1* because of communications complications that would be posed by the arrival of *Viking 2*. Once the second craft came close to the planet, Earth-based controllers would have to ignore *Viking 1* temporarily. According to Martin, the first new milestone would come on 29 June. By that time, they would have the P8 and P10 photos from the northwestern portion of Chryse. If those images indicated an impossible terrain, the orbiting *Viking 1* would be commanded to move over the A-2 region for landing release on 20 or 22 July. Martin pointed out that it was now essential to get more coverage at the B-1 and C-1 sites. The delay in landing meant they would know less about the surface than they had planned when it came time to find a landing site for *Viking 2*. The Landing Site Staff had hoped to have orbital and surface photography that would establish "ground truth" for the orbital images. According to Masursky, "ground truth" simply meant that you could trust the 100-meter photographs to tell you there was nothing at a smaller scale that would hurt the lander. There was no time for determining such truth now.

Jim Martin especially wanted a couple of passes over the C-1 region so the photographs could be compared with the radar observations made earlier. He believed there would be suitable landing areas northwest of the A-1 site. If better targets did not materialize, they would move *Viking 1* over A-2 after photographing the C-1 area. Martin tersely explained:

> The risk we are running in this change in plans is that we may have 2 landers in orbit at the same time. The last date in July we can land VL-1 is 24 or 25 July. From July 26 through August 8, we can't land VL-1 due to Mission 2 work. If we have any problems in any of this new plan, we will have 2 landers in orbit and we may have to land one after conjunction.

Furthermore, after the November-December conjunction, Viking could not land at the C site because the region would have begun to warm up and the biologically important water would have dissipated. However, they could land the second craft at C-1 before conjunction, leaving the first orbiter and lander circling the planet, temporarily inactive. At the 26 June meeting, Martin asked the group to vote on three options:

  a. Do you want to land at A-1S (19.35°N and 32.5°W) on July 4, 1976?
  b. Do you want to observe NW of Chryse and plan to land there July 21 with the contingency to go to A-2 and land after August 8 if anything goes wrong?
  c. Do you want to go to A-2 as soon as possible, keeping B1 and C1 observations and landing about July 22?

The votes were 24 for option a, 17 for b, 2 for c.

Len Tyler reminded the group, however, that the facilities at Goldstone, Haystack, and Arecibo had already made radar observations in the 22.5° north, 36.1° region that indicated the terrain became rougher to the northwest. He expected the upcoming Arecibo observations to confirm this evaluation, and he predicted slopes of up to 8 degrees. Mike Carr argued that the Viking lander was extremely tolerant of slopes up to 25 degrees and less significance should be given the radar results. The differences in outlook between the radar specialists and the photogeologists were becoming more apparent.[32]

Closing the meeting with, "Safety is the only consideration," Martin went to telephone Washington: there definitely would be no Fourth of July landing. It was after midnight on the East Coast when Nick Panagakos, the NASA Headquarters public affairs officer at JPL, began trying to find Administrator Fletcher. After several hours, Fletcher was reached in San Diego, where he was addressing the American Academy of Achievement. Although disappointed to hear that Viking would not land on the Fourth as planned, he immediately agreed that a safe landing was the paramount concern. He authorized a news release that could be delivered to the eastern newspapers and television networks on Sunday before the media representatives left for California to cover the landing.[33]

---

**VIKING MISSION STATUS REPORT**
12:01 a.m., PDT, Sunday, June 27, 1976

NASA has decided to delay the Mars landing date beyond July 4, pending a further investigation of likely sites on the Red Planet.

Project officials feel that the terrain in the pre-selected landing area, called Chryse, may be too hazardous. Orbiter photographs taken during the past few days reveal a much more cratered and rougher area than previously shown.

Officials want to study an area to the northwest of the primary landing site, called Chryse Phoenicia, which may be more suitable than the previously selected site.

A new landing date will be selected in the next several days, depending on what new information is revealed by further site investigation, officials said.

Additional details concerning the rescheduled landing of Viking-I will be discussed at a news briefing at the Viking News Center at 9:00 a.m., PDT, Sunday.

Viking-I has been orbiting the planet since June 19, taking photographs of potential landing sites.

---

Jim Martin met with the press Sunday morning, 27 June. "After careful examination of the landing site pictures that we have been taking for the last several days, we have decided that the A-1 area . . . appears to have too many unknowns and could appear hazardous." He had decided, and the NASA leadership had agreed, to postpone the touchdown while other areas were examined. He explained the A-1 northwest strategy, which if unsuccessful would be followed by a look at A-2. By going northwest, they hoped to get out of the channel, or "river bed," and into a basin, or "river delta," region. "It has been suggested that the fine material that has been washed out of the river bed . . . has been swept downstream and maybe has collected in this basin. If so we might expect to see sand dune fields, we might expect to see craters filled with sand or dirt." He hoped this could be a better landing site. Noting they had always planned for such a contingency, he outlined the steps they would take during the next week.

After the C-1 photos became available, Martin thought the project team could draw some more decisive conclusions, but he warned the Viking specialists that even after their examinations they still might not come to understand Mars. "Things completely unknown to us" might be going on there, Martin said at the press conference. Available *Mariner 9* photography indicated that A-2 was likely to be rougher than the parts of Chryse seen thus far. Unhappy as they might be, they might have to land at a point in A-1 that they did not like. "If that were to happen, we would land some time between the 8th and 12th of July." A landing at A-2 would take place on the 22d or 23d. At this time, Martin could not be more specific; he and his advisers needed more data.

During the question and answer session that followed the press briefing on the 27th, Martin was asked if any single factor had caused him to decide against A-1. He replied that he had been concerned since he saw the first pictures and a great deal of analysis had been done since then. Hal Masursky had been working 20 hours a day; others on the team had been putting in 16 to 18 hours daily. The telling points came at the meeting on Friday evening and the long session held that day. "I came to the conclusion last night that I had enough concerns about the safety of the landing site that I thought we must go examine additional sites."

No one understood how the Martian "river bed" had been formed. Masursky added to Martin's remarks that the geologists just did not have enough data to make judgments. With just one site, it was hard to say what the surface was like; they needed comparative data. The P9 photos of B-1 and the P12 coverage of C-1 might help. Meanwhile, detailed analysis of existing data, including the reprocessing of photos using the computers, would give them a better idea of the terrain they were up against.[34]

*Men and Machines*

Behind the scenes, much hard work and intense activity was under way. Viking's cameras had taken some 200 photographs by 27 June 1976, and an additional 40 covering the B-1 site were taken on the 28th. Getting these images was no simple job. Masursky commented that many of the young people he had talked with had thought NASA's unmanned space projects were controlled by one great computer with no human beings involved. For Viking, the computers were essential, but they were only a tool to aid the scientists and engineers. As Masursky put it, "Computers are just like wearing shoes. You need them when you are walking on gravel, but they don't get you across the gravel." Viking was people interacting with the computers and with one another, and according to Masursky it was "an intensely human experience. It was young college undergraduates counting craters. Grunt work is what the photogeologists called it, but it was essential." Hour after hour, they peered through magnifying glasses, counting large craters and those no bigger than a pin prick. It may have been grunt work, but to someone 17 or 18 years old it was exciting to be at the center of a major space project and know your work really counted.[35]

Gentry Lee also talked of the persons who worked outside the lime-light. Many Viking team members thought of themselves as Earth-bound sailors guiding their ships across the vastness of space. Jim Martin and Tom Young stood in the command center, surrounded by their technical and scientific advisers. Many of these men became known to the press as they went before the microphones and cameras to explain the problems and progress of the day. Lee, Masursky, Carr, and other members of the science team became familiar faces on the evening news. Even Pete Lyman, head of the Spacecraft Performance and Flight Path Analysis Directorate, took time out from his busy schedule to brief the media on Viking's status. But many others working "in the bowels of the organization" the reporters did not see. Akin to the boiler room crew on a ship, they did all the work necessary to enable the men at the top to pick from several options; they did all the paperwork, computer programming, and system checkouts. Lee noted that he and others toward the top of the project hierarchy got positive re-inforcement for their efforts; they got their names in the paper, they got their faces on television. But the stokers in the boiler room just got groused at and told to work faster and harder. At least 60 persons reworked mission blueprints every time a change was made in the proposed landing zone or in the date of the landing. But the esprit de corps was excellent because each person was doing the job for which he had trained. They were doing more than they had expected, but pride being part of the Viking team made the extra effort a matter of honor.[36]

As things worked out, the hard work had just begun. Landing Site Staff members had to schedule their duties around noon status briefings for the press and their evening staff meetings. Sometimes working copies of mosa-ics were spirited off to the photo lab so that composite pictures could be released to the media. Given the strong interest and positive attitude of the news people, the photogeologists could not really complain, but such incidents were trying. It was not uncommon for new data to be delivered during staff meetings, and Masursky and his colleagues would be called on to make instant analyses before a group of several dozen specialists. Instant science became a way of life during the last days of June and early July. There was no time for idle speculation, no respite for reflection. Decisions had to be made against the clock and the mission schedules. And only human beings could make these decisions.

### Which Option?

"If one sets off as Columbus did to find a new world, he would not apologize for looking for a safe harbor," Jim Martin commented to the press on 28 June. To give them some idea of the complexity of the decision-making process, the next day Martin distributed to the press a "logic flow diagram so each of you can be your own judge of where we should go based upon the evidence." After some laughter, he added, "I kid you not. You see almost as much evidence as I do. So at each milestone, you can decide which way you would go." At the two-hour Landing Site Staff meeting that evening, the attendees considered that same logic flow chart as they prepared for their big session on the 30th.[37]

Meeting for the 13th time, for four and a half hours on 30 June, the Landing Site Staff wrestled with the three potential landing targets. During the facilities report, the men in charge noted that fatigue was beginning to catch up with some of their people. In turn, 10 specialists reported their latest information and opinions.

Masursky synthesized the A-1 site selection and certification process: Two and a half years earlier they had put the ellipse at the channel fronts "in the hope of getting wet sediments." This spring they had added one further northwest site, to avoid channel-borne boulders, and one northeast site, to include the best radar location. "In all of this, we did not anticipate that the channels would be incised deeply in the A-1 site region." *Mariner 9* had shown gentle workings in the areas where they could now see stream cratering. Comparative crater counts showed that if they had gone to "A-1 biased" (A-1 revised, formerly called A-1S, 19.35°N, 32.5°), "we would have been in reasonable shape. There is not a significant difference between A-1 and A-1NW. We did have to go further NW to avoid incised channels, but this is not a marked change. Our course is not dramatically different from what we set out on" two and a half years earlier. A-2 seemed less safe at the moment; knobs and craters were more predominant, according to *Mariner 9* findings. In the A-1 region, Masursky said, "the ranking is straightforward. Jim Cutts' crater counts clearly show that we are moving in the right direction. The choice is heavily weighted to the NW." The sun elevation, however, was somewhat higher in the P8 and P10 pictures than in P4 and P6, and lunar experience indicated that as the sun went up craters disappeared from photos.

Overall, Masursky rated the sites: A-2, worst. A-1 revised, next. A-1 NW, best from available data, the most favorable site at the moment.

When Masursky finished his summary, Tom Young asked for a vote by those who had experience or a feel for the factors. No one present was ready to land at A-1 revised. Thirty specialists favored A-1 northwest, while two wanted to go to A-2. Three abstained—Carl Sagan, Len Tyler, and Henry Moore—because there were not enough data to make a decision. Under questioning, Sagan said if he had to he would vote for science and choose A-1 revised. Tyler was strongly negative to A-1 revised and said that A-1 northwest and A-2 looked equally good. Moore favored A-1 northwest and A-2. If Arecibo radar confirmed the site, he would vote for A-1 northwest, but he just did not feel comfortable about trying to land without radar. Jim Martin closed the meeting by reminding the group that he had not made his decision, because not all the P10 photos were available yet.[38]

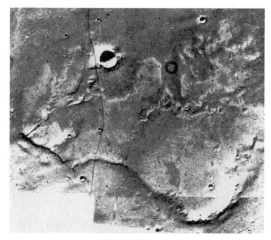

*Views on 27 June 1976 of the A-1 NW Chryse Planitia site, an alternative to the A-1 landing site on Mars, did little to relieve Viking team worries.* Viking 1's close look, in the two orbiter frames at left, reveals an impact crater, ejecta blanket, many small craters with wind tails (probably dunes), fractures, and knobs of rock. Above, the irregular south edge of a plateau appears to have been shaped by the flow of water.

Fifty-three persons attended the project manager's 8 a.m. landing site meeting on 1 July. Without "rehashing" the previous night's meeting, Martin wanted to hear any new information from those who had been working all night, and then he wanted another vote. In the course of the discussion, it became apparent that the worrisome factor was the Arecibo radar observations scheduled for 4 and 5 July. John Naugle from headquarters asked the assembled specialists how bad the radar at A-1 northwest had to be "to make us go to A-2." Len Tyler said it was difficult to anticipate the results. If the data were similar to the Goldstone results for A-1, he thought they should go to A-2. Carl Sagan thought that two different situations were possible at A-1 northwest—good quality pictures with mediocre quality radar, or good quality radar with mediocre pictures. For safety, he wanted to see both good-quality radar and good-quality photographs. After some additional discussion of radar by Von Eshleman, Martin called for a show of hands. All present were asked to vote, and to discourage fence-sitting he told them that anyone who abstained would have to explain his position. Should they try to land at A-1 revised? No hands were raised. Should they try for A-1 northwest? All but one voted for this option; he favored the third possibility, A-22.[39]

At the noon briefing, Administrator Fletcher and Jim Martin talked with the press. Fletcher congratulated Martin and his people for their hard work and their apparent success in finding a safe landing site in what the

Administrator called the "Northwest Territory" (A-1NW). Martin reported that the flight team was working toward a landing on 17 July at 3 a.m. PDT. Looking back later, Martin said it had been very troubling to find geological features that the specialists neither expected nor understood. On Friday, 25 June, when the landing team was still struggling with the question of what was going on in that area, Dr. Bob Hargraves from Princeton had made a suggestion that caught Martin's attention. If you do not like this river bed area from which material has been excavated, look for the area where that debris has been deposited. "We started thinking about where that river went." Since the river bed seemed to be going off to the northwest, "that prompted our decision against landing at A-1.[40]

Gentry Lee and Hal Masursky had vivid recollections of that event. Lee commented on the decision to go northwest: "I distinctly remember the point where I believe Jim Martin changed his mind because I may have been an hour or two ahead of him. . . . It wasn't exactly as if light bulbs went off, but Bob Hargraves has a way at times of explaining things in such a— especially big geologic things—in such a way that it becomes very clear." Masursky thought it ironic that the leader of the magnetic properties team could make so easily a point that he, the leader of the landing site certification team, had been trying to get across for 18 months. But when Hargraves had said, "Let's go downstream," it had come at precisely the proper moment for Martin to react. Since Bill O'Neil and his navigators had already worked out the procedures beforehand at Masursky's suggestion, the Landing Site Staff had been able to move quickly once it had decided in which direction to move.[41]

Between 1 and 6 July, as the project team waited for the Arecibo results, the Landing Site Staff continued to evaluate existing information. During an afternoon meeting on the 6th, Tyler telephoned from Stanford, where the Arecibo data were still being analyzed. Generally, the results seemed to corroborate earlier observations that the topography between 30° and 50° longitude was rough—A-1 northwest was at 43°. Farther to the west, it seemed to get smoother. The news was not reassuring.[42]

### Renewed Crisis

Two crucial meetings were held 7 July. At 8 a.m. PDT, Len Tyler presented the results of the Arecibo radar observations of early July. His remarks were essentially the same as those given over the telephone the previous afternoon:

Observations—

1) Good data obtained from Chryse Planitia July 3, 4. Data to West and to the East obtained July 2 and 5 respectively.

2) July 3 and 4 provide detailed repeatable results from 41° to 46°W with integrations as short at 0.7° in longitude.

341

Results—

3) One-half power widths generally corroborate Carpenter['s 1967 observations]. Generally rough between 35° and 50°W, smoother to the East and to the West; with general quantitative agreement.

4) Chryse Planitia is a complex radar area, generally of roughness comparable to area observed by Arecibo SW of A-1 (33°–37°W, 17.5°N). On the average, Chryse Planitia and SW A-1 are not distinguishable by the current observations.

5) Spectra from Chryse Planitia on both July 3 and 4 show a sharp drop (2:1) in total reflectivity at about 44°W (23.2°N). This is interpreted as a marked increase in roughness and/or decrease in reflectivity at that location. However, the apparent abruptness of the change cannot be understood in terms of a simple two-unit model for the scattering, indicating the complexity of the area. (One needs to build specialized models to explain such abrupt behavior on spectra averaging over wide areas.)

6) Spectra from Chryse Planitia on both July 3 and 4 show a "spike" corresponding to *approximately* 42°W, suggestive of a smoother area near that longitude.

7) There is no area within the regions probed in Chryse Planitia, of size greater than about 3° in diameter, as smooth as the Martian average (assuming that the reflectivity is not also anomalously low).

Radar was saying that the surface was rough where the photographs had indicated it was smooth. The question was which to believe—whether photos you can see, but at a scale larger than the lander, or radar, which produces only spectral lines on graph paper but which supposedly has "felt" the surface. Tyler's conclusions were that the southwestern and northwestern regions of A-1 were twice as rough as the Martian average and that west of 50° the surface was back to average. Tom Young closed the morning session by summarizing their choices: (a) Go to A-1NW. (b) Go to A-1WNW—because of new radar results, no Viking visual imaging. (c) Go to A-2—because of old radar, no Viking visual imaging. "We may be surprised at A-2 and there is a timeline problem." Young sent the Landing Site Staff off to study these options before they reconvened at 4 p.m.[43]

Jim Martin summed up the new situation for the press at noon: "The visual images are only really telling us what is observable at . . . 100 meters and up, . . . Rose Bowl size hazards." Tyler and his colleagues believed that radar "feels slopes, boulders, in the order of a meter or a few meters in size." Martin and his men had a decision to make that night—go ahead with the plans for a 17 July landing or use the next day's maneuver to look for a new site in the 50° longitude area. The map looked good, but no detailed photographs had been taken in that region. Should a decision be made to look farther west, any landing would be delayed another three to five days. He believed that the radar data looked good; the problem was one of interpretation, and he had to admit that there were differences of opinion as to what the radar was telling them.[44]

Martin, presiding at the 18th Landing Site Staff meeting the night of 7 July, opened by saying: "We must move forward, if not to land, to do other things. We must today tell" the Spacecraft Performance and Flight Path Analysis Directorate what direction (east or west) to go tomorrow. "The outcome of these discussions will be to continue to 23½°N and 43½°W (the A-1NW site), or go over and observe farther west. . . . We must be prepared to continue beyond 6 p.m. tonight to air all viewpoints."

Tom Young took the floor. The fifth Mars orbit trim maneuver had to be executed at 5 the next evening. "We will correct latitude and walk [move the spacecraft] westward. If we decide tonight for A-1NW, we can land on July 17. If we decide to go to WA-1NW [west A-1 northwest], we keep on walking, and land at WA-1NW about July 20 plus one or two days." The calendar of events was a full one:

| | |
|---|---|
| Thursday, 8 July | P-19, Mars orbit trim 5. |
| Friday, 9 July | P-20, take 80 frames of monoscopic reconnaissance coverage of WA-1NW to 55°, contiguous to the P-10 coverage. |
| Saturday, 10 July | P-21, photo coverage. |
| Sunday, 11 July | P-22, 80 frames contiguous to P-20 coverage. |
| Wednesday, 14 July | P-24, accept or reject area covered in P-20 and P-22. |
| Thursday, 15 July | P-25, Mars orbit trim 6. |
| or | |
| Friday, 23 July | P-33, Return to A-1NW and land there. |

And photography versus radar continued to be a dilemma. Site A-1NW assessment by radar was that it was "bad"; by photos, "good." Site WA-1NW radar assessment was "good," but no photos would be available until P-20 and P-22. Good photos to accompany the good radar of the "far west" would mean a landing there. Bad photos and good radar would mean going back to A-1 northwest. It was obvious that it was difficult to say exactly what the various radar signal returns meant. Sometimes the Landing Site Staff could say with assurance that a particular signal reduced to spectral lines on a graph equaled a specific terrain. Other spectra were just not fully understood. Tyler was the first to say that he did not "want to land without images" of the landing site. Young gave the group two choices—A-1 northwest on 17 July, or go west to west A-1 northwest and try for a landing on the 20th. If the pictures there were bad, return to A-1 northwest and land on the 23d.

The vote, when it came, totaled 23 for site A-1NW and 12 for going west. Essentially, the voting indicated that the scientists were ready to land

343

anywhere and get on with the mission. Landing site and project staff members, surprised by this vote, were still playing it cautious and wanted to look at another location before chancing a landing. Martin told the group he would make public his decision that night.[45] That evening the Viking news center at JPL released a mission status report:

> NASA officials have decided to study a possible new landing area on Mars, some 575 kilometers (365 miles) further west than the previously planned site. This will delay the landing of Viking 1 at least until July 20.
>
> New radar results obtained July 3 and 4 at Arecibo Observatory indicate that a more westerly area of Chryse Planitia may be smoother than the previously selected northwest site. This area . . . has not yet been photographed by Viking.
>
> Viking 1 will perform an orbital trim maneuver at approximately 5 p.m. PDT, Thursday, July 8, to begin moving the spacecraft over to the western region, where high resolution photographs will be taken Friday, July 9, and Sunday, July 11.
>
> If these photographs indicate agreement with the recent radar data, the landing can occur as early as July 20. . . .[46]

Martin gave further details the next day at noon, and Len Tyler briefed the press on the complex business of radar observation and the interpretation of data. He tried to explain such terms as *rms slope*, "root mean square" being a specific kind of mathematical average. He talked about sending a radar beam out to Mars and then 36 minutes later measuring the nature of the reflected signal. Using the analogy of a spotlight he said:

> If Mars were perfectly smooth, one would see a single spot . . . that's about one kilometer in size. That spot would be bright; otherwise Mars would be dark. As you roughen the surface of the planet this single spot breaks up into a multitude of smaller spots so that one sees a speckle pattern around the . . . radar point. . . . This pattern would be bright and otherwise the planet would be quite dim. . . . As you increase the roughness . . . the size of the speckle pattern increases. So a very smooth location on Mars will produce a very tight pattern, and a very rough location produces a broader pattern.

While roughness affected the pattern of the reflected signal, it did not affect its strength. On the other hand, the nature of the surface—hard to soft—influenced the returned signal's power. The Arecibo data indicated a rougher-than-average surface beneath the radar spot when it was aimed at A-1 northwest. With these results the same as Haystack's radar findings of nine years earlier, Tyler had voted to go farther west.[47]

When it was his turn to speak, Hal Masursky frankly indicated that he was puzzled at the discrepancy between the photographs and radar observations. He noted that "if our backs were to the wall we would have . . . [taken] the increased risk of attempting to land in this small area embedded between the radar rough areas" at A-1 northwest. "But since we have the chance of looking just to the west . . . where the radar spectra show a much sharper, cleaner echo, then it seemed prudent to take the additional series of pictures." To put a landing ellipse in a safe part of the A-1 northwest region, "we'd have to put at least half of it outside the photographic coverage and again that didn't seem like a good idea." So Masursky also wanted to look further before committing *Viking 1* to a final landing place.[48]

### Viking 1 *Landing Site Decision*

The fifth-orbit trim maneuver was executed just before 6 p.m. PDT on 8 July. After loading the spacecraft's computer memory with the maneuver command, ground controllers had temporarily lost contact with Viking for an expected blackout period, from 4:40 to 6:13 p.m. *Viking 1* performed the 40.77-second engine burn flawlessly and was on its way to look at the "far west." The next day for important decisions would be 14 July, by which time the P20 and P22 photos would be fully evaluated. Then the project team could choose between west A-1 northwest and A-1 northwest.

Meanwhile, the Landing Site Staff was trying to draw conclusions about the B and C sites for *Viking 2*. On 8 July, Barney Farmer reported on his atmospheric water studies over the Capri region (C-1 at 5°S); at this season of the year it appeared warm and dry, not biologically promising. Farmer remarked that if he had his choice free from all other constraints he would land at Hellas Planitia, which because of its low elevation (high pressure) offered the probability of more water and higher temperatures than other sites. His description of Capri led to a phrase popular with the specialists—it was "hotter than Hellas."[49]

On the 12th, the site staff met to consider the insights for the *Viking 1* site gained up to that point from the P10, P20, and P22 photographs. John Guest of the University of London had reviewed the revised and updated geology hazard map and found that neither textured surface nor grooved plains existed in the landing ellipse, except possibly some fine grooving below the resolution limit of the cameras. Additionally, channels disappeared or stopped rather suddenly, and Guest thought this indicative of their being covered over by wind- or water-borne dust or larger particles (a process called mantling) rather than their being below the resolution limit. Hal Masursky believed that existence of this younger, thicker mantling was consistent with the drop in radar reflectivity in that direction. Norm Crabill noted in the meeting minutes, "As we go west, we get into older geologic units and sharper reflectivity, with sharper features appearing further west." Masursky believed that they had reached the best location for a landing. He reported that although the slopes in the new ellipse (47.5° longitude) were as bad as those in A-1 the radar reflectivity was better. Significantly, this region seemed to have relatively few young impact craters, which meant that the area was probably covered with weathered materials that would pose less of a hazard to the lander.

Len Tyler presented findings from the continued radar analysis. Tongue in cheek, he suggested that the reflected signals dropped off significantly either because of scattering caused by the surface or because of a hole through the planet. But, despite the "Chryse Anomaly," he noted that the surface looked better at 47° to 48°. Radar data were once again the subject of considerable discussion among the specialists, but after a couple of hours Martin closed the session. They would reconvene that night to consider the additional P22 pictures processed by then and reach a decision. If they could not do so quickly, they would meet at 3:00 the next morning and continue to meet until they selected a landing site. Some tempers and senses of humor were wearing thin, but Martin continued to display his steady, firm, authoritative manner. A decision needed to made, and he intended to see it through.

Hal Masursky opened that night's session. He saw three possible landing areas: alpha, at 22.4°N, 47.5°; beta, at 22.5°N, 49.0°; or gamma, at 22.0°N, 51.0°. After the staff had moved ellipses around the photomosaics (playing what Masursky called "cosmic ice hockey"), counted hazards, and evaluated radar, alpha looked best. Mike Carr expressed an opinion held by several attendees at that late meeting on 12 July, "Don't prolong the debate, the choice is clear." Too often, he thought, meetings had lasted a specific number of hours simply because it was traditional for them to last that long. He was ready to force the vote.

The alpha site would be a compromise between the hazards visible in the photographs, primarily impact craters and the blocks ejected from them, and the small-scale surface properties "felt" by the radar. A vote was called for, and alpha, at 22.5° north latitude, 47.5° longitude, was the unanimous choice for the spot to land *Viking 1*. The 22d meeting of the Landing Site Staff adjourned at midnight.[50] Mike Carr reflected that he never had any second thoughts once the decision had been made. "I didn't realize how great a strain it had been on me. . . . When the decision was finally made it was as though a tremendous load went off."[51]

With site certification completed on the 12th and the spacecraft's orbit adjusted on the 16th, the project focused its attention on preparing for a 20 July landing. Final descent trajectory information and minor sequence changes were sent to the orbiter, and a set of commands for entry, landing, and the preprogrammed mission was transmitted to the lander. The same set of commands was transmitted to the Lander Support Office at Martin Marietta, where a computer-simulated mission was being flown.

At 5:12 a.m. PDT on 20 July 1976, the seventh anniversary of man's first steps on the moon, the Landing Site Staff learned that the *Viking 1* lander had touched down safely on Mars 19 minutes earlier. The job half done, the staff continued to evaluate sites for the second spacecraft.

## SECOND SITE NO EASIER

On 14 July, after one day off, the Landing Site Staff renewed its work. As part of the project's open policy, several reporters and photographers were permitted to watch the group in action. This meeting concentrated on

*Hal Masursky (above) on 12 July 1976 explains the geology apparent in the P22 photos. Jerry Soffen stands on his right. Across the mosaics, standing from right, are Tom Young, Jim Martin, Carl Sagan, Mike Carr, and Joe Boyce. Photo taken by Hans-Peter Bieman. Target for* Viking 1 *in Chryse Planitia at 22.4°N latitude, 47.5° longitude is shown in a photo taken 17 July. The area, photographed from 1551 kilometers away, is a smooth plain with many impact craters.*

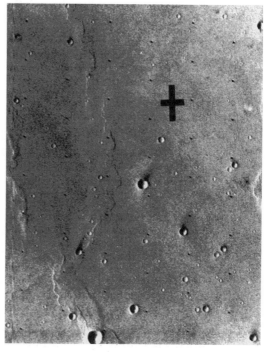

engineering considerations affecting latitude selection for *Viking 2*. The tenor of the discussions indicated that the C site latitudes were less favored than the B site region to the north, because the former were too high, too hot, and too dry. Debate continued as the Landing Site Staff once again tried to evaluate a landing zone without having seen the surface.[52]

On the 16th, Hal Masursky led off the day's session by asking what they would learn if they could put a spacecraft down in each of the sites they had considered. A-1 was where the largest Martian channel complex opened onto Chryse Planitia. According to Masursky and Crabill, it was "the best area to observe where water and possibly near-surface ice had occurred in large quantities in the past—the optimum place to look for complex organic molecules." B-1 had been selected because it was in a region where high water vapor concentration might be expected; also, at this longitude the two orbiters could provide relay support for either lander. C-1 and C-2 had been chosen because they had appeared safe to the radar team. The landing site team had to choose between B and C for *Viking 2*.[53]

Among others, John Guest, Mike Carr, and Ron Greeley presented their thoughts on the nature of Martian geology in these regions. Harold Klein gave his reasons for preferring B; central to his argument was water. And Josh Lederberg, also believed that there was a better chance for water at the northern latitudes. Vance Oyama dissented, saying that the B region was too cold; the temperature was always below freezing. The next day, only 4 members of the group were in favor of the C sites (5°S), while 30 wanted to go to the B region (44°N). Jim Martin reminded them he would have to make the final choice for the second mission by 3 p.m. 24 July.[54] A poll was taken again on the 21st after the first *Viking 1* pictures of the surface had been studied—3 favored the C latitudes, while about 40 voted for B.[55]

### Worries about the B Sites

Cydonia (B-1, 44.3°N, 10°) had been chosen as the *Viking 2* prime site because it was low, about five to six kilometers below the mean Martian surface, and because it was near the southernmost extremity of the winter-time north polar hood. B-1 also had the advantage of being in line with the first landing site, so the *Viking 1* orbiter could relay data from the second lander while the second orbiter mapped the poles and other parts of Mars during the proposed extended mission. While this was a good spot to find water,* Masursky was worried about the geology of the region. He asked David Scott, who had prepared the geology maps, to work up a special hazard map for B-1. After studying the map, Masursky came to the conclusion that the area was not "landable." This analysis, of course, was made with maps based on *Mariner 9* photographs. He told Tom Young and Jim Martin, however, that there was one hope; wind-borne material may have mantled the rough terrain and covered "up all those nasties we see."

The first pictures of B-1 were taken on periapsis 9, and it was worse than Masursky had imagined. "But it was not particularly a shock because I was scared to death of that site before it happened." Masursky proposed a big swath of pictures heading off to the northeast to about 57° north. Somewhere in the "Northeast noodle," he hoped that they would see the mantling develop and cover the rough terrain. Because of engineering constraints, however, *Viking 2* could not land above 50° north. "So," according to Masursky, "we cut off the noodle and . . . called it the 'Northeast rigatoni'—that's a short noodle."[56]

Meanwhile, the lander science team was having some worries of its own with the lander on the surface. The seismometer had failed to respond

---

*If the pressure was as high as 7.8 millibars and the temperature rose above freezing, liquid water was possible at Cydonia.

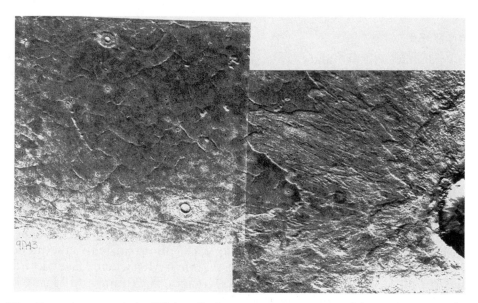

*The first pictures of the* Viking 2 *site were not very promising. Photos of the preselected primary landing site, Cydonia, were obtained 28 June by* Viking 1 *orbiter cameras from a range of 2050 kilometers. The rim of the crater Arandas is on the right edge. Rocks outcrop in the inner well, and ejecta form a lobate pattern with surface ridges and grooves. Small pedestal craters may have been caused by impact and etched out by wind. The center of the left photo is at 43°N latitude and 7.6° longitude; the right, 42.4°N latitude, 7.3° longitude.*

to actuation commands from Earth, and some transient communications difficulties had to be corrected. But most significant for the Landing Site Staff, the sampler arm on the lander had stuck on the third day of the landed mission (sol 3). Although the Viking team was able to diagnose the difficulty, devise and test a solution, and free the sampler boom assembly in time to collect soil samples on schedule (sol 8), the problem emphasized the importance of safely landing *Viking 2*. Simply put, if *Viking 1* failed, then no chances could be taken in choosing a landing site for the second mission. All the tension and pressure experienced during the past month reappeared.[57]

With the second spacecraft about 2.7 million kilometers from Mars, Martin told the press on Sunday, 25 July, that the navigators were going to make an approach midcourse maneuver at 6 p.m. PDT on the 27th to position the craft for orbit insertion on 7 August. Targeting the spacecraft for about 46° north, "we're going into an orbit which will allow us to spend some time observing three possible north latitudes. Two of them are known as B-1 and B-2. . . . We've spent a fair amount of energy looking for landing sites in B-1; so far we haven't seen anything I would like to put an ellipse in." Continued observation of the B-1 "rigatoni" identified a complex history of "aeolian deposition interleaved with erosion stripping." As Masursky and Crabill reported in *Science*, in some places secondary crater clusters peppered the small plains. In other areas, stripping of the uppermost aeolian mantle had left secondary craters protruding as hummocks where crater ejecta had inhibited stripping. "In any case, the small areas of aeolian mantle were not large enough to locate an ellipse in, and the entire B-1 region was rejected."[58] During the 27 July Landing Site Staff meeting, held just before the successful execution of the midcourse correction maneuver, Masursky indicated that there were several nice little spots in B-1 but they were only one-tenth the size of the necessary landing ellipse.[59]

A suitable landing site had still not been found in B-1 when *Viking 2* went into orbit on 7 August at 5:29 a.m. PDT. Project strategy now called for using the second orbiter in the search for a site at Alba Patera (B-2) or Utopia Planitia. Masursky and Crabill noted that coverage of Alba Patera during P4 and P7* "raised and then dashed hopes that it would prove to be a suitable landing site." On revolution four, the photo calibration sequence indicated a possible site in the northern region of Alba where Masursky had located his first ellipse using *Mariner 9* images. As Young and Masursky watched the P7 pictures come in on the digifax machine, the region still looked smooth. Thinking they had found a likely prospect for a landing zone, they decided to get more pictures during the 15th orbit.[60] During the night, the mosaic team pasted up all the P7 photos, and Masursky was shocked by what he saw. Where the individual photos had shown a smooth terrain, the mosaic revealed a territory that was "rougher than hell." The difference was in the computer processing; under the proper processing, the region appeared very rough, covered by textured lava flows. Masursky thought to himself, "I think we've got a problem." He called for a special meeting of the landing team on 17 August.

Gentry Lee opened the 42d meeting with a review of the "fast breaking events" since the previous day. After closer examination of Alba Patera's "smooth" spots, Masursky had concluded that it was not smooth enough. "Nothing in B-2 looks comfortable." Lee also noted that "we have an exhausted crew, as evidenced by the high frequency of errors in the products, from trying to maintain all four options." Those choices had been labeled B-2 early, the earliest site at which they could land; B-2 late, the latest site at which they could land; B-1; and B-3. Earlier that day, the Landing Site Steering Committee, an independent group of scientists advising Martin, had met and decided to drop the first three options and shoot for B-3. This decision was the most controversial action taken during the entire site selection-certification process.[61]

Masursky recalled the events that led to that decision. At the 17 August steering committee meeting, he told Young and the others that B-2 was just no good; they did not need to break their backs getting the P15 photographs. Since B-2 was out, everyone also agreed to drop B-2 early, and Masursky concurred. But then someone suggested getting rid of B-2 late, as well. Masursky protested; they had not "even looked at the rest of the pictures in the B-2 area." However, they dropped the B-2 late option since it

---

*These orbit numbers (periapsides figures) are for the second mission.

would save work and time. The next recommendation was that a candidate they had been calling "B-1 awful" be scrapped, too. "We were taking six more sets of pictures there . . . to see if we could find a site," Masursky remembered. Dropping "B-1 awful" cut out another option. The meeting had lasted only 15 minutes, when others had lasted hours. Masursky left the session stunned: "We had committed the project to landing at B-3 where we had zero data."[62]

What happened at that meeting? Gentry Lee reflected that each man in the Viking management had a different perspective and a different worry, even though they were all directed toward the same goal. From his perspective as manager of the Science Analysis and Mission Planning Directorate, he worried about his team; by mid-August he had a near mutiny on his hands. His people had to preserve four mission alternatives. To do that, "They had to do every day four times as much work as normal. They had to have a plan for what was going to happen eight days in the future on each of those options and so those poor people were just about to bite the dust." Norm Crabill also noted that the flight team was too tired to jump through any more hoops. Young and Martin saw the signs as well. Masursky was not happy with the decision, but being a team member and a team player he agreed to try for B-3.[63]

B-3 called for a landing late on the afternoon of 3 September. From that location, called Utopia Planitia (47.9° north, 225.9°), 186 photos would begin coming in the night of the 17th and continue being played back until 2 a.m. PDT on the 21st. All observations previously planned through P15 would still be made and processed, but no operational planning would be done for any of the B-1 or B-2 areas.[64] According to Masursky, the B-3 pictures looked terrible. While he was pondering the situation in the photomosaic room at JPL, he was visited by Henry Moore, who picked up a recently completed mosaic of B-2. Moore found what looked like "sand dunes all over that area" and called over his colleague. "Hank, I think you're smoking pot!" was Masursky's first reply, but when he looked at the mosaic he had to concede that there might be dunes. Because of poor exposure, it was difficult to tell, so they worked up special enhancements. "My God, that really looked good. That looks like that area is really covered by dunes." B-2 west was a promising area. Next they spotted smaller dunes in the B-3 region (48–49°N, 220°) that covered the ejecta blanket outside the large crater Mie (100 kilometers in diameter) and actually went into the crater. Farther to the west, some faint marks could be interpreted as the beginning of aeolian-deposited mantling material. This discovery led to a whole new debate: " Do dunes cover rocks, blocks and other hazards created by the erosional and cratering processes that might otherwise menace a lander?"[65]

The dune controversy began on the afternoon of 18 August and continued through the final site review on the 21st. Openly admitting his preference for the B-2 area where they had spotted the dunes, Masursky developed a new argument for landing there. In the absence of lunarlike mantles, dunes offered reasonable protection from rocks and blocks. Big dunes, as seen at B-2, offered better protection than the small dunes seen near the crater Mie. Before the meeting on the 21st, one of the landing site team members told Masursky that they would have to land at B-3. They could not "announce on Tuesday that all options are closed off except B-3, and then on Saturday decide to go back to B-2." Masursky believed they would "do the right thing."[66]

On Saturday afternoon, 21 August, a formal review of two candidate sites—B-2 west and B-3 east—was held. During the first mission search, individual rock units had been mapped but, during the *Viking 2* analyses, hazards had been defined in terms of debris ejected from craters, steep slopes, or areas subjected to different processes (stripping, mantling, and texturing). All these features had been mapped to determine favorable areas, and those mapped features were the center of discussion. Tom Young reminded them that safety was the fundamental issue but that they must try to keep "the science factors visible." And Hugh Kieffer reported on his infrared thermal-mapping instrument (IRTM), which was being used as a substitute for radar since there was no radar information for the northern sites. Masursky identified the five candidate ellipses:

| | | | |
|---|---|---|---|
| B-3 East | alpha | 47.2°N lat. | 224.9° long. |
| B-3 East | beta | 48.0°N | 228.0° |
| B-2 West | I | 44.1°N | 154.9° |
| B-2 West | II | 47.3°N | 156.6° |
| B-2 West | III | 43.5°N | 153.0°[67] |

In their report in *Science*, Masursky and Crabill evaluated these areas. The dunes in the B-2 region appeared to be bigger and apparently thicker than those in B-3. In B-3 there seemed to be some favorable aeolian mantle, even if it showed signs of being pitted. The northern part of B-3 looked better; cracks became shallower and craters were less abundant. Because of defects in some of the B-2 photos and an atmospheric haze that had obscured the surface, the interpreters were cautious in their estimates of the dunes in that zone. Some of the photogeologists believed that the decreasing number of small craters in B-3 would continue down to the scale of the lander. Some felt that the B-2 craters and ejecta blocks, being smaller than those in B-3, were better covered by B-2's bigger dunes and that B-3's smaller dunes might not cover the ejecta from the larger craters there as well. Still others favored site B-3 because they thought it appeared smoothed by uniform mantling. It was Tom Young's opinion that, although the geological conditions at B-2 east and B-3 west were different, the hazards gave them about the same safety ranking.

Continued discussion at the meeting on 21 August centered on the size of the block hazards that they might encounter. Their best analyses predicted that the wind-borne mantling material was sufficient to cover the small blocks thrown out of the craters by meteoritic impact. In the region of the crater Mie, it was hard to project what size blocks could be anticipated. Rocks up to 10 meters in diameter were not ruled out, because of the ejecta measured in the *Surveyor 7* lunar landing site and *Apollo 15, 16,* and *17* high-resolution photographs of the moon. According to Masursky and Crabill, "The block populations depend on the number of small craters below the resolution limit that may excavate blocks from below the wind-laid mantle and the number exposed by deflation. Slopes were deemed acceptable based on Earth analogs, except on the inner margins of craters."[68] Hugh Kieffer's infrared thermal-mapping device did not provide conclusive assistance in selecting a site. When the pictures looked good, there were no IRTM data. Where the IRTM gave good results, there were no photographs. For no site were there both photos and IRTM information. From the IRTM, B-2 looked less blocky than B-3, but with the IRTM the latter looked a good deal like *Viking 1*'s landing point. After nearly two hours of discussion, a dinner break was called.

The meeting reconvened at 6:35 p.m. "Is there demonstrated evidence that a significant increment in safety exists in going to B-2 that warrants changing the current plan to go to B-3?" asked Tom Young. Mike Carr thought the larger dunes in B-2 argued for safety, but then so did the effects of wind in B-3. Klein and Biemann favored B-2 because more water might be present for the biology experiment. After a number of other opinions had been expressed, Young called for a show of hands on two questions: "Should we select B-3E/B-2W or continue to search? If we decide on a specific site tonight, should it be B-3E or B-2W?" Of those voting, 28 wanted a site; 9 wished to continue searching. The B-2 site would be chosen by 20, B-3 by 10.

Jim Martin concluded the open part of the meeting by requesting Tom Young, Gentry Lee, Jerry Soffen, Carl Sagan, Hal Masursky, Norm Crabill, Mike Carr, Hugh Kieffer, Conway Snyder, Brad Smith, Tim Mutch, and Bob Hargraves to attend an executive session of the Landing Site Staff. Young favored B-3 and enumerated his reasons: (1) Safety—B-3 appeared to be mantled, muted, and filled. With all that cover, it was hard to believe that there could be serious hazards to the lander. Since Carr and others were not particularly confident about B-2 because of the visibility problems, Young liked B-3 better. (2) Science—Young saw limited distinction between the two. B-2 might have a slight edge because of more water and higher temperatures, but those elements did not outweigh the safety differential between the two sites. As Masursky and Crabill pointed out, "The most significant scientific distinction had already been realized when the northern latitude band was selected." (3) Operations—The landing would be more straightforward at B-3. To land at B-2 would require additional data analysis, and that would delay the landing significantly (according to Martin, the delay could be as long as two or three weeks). Such a delay, attended by greater operational complexity, did not seem to be justified by his readings of the two sites.

Project Manager Martin agreed with his mission director. He noted that several of the scientists wanted to do more ambitious exercises with the second lander, but he believed that additional observations of B-2 would work an already tired team into the ground. He could not see imposing such a killing load on the flight team. After some brief comments from others, Martin said they would go with B-3. It was safe enough; it had good enough science. There was no radar, but he was willing to take that risk. Gathering around the B-3 east stereo mosaic, the group determined the preliminary coordinates—48.0° north, 226.0°.[69] Final coordinates were chosen on 30 August after reviewing the P20 photographs: 47.89°N, 225.86°.

*Renewed Drama*

An orbital trim maneuver on 25 August 1976 ended *Viking 2*'s walk around the planet. Two days later a final trim synchronized the periapsis point relative to the landing site, which was centered 200 kilometers from the crater Mie. Before preseparation checkout of the second spacecraft, mission control put the first lander into the "reduced mission mode," permitting the flight team to concentrate on the second craft. At 9 a.m. PDT on 3 September, the Viking Flight Team met for the "go/no-go" separation meeting. With the exception of one of the terminal descent radar beams and a gyroscopic stabilizer that had given them some trouble, there were no problems with the spacecraft; all systems were "go" for separation. The radar problem was solved by locking out the troublesome unit, since the lander could touch down with only three of the landing radar beams functioning. And further analysis of the Y-axis gyro led the specialists to believe that it would not give them any trouble. *Viking 2* was ready for the big moment—separation and descent to the surface.

When asked how he assessed the risks and dangers of the Utopia site as compared to the Chryse site, Martin replied that he believed Utopia was safe. Even without ground-based radar information, he believed the processing of the planet had laid a thick mantle of sand or soillike material over any rocks and obstacles such as seen around lander 1. The Utopia area appeared to have perhaps more undulations, hills, and valleys, but he thought the slopes were gentle, and only 10 percent as many craters were visible. To the query, "Do you call 155 foot [47-meter] high sand dunes a better landing area?" Martin replied:

> Well let me say that there was not unanimity in the selection of this landing site. My job is much easier when everybody gets up and says let's go this direction. Well, here we had a case where people were wanting to go in a couple of different directions. I still believe that from my own

*A landing site for the second Viking lander is chosen in the eastern end of Utopia Planitia, 48°N latitude, 226° longitude. These three photos were taken by the Viking 2 orbiter 16 August 1976 from 3360 kilometers away. Rough ground and craters appear blanketed by dunes.*

knowledge of big sand dunes, that we can land on essentially any sand dune in the United States. I think the Lander is very tolerant to this kind of hazard. I think it is very intolerant to big rocks. So I would trade sand dunes for big rocks any day.[70]

On 3 September, the world would be able to judge the wisdom of the landing site team's decision.

But there were some heart-stopping moments before Viking mission control knew that the lander was on the surface. Confirmation of separation came as scheduled at 12:39:59 p.m. Three seconds later came an indication that the orbiter had been upset. Twenty-six seconds later the power supply to the gyros on the orbiter cut out; the second power unit went out at 12:41:19. Without power, the inertial reference unit, which kept the orbiter aligned properly in space, could no longer control *Viking 2*. As the spacecraft began to drift off course, its high-gain antenna lost contact with Earth. Within minutes of the failure, the orbiter's computer sensed the problem and commanded the backup inertial reference unit to take over and stabilize the attitude of the spacecraft.

While the men in the Deep Space Network worked to regain contact with *Viking 2*, the lander was on its way to the surface. To monitor the progress of the descending craft, the flight team tensely watched a small stream of engineering data coming down through the low-gain antenna. Throughout the Jet Propulsion Laboratory, project personnel, news peo-

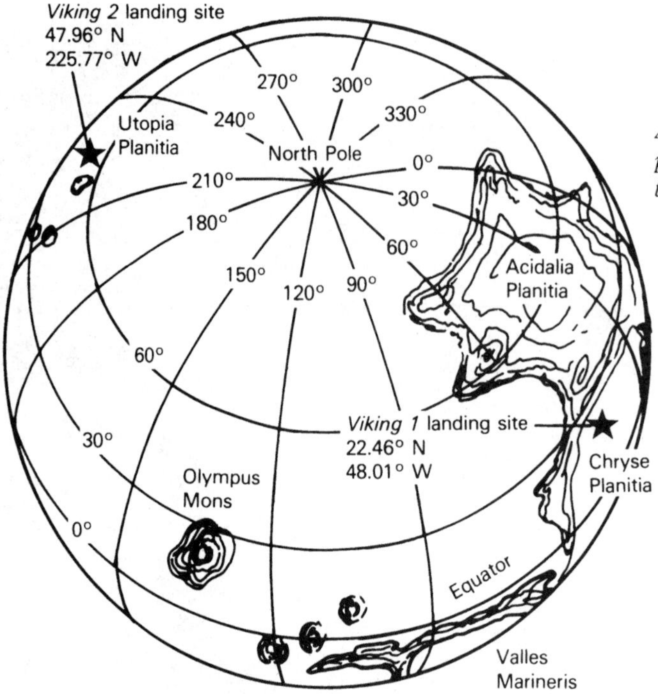
Viking 2 landing site
47.96° N
225.77° W
Utopia
Planitia
270°
300°
330°
240°
North Pole
0°
210°
30°
180°
60°
150°
120°
90°
Acidalia
Planitia
60°
Viking 1 landing site
22.46° N
48.01° W
30°
Chryse
Planitia
Olympus
Mons
0°
Equator
Valles
Marineris

*A view from the Martian north pole shows the location of the two Viking sites.*

ple, and guests waited, subdued, for each little clue that would tell them all was going well.

At 3:58:20 p.m. PDT (9:49:05 a.m. local Mars time) on 3 September 1976, the second lander touched down safely. Cheers mixed with sighs of relief, though the crisis was not over yet. The Deep Space Network worked with the flight team to get the proper commands to the orbiter. Once the spacecraft locked back onto its celestial reference point—the star Vega—Earth control again began to recover mission data, including the first two photographs taken by the lander's cameras immediately after the craft had reached the Martian surface.[71]

*Viking 2*'s landed photos illustrated a much rockier terrain than even the first site. One rock near the lander's footpad in the first picture looked as if it had been moved during landing. Martin and Young reported that the panoramic second picture revealed "a flat horizon and a landscape strewn with many rocks of various types. The tilt of the horizon indicates that the spacecraft may have landed on a rock." They also noted, "As a surprise, the panorama shows none of the sand dunes expected from the observations from orbit. A generally featureless terrain spreads flatly toward the horizon, more so than at the site of *Viking 1*."[72]

## Lessons Learned

After the second landing, seven key Viking team members talked about the landing site selection process. Of the lessons they had learned, had they labeled any as especially significant? If later there were a third mission, what would they do differently? All these men had worked toward the same end—safely landing two spacecraft on the Martian surface—but they had viewed the experience from seven different perspectives.

*James S. Martin.* Support for his decisions from the space agency's top management in Washington figured highly in Martin's recollections. Postponing the Fourth of July landing had probably been one of his most difficult moves. Martin had been "quite horrified" by the first photographs of Chryse, remembering the rough dry river beds and uneven washes he had walked over in Death Valley. And the river bed they had seen on Mars was many times larger, with cliffs hundreds of meters high. It has been his choice to make, and he had wanted a safer site. Martin remarked that even if Viking had safely touched down on the Fourth of July, the landing "would have been lost among the Tall Ships," a reference to the publicity given the bicentennial parade of ships in New York harbor. That historic date had been chosen in 1970 after a preliminary trajectory analysis singled out the first week in July 1976 for a landing, but the Red Planet had not cooperated. The bicentennial was celebrated without a Viking on Mars.

If Martin were going to land a third Viking, he would make some changes. He was unhappy with the data with which they had had to work. If there were to be a next time, he wanted to equip the lander with a terminal-hazard-avoidance device or a computer-controlled laser guidance system that could evaluate the surface and pick the safest part of a general area in which to land. Both kinds of hardware were available; the latter concept had been used in "smart bombs" in Vietnam. Martin and all his colleagues wanted more information guiding the next Viking on its final approach. A terminal guidance system would eliminate any radar versus photography controversy, Martin suggested, still skeptical about the use of radar.

> I'm not convinced that the radar told us anything useful at all. But on the other hand, I believe that it provided an input and a source of information that [we] could not ignore. . . .
>
> I looked at the radar as a source of data. I frankly never did . . . accept it as an absolute. . . . But I've got to believe that when they get a pass, like at that Northwest Site, and there's something screwy right in the middle of a place that looks just like everything else [in the photographs], the radar is seeing something. For all I know, it was seeing sand dunes . . . it could have been seeing something perfectly safe, but the fact that it was so different scared me off.[73]

*A. Thomas Young.* Radar played a useful role, Young believed, as he reviewed the background of using radar as an aid in landing site certification. "When we went through the initial selection [process], radar played no role, because we weren't smart enough to know how to use it." But Young and Gerald Soffen had gone to Stanford University to confer with Von Eshleman and Len Tyler.

NASA provided the funds for Tyler and his colleagues to develop the means of interpreting radar data so this tool could be used to evaluate the nature of proposed landing sites on Mars. Young's basic philosophy had been: "Use whatever tools we had available to the maximum extent we could, recognizing that none of them was good enough for absolute site certification." He thought they had probably used this tool before they fully understood what the signals meant. While the technique for interpreting radar data had not matured to the extent that they had absolute confidence in the results, he believed that the radar signals received from the A-1 northwest site did indicate that it was unsafe. Above all, Young commented, they had to be responsible in how they used the data provided them.[74]

*Gerald A. Soffen.* As project scientist, Soffen was interested in the process of scientists at work and concerned that that work be consistently credible. Caught in a philosophical frame of mind after a few days' rest, Soffen said he believed that the crisis over the landing sites had forced them to study the planet with an intensity that would not have existed if Mars had been as bland as *Mariner 9* had led them to believe. Talking about the days between 23 June and 21 August, Soffen said:

> We learned about Mars in that period. And it is sad to say we will probably never learn as much from the Orbiter pictures . . . as we did during that intensive period—because we had to. Because people were forced around the clock to do work and integrate their efforts in a way that unfortunately they don't do simply because they are inspired. Inspiration works to a very small extent on any person. What drives us is necessity. . . .

Soffen's observation was that, since only so much data could be collected and since they were working against the clock, the scientists could not retreat into the familiar excuse "I need more data."

> Because time was an element that we could not sacrifice, the energies of the people and the brilliance, deduction, the thought, the concerted effort, was as intensive as anything I have ever seen. . . . It was most remarkable. Remarkable because I saw people who otherwise have to take days off, have to take time off, have to relax. Their adrenalin kept them going in a way that I have never seen. . . . That was the moment in which the true concept of a team met its test. It was like an army that was desperately fighting for its life. It was either going to win or it was going to lose. It is not a question of "Maybe I'll survive and they won't." We're all in the same space program.

Soffen believed some important lessons were learned during the search for sites. First of all, they had erred in trusting the Mars maps based on *Mariner 9* findings. He suspected that if someone had shown the Landing Site Staff the actual photographs or had verbally described the surface of the planet to them using the raw data, they would not have had such confidence. "But seeing the U.S. Geological Survey maps, the straight lines and real numbers and real elevations, gave it an air of credence. . . ." A second lesson was that real-time decision-making had to be a combination of effort between mission specialists and managers. Before Viking arrived at Mars, Soffen had formed a four-man landing site advisory group—Josh Lederberg, Brad Smith, Toby Owen, and Carl Sagan—to listen to the site certification deliberations and advise Soffen, Young, and Martin. But events moved too fast. There was no time to reflect and cogitate; decisions had to be made; Landing Site Staff meetings never followed a neat pattern.[75]

Hal Masursky pointed to the same problem. Mosaics, just recently pasted together, were often brought into meetings in progress. There he was, faced with interpreting completely new information on the spot, with everyone waiting for his words of wisdom so a decision could be made. "That's hard to do," he noted. "Emotionally and managerially it's not the right way to do it. You need to work, digest, come to conclusions by arguing and then pass on a recommendation." But there was never time. For example, the team would schedule 8 days for the analysis of a particular issue, but the specialists might be able to devote only 10 seconds of wits-gathering to the problem. Pressure from the mission schedule made the scene tense, and too often general scientists, members of the landing site advisory group, for example, had to defer to specialists. Masursky, among others, was not as concerned about the pace as he was with the precipitous nature of their decision-making. But they had committed themselves to the real-time game, and decisions had to be made on schedule.[76]

One other observation Soffen made dealt with spheres of influence. Position in the project heirarchy had little to do with power of influence over Jim Martin, "an absolute dictator," in Soffen's words. If any one person—regardless of rank—had an idea that made good sense to Martin, he listened and acted accordingly. During early July, it had been Tyler who had held center stage with his radar data. "A week earlier we dismissed what Len Tyler had to say, as though we weren't interested," Soffen recalled. The activity of so many intense individuals working closely together gave the site selection–site certification process a dynamism typical of the entire Viking effort.[77] Such a human endeavor needed discipline.

*B. Gentry Lee.* If Jim Martin were the dictator, as many had suggested, Gentry Lee was the intellectual disciplinarian. From his vantage point as science analysis and mission planning director, Lee noted that "we went into the site certification process with two distinctly different views of how it was going to operate." Many project personnel members—Lee, Martin, and Young among them—did not want to deviate from the previously selected sites unless it were absolutely necessary, since they were relatively certain they had found the best sites available. All the mission operating plans were designed for those targets, with time and money arguing against changes. But a second group, primarily scientists, wanted to search for even better sites during the certification process. Caught between the two were men like Mike Carr and Hal Masursky, who simply wanted to see that the spacecraft landed safely in a scientifically valid location. Probably the only thing that averted open controversy was the terrible nature of the prime Chryse region.

Lee found himself in the role of interlocutor. Even more important, it was his task to ensure that the operational people and the scientists understood each other's needs and limitations. Lee participated in the Landing Site Staff meetings not only to translate for each group the other's goals and problems, but also to make certain that the discussions remained germane to the issues at hand. When they did not, he turned disciplinarian, "trying to get people back where they belonged."

After Lee looked over the photographs of the Martian surface sent back from the two landers, he concluded that if the flight team had seen the sites that clearly before the landings, it would not have certified them as safe. The team had sought zones that were 99 percent "landable," and to Lee the sites they had chosen now appeared to be hazardous at best. "But the thing that we don't know is how much worse the areas may have been that we rejected." Without the rigorous search-certification operation, they would have had no hope of a successful landing. Lee and Martin agreed that a third Viking landing using the same certification tools would have no guarantee of touching down safely. Before another craft was sent to Mars, Lee hoped they would have a better understanding of radar and infrared thermal mapping. He also had hopes that the low-altitude photography planned for the extended Viking mission, with periapsides as low as 300 kilometers, would give them a totally new look at the surface, including hazards of the 15- to 20-meter scale.[78]

*Michael H. Carr.* Mike Carr also had something to say about low-level photography. Commenting on the gap between the 100-meter-resolution orbiter imaging photographs and the lander photographs, the leader of the orbiter imaging team said:

> We've got to bring the orbiter down in the extended mission to 200, 300, 400 kilometers and use the scan platform for image [motion] compensation. [We must] squeeze the maximum resolution we can out of the orbiter cameras over significant areas, so that we're getting data at a much finer scale in anticipation of the next [mission]. . . . There will only be one next one—a rover. We just can't afford to have it crash.

Even though cameras would be on board any future spacecraft, Carr believed that the site selection team ought to be armed with data at a scale relevant to the lander. Better photography and a clearer understanding of radar and the infrared system would make the job easier.[79] Both Carr and Masursky thought that image-motion compensation was necessary for any future low-altitude orbital photography of Mars, to prevent the images from smearing.

*Harold Masursky.* The leader of the landing site certification team said he would like to attach a mechanical image-motion compensator to the Viking cameras. With this device, he knew how he would fly a third mission. The spacecraft would be inserted into low Mars orbit, to take higher resolution site certification pictures. From these low-altitude images, the flight team would be able to avoid hazards of a smaller scale and select key topographical features to which the lander could guide itself, using either laser or television. After a site was approved, the orbiter would obtain a higher altitude and release the lander. All the technology Masursky wanted to use was available. What they did not have was NASA's approval for a third flight—and the funds.[80]

Coming down to Earth, Masursky commented on the effort made by Martin and Young to maintain the scientific integrity of the Viking missions. No matter what problems came up, the management kept reminding everyone that the primary objective was scientific investigation after a safe landing. With many critical issues facing them daily, Martin and Young never forgot the main goal.[81]

*Carl Sagan.* As a scientist, Sagan was impressed by how "remarkably willing to listen the project manager was." If anything, Sagan had been prepared for resistance to such items as postponing the 4 July landing and taking a closer look at radar data. But Martin had kept an open mind. "It sounds like a reasonable thing for a project manager to do, but that's not always been the case in past missions."[82]

Martin and Young had listened. They had not always accepted the advice given them, but considering the immense task they had faced and their success they must have made the right decisions. They had safely landed two out of two spacecraft, and luck had had very little to do with it. Martin would continue to believe that hard work and discipline were better bets than luck. The site selection–site certification process had been time-consuming, tension-filled, and seldom an "exact" science, but it had worked—and worked on a planet 348 million kilometers from Earth. With two successful landings behind them, the Viking team could turn to the real reasons for its labors—the scientific examination of the Martian surface and the search for possible life forms on the Red Planet.

# 11

# On Mars

As anticipated, the information relayed to Earth by the Viking space-craft has greatly affected man's perceptions and understanding of the planet Mars. The increase in basic, directly confirmed knowledge of the Red Planet began even before the landings. Once in orbit, the spacecraft began transmitting the first of tens of thousands of images of the planet and its satellites.

## IMAGES FROM ORBIT

Heterogeneity was the most striking aspect of Mars as scientists iden-tified a greater variety of terrains than known to exist on the moon or Mercury. Conway B. Leovy, a member of the meteorology team, noted: "Unlike the moon, whose story appears essentially to have ended one or two billion years ago, Mars is still evolving and changing. On Mars, as on the earth, the most pervasive agent of change is the planet's atmosphere, itself the product of the sorting of the planet's initial constituents that began soon after it condensed from the primordial cloud of dust and gas that gave rise to the solar system 4.6 billion years ago."[1]

Some information about the nature of the Martian atmosphere had been derived from telescopic observations and from earlier Mariner mis-sions, but those sources of data were "unverifiable and subject to misinter-pretation." With the exception of its significantly different composition and its being "less than a hundredth as dense as that of the earth," the atmosphere of Mars behaves much like that of our own planet. "It trans-ports water, generates clouds and exhibits daily and seasonal wind pat-terns." Responding to seasonal changes in the heat generated by solar radiation, localized dust storms occur and sometimes grow in strength until they cover the entire planet, a fact with which Mariner and Viking special-ists were familiar. Global dust storms appear to be a phenomenon unique to Mars, which lacks large bodies of water that would prevent their buildup.

Atmospheric weathering of the primitive crystalline rocks on Mars has reduced them to fine particles that have oxidized and combined chemically with water to produce the reddish minerals so apparent in the color images returned from the Viking landers. Whereas on Earth the dominant weathering process has been from the movement of liquid water, on Mars the primary agent of change has been the wind. It erodes the landscape, transports the dust, and deposits it elsewhere on the planet. The Viking landing sites appear to have been "severely scoured by winds." In addition, pictures taken by the orbiter cameras reveal deep layers of wind-borne sediment in the polar regions, while dunefields of Martian dust and sand much larger than those on Earth were observed near the north pole.[2]

The geologic history of Mars, according to orbiter imaging team leader Michael H. Carr, "shows evidence of floods and relatively recent volcanic eruptions, at least in the hundreds of millions of years that geology uses as a measure." There are also features that resemble terrestrial river systems. "Apparently tremendous floods occurred many times over Mars' history, indicating that the planet must have been drastically different in the past."[3]

Earlier Mariner flights indicated the presence of volcanoes on Mars; Viking measured their extent and variety. A large portion of the northern hemisphere is covered by volcanoes, some spreading broad lava fields for hundreds of kilometers. Others, such as Olympus Mons and Arsia Mons, rise some 27 km above the reference surface level of the planet. Distinct lava flow patterns can be seen 300 km from their source in Arsia Mons, with the general pattern of the terrain indicating that the lava may have traveled up to 800 km, the distance from Washington, D.C. to Cincinnati, Ohio.[4] Geologists who have studied the Viking photographs believe that the nature of volcanic activity on Mars is essentially the same as that on Earth—the movement of a basaltic, low-viscosity lava. One kind of volcano appears to be unique to Mars: the patera, or saucer-shaped, volcano with a low profile covering a vast area. Alba Patera, with a maximum diameter of 1600 km, is probably the largest such volcano on the planet. A similar volcano centered on Denver would have spilled its lava across all of Colorado, Wyoming, Utah, large parts of New Mexico, Kansas, Nebraska, South Dakota, and corners of Montana, Idaho, Arizona, Texas, and Oklahoma. Scientists think that the caldera—the crater formed by the collapse of the central part of the volcano—of a patera is the result of simultaneous lifting and collapsing of the sides of the volcano, probably repeated many times over a long period. According to Carr, "the total volumes of lava erupted to produce single flows are orders of magnitudes greater than they are in terrestrial lava flows, and the total volumes of lava erupted from essentially a single-vent volcano are enormous."[5] Production of sufficient magma (molten rock) for such lava flows cannot be explained, but as Carr pointed out, the plains regions appear to have been formed several billion years ago by this movement of lava.[6]

In addition to lava, the movement of water also has affected Martian topography. The large riverlike channels are one of the big Martian puzzles. Carr and his colleagues believe there are two major kinds of water features:

*Two variations of the first color photo from the Viking 1 lander, taken on the Mars surface 21 July 1976. The blue-sky version above was released the same day. Below is the true red-sky version released 26 July. The red cast is probably due to scattering and reflection from sediment suspended in the lower atmosphere. To assist in balancing the colors, a photo was taken of a test chart mounted on the rear of the spacecraft and the calibration then applied to the entire scene.*

*The two photographs above were taken with the Viking lander camera during tests in the summer of 1974. At the top is a panoramic shot from a site overlooking the Martin Marietta Corporation factory in Denver. The lower photo was taken at the Great Sand Dunes National Monument in southwestern Colorado. The lander camera is a facsimile camera, different in design from the television and film cameras which have been used on many space missions. The field of view is not imaged simultaneously. Instead, adjacent vertical lines are successively scanned. Reflected light from each of the "picture elements" in the line is recorded on a very small photodiode in the focal plane of the camera. Twelve diodes are available for use, each optimized for a different distance and a different part of the visible near-infrared spectrum.*

*Photos permit comparison of the color of the Viking lander on Mars (at left) and Earth (above)—especially the orange cables. Tim Mutch used this guide to show that the red-sky rendition of the Mars landscape was the correct one. In the Earth photo, Jim Martin stands beside the science test lander in the Von Kármán Auditorium at Jet Propulsion Laboratory.*

Photos taken by the Viking lander camera provide comparison of an Earth scene (above) and one on Mars (below). In a photo taken near the Martin Marietta Denver facility during tests in 1974, tan and reddish sedimentary rocks have been tilted and eroded to form prominent cliffs. Data from three diodes (blue, green, and red) were combined for the color picture. Colors have not been balanced; the blue contribution is unnaturally large. For mission photos, colors were carefully calibrated. The Martian horizon stretches across nearly 200° in the composite of three color photos taken 4 September 1976 (center), 5 September (right), and 8 September (left). A thin coating of limonite (hydrated iron oxide) colors the surface predominantly rusty red, although some dark volcanic rocks can be seen. The horizon is flat because the photo has been rectified to remove the effects of the 8° tilt of the spacecraft.

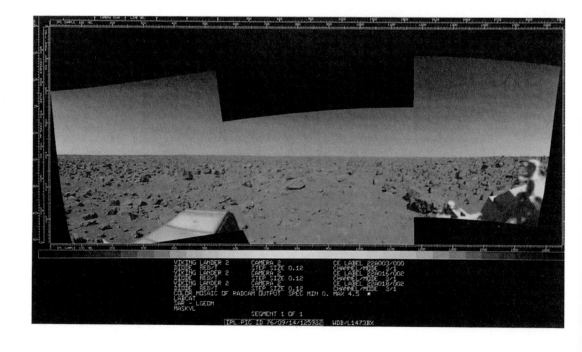

There are the large flood features and then there are dendritic or branching drainage features that resemble terrestrial river systems. It appears from the crater counts that the fine terrestrial-like river channel systems are older than the flood features. It appears that the large flood features came in middle Mars history. There was a period of vast floods, then the flooding for some reason ceased or became less frequent because we don't have flood features with crater counts comparable to those we find on the Tharsis volcanoes. Very early in Mars' history, dendritic drainage patterns developed; in Mars' middle history it had a period of flooding, and then mostly after that the volcanics of Tharsis accumulated. This general picture has come out of the Viking data.

A lot of skeptics didn't believe there had been any period of surface drainage. Some said all those things could easily have been formed by faulting and so on. The Viking pictures are full of examples of dendritic channels. I can't believe there are many skeptics left. I think we have really established that there was this early period of surface drainage. There can be very little doubt about that.[7]

The scientists are still left with explaining where all the water for the floods and rivers came from. More important, where did it go?

Because of low atmospheric pressure at the surface, there are no contemporary large pools, rivers, or collection basins filled with water, and because of low temperatures the atmosphere cannot contain much water. However, there is probably a great quantity in the permanent polar caps and within the surface. The low pressure permits water to be present only in the solid (ice) or gaseous (water vapor) state. One possible explanation for the apparently contradictory vision of rushing rivers on Mars was presented by Gerald A. Soffen: "Broad channels formed when subsurface water-ice (permafrost) was melted by geothermal activity from deep volcanic centers. When the melting of the permafrost reached a slope the interstitial water suddenly released great flows, sometimes a hundred kilometers wide that modified the channels."[8] Seasonal heating of the permafrost may have occasionally released large flows of water, as well—a possible explanation for the channels that originate in box canyons and spill onto the plains. The easiest method of accounting for the dendritic channels is to conjure up a Martian rainstorm, but that suggestion raises many problems, all of which hinge on the basic question: "How is it possible that these ancient rivers could [have] existed and there be none today?" Obviously, atmospheric pressure would have to have been different during such a period. This hypothesis seems to be supported by studies of the Martian atmosphere encountered by Viking.

If the atmospheric pressure once was sufficient to permit the formation of liquid water, how long ago was that? This is still a subject of some debate. Harold Masursky and his colleagues estimated the relative age of the channels by counting the number and judging the age of the craters in and near the channels. The different kinds of channels appear to have been created in

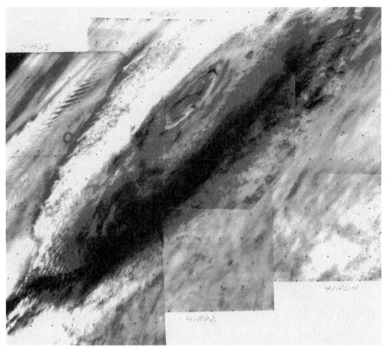

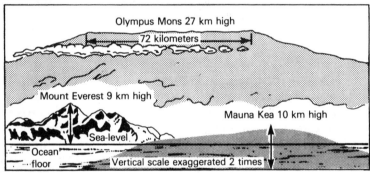

Olympus Mons 27 km high

72 kilometers

Mount Everest 9 km high

Mauna Kea 10 km high

Sea-level

Ocean floor

Vertical scale exaggerated 2 times

*The Martian volcano Olympus Mons, at top, was photographed by the* Viking 1 *orbiter 31 July 1976 from a distance of 8000 km. The 27-km-high mountain is wreathed in clouds extending 19 km up its flanks. The clouds are thought to be principally water ice condensed as the atmosphere cools. The crater is some 80 km across. At left, Arsia Mons, called South Spot during the* Mariner 9 *mission, is shown in a mosaic of photos taken 22 August. The crater is 120 km across, and the peak rises 16 km above the Tharsis Ridge, itself 11 km high. Vast amounts of lava have flooded the plains.*

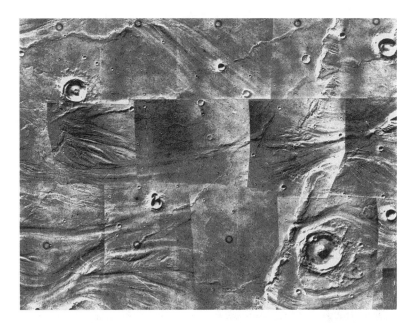

A 9 July mosaic of Viking 1 orbiter photos above shows lava flows broken by faults forming ridges. Apparently a small stream once flowed northward (toward upper right) from Lunae Planum, crossed the area, and descended toward the east. In places water may have formed ponds behind ridges before cutting through. At right, a fresh young crater about 30 km across, in Lunae Planum, is near a dry river channel running alongside a cliff in possible lava flows (Kasei Valley). Below, an oblique view across Argyre Planitia (the relatively smooth plain at top center of the photo) shows surrounding heavily cratered terrain. Brightness of the horizon to the right (with north toward upper left) is due mainly to a thin haze. Above the horizon are detached layers of haze 25 to 40 km high, thought to be crystals of carbon dioxide (dry ice). Both the lower photo mosaics were taken 11 July.

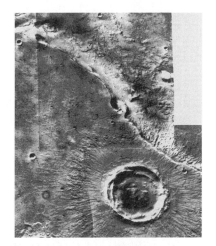

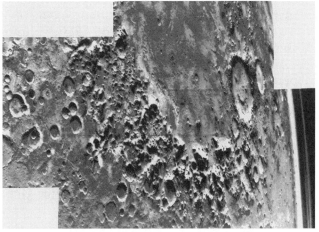

different epochs, or episodes, and all of them at least 50 million years ago and perhaps as long ago as several billion years.[9]

In addition to the effects of lava and water, shifting of the permafrost also is believed to have influenced the texture of the planet's surface. Investigators assume the existence of permafrost, sometimes to the depth of several kilometers and generally thought to have been present for billions of years. Carr stated:

> To me one of the more exciting things we've observed is the abundant evidence of permafrost. The most striking features indicative of permafrost occur along the edge of old crater terrain. They form by mass movement of surface material probably aided by the freezing and thawing of ground ice. Another possible indicator of ground ice is the unique character of material ejected from impact craters that is quite different from the pattern on the Moon and on Mercury. We interpret the difference as due to ground ice on Mars. The impact melts the ground ice and lubricates the [ejecta] that is thrown out of the crater so when it lands on the ground it flows away from the crater in a debris flow and forms the characteristic features we have observed.

Slow movement and a freeze-thaw cycle could account for the chaotic, jumbled terrain seen over vast stretches of the Martian surface. Irregular depressions caused by localized collapsing of the crust when permafrost thawed could have formed the flat-floored valleys in Siberia and the table-lands of Mars. Large polygonal-patterned regions on Mars resemble the ice wedges in terrestrial glacial areas.[10]

The Martian class of lobate craters is distinct. Unlike lunar craters and those photographed on Mercury, which have radial sunburst patterns caused by ejected debris, on Mars debris apparently flowed smoothly away from the points of impact of many craters. Craters on the moon and Mercury typically had a coarse, disordered texture close to the rim that became finer farther out, grading almost imperceptibly into dense fields of secondary craters. "The most distinctive Martian craters have a quite different pattern. The ejecta commonly appears to consist of several layers, the outer edge of each being marked by a low ridge or escarpment." Recognized in *Mariner 9* photographs, the shape was attributed to erosion caused by the wind. With improved-resolution Viking photographs, the geologists have changed their minds; they theorize that on Mars objects also struck the surface with explosive force, but the difference lay in the heating of the permafrost. Resulting steam and momentarily liquid water transported surface materials away from the point of impact and created the distinct lobate flow patterns around the central point. Where the crater ejecta patterns do resemble those on the moon and Mercury, geologists believe that the permafrost was too far below the surface to have been heated, or else possibly absent.[11]

On a planet that has many spectacular features, one of the most interesting is the Valles Marineris, the Grand Canyon of Mars. First observed by *Mariner 9* cameras, only the gross proportions of the canyon system were appreciated at the 1- to 1.5-km resolution. A small sample of higher resolution *Mariner 9* photographs (100–150 meters) hinted at the huge landslides and related features that would be seen on the canyon walls and floors. The images from Viking were much better (resolution of objects as small as 40 meters), and many parts of the 4000-km-long canyon system were photographed in stereo, the combination permitting geologists to understand more precisely the processes that formed it. Significantly, neither volcanic activity nor erosion caused by flowing water seems to account for the changes in the Valles Marineris. After examining the Viking photos, Karl R. Blasius and his colleagues believe that tectonic shifting of the planet's crust may have enlarged the canyons. Volcanism was not seen in the Viking images, they point out, and evidence of fluvial activity was only indirect, from chaotic terrain. But tectonic activity appeared to have been prolonged, deepening canyons and offsetting erosion and deposits that would have broadened and filled them. Vertical adjustment of crustal blocks under north-south and east-west extensional stresses appeared to have been the primary process. Some blocks may also have tilted, forming "peculiar slopes near canyon rims and on the intratrough plateau and possibly causing the formation of strings of collapse pits." The history of canyon erosion and deposits was also more complex than had been realized. "Layered materials, including some very regularly imbedded sediments first recognized in the Viking images," were highly diverse and wide-spread.[12]

One of the basic reasons for studying the Valles Marineris was an interest in the interrelations through time of the volcanic and tectonic forces that produced the large volcanoes to the west—Olympus Mons and the Tharsis craters, which include Arsia Mons—and the development and

*Material appears to have flowed out of the Arandas crater on Mars, rather than being blasted out by the meteorite impact. Radial grooves on the surface of the flow may have been eroded during the last stages of the impact process. Photographed 22 July 1976 by the Viking 1 orbiter at 43°N latitude, 15° longitude, Arandas is about 25 km in diameter.*

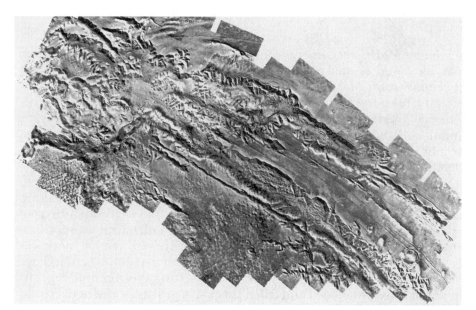

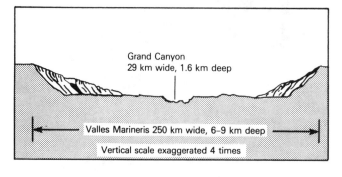

Grand Canyon
29 km wide, 1.6 km deep

Valles Marineris 250 km wide, 6–9 km deep

Vertical scale exaggerated 4 times

*More than 100 photos form the top mosaic mapping Valles Marineris, huge Martian complex of canyons. Taken by the* Viking 1 *orbiter 23–26 August 1976, they are centered at 5° south latitude, 85° longitude, with north at the top. Ten photos taken 22 August form the center mosaic of the western end of the canyon. The volcanic plateau is deeply dissected into connected depressions.*

evolution of the canyon lands. Both geological regions are young in terms of the life of the planet, and changes in both areas likely have continued to the present. Mars and Earth may thus be more alike in geological terms than previously expected. The Viking images have contributed to a new field of study called comparative planetology. Undoubtedly, the wealth of new information gathered by the cameras on the orbiter was ample reward to the people who had fought so strongly to send an improved imaging system to Mars to complement the scientific instruments. As Mike Carr and his associates had predicted in October 1970, "The high-resolution imaging system may be considered as the 'meat and potatoes' low-risk but guaranteed-significant-gain experiment in the mission."[13]

Further analysis of the photographs taken over the Chryse and Cydonia regions during and after landing site certification had indicated that many of the assumptions specialists had made on the basis of *Mariner 9* photography had to be changed. Viking science investigators benefited from approaching the planet at a time when it was far from the sun, since lower solar radiation nearly eliminated the worry about dust storms.[14] The clarity of the Viking orbiter images indicated that the Martian atmosphere probably had never cleared during the *Mariner 9* mission. *Viking 1* arrived at Mars just before the beginning of summer in the northern hemisphere and soon after aphelion. Every Viking scientist reaped benefits from the clear orbiter images, and Ronald Greeley and his geologist colleagues had specific comments about the importance of the Viking orbital pictures in the Chryse and Cydonia regions: "High-resolution Viking orbiter images show Chryse Planitia to be much more complex than had been suspected from Mariner 9 images. Ancient heavily cratered terrain appears to form the basement for the basin. Much of its heavily cratered terrain is mantled with deposits that may be of aeolian, fluvial, or volcanic origin."[15] They were certain that the *Mariner 9* view of Mars had been "simplistic." From a close examination of the southern hemisphere, scientists had made some false assumptions about the northern half of the planet. "From Viking photography it is suggested that not only is the northern hemisphere more complicated than was expected, but as . . . predicted, although the present surfaces are young, some of the rocks exposed at the surface may be old."[16]

Orbiter photographs coupled with data from the infrared thermal mapper (IRTM) gave scientists a new understanding of the polar caps, too. The Martian poles change dramatically with the seasons. When the Viking craft arrived at the planet, the northern cap had shrunk to its minimum size, revealing the permanet cap, which—contrary to some expectations—consisted of water ice. The part that had dissipated had been made of solid carbon dioxide, dry ice. Meanwhile, the southern ice cap expanded. The northern polar region displayed terraced deposits, indicating an episodic pattern of rapid erosion and deposition of materials. "An unconformity within the layered deposits suggests a complex history of climate change during their time of deposition."

*Table 52*
*Geological Evolution*
*of Martian North Polar Region*

| | |
|---|---|
| Stage 1 | Onset of polar activity.<br>Moderate aeolian modification of ancient volcanic terrains. |
| Stage 2 | First depositional period.<br>Layered deposits of silicate dust and possibly interbedded ice accumulate to thickness of several kilometers. |
| Stage 3 | First erosional period.<br>Erosional attack of layered deposits results in landscape of gently curving scarps and channels with terraced slopes. |
| Stage 4 | Second depositional period.<br>More layered deposits accumulate unconformably on top of units formed in first depositional period. |
| Stage 5 | Second erosional period.<br>Further erosional attack of layered deposits results in exhumation of earlier formed landscapes and reveals unconformable contacts between deposits of first and second depositional period. Some eroded material reaccumulates as girdle of sand dunes between 75°N and 80°N. |
| Stage 6 | Recent period.<br>Ice in permanent polar cap assumes its present form and distribution. |

While this scenario might not represent a completely accurate explanation of the manner in which the polar terrain evolved, James A. Cutts, Karl Blasius, and associates argue that "it does offer a credible framework . . . against which further observations and theoretical models may be tested."[17]

Meanwhile at the south pole, the infrared thermal-mapping team had observed some interesting temperatures. In their first report in *Science*, Hugh H. Kieffer and his colleagues noted that "areas in the polar night have temperatures distinctly lower than the $CO_2$ condensation point at the surface pressure." From the atmospheric pressure of 6 millibars at the south pole, the mapping team had anticipated temperatures of about –125°C, the equilibrium temperature for carbon dioxide at that pressure, but, when initial results came in, temperatures as low as –139°C were recorded. The infrared specialists decided that this extra cooling was attributable to a freezing out of the carbon dioxide, leaving a higher concentration of non-condensable gases (such as nitrogen and argon) than is normal for the atmosphere elsewhere. Since these gases would not condense into solid form at –139°C, that could explain the cooling, but other questions were raised by this theory.[18] How did the noncondensable gases concentrate in the polar region? What did this phenomenon mean for global circulation patterns? What did it tell scientists about the movement of carbon dioxide and other gases from one pole to the other during the change of seasons?

Once again, new knowledge raised as many interesting questions as it answered.

By the end of the primary mission, the infrared thermal-mapping team had begun to devise theories to answer some of the questions. Large-scale patterns in the temperatures of Mars appear to be similar in size to continental weather patterns on Earth. Viking scientists believe that these patterns may be associated with cloud patterns. As team leader Hugh Kieffer put it, "It's possible we're seeing what I call continental scale weather." Temperatures shortly before dawn in some places are much cooler than expected. Over the Valles Marineris, the temperatures were unexpectedly quite warm before dawn. Kieffer noted that "the temperatures just before dawn are more directly related to the physical properties of the surface because there is no solar energy being absorbed during the 12 hours of night. This means the temperatures are a good indication of how well the surface can hold its heat."[19]

Infrared thermal-mapping measurements indicated wide daily temperature variations on Mars. The typical day-night variation on Earth is 5° to 10°C, but on Mars the temperature can go from a low of –133° to a high of 4°C. The reason for this wide range is not yet fully understood, nor is the tendency of the temperatures in the afternoon to drop much more quickly than expected. Keiffer reported that in several regions on Mars temperatures begin toward the middle of the afternoon to drop more rapidly than predicted until just before dusk. They may be 10 to 15 degrees cooler than expected. Then they "cease to drop so rapidly and slowly merge with the predictions for the evening." In the afternoon, "the only atmospheric regions that are cooler than the surface are very high and thus we don't know what process at the moment is causing this rapid surface cooling." The process "may be related to clouds in some way, but most of the atmosphere near the ground, where one expects clouds to form, is, in fact, warmer than the surface just before sunset."[20]

A more important contribution from the infrared thermal-mapping experiment was the discovery of the nature of the polar ice cap. One of the major questions posed by the *Mariner 9* data was the composition of the residual polar cap left when the winter polar cap, made of frozen carbon dioxide, retreated in midsummer. A major controversy existed over whether this summer cap was also frozen carbon dioxide or was frozen water. According to Viking data, the temperatures of the residual cap are near –68° to –63°C, making a case for water frost. Also, the brightness of the frost "indicates it has a lot of dirt mixed in with it. The dirty nature of the ice had also been seen now by the orbital imaging system." Apparently there is no permanent reservoir of carbon dioxide in the polar regions of Mars, a finding that tends to rule out the theory of a rapid climate change induced by the instability of the carbon dioxide on the planet. "This means we still don't have an adequate explanation of how the atmosphere could have been of sufficient density to sustain the liquid water that appears to have flowed at one time in streams and rivers on the surface of Mars," said Kieffer.[21]

## MEASURING THE ATMOSPHERE

The water-vapor-mapping investigation was designed to map the distribution of water vapor over the planet and to determine the pressure of the atmosphere at the level where vapor is present. Understanding the distribution of water vapor is crucial to understanding the geological features of Mars and the possibility of the existence of life. Viking's measurements of water vapor varied, depending on the location, season, and time of day.

Specialists discovered a direct correlation between elevation and the amount of water vapor present, with the lowest points on the planet having the greatest concentrations and the highest features the minimum. More water vapor was found during the summer season than during winter, when it was barely perceptible. In regions of rough terrain, there were marked daily variations in water vapor, and C. Barney Farmer and his team believed the variations were attributable to local phenomena—shifting wind patterns, dust, or a thin cloud or haze that is present at dawn but dissipates by noon. For example, early in the first mission one site was monitored over a six-hour period. The water vapor content in the atmosphere rose steadily from dawn until noon. This water could have been brought into the area from another region by the wind, or the haze or dust in the air could have affected the instrument's measurements. Whatever the cause for the change, the increase would be considered minute when compared to Earth's atmosphere with 1000 times as much moisture.[22]

During the Viking primary mission, the Martian water vapor underwent a gradual redistribution, the latitude of the maximum amounts moving from the north polar region toward the equator. Interestingly, while the amounts of vapor at some latitudes changed dramatically, the total global water remained almost constant at the equivalent of about one cubic kilometer of ice. The largest amounts observed were found over the dark polar region, which is inaccessible to Earth-bound observers. Maximum vapor column abundances of about 100 precipitable micrometers were measured adjacent to the residual cap itself—"a very large amount considering the temperature of the surface and atmosphere in this region." The Mars atmospheric water detector also confirmed the conclusion that the residual cap is made of frozen water and that the atmosphere above it is saturated with vapor during the polar summer.[23]

Orbital science investigations had given a better grasp of the global nature of Mars, and the entry science experiments provided the first direct measurements of the physical and chemical composition of the planet's atmosphere. The scientists were for the first time "getting their hands on" some more tangible data. Entry science investigations consisted of measurements by the retarding potential analyzer, the upper-atmosphere mass spectrometer, lander accelerometers, the aeroshell stagnation-pressure instrument, and the recovery temperature instrument. The analyzer had been designed to study the nature of the ionosphere. The mass spectrometer was to provide mass spectra for the constituents of the upper atmosphere.

Three of the instruments—the lander accelerometers, the aeroshell stagnation-pressure instrument, and the recovery temperature instrument—made up the lower-atmosphere structure experiment, which measured the density, temperature, and pressure profile of the atmosphere as the lander approached the surface. As with other experiments and Viking hardware, the entry investigations had been based on the common "Mars engineering model" adopted early in the project. That model described the nature of the planet as it was believed to be, from the best knowledge then available. As Jerry Soffen recounted, the model was developed to set the boundaries for design, prescribing the atmospheric envelope, the variety of possible surfaces, range of textures, radiation environment, etc. This "working manual" was constantly reviewed by scientists both within and outside the project and used by all the engineers. The Mars engineering model "was an excellent crossroads for scientists and engineers." With the mission definition, it "truly spelled out what we were trying to do and the planetary constraints we believed existed."[24]

The lander's mode of descent altered several times before touchdown, and the entry instruments operated during different phases of the entry process. At separation, the lander capsule—consisting of the aeroshell and basecover surrounding the lander—was deorbited by ignition of the deorbit engines. The capsule began the first part of its descent trajectory through the undisturbed interplanetary medium of ions and electrons. The interplanetary medium streams away from the sun at hypersonic velocities in what is called solar wind. Closer to the planet, the lander capsule passed through a disturbed region where the solar wind is diverted to flow around and past Mars. Beneath this zone of interaction lay the Martian ionosphere, a region of charged atomic particles. It was in the ionosphere, 3 minutes after the completion of the deorbit burn, that the retarding potential analyzer began 18 sampling sequences, during which 71 seconds of data were collected.

*Entry* has been arbitrarily defined as starting at 250 kilometers, although the atmosphere is only readily apparent from about 91 kilometers. From separation to entry required about 3 hours. At entry, the lander capsule was oriented with the aeroshell and its heatshield facing the direction of travel; before the atmosphere exerted an appreciable drag, the capsule would accelerate to about 16 000 km per hour. Almost 1 hour before the lander reached the 250-km mark, the upper-atmosphere mass spectrometer was turned on for a 30-minute warmup period. The spectrometer and the retarding potential analyzer would continue to take measurements until the capsule system sensed 0.05 gravity, at which time they would shut down. The capsule-mounted temperature sensor was then deployed. With pressure sensors (deployed 10 minutes before entry), it would continue to function until the aeroshell was jettisoned (12 seconds after the radar altimeter sensed an altitude of 5.9 km).

At about 27 km above the surface, the capsule reached its peak deceleration and for a time its path leveled off into a long glide, because of the aerodynamic lift provided by the aeroshell. As the effects of atmospheric friction and gravity overcame the lift, the capsule resumed descent. By the time its radar altimeter indicated an altitude of 6.4 km, the capsule was traveling slowly enough (an estimated 1600 km per hour) to deploy the parachute. Seven seconds later, the aeroshell separated from the lander, and the remaining lift in the lightened aeroshell permitted it to drift well away from the landing site. Twelve seconds after aeroshell separation, the lander legs were deployed, at which time the footpad temperature sensor began collecting data, doing so until touchdown.[25]

From the retarding potential analyzer, new information about the Martian ionosphere was collected through measurements of the solar wind electrons and ionospheric electrons, the temperatures of the electrons, and the composition, concentrations, and temperatures of positive ions. At the higher altitudes, the analyzer examined the interaction of the solar wind and the upper atmosphere. The planet's weak (or nonexistent) magnetic field permits the solar wind to penetrate closer to the surface of Mars than it does to Earth's surface. Data obtained during descent indicates that singly ionized molecular oxygen ($O_2$+) is the major element of the upper atmosphere, with peak concentration at an altitude of 130 km. Singly ionized molecular oxygen is about nine times as abundant as singly ionized carbon dioxide ($CO_2$+), the primary ion produced by the interaction of sunlight with the Martian atmosphere. This new finding lends support to theoretical analyses by M. B. McElroy and J. C. McConnell, which call attention to the reaction of atomic oxygen with $CO_2$+ that would produce carbon monoxide and the more stable ion $O_2$+. The temperature of the observed ions at 130 km was about $-113°C$.[26] Viking measurements of O+ ions moving away from the planet coupled with *Mariner 9* observations of hydrogen escaping from the planet's upper atmosphere suggest that the planet has been losing the basic ingredients for water for billions of years. Perhaps some of the water that once carved the massive channels on the surface of Mars slowly escaped in the form of ionized hydrogen and oxygen.

The upper-atmosphere mass spectrometer obtained data about the identities and concentrations of the various gases from 230 to 100 km. As expected, the main constituent of the upper atmosphere is carbon dioxide, with small amounts of nitrogen, argon, carbon monoxide, oxygen, and nitric oxide. Taken together, what do these upper atmospheric measurements suggest? The discovery of nitrogen was a particularly pleasant surprise. As Tobias Owen of the molecular analysis team commented, the search for nitrogen in the Martian atmosphere goes back several decades, and he was "delighted" that they finally had found it. When he first became interested in Mars during the 1950s, "it was an established doctrine that the pressure on Mars was eighty-five millibars, plus or minus three millibars, and that the atmosphere was well over ninety-five percent nitrogen." As time passed, predictions changed; both the surface pressure and the amount of nitrogen decreased. As the estimated amount of carbon dioxide grew to more than 95 percent of the gas in the atmosphere, detection of any nitrogen seemed unlikely. This outlook was disheartening to the exobiologists who believed that nitrogen was an essential ingredient in any environment in which life might have evolved. But the upper-atmosphere mass spectrometer did detect nitrogen. Happily, Toby Owen said, "And now we finally got it; it's really there."[27]

Michael McElroy of the entry science team went even further. According to him, Mars was a very "cooperative" planet, and it had given the Viking scientists some bonus information. Beyond defining the chemical composition of the atmosphere, they discovered some "clues as to the evolution of the planet from its isotopic abundance." Mars has more of the heavy form of nitrogen than does Earth, which allows specialists to theorize that Mars is "remarkably Earth-like although it has gone through a different evolutionary history." McElroy explained that there are two abundant isotopes of nitrogen: Mass 14, which is the common form, and Mass 15, which is less common. They are both present in Earth's atmosphere and in the Martian atmosphere, but Mars has rather more of the heavy component than does Earth. The implication is that Mars must have lost the light material over time. The initial amount of nitrogen on Mars was apparently similar to the initial amount on Earth, but slightly lower gravity on Mars allowed the lighter nitrogen to escape. Perhaps Mars has "evolved to a larger extent than the Earth because of this escape process."[28]

While the presence of 2.5 percent nitrogen in the atmosphere opened the door for speculation about possibilities of organic material, the levels of argon led to other theories, some of which were contradictory to the one used to explain the presence of nitrogen. Argon was measured at 1.5 percent, considerably less than indicated by the indirect measurements made by the Soviet Union with its *Mars 6* mission in 1974. The discovery that Soviet scientists were mistaken was welcome to Klaus Biemann and his colleagues on the molecular analysis team, because it relieved their worry that argon might choke the gas chromatograph–mass spectrometer. The low amount of argon in the atmosphere would not prevent that instrument from performing a series of atmospheric analyses on its way to the surface before it could be contaminated by organic compounds from the Martian soil.[29]

A low concentration of argon also had significant implications when it came to reconstructing the early Martian atmosphere. The two common isotopes of argon are argon-36 and argon-40. The former is an inert element produced in the interior of stars such as our sun, and the latter is created during the radioactive decay of potassium-40. Both isotopes have been released over time from the rocks of planets, and it is generally held that the relative amount of the two says something about how the atmosphere evolved. For Mars, this theory poses some interesting problems and questions. Toby Owen proposed the following scenario during a 28 July 1976 Viking science symposium at JPL. Using the Earth's atmospheric history as a guide, Owen argued that one could by analogy plot the evolution of the Martian atmosphere back over time. One way to make this analysis for the two planets was to use argon-36 as the common piece of information. It was assumed that Earth and Mars were formed at the same time and from the same inventory of gases in the solar nebula. If that is true, then Earth and Mars should have about the same ratio of argon-36 and argon-40 in their atmospheres. They do not. Earth is relatively poor in argon-36; it is held that this gas was lost early in the evolution of the terrestrial atmosphere. Scientists thought that they could deduce from the amount of argon-36 in the Martian atmosphere the gases that have been lost. Viking measurements indicate that the planet should have lost 10 times the amount of carbon dioxide and nitrogen now measured in the atmosphere. But the loss was not out into space; it was hidden in some form on the planet itself. Ten times the present amount of carbon dioxide constitutes a considerable amount of material to hide. Owen reported: "I'm suggesting that somewhere between 1 and 10 times the present amount of $CO_2$ is missing on Mars . . . and some fraction could still be present in the form of $CO_2$ trapped in the [polar] caps. The other part of this reconstruction, which is interesting, is that it implies a couple of tens of meters of water on the surface which must also be sequestered somewhere."[30] The water could have become permafrost, but this explanation disagrees with the theory that the water left the planet in the form of ionized hydrogen and oxygen.

Although no general agreements have been reached on how the upper atmosphere of Mars was formed, one point seems certain: that atmosphere was significantly different in the past. Just as the evolution of Earth's atmosphere helped determine the nature of its environment, the evolution of Mars is linked with the development of its atmosphere. As Jerry Soffen concluded: "It appears that there was a considerably denser atmosphere in the past, somewhere between 10 and 50 times the present value of 7.5 millibars at the surface. This denser atmosphere would account for the possibility of the ancient river [beds] seen from the orbiter."[31] Whatever explanation the scientific community comes to accept, Viking has made two points very clear—the Red Planet's environment has not been static, and in the past was very dynamic.

The lower atmosphere structure experiment provided vertical profiles of the density, pressure, and temperature of the atmosphere from an altitude of 90 km to the surface. Accelerometers, part of the lander's inertial reference unit, acted as sensors for the initial measurements from which the density profile was derived. The profile was determined by observing the retardation of the capsule's descent by atmospheric drag. Pressure and temperature measurements came at first from the two instruments in the aeroshell. Because of the high initial velocities of the lander capsule, the pressure sensor determined the pressure of the atmospheric molecules against the aeroshell surface; the actual pressures were determined analytically later. In a similar fashion, the temperature probe, near the outer rim of the aeroshell, measured the temperature of molecules flowing around the aeroshell. During the parachute phase of the descent, after the aeroshell had been jettisoned, the lander's pressure and temperature sensors provided this information.

Altitude data for construction of profiles came from the radar altimeter. A by-product of the radar altimeter measurements was information about the terrain beneath the lander. The terminal descent and landing radar system, which controlled the very last stage of the landing, also measured the extent to which the lander drifted because of winds above the point of touchdown. Pressure and temperature variations were measured by the two landers at selected intervals during the descent (table 53). The temperature in the region between 200 and 140 km above the surface averaged about -93°C; for the region between 120 and 28 km it was -130°C. At touchdown, the *Viking 1* atmospheric temperature was about -36°C, and *Viking 2*'s reading was -48°C.[32]

*Table 53*
*Structure of Martian Atmosphere*

| Altitude (km) | Viking 1 | | Viking 2 | |
|---|---|---|---|---|
| | Pressure (mb) | Temperature (°C) | Pressure (mb) | Temperature (°C) |
| 120.0 | 0.000 004 14 | -136.85 | 0.000 001 99 | -157.15 |
| 108.0 | 0.000 018 40 | -126.75 | 0.000 013 00 | -152.05 |
| 96.0 | 0.000 080 20 | -127.25 | 0.000 066 00 | -122.95 |
| 84.0 | 0.000 387 00 | -128.95 | 0.000 288 00 | -131.75 |
| 72.0 | 0.002 050 00 | -134.05 | 0.001 680 00 | -142.25 |
| 60.0 | 0.009 110 00 | -127.65 | 0.008 540 00 | -135.85 |
| 48.0 | 0.044 500 00 | -124.55 | 0.039 200 00 | -102.45 |
| 36.0 | 0.198 000 00 | -107.05 | 0.158 000 00 | -108.75 |
| 28.0 | 0.483 000 00 | - 89.35 | 0.404 000 00 | - 99.95 |
| 4.5 | 5.160 000 00 | - 51.05* | 5.222 000 00 | - 51.95 |
| 4.0 | 5.390 000 00 | - 50.53 | 5.483 000 00 | - 51.55 |
| 3.5 | 5.635 000 00 | - 48.45 | 5.747 000 00 | - 51.05 |
| 3.0 | 5.885 000 00 | - 46.65 | 6.015 000 00 | - 50.55 |
| 2.5 | 6.150 000 00 | - 44.85 | 6.282 000 00 | - 50.05 |
| 2.0 | 6.427 000 00 | - 43.05 | 6.564 000 00 | - 49.55 |
| 1.5 | 6.707 000 00 | - 41.35 | 6.853 000 00 | - 49.15 |
| 1.0 | 6.994 000 00* | - 39.45* | 7.160 000 00* | - 48.55* |
| 0.5 | 7.301 000 00* | - 37.65* | 7.480 000 00* | - 48.05* |
| 0.0 | 7.620 000 00* | - 35.85* | 7.820 000 00* | - 47.55* |

*Extrapolated.
SOURCE: Alvin Seiff and Donn B. Kirk, "Structure of the Atmosphere of Mars in Summer at Mid-Latitudes," *Journal of Geophysical Research* 80 (30 Sept. 1977): 4367, 4371.

Compared to the scientific instruments aboard the orbiter or the lander, the entry experiments were very short-lived. They operated only during the descent to the surface. Still, these instruments provided investigators with several new insights into the Martian environment and clues that, when coupled with orbital and landed data, would help frame new hypotheses about the evolution of the planet.

As interesting as the orbital pictures and measurements were and as informative as the entry data instruments were, the best was to come. Science aside for a moment, the reception of the first pictures from the lander cameras had to be the most exciting event for many project participants, scientists and engineers alike. For the public, the surface pictures were certainly the main event.

## ON THE SURFACE

The first lander's first picture, of footpad 3 (a 60° high-resolution image), demonstrated to everyone that the craft was safely down on the surface. Minutes later, camera 2 began taking a real-time picture, a 300° panoramic view of the scene in front of the lander. These shots had been planned to provide the maximum amount of immediate information so that images of value would already have been collected should something unforeseen terminate the operation of the lander. Thomas A. Mutch, lander imaging team leader, recalled, "The planning for these first two frames was exhaustive." Characteristically, everyone had some advice about the best photographs to take. More than a year before the landing, team members had been called to Washington to brief NASA Administrator James Fletcher on camera strategy. "In the event of a botched landing, the first two images might constitute our only pictorial record of Mars." The pictures would be sent to the orbiter in the first 15 minutes after landing and thence to Earth. Not for 19 hours, including the first night on Mars, would it be possible to communicate again with the lander.

Some of Mutch's associates argued with the decision to photograph the footpad and then the view in front of the lander. One challenged, "If you were transported to an unknown terrain, would you first look down at your feet?" Mutch had to agree that the common mental image of the explorer was that of an individual shading his eyes with his hand looking far away to the horizon. He records that his counter argument was rather pedestrian. He thought—in the terms of a photogeologist—that the first picture of the footpad would be technically the better of the two:

> A primary photogeologic goal, perhaps because it is so easily quantifiable, is increase in linear resolution. Looking straight down, the slant range was abut 2 m, yielding a linear resolution of approximately 2 or 3 mm. Looking toward the horizon, nominally 3 km distant, the linear resolution would have been reduced toward two or three orders of magnitude.
>
> Our logic would have been persuasive if the surface of Mars had been generally flat, but covered with small objects of unusual form. As it turned

out, this was not the case. The rock-littered surface in the near field is relatively undistinguished, but the undulating topography and diverse geology of the middle and far field is spectacular. From both an exploratory and scientific perspective, the panorama to the horizon is the more impressive of the first two pictures.[33]

This self-effacing evaluation is characteristic of many of the Viking scientists, but especially of Tim Mutch. Seated in the "Blue Room" as the first electronic picture data began appearing on the television monitors throughout the Jet Propulsion Laboratory facilities, Mutch in almost a boyish manner commented, "The neat thing about pictures is that everyone can do their own analysis. We're really quite superfluous here." The images from the lander were reconstructed, picture element (pixel) by picture element from left to right, just as they had been taken by the camera on Mars. After going through the decoding process in the ground reconstruction laboratory, the image was shown throughout JPL a few lines at a time. From left to right, the first pictures of Mars began to evolve on the monitors. Reactions were varied, but nearly all were happy ones. For Tim Mutch, it was "a geologist's delight." Jim Martin saw the first picture in very practical terms—Viking was so far a success. He expressed his appreciation to the entire Viking team and to the "10 000 people across the country who deserve a part of the credit given to me." Mission Director Tom Young was also pleased with the performance of his spacecraft. As for the pictures, he said, "quality was consistent with what we should get, but they have exceeded my expectations." The quality was very good, and Young added that "Mars has demonstrated that it is photogenic!"[34]

### The Colors of Mars

The first two photos of Mars received on 20 July 1976 were followed by a color photograph on the 21st. A lot of people would not forget that first color picture. Mutch tells the tale as well as anyone. During the first day following the early morning landing of *Viking 1*, his team was preoccupied with analysis and release of those first two images, "which, in quality and content, had greatly exceeded our expectations." So much were they concentrating on the black and white pictures, that they were "dismally," to use Mutch's word, "unprepared to reconstruct and analyze the first color picture." Mutch and his colleagues on the imaging team had been working long hours, along with everyone else, during the search for a landing site. Despite enthusiasm, people were tired. Many of the Viking scientists in the upcoming weeks would have to learn to present instant interpretations of their data for the press. For the first color photograph, haste led to processing the Martian sky the wrong color.

In a general fashion, Mutch and his team understood that a thorough preflight calibration of the camera's sensitivity to the colors of the spectrum was necessary. They also knew that they would need computer software programs to transform the raw data efficiently into an accurate color

*The first photograph (above) from the surface of Mars, taken minutes after the Viking 1 lander touched down on 20 July 1976. Center of the image is about 1.4 meters from the lander's camera no. 2. Both rocks and finely granulated material are visible. Many foreground rocks are flat with angular facets. Several larger rocks have irregular surfaces with pits, and the large rock at top left shows intersecting linear cracks. A vertical dark band extending from that rock toward the camera may have been caused by a one-minute partial obscuring of the landscape by clouds or dust. The large rock in the center is about 10 centimeters across. At right is a portion of the spacecraft's footpad, with a little fine-grained sand or dust deposited in its center at landing.*

*Below is the first panoramic view by Viking 1 on the surface. Horizon features are about 3 km away. A collection of fine-grained material at left is reminiscent of sand dunes. Projections on or near the horizon may be rims of distant craters. Some of the rocks appear to be undercut on one side and partially buried by drifting sand on the other. The housing of the sampler arm, not yet deployed, and the low-gain antenna are at left. In the right foreground are the color charts for camera calibration, a mirror for the magnetic properties experiment, and part of a grid on top of the lander body. At upper right is the high-gain antenna for direct communication between the lander and Earth.*

representation. "What we failed to appreciate were the many subtle problems which, uncorrected, could produce major changes in color. Furthermore, we had no intimation of the immediate and widespread public interest in the first color products—for example, intuitively corrected color images were shown on television within 30 minutes following receipt of the data on Earth." Although they resisted at first, the lander imaging team was obliged to release the first color prints within 8 hours of having received the image.[35]

Instinctive reactions and intuition can lead to mistakes when dealing with an alien world. Here is Tim Mutch's first public reaction to the color photograph:

> Look at that sky—light blue sky—reddish hue. It's a very exciting thing to see this distinct reddish coloration to the surface. These are subtle hues. It's a geological scene, a natural scene. Even in the deserts here on Earth the reds are not crayon reds as painted by a child. This is a surprisingly terrestrial-like desert scene.[36]

But to borrow Carl Sagan's phrase, to see this picture in terms of deserts on our own planet was an "Earth chauvinism." The photo was of Mars, not of Earth; the sky should have been red. When James A. Pollack of the imaging team told a press conference on July 21 that the Martian sky was pink, he was greeted with some friendly boos and hisses. Sagan, in a way that only he could, chided the newspeople the following day: "The sort of boos given to Jerry Pollack's pronouncement about a pink sky reflects our wish for Mars to be just like the Earth."[37]

There were three sensors with blue, green, and red filters in the focal plane of the camera to record the radiance of the scene in blue, green, and red light. The multilayer, interference filters used in the lander cameras (filters that could withstand the rigors of sterilization) have an irregular spectral response. The blue channel, for instance, responds slightly but significantly to light in the infrared portion of the spectrum. The unwanted part of the signal must be subtracted, "so that the absolute radiances at three specific wavelengths in the blue, green and red are represented." Subsequently, color prints were produced by exposing conventional color film to

individually modulated beams of blue, green, and red laser light, scanning the film with the same geometry employed in the camera.

Before the flight, the cameras had been calibrated and the sensitivity of each sensor-filter combination determined. "Qualitative tests indicated that simple normalization of the voltages for the three color channels . . . was sufficient to produce reasonable color images. In making that judgment our attention was generally directed to saturated colors in the natural scene and test target." When the first color data were received, Mutch's specialists used the same normalization techniques to calibrate the image. "The result was surprising and disquieting. The entire scene, ground and atmosphere alike, was bathed in a reddish glow. Unwilling to commit ourselves publicly to this provocative display, we adjusted the parameters in the calibration program until the sky came out a neutral gray." The soil and rocks demonstrated good contrast, and the colors "seemed reasonable." This was the picture released eight hours later. "But to our chagrin," Mutch recalled, "the sky took on a bluish hue during reconstruction and photo-reproduction. The media representatives were delighted with the Earth-like colors of the scene."

While the television and newspaper reporters hurried to get this color print before their respective audiences, continued analysis supported the reality of an orangish tint throughout the scene. The atmospheric coloration was due to the presence of suspended soil particles in the thin air. Mutch recalled: "Several days after the first release, we distributed a second version, this time with the sky reddish. Predictably, newspaper headlines of 'Martian sky turns from blue to red' were followed by accounts of scientific fallibility. We smiled painfully when reporters asked us if the sky would turn green in a subsequent version." Experience with color imaging over the next year indicated that the colors of Mars might vary, but the sky would retain its reddish hue. "In summary," Mutch said, "the color of the Martian scene, perceived by the necessarily abnormal eyes of Viking, is elusive. In response to the inevitable question: 'Is that exactly how it would look if I were standing on Mars?' a qualified 'yes' is in order."[38]

## A Real World

No matter what the color of the sky, the Viking pictures created a new reality for many people. Jerry Soffen said that, if any one thing stood out in his mind, "Mars had become a place. It went from a word, an abstract thought, to a real place." Soffen doubted that he would ever have an adventure like climbing Mount Everest, but he knew that it existed because other people had been there and had taken pictures of it, just as people had been to other extraordinary places on Earth. And now, their "guy" had made it to Mars. "He was not a person, but he was a close friend." For many associated with the Viking project, the lander had become personified. "It is like a person invented by a committee. And we sent him there and he did his thing. . . ." Before the Viking missions Mars was a fictional or fantasy

*Two variations of the first color photo from the* Viking 1 *lander, taken on the Mars surface 21 July 1976. The blue-sky version above was released the same day. Below is the true red-sky version released 26 July. The red cast is probably due to scattering and reflection from sediment suspended in the lower atmosphere. To assist in balancing the colors, a photo was taken of a test chart mounted on the rear of the spacecraft and the calibration then applied to the entire scene. (See color plates between pages 364 and 365.)*

*The two photographs above were taken with the Viking lander camera during tests in the summer of 1974. At the top is a panoramic shot from a site overlooking the Martin Marietta Corporation factory in Denver. The lower photo was taken at the Great Sand Dunes National Monument in southwestern Colorado. The lander camera is a facsimile camera, different in design from the television and film cameras which have been used on many space missions. The field of view is not imaged simultaneously. Instead, adjacent vertical lines are successively scanned. Reflected light from each of the "picture elements" in the line is recorded on a very small photodiode in the focal plane of the camera. Twelve diodes are available for use, each optimized for a different distance and a different part of the visible near-infrared spectrum. (See color plates between pages 364 and 365.)*

Photos permit comparison of the color of the Viking lander on Mars (at left) and Earth (above)—especially the orange cables. Tim Mutch used this guide to show that the red-sky rendition of the Mars landscape was the correct one. In the Earth photo, Jim Martin stands beside the science test lander in the Von Kármán Auditorium at Jet Propulsion Laboratory.

(See color plates between pages 364 and 365.)

*Photos taken by the Viking lander camera provide comparison of an Earth scene (above) and one on Mars (below). In a photo taken near the Martin Marietta Denver facility during tests in 1974, tan and reddish sedimentary rocks have been tilted and eroded to form prominent cliffs. Data from three diodes (blue, green, and red) were combined for the color picture. Colors have not been balanced; the blue contribution is unnaturally large. For mission photos, colors were carefully calibrated. The Martian horizon stretches across nearly 200° in the composite of three color photos taken 4 September 1976 (center), 5 September (right), and 8 September (left). A thin coating of limonite (hydrated iron oxide) colors the surface predominantly rusty red, although some dark volcanic rocks can be seen. The horizon is flat because the photo has been rectified to remove the effects of the 8° tilt of the spacecraft. (See color plates between pages 364 and 365.)*

place—the planet of Flash Gordon or some world peopled by Edgar Rice Burroughs. School children learn about the orderly progression of planets, and one of them has the same name as the world of many science fiction dramas. One Mars had physical, scientific properties like Earth; the other was a fantasy land. Now they could think of Mars as a genuine world. The shift from an object to be studied to a real place might not have been important scientifically, but it was a big change intellectually.[39]

Soffen pointed out that his personal involvement with the planet was not unlike that of the other Viking scientists. It had been eight years since the beginning of Viking. With the landing, the investigators were hungry for every bit of knowledge, any new speculation that would lead to a better understanding of the nature of Mars. Before the first photographs were received from the lander, Mars was more a scientific problem than an actual planet. When scientists talked about atmospheric conditions, they were describing numerical quantities that had an engineering significance for the designers. But it was difficult to think in terms of real clouds, real winds, real temperatures in the way we discuss our own weather. As the science fiction writers had built imaginary worlds on which their stories could take place, the scientists too had created a Mars that seemed to fit their assumptions. But the planet created from earlier known scientific facts had very little similarity to the Mars that the orbiter and lander cameras portrayed. Mars as a real place was much more complex and interesting than any that had been conjured up in the minds of scientists. The new Mars of Viking has as many complicated processes at work as does Earth.

Geologist Tim Mutch also had some personal reflections on what they had found awaiting them on Mars:

> If you were to tell a geologist that you were going to go out to two places on Earth with your little Brownie to take one or two rolls of film at each locality and then were to come back and from this interpret the history of the planet, he would think you were out of your mind, the most absurd thing he had ever heard of. In a sense it is. So one should not overestimate the exclusive model that you can generate from pictures.

But one thing that could be said definitively was that the terrain of Mars was not bland. A complicated history is exposed particularly in the photographs taken at the *Viking 1* site. "From a geological point of view, there is clearly a sequence of events represented. . . . involving fundamentally different processes—for example, impact, wind, volcanic activity, possibly fluvial activity and possibly ground ice."

The specialists confirmed a diversity of rock types on Mars, indicating several petrographic types; that is, rocks that probably have different mineralogy and at least have different texture. More boulders seem to be on the surface than can be accounted for by impact processes; perhaps the weathering of bedrock or the deposition of rocks by fluvial mechanisms account for them. And the bedrock visible in the Viking images indicates that some process, either fluvial or alluvial, is stripping off the soil to reveal the rock. At the *Viking 2* site, the rocks are more homogeneous. "They are highly pitted, due either to volcanic vesiculation or to some peculiar process we simply do not understand," reported Mutch. Some scientists think *Viking 2* landed on a widespread, fine-grain sediment mantle—the polar mantle. The boulders littering the scene were probably imposed, either as broken lava flows or as ejected boulders from a nearby crater.[40]

Seeing another planet up close opened the way for a comparison of two evolving worlds. With the passing of the romantic Mars and the gradual acceptance of the new Red Planet has come both excitement and disappointment. Looking at a tangible place is far more exciting than ruminations about abstract places, but the absence of life was a blow to many who had hoped to discover life or who had hoped that life might have had a chance to evolve. The biological and organic investigations indicated that the prerequisites for life on Mars were not evident at either landing site.[41]

## SCIENCE ON MARS

### Weather

When Viking touched down on the surface, weather reports started streaming their way to Earth. Martian weather was clear, cold, uniform, repetitious. Seymour L. Hess, meteorology team leader, reported on conditions at Chryse Planitia on sols 2 and 3:*

> Winds in the late afternoon were again out of a generally easterly direction but southerly components appeared that had not been seen before. Once again the winds went to the southwesterly after midnight and oscillated about that direction through what appears to be two cycles. The data ended at 2:17 PM (local Martian time) with the wind from the ESE, instead of from the W as had been seen before. The maximum mean wind speed was 7.9 meters per second (18 mph) but gusts were detected reaching 14.5 meters per second (32 mph).
>
> The minimum temperature attained just after dawn was almost the same as on the previous Sol, namely –86°C. . . . The maximum measured temperature at 2:16 PM was –33°C. . . . This [was] 2° cooler than measured at the same time on the previous Sol.
>
> The mean pressure was 7.63 mb, which is slightly lower than previously. It appears that pressure varies during a Sol, being about 0.1 mb higher around 2:00 AM and 0.1 mb lower around 4:00 PM.[42]

During the course of the Viking lander experiments, Hess and his fellow meteorologists discovered two interesting facts about Martian weather patterns. One was the extreme uniformity of the weather, presum-

---

*Sol is used to designate the Martian day, which is 39.6 minutes longer than an Earth day; 20 July was listed as sol 0 because just a few hours were left in the sol (local lander time) at the time of landing. Sol 1 began late on 20 July, at the first lander 1 midnight.

*Table 54*
*Mars and Earth Temperatures*
*21 July 1976*

|                        | Mars      | Earth                                               | United States                        |
| ---------------------- | --------- | --------------------------------------------------- | ------------------------------------ |
| Lowest temperature     | −85.5°C   | −73°C (Soviet Vostok Research Station, Antarctica)  | 2.7°C (Point Barrow, Alaska)         |
| Highest temperature    | −30.0°C   | 47.2°C (Timimoun, Algeria)                          | 42.7°C (Needles, California)         |

ably due to the Martian atmosphere, which is much simpler than Earth's. The Red Planet has only very, very small amounts of water vapor and no oceans—makers of extreme weather on Earth. Earth's atmospheric and surface water contribute substantially to the variability of its weather. The second discovery was the seasonal variation of pressure. When Viking first landed, its instruments detected a steady decrease in the mean pressure from day to day. But in the extended mission, the pressure at both landing sites reached its lowest value seasonally and began to rise again. The Viking meteorologists think this variation is due to the condensation of carbon dioxide on the winter cap and its release as spring comes to the northern hemisphere. This process would remove a major constituent from the atmosphere at a certain rate, changing the pressure accordingly.

At the second Viking site, 48° north, the temperatures dropped as expected during the Martian winter. Early in the mission, the minimum temperature was about −87°C, but during winter the minimum temperature at dawn was −118°C. Frost on the surface was first observed in mid-September 1977. At the time, the second lander was recording nighttime temperatures of −113°C, and a photo of the frost was taken at −97°C. With winter, the wind speeds increased slightly, especially at the *Viking 1* site, with several interruptions in what had been a regular pattern of wind

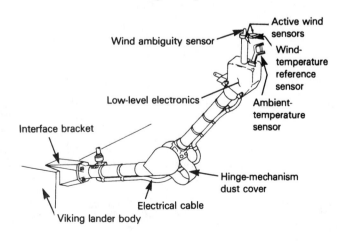

*Viking lander's meteorology boom and sensors in deployed configuration.*

Active wind sensors
Wind ambiguity sensor
Wind-temperature reference sensor
Low-level electronics
Ambient-temperature sensor
Interface bracket
Hinge-mechanism dust cover
Electrical cable
Viking lander body

direction. There were several periods of northerly winds all day for several days in a row, associated with temporary drops in temperature—Martian cold fronts. Hess and his colleagues had thought winter, with the fallout of carbon dioxide, would greatly increase wind speeds and variability. There were some wind directional changes and gusts, but no noticeable changes of patterns in wind direction or speed were recorded.[43]

## Hardware Problems

While Hess and his meteorological colleagues began to compile weather data for the *Viking 1* landing site, other experimenters were having their difficulties. First, the seismometer was not functioning. Its seismic sensor coils had been "caged" mechanically to prevent damage to these sensitive components during the shock of landing. Following touchdown, a fusible pin-pulling device was to have detonated, unlocking the seismometer so it could begin full operation. For some reason, perhaps a broken or misconnected wire, the fusible device failed to work, and the instrument remained in the caged position. While the *Viking 2* seismometer performed satisfactorily, the *Viking 1* failure prevented the seismology team from locating the approximate origin points of recorded seismic activity.[44]

Don Anderson and his colleagues on the seismology team was afraid that the sensitive seismometer on Viking would be hampered by the high winds on Mars. But during the night from about 6 p.m. through the next morning, the winds die down to about virtually zero and there are essentially no seismic background noises. During that time, the seismometer can be operated "at a very high sensitivity." Marsquakes as small as a magnitude of 3 at a distance of about 200 kilometers can be recorded. By comparing a Marsquake with a similar Earthquake, the specialists estimated the mean crustal thickness at the *Viking 2* landing site to be about 14 to 18 km, about half the thickness of the crust in the continental parts of Earth and about 50 percent greater than the average thickness of the oceanic crust. Viking scientists think the crust on Mars may be as thick as approximately 80 km, much thicker than the crust under continental regions on Earth.

An unexpected result of the seismic experiment was a great amount of information about the winds on Mars. A very sensitive wind detector, the seismometer picks up the wind pressure on the lander, from which characteristics of the wind can be determined. Like the meteorologists, the seismology team detected the cold fronts. The wind pattern "changed very rapidly on the 131st Martian day. The winds . . . started to blow all night until 2 or 3 a.m. indicating a substantial change in the weather patterns. If very high winter winds had continued at night, they could have generated the massive dust storms we have observed in the winter time." However, orbiter photographs have shown only a few isolated dust storms, with none reaching the magnitude of the planetwide dust storm of 1971.[45]

Another cause for concern for the Viking team appeared on the second day of landed operations. The lander's UHF transmitter had been designed to operate at three different power levels—1 watt, 10 watts, and 30 watts—depending on the rate of data transmission required. During the relay-link portions of the mission, the 30-watt power level was scheduled for use, to permit the transmission of the maximum amount of scientific data. From the observed performance of the initial landed relay link, confidence in the system was high. During the first relay, approximately 30 million bits of data were transmitted to the orbiter, recorded, and subsequently transmitted to Earth, all within a few hours after the information left the surface of Mars.* Success, however, was short-lived.

On 22 and 23 July, the UHF transmitter switched over to the 1-watt power level without instructions to do so. Tom Young told the press, "In the one-watt mode you can get slightly over seventeen minutes' worth of data from the Lander to the Orbiter." The mission had been designed so that slightly more than 18 minutes of data would be transmitted to the orbiter as it passed overhead, so the problem was not a critical one, but it did pose a vexing limitation. At the 30-watt level, the lander could transmit telemetry to the orbiter for 30 to 32 minutes.[46]

On the morning of 24 July, the UHF transmitter switched back to the 30-watt power level. Tom Young reported this second mysterious power change at the news briefing that day: "When we had the relay [of information] today, lo and behold, it came up in the 30-watt mode, operating as we would like for it to. So our statistics, to date, are two relay periods in the 1-watt mode, two periods in the 30-watt mode. We are continuing the analysis of this particular anomaly.[47] The radio specialists suspected that the problem lay in the power-mode control-logic subassembly of the UHF transmitter. To counteract this trouble, commands had been prepared to order the guidance control and sequencing computer to eliminate the electronic "noise" causing the problem. Before this command was sent up, the tansmitter switched back to the 30-watt power level. The change supported the theory that the problem was associated with noise susceptibility. Following the self-correction, the UHF transmitter performed as expected until one week before the end of *Viking 1*'s primary mission. At that time, telemetry indicated that there were potentially new problems with the 30-watt level. To avoid a catastrophic failure and to extend the transmitter's life for use in the "follow-on" mission, the lander performance analysts decided to use the 10-watt power mode for the last sols of the basic mission.[48]

The landed relay communications for *Viking 2* did not demonstrate any anomalies. On sol 21 of the second landed mission, orbiter 1 was moved into position over lander 2 to provide a relay link. This maneuver permitted mission planners to send orbiter 2 on an extended "walk" around the planet, to photograph the poles and other regions of Mars and scan them

---

*The relay links for the first 11 sols were pre-programmed for redundant playback and transmission to Earth of the lander-recorded data so as to prevent loss of any important information.

with the infrared thermal mapper and the Martian atmospheric water detector. Orbiter 1 continued to provide the communications link for the second lander during the remainder of *Viking 2*'s primary mission.[49]

A more serious problem emerged in the first days after *Viking 1*'s touchdown when the surface sampler arm became stuck. On Thursday, 22 July, the surface sampler assembly was rotated so that the protective shroud covering the sample collector head (scoop) could be jettisoned. During this operation, the sampler boom was to be extended a few centimeters and then returned to the stowed position. Extending the boom was no problem, but on retraction it stuck. At first, Jim Martin and crew thought the problem was one of electronics. At 6:30 p.m. on the 22d, Martin told reporters preliminary indications were that perhaps the soil-sampler control assembly—the receiver for computer commands—had "some kind of an electronic problem." He could switch to a redundant soil-sampler control assembly if that was the problem, but, "the concern I have at the moment is that unless we can solve or understand this problem and solve it in fairly short order we are likely to run the risk of impacting the soil acquisition sequence on Sol 8."[50]

By 10 p.m. on the 22d, Martin's team had arrived at a new theory. Prefacing his remarks to the media with, "It has been a very busy day," Martin addressed the problem of the sampler. Everyone knew that loss of the sampler would be a major setback for Viking science activities. Without it, no samples would be delivered to the biology instrument, the gas chromatograph–mass spectrometer, or the x-ray fluorescence spectrometer. Martin believed that his people, who had worked all evening, had "isolated the most probable cause of the problem. It turns out, contrary to my expectation, not to be an electrical problem." Instead, it was apparently a simple— if anything can be simple when working with a piece of equipment millions of kilometers away—mechanical hangup. Martin pointed out "that there is a locking pin that is part of the shroud latching system"; that pin "was supposed to drop to the Martian surface during the boom extension. . . . It now appears that the extension that had been commanded in the sequence was not long enough to allow this pin to drop free."

Martin had observed a duplication of the difficulty on the science test lander, which was housed in a glass-walled room next to the auditorium in which the press briefings were held at JPL. Commenting on the fishbowl atmosphere in which his people had been working, Martin told the reporters, "I went in and looked at it myself when some of you weren't looking." The stuck pin was "certainly a plausible and possible failure mode." To test this theory, "we plan to send up a new command sequence on the Sol 5 command load which will go up at around midnight Saturday night," 24 July. Mission analysts thought that extending the boom to about 35 centimeters would let the pin fall. Martin added, "If by some chance the pin was retained within the mechanism, which I really believe is doubtful, we don't ever intend to retract it as far as we did in the original sequence." That way, they would avoid another difficulty; at a certain point the boom extraction motor would clutch on purpose and then shut itself off to avoid damage to the motor. If the pin did not drop free this time, the boom would be ordered to extend far enough so that the "no-go" signal would not be given.[51]

Two photographs taken by the lander camera on sol 5, 25 July, showed that the retaining pin did fall free, landing on the ground in front of the craft.[52] The apparent ease with which this problem had been diagnosed and corrected hid the months of training and preparation for such mission operations. Subsequently, more serious troubles were to plague the soil-sampler assembly, but each time training and ingenuity permitted the team to work out solutions and keep the mechanism functioning. Adaptability was one of the key elements of Viking's landed operations.

## Communicating with the Spacecraft

Before separation from the orbiter, the lander had been given an initial computer load (ICL, or "ickel"), which contained all the computer commands necessary for a basic 60-day mission, even if there were no further communications from Earth. With normal communications between the spacecraft and mission control, the mission programmers could modify the initial computer load as needed to get the most out of the lander. Commands were "uplinked" to the lander from JPL through the stations of the Deep Space Network to the orbiter and then to the guidance, control, and sequencing computer. The command uplinks, made in three-day cycles, were the responsibility of the lander command and sequencing team of the lander performance and analysis group.

Agreeing on the commands to be sent to the lander, programming them, and checking them out through simulations was a complex series of tasks, which required a great deal of work and interaction among many persons. An example is the decision to photograph the sampler boom immediately after acquisition of a sample. The requirement would first be sent to the lander imaging team, which had three three-person squads who handled such requests. These uplink squads, plus a "late-adaptive squad" responsible for last-minute alterations, would investigate the picture called for and determine if it could be combined with others or if it had to be taken by itself. The series of pictures for a given sol was then described and combined into a science requirement strategy that was passed on to the Lander Science Systems Staff, which had the difficult task of matching wants (requirements) with the constraints imposed by the lander systems and the other tasks that had to be accomplished.

The Lander Science Systems Staff received the uplink plans in the form of computer printouts called science instrument parameters—specific commands to the guidance, control, and sequencing computer. Lander imaging had 56 commands available, and each could be adapted to special requirements. Once approved by the Lander Science Systems Staff, the parameters were passed on to the lander computer simulations personnel, who ran through the commands to see if there were any software or hard-

## Viking Surface Sampler

The Viking lander's chemical and biological investigations all used samples of surface materials excavated by the surface sampler. In addition, as the experience with lunar Surveyor spacecraft demonstrated, there was much to learn about the surface simply by digging in it. In the Viking mission, digging was part of the physical properties and magnetic properties investigations.

The surface sampler consisted of a collector head attached to the end of a three-meter retractable boom. The arm housing the boom could be moved both horizontally and vertically. The boom itself was constructed from two ribbons of stainless steel welded together along the edges. When extended, the two layers opened to form a rigid tube. When retracted, the boom flattened. A flat cable sandwiched between the boom layers transmitted electrical power to the collector head.

The collector head was basically a scoop with a movable lid and a backhoe hinged to its lower surface. Where the scoop is attached to the end of the boom, a motorized rotator acted as a mechanical wrist to permit manipulation of the collector head. To fill the scoop, the lid was first raised and then the boom was extended along or into the surface. Once full, the lid closed. The top of the lid had holes two millimeters in diameter, which formed a sieve. When the collector head was positioned over one of the inlets for the instruments, it was inverted and vibrated. Only particles smaller than two millimeters were delivered to the instrument inlets. Coarser samples could be delivered to the x-ray fluorescence spectrometer, if desired. The gas chromatograph-mass spectrometer and the biology instruments had their own filters to control the size of material introduced into their sample processing assemblies.

The surface sampler could also dig trenches, by lowering the backhoe to place the sampler head on the surface, and then retracting the boom. Excavated materials could be scooped up for sampling. A brush, magnets, temperature sensor, and other instrumentation also provide data concerning the physical properties of the materials.

ware conflicts. Considerations such as electrical energy required or the thermal impact of a command were also determined. Following simulations, the request was codified into a "lander sequence." After all the necessary changes (massaging) were completed, the command was entered into the ground-based computer and relayed to the Deep Space Network for transmission to Mars.

Uplink teams preparing lander sequences worked about two weeks ahead of the time the command was to be executed. Changes could be made in the planned uplink until about 48 hours before it was loaded into the

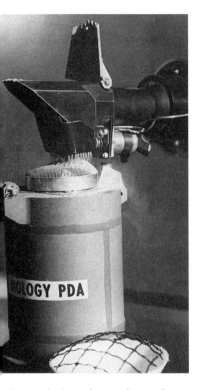

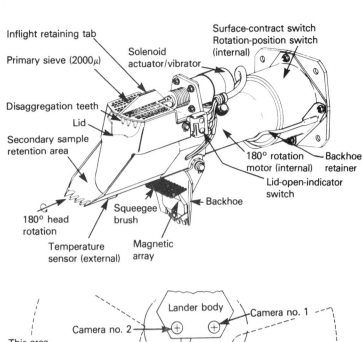

Inflight retaining tab

Primary sieve (2000μ)

Solenoid actuator/vibrator

Surface-contract switch
Rotation-position switch
(internal)

Disaggregation teeth

Lid

Secondary sample retention area

180° rotation motor (internal)

Backhoe retainer

Lid-open-indicator switch

180° head rotation

Squeegee brush

Backhoe

Temperature sensor (external)

Magnetic array

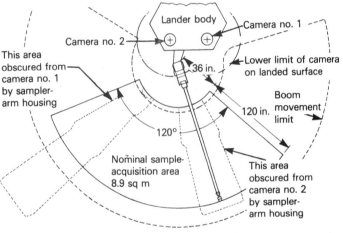

Lander body

Camera no. 2

Camera no. 1

This area obscured from camera no. 1 by sampler-arm housing

36 in.

Lower limit of camera on landed surface

Boom movement limit

120 in.

120°

Nominal sample-acquisition area 8.9 sq m

This area obscured from camera no. 2 by sampler-arm housing

**Sampler-Arm Area of Operation**

*A premission photo, above, shows how the surface-sampler collector head deposits its contents into the biology-instrument processor and distributor assembly. The collector head and the area of the sampler arm's operation are sketched at right. Below, project scientist Gerald Soffen examines the collector head on the science test lander at JPL.*

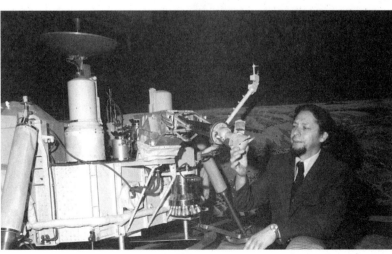

computer. Obviously, uplinking was a precise, demanding business. Mistakes were totally inadmissable. Although out of the limelight, the people responsible for talking with the lander had a difficult task. Occasionally, nerves wore thin when the requirements of different science teams conflicted. The uplinkers were expected to satisfy everyone's needs, and for the most part they did.[53]

### Sampling the Martian Surface

Scientifically, the most important experiments aboard the lander were those which sampled the planet's surface. Of these, the chemical analyses were interesting, but the biological experiments were a disappointment. As with other investigations, Mars again turned out to be a more complex riddle than anticipated and, while there is still disagreement over the exact causes of some of the reactions observed, most—but not all—of the Viking scientists have come to the opinion that detection of life on Mars is a very unlikely prospect.

The first soil samples were acquired on sol 8, 28 July. Four samples were dug, with the first being deposited into the biology instrument distributor assembly, the next two into the GCMS processor, and the fourth into the funnel of the x-ray fluorescence spectrometer. All the commands were successfully executed, but there was no positive indication that the gas chromatograph–mass spectrometer processor had been properly filled. A second acquisition attempt still did not provide a "sample level detector 'full' indication." The sampler system, having completed its programmed sequences in a normal manner, parked the boom as planned. On Earth, the lander performance specialists began to analyze the possible causes of the anomaly: (1) insufficient sample acquired in the collector head because the same sample collection site had also been used for the biology sample; (2) insufficient time allowed for the sample to pass from the funnel through the sample grinding section and then through the fine (300-micrometer) sieve into the metering cavity of the instrument; (3) grinder stirring spring not contacting the sieve; or (4) sample-level-detector circuit faulty. Since the "level-full" detector consisted of a very fine wire stretched across the cavity to which the sample material was delivered, it was also possible that it had broken when the soil was dropped into the funnel.[54]

An anomaly team headed by Joseph C. Moorman, who had worked closely with the builders of the GCMS, went to work on this problem. While preparations were made for another sample to be collected on sol 14, 3 August, Martin and Young had to decide whether to proceed on the assumption that the GCMS had actually been filled and chance wasting one of the two remaining ovens on an empty chamber (the specialists had determined that one of the ovens was inoperable during the GCMS in-flight checkout) or pick up another sample on sol 14. Conservatism and caution argued for the latter decision, and the managers chose that option. But the boom did not cooperate. It jammed.

The surface-sampler control-assembly sequences performed normally through the 12th command. During the execution of the 13th (boom retraction to 26.7 centimeters), trouble showed up; when the computer issued the 14th command, the assembly would not respond. Examination of photos taken on sol 14 revealed that the sampling trench had been dug as ordered, but the collector head was not over the GCMS funnel where it was supposed to be. An image received on sol 15 showed the back of the boom. Three possible reasons for this new anomaly were considered: (1) failure of the surface-sampler control-assembly electronics; (2) failure of the boom motor or related equipment; or (3) jamming of the boom, precluding proper retraction. Causes 1 and 2 were rejected after analyzing the proper performance through the first 12 commands. Jamming had most likely caused the difficulty since the failure appeared to be similar to the "no-go" response encountered with the sol 2 shroud-pin jam.

Frozen carbon dioxide or surface material were rejected as possible causes of jamming the boom, because of the absence of a slowly increasing motor load, which the investigators would have detected. Discussion of the anomaly with the boom designers revealed that a similar problem had occurred during early test phases, and they believed it was caused when a series of successive retract (or extend) commands had been issued. In testing, the successive commands tightened the boom element on the storage drum, and the boom element tended to wind around the drum in a 5- or 6-sided configuration rather than in a perfect circle. This arrangement caused intermittent high loading when the "points of the hexagon" passed under the boom restraint brake shoes. The reliability of the system was further weakened when operated at low temperatures; the motor torque limiter finally decoupled, and movement of the boom ceased. Two major operating procedures were proposed to meet the problem: (1) All sequences were to be revised to eliminate successive extend or retract commands, avoiding excessive tightening of the boom element on the drum. The command reversals would cause the extend or retract "flip-flop" gear to disengage the load during each cycle, allowing the motor to attain full speed and operating torque before it reengaged the load in the opposite direction. (2) Future operations were to be performed within one to two hours of the peak temperature during the Martian sol. An uplink diagnostic sequence was designed for sol 18; the boom would be used in each axis of operation— extend, retract, up elevation, down elevation, clockwise, and counterclockwise. The sequence was executed properly and no anomalies were met. Following Martin Marietta's instructions, all activities of the sampler arm were redesigned "to exclude, wherever possible, successive extend or retract commands, and to perform these operations during the warmest part of the sol." The Viking team had no further problems with the sampler boom on either lander, and operating temperature restrictions were eventually waived because of the need to acquire early morning biology samples. Preflight testing and the documentation of those procedures had paid off.[55]

The sol 14 anomaly forced Martin and Young to reconsider their decision not to analyze the "possible" sample acquired on sol 8. Influenced by early results from the biology experiments, the molecular analysis team urged that the contents of the gas chromatograph–mass spectrometer be analyzed. Jim Martin and Tom Young agreed.

*Biology.* At the 1:30 p.m. news briefing on 31 July 1976 (sol 11), Jim Martin made an announcement. Prefacing his remarks with, "I wanted to state that it's been project policy for seven years to make data available to the media when we have [them]," Martin noted that this day was "no exception. We have received biology data that we believe to be good data." Engineering telemetry indicated that the biology instrument was performing "extremely well," perhaps too well, since early reactions from the gas-exchange and labeled-release experiments were very positive. That could possibly be the consequence of biological activity, but Martin was cautious: "I think Chuck Klein will continue to caution you that the biology experiment is a complex one. We've seen that Mars is a complex planet. There are many things that we do not understand." The scientists were proceeding systematically and methodically.[56]

Biology Team Leader Harold P. Klein and his colleagues had already conducted a number of tutorials for the news people covering the Viking mission, and at each session where they presented analytical details they took time to explain the experiment in question. The biologists started with the basics. Each Viking lander carried an integrated biology instrument, which contained three experiments designed to detect the metabolic activity of microorganisms should they be present in the soil sampled. First, the gas-exchange experiment would determine if changes caused by microbial metabolism occurred in the composition of the test chamber atmosphere. Second, the labeled-release experiment, also known as Gulliver, would determine if decomposed organic compounds were produced by microbes when a nutrient was added. Third, the pyrolytic-release experiment would detect, from gases in the chamber, any synthesis of organic matter in the Martian soil. A change could be the result of either photosynthetic or nonphotosynthetic processes.

On 31 July, Klein told the press: "What we are proposing to do for you today [is] to give you a status report on the three experiments and we'd like to then focus on one of the experiments, the labeled release experiment, a little more closely since some of that data is exciting and interesting." First, all three instruments were working normally. "We have no anomalies, no problems despite what some of the press or other news media have said." He had heard rumors that the biology instrument was "sick, dead in the water." The truth was that the instrument was in good shape, and he had two important, unique facts.

First, the gas-exchange experiment had given them reason to believe that "we have at least preliminary evidence for a very active surface material. . . . We believe that there's something in the surface, some chemical or physical entity which is affording the surface material a great activity." But, adding a word of caution, he noted that the reaction observed in the gas-exchange experiment might be mimicking some aspects of biological activity. Second, the labeled-release experiment's radioactivity counters were measuring "a fairly high level of radioactivity which to a first approximation would look very much like a biological signal." The highly active nature of the soil, however, caused the biology team members to be cautious. "That second result must be viewed very, very carefully in order to be certain that we are, in fact, dealing with a biological or non-biological" phenomenon.

Klein reported on the sequencing of the three biology experiments. Norman Horowitz's pyrolytic-release experiment had been started first. After the soil had been injected into the test chamber and carbon 14-labeled carbon dioxide added, the xenon lamp had been turned on; incubation would last until at least sol 14, when the first results might be available. Vance Oyama's gas-exchange experiment had also received its soil sample on 28 July, but the incubation process was not begun until the morning of the 29th, when the chamber containing the soil and Martian atmosphere was injected with a mixture of carbon dioxide, krypton, and half a cubic centimeter of nutrient. About two hours later, gas in the chamber was analyzed—a calibrating measurement against which all subsequent analyses would be measured. Calling for the lights in the Von Kármán Auditorium to be turned off, Klein had a chromatogram based on the first gas exchange results projected on the screen behind him:

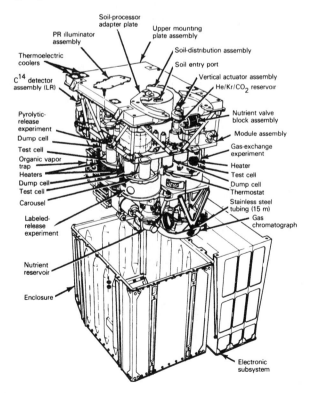

*Biology instrument*

> What we saw were five peaks—little tiny peaks: neon, over here on your left and that's explainable by the neon that we used in the nutrient chamber itself and that's our indication that we, in fact, injected nutrient and that's fine—there's nothing unusual about that. Then you see nitrogen and that amount of nitrogen can be accounted for by the nitrogen in the atmosphere and a small amount of nitrogen that we know was contaminating our $CO_2$ krypton mixture. Then we see this oxygen peak which I will come back to in a moment. And then as a shoulder beside the oxygen, you see a small peak and that's a combination of argon and carbon monoxide and that amount of gas would be consistent with current estimates of argon and carbon monoxide in the atmosphere.

A large krypton peak, Klein explained, was present because they had added krypton in a specifically known amount to provide a standard reference for determining the amount of other gases that might be present. He turned back to the oxygen peak: "You will see at the base of that oxygen peak, a little bar—that's the amount of oxygen down there that we can account for, or could account for from all known sources in the atmosphere or in the contamination of our gas mixture." But the instrument on *Viking 1* was indicating 15 times more oxygen than the scientists could account for from known sources. The results from the second measurement made 24 hours later showed that all the gases had remained the same except oxygen. It had increased by 30 percent. After ruling out all other possible causes, the scientists concluded that the oxygen had to be coming from the soil itself. While one possible explanation for the increase was biological activity, other explanations were possible, too.[57]

A possible alternative answer to why the initial amount of oxygen had been released lay in the desert area of landing site; the Martian samples contained peroxides and superoxides, which when exposed to abnormal (non-Marslike) humidity in the instrument quickly released oxygen. The related release of carbon dioxide suggested that the samples had an alkaline core. Although such reactions had not been witnessed on Earth, the scientists believed that the intense ultraviolet radiation bombarding the surface of the Red Planet could have produced unique photocatalytic effects. Still, there was much to be explained, including the reactions observed from the labeled-release investigation.

Gulliver was sending back some surprises. As with the gas-exchange experiment, the labeled-release experiment added a small amount of nutrient to the soil sample. It also produced a large amount of gas after that injection. Where the gas-exchange produced a spectrum of the gases, the labeled release measured the amount of radioactivity produced by the carbon-14 "labeling" material in the nutrient. Shortly after the addition of the nutrient, the radiation counts rose sharply, leveling off at about 10 000 counts per minute.

Gil Levin gave the audience at JPL a brief resume of the activities since the injection of the nutrients, which had occurred at about 1:45 p.m. PDT on

30 July. That injection had consisted of about 0.1 milliliter, or about 2 drops, of liquid. As Levin noted, "If any organisms are present that can utilize the nutrient and if these organisms behave biochemically—roughly as terrestrial organisms do—they should imbibe the nutrient and exhale a radioactive gas." Resulting radioactivity was measured periodically by a radiation detector. The result on Mars was very interesting. It was similar to ones encountered with living organisms detected in terrestrial soil, but Levin warned, "We are far too early in the game to say that we have a positive response." There were too many factors that had to be weighed and tested. "All we can say at this point is that the response is very interesting, be it biological or non-biological, it is unanticipated."

As in the gas-exchange experiment, there was a possibility that the soil itself contained catalysts, minerals, inorganics that produced some breakdown of the radioactive compounds. "The effect of water introduced into the dry Mars soil may cause violent chemical reactions that would disintegrate a portion of our medium." As a consequence, Levin thought that any speculation about the biological or nonbiological nature of the response would have to await further data.[58]

By 1 August, the production of oxygen in the gas-exchange experiment had decreased considerably, thus supporting the belief that the release was the function of oxides in the soil. In a 2 August update on the labeled-release experiment, Levin noted that they had examined the radioactivity curve very carefully. They had found no evidence of any doubling of cells. No growth appeared to be taking place, but the curve did not seem to behave as scientists would have expected it to for chemical reactions either. "We find that the chemical reaction took place at a very rapid rate initially, and then uncharacteristically slowed down and took a long time to plateau." The curve detected with the labeled-release experiment did not agree with known responses for either chemical or biological reactions.[59]

Data returned by the pyrolytic-release experiment and reported by Norman Horowitz on 7 August were equally confounding. Once again, the specialists had detected a reaction, but they did not know what it meant. "There's a possibility that this is biological," Horowitz said, but "there are many other possibilities that have to be excluded." The results obtained the night before were interesting but he emphasized that they were not ready to say that they had discovered life on Mars. "The data point we have is conceivably of biological origin, but the biological explanation is only one of a number of alternative explanations." He told the press:

> We hope by the end of this mission to have excluded all but one of the explanations, whichever that may be. I want to emphasize that if this were normal science, we wouldn't even be here—we'd be working in our laboratories for three more months—you wouldn't even know what was going on and at the end of that time we would come out and tell you the answer. Having to work in a fishbowl like this is an experience that none of us is used to.

He also cautioned the reporters that they were being included in the analysis phase of the experiments. They were "looking over the shoulder of a group of people who are trying to work in a normal way in an abnormal environment."[60] The scientist's caution was prompted by his knowledge that "we well might be wrong in anything we say. Anyone who has carried out a scientific investigation knows that the pathway of science is paved not only with brilliant insights and great discoveries, but also with false leads and bitter disappointments. And nobody wanted to be wrong in public on a question as important as that of life on Mars."[61]

Later in a November 1977 *Scientific American* article, Horowitz was able to speak more authoritatively about the results that had been observed in all three experiments. In the gas-exchange experiment, "the findings of the first stage of the experiment were both surprising and simple." Immediately following the addition of the moisture to the sample chamber—the soil sample was not directly wetted—carbon dioxide and oxygen were released. The evolution of gases was short-lived, but the pressure in the chamber increased measurably. At the Chryse site, the amount of carbon dioxide increased by about 5 times, and the amount of oxygen increased by about 200 times in little more than one sol. At the landing site in Utopia, the increases were smaller but still "considerable." Upon reflection, Horowitz stated that "the rapidity and brevity of the response recorded by both landers suggested that the process observed was a chemical reaction, not a biological one." Horowitz felt that the appearance of the carbon dioxide was readily explainable: "Carbon dioxide gas would be expected to be adsorbed on the surface of the dry Martian soil; if the soil was exposed to very humid atmosphere, the gas would be displaced by water vapor." The presence of the oxygen was logical but harder to account for, since so much oxygen would seem to require an oxygen-producing substance, not just the physical release of preexisting gas. There was just not that much oxygen available in the atmosphere—past or present—to account for the quantities measured. Horowitz argued that it was "likely that the oxygen was released when the water vapor decomposed an oxygen-rich compound such as a peroxide. Peroxides are known to decompose if they are exposed to water in the presence of iron compounds, and according to the X-ray fluorescence spectrometer . . . the Martian soil is 13 percent iron."

At both sites, the second phase of the gas-exchange experiment was "anticlimactic." When the sample was saturated with the aqueous nutrient, more carbon dioxide and oxygen were produced. The additional evolution of carbon dioxide was probably a continuation of the reaction observed in the humid stage of the experiment. Horowitz believed that the amount of oxygen then diminished because of its combination with the ascorbic acid in the nutrient medium. "And so . . . it became clear that everything of interest happened in the humid stage of the experiment, before the soil came in contact with the nutrient!" Thus, in November 1977, Horowitz confidently stated that the gas-exchange experiment had detected "not metabolism but the chemical interaction of the Martian surface material with water vapor at a pressure that has not been reached on Mars for many millions of years."[62]

In the labeled-release experiment, there was a similar rapid surge of gas into the test chamber when the nutrient solution was added to the soil. This release tapered off shortly after the passage of one sol. Horowitz noted,"The gas, undoubtedly carbon dioxide, was radioactive, showing that it had been formed from the radioactive compounds of the medium and not from compounds in the Martian soil." He also believed that other nonradioactive gases were evolved when the water in the nutrient medium came in contact with the sample, but that these could not be detected by the instrument. "The production of radioactive carbon dioxide in the labeled-release experiment is understandable in light of the evidence from the gas-exchange experiment suggesting that the surface material of Mars contains peroxides." Formic acid, which was one of the compounds in the labeled-release nutrient, is oxidized with relative ease. "If a molecule of formic acid ($HCOOH$) reacts with one of hydrogen peroxide ($H_2O_2$), it will form a molecule of carbon dioxide ($CO_2$) and two molecules of water ($H_2O$)." The amount of radioactive carbon dioxide produced in the experiment was only slightly less than would have been predicted if all the formic acid in the nutrient had been oxidized in this manner.

Going a step further with his analysis, Horowitz said that if the source of the oxygen in the gas-exchange experiment was peroxides in the soil decomposed by the water vapor, then the labeled-release experiment should have decomposed all of the peroxides with the first injection of nutrient. The second injection should have produced no additional radioactive gas. That was what happened. "When a second volume of medium was injected into the chamber, the amount of gas in the chamber was not increased; indeed, it decreased. The decrease is explained by the fact that carbon dioxide is quite soluble in water; when fresh nutrient medium was added to the chamber, it absorbed some of the carbon dioxide in the head space above the sample."

In the labeled-release experiment, the stability of the reaction to heating at various temperatures was examined. Heating reduced and subsequently stopped the reaction. This result has been interpreted by some to be evidence in favor of biological activity, but Horowitz, although conceding that the effects of heating could be explained by biological activity, said that these results were also consistent with a chemical oxidation in which the oxidizing agent is destroyed or evaporated at relatively low temperatures. "A variety of both inorganic peroxides and organic peroxides could probably have produced the same results."[63]

The third biology experiment, pyrolytic release, differed from the others in two basic respects. First, it attempted to measure the synthesis of organic matter from atmospheric gases rather than the decomposition of that matter. Second, it was designed to operate under pressure, temperature, and atmospheric composition that were nearly the same as those on the planet. During the actual operation of the pyrolytic-release investigation, the temperatures ran higher than those normally encountered on Mars because of heat generated within the lander. A sample of the soil was sealed in the test chamber along with some of the planet's atmosphere. A xenon arc lamp simulated the sun. Into this Martian microcosm, small amounts of radioactive carbon dioxide and carbon monoxide were introduced. After five days, the xenon lamp was turned off, and the atmosphere was removed. The soil was then analyzed for the presence of radioactive organic matter.

Analysis of the soil began with heating it in the pyrolyzing furnace—hence, the name pyrolytic release—to a temperature high enough to reduce any organic compounds to small volatile fragments. Those "fragments were swept out of the chamber by a stream of helium and passed through a column that was designed to trap organic molecules but allow carbon dioxide and carbon monoxide to pass through." In this process, radioactive organic molecules would be transferred from the soil to the column while being separated from the remaining gases of the incubation atmosphere. Any organic molecules would be released from the column by raising the column's temperature. Simultaneously, the radioactive organic molecules would be decomposed into radioactive carbon dioxide by copper oxide in the column and transported to the radiation counter by the helium carrier gas. If, as a result of this process, organic compounds had been formed, there would be detectable radioactivity; if there were no organics, there would be no radioactivity.

Horowitz noted that, surprisingly, "seven of the nine pyrolytic-release tests executed on Mars gave positive results." The negative results occurred with samples obtained at *Viking 2*'s Utopia site. The amount of radioactive carbon dioxide obtained by the experiment was small; still, it was enough to furnish organic matter for between 100 and 1000 bacterial cells. Significantly, "the quantity is so small . . . that it could not have been detected by the organic-analysis experiment," the gas chromatograph–mass spectrometer (see below). Though small, the quantity was important, because as Horowitz expressed it, "it was surprising that in such a strongly oxidizing environment even a small amount of organic material could be fixed in the soil." Even more important to him was the fact that "the pyrolytic-release instrument had been rigorously designed to eliminate non-biological sources of organic compounds." To encounter positive results from the Martian soil in spite of all the precautions was in the biologist's word "startling."

However, on reflection, it appeared that the findings of the pyrolytic-release experiment had to be interpreted nonbiologically. The reaction did not respond to heat in a manner consistent with a biological reaction. Martian microbes, accustomed to the very low temperatures on that planet, would have been killed by the elevated temperatures experienced during the test, the investigators thought. "On the other hand, it is not easy to point to a non-biological explanation for the positive results." Investigations into this curious reaction have continued in terrestrial laboratories, and until "the mystery of the results . . . is solved, a biological explanation will continue to be a remote possibility."[64]

*Gas Chromatograph-Mass Spectrometer (GCMS).* While the results of the biology experiments did not seem as bleak in the summer of 1976 as they have appeared subsequently, there was considerable concern during the missions about the proper interpretation of the reactions being witnessed. During August 1976, the Viking scientists believed that the GCMS was one possible tool for deciding if the reactions observed in the biology instrument were biological or chemical in origin.

As one observer noted, the gas chromatograph-mass spectrometer was the court of appeals in the event that the biological experiments did not present a clear verdict.[65] With the initial uncertainties from the biology experiments, the molecular analysis team decided to gamble that the GCMS had received its sample on sol 8 (see pages 398-400) and made the first analysis on 6 August (sol 17). Klaus Biemann reported to the press on the molecular analysis—"the first half of the first sample experiment of the organic analysis"—the following day. The soil sample was there! And the oven had worked as planned. There was always speculation among the news representatives about what new hardware problems might appear, but this time the scientists could report, "It did work as predicted, heated to 200° and stayed there for thirty seconds. The entire gas chromatograph mass spectrometer worked well like all gas chromatograph mass spectrometers do." Although the molecular analysis team was obviously pleased that its instrument was working well, the results from the GCMS would be the source of the most frustrating data for those exobiologists who were hoping to find life on the Red Planet.

About 300 mass spectra, electronically provided graphs identifying the molecules detected in the Martian soil sample, were returned by the first run of the GCMS. The molecular analysis specialists were particularly interested in determining if carbon compounds were in the sample, since biochemistry is largely the chemistry of carbon. The basic structure of the carbon atom enables it to form large and complex molecules that are very stable at ordinary temperatures. While no carbon compounds were detected in the first sample analysis, there was no great concern, since it was believed that the sample would have to be heated to 500°C before the organics would be broken down and detected by the instrument. The only surprising aspect of the first data was the very small amount of water released by the sample.[66]

On 12 August, the GCMS experiment was run again with the first sample being heated to a maximum temperature of 500°C. Biemann reported that this analysis "to our surprise, evolved a large amount of water. Indeed so much that it gives us trouble in analyzing the data." Still, the critical point of this analysis was that there were probably no organics. If the reactions observed in the biology instrument were the consequence of life, then it was expected that the GCMS would detect organic compounds

407

in the same soil. Neither this analysis nor the subsequent one at the *Viking 1* site, nor those carried out at the *Viking 2* landing area, produced traces of organic compounds at the detection limits (a few parts per billion) of the GCMS.[67]

Failure of the gas chromatograph–mass spectrometer to detect organic compounds was devastating for those who believed that life on Mars was possible. For Jerry Soffen, the GCMS results were "a real wipe out." Once he assimilated the fact that the GCMS had found no organic materials, he walked away from where the data were being analyzed saying to himself, "That's the ball game. No organics on Mars, no life on Mars." But Soffen confessed that it took him some time to believe the results were conclusive. At first, he argued with Tom Young that there must have been no sample present in the GCMS, because there had to be organics of some sort on the planet. Soffen bet Young a dollar that the second analysis would prove that the instrument had been empty. To his dismay, the data indicated instead that there was a sample in the instrument and that the sample was devoid of organics.

Klaus Biemann, the molecular analysis team leader, had some reflections on the search for organic compounds. Looking in the soil for compounds made of carbon, hydrogen, nitrogen, and oxygen at the level of a few parts per billion, they found none. The gas chromatograph–mass spectrometer could have detected smaller concentrations of organic materials than are present in typical antarctic soil, which is low in organic compounds because there is little vegetation and animal life on that part of Earth. Compared to Antarctica, Mars is devoid of organic material, and a number of conclusions could be drawn from that finding. First, no synthesis of organic compounds is occurring on the surface, at least where the two Vikings landed. Second, if millions of years ago organic compounds did exist, they must have since been destroyed. Third, since organic compounds must be arriving on Mars in the form of meteorites, that material must have been imbedded in the surface very deeply or, more likely, destroyed by the planet's harsh environment. Finally, says Biemann, "if we use terrestrial analogies, we always find that a large amount of organic material accompanies living things—a hundred times, thousand times, 10 thousand times more organic materials than the cells themselves represent." Since the Viking instruments did not detect any large amounts of organic waste material, it is difficult to see how microorganisms could be living at the areas investigated "if they behave as terrestrial organisms do."

Of course, reminded Biemann, "this does not rule out a different kind of living mechanism that would protect its organic constituents very well and, therefore, avoid this waste of a scarce commodity." Martian organisms could have evolved along those lines, and as the environment became harsher and harsher they could have become more and more efficient in using the organic materials they needed. Viking looked at only two samples at each of the two landing sites from depths of 5 to 10 centimeters. If organic materials were produced millions or hundreds of millions of years ago, they could be present at greater depths and protected there from the damaging ultraviolet radiation. The Viking spacecraft could be sitting on an area containing a deposit of organic material a few meters down. There could also be other areas on the planet where the surface material is more protected or where organic material is now being synthesized and not destroyed. To help answer these puzzling questions, Biemann and his colleagues had plans to study in their laboratories the rate of decomposition of certain typical organics under Martianlike conditions, to determine how fast organic materials might be destroyed at the surface.[68]

## LIFE OR NO LIFE?

Soffen's disappointment was shared by others on the biology team. For years, they had discussed the scientific possibilities of discovering life or the prerequisites for life on the Red Planet, and Soffen recalled the long debates with his colleagues on the subject. Some, like Wolf Vishniac, had argued that a negative result—that is, no life—was as important scientifically as the discovery of life. But such a discovery had not proved very exciting. Before the Viking landings, Soffen had been very careful in all his public statements to say that they would likely find nothing on the planet, but personally he had wanted to find life.

While Soffen believed that it was possible for life to have developed on Mars, he also thought it likely that the biology instrument, for a host of reasons, had not been designed properly to detect it. However, he was also very confident that if organic compounds had been present, the GCMS would have detected them. For that reason, he had fought for the instrument throughout the evolution of the Viking project. Soffen could have accepted a negative biology result, if there had been a positive measurement of organic compounds. But positive biology results could not be interpreted as indicating the existence of life in the absence of organics. Others have argued that perhaps Viking landed at the wrong places on the planet. Nearer the poles where there was a higher moisture content in the soil and atmosphere, life might exist. Or perhaps, as suggested by Carl Sagan and Joshua Lederberg, there are Martian microenvironments where in small oasislike areas life has evolved and survived. Soffen thought this unlikely since the homogenizing effects of wind and dust storms would have likely distributed any organic material all over the planet. He reluctantly concluded that life on Mars was unlikely.[69]

The apparent absence of life on the Red Planet had a far-reaching philosophical and emotional impact on members of the biology team. The team had never been a cohesive group of investigators, and the results of the biology and GCMS experiments served to accentuate their differences. Norman Horowitz came to the opinion that there is no life elsewhere in the solar system. While he did not rule out the possibility in theoretical terms, he believes, practically speaking, that scientists will never be able to prove the existence of life on another planet. Horowitz noted:

There are doubtless some who, unwilling to accept the notion of a lifeless Mars, will maintain that the interpretation I have given is unproved. They are right. It is impossible to prove that any of the reactions detected by the Viking instruments were not biological in origin. It is equally impossible to prove from any result of the Viking instruments that the rocks seen at the landing sites are not living organisms that happen to look like rocks. . . . The field is open to every fantasy. Centuries of human experience warn us, however, that such an approach is not the way to discover the truth.[70]

One man who is still not convinced is Gil Levin. He cannot rule out the biological interpretation of the Viking biology experiment results. "The accretion of evidence has been more compatible with biology than with chemistry. Each new test result has made it more difficult to come up with a chemical explanation, but each new result has continued to allow for biology." Furthermore, Levin believed that all of the life-seeking tests showed reactions that "if we had them on earth, we would unhesitatingly have described as biological."[71] But other members of the biology team were not as easily convinced.

Vance Oyama, who fathered the gas-exchange experiment, publicly stated in early 1977 that "there was no need to invoke biological processes" to explain the results obtained from the experiments. While far from being accepted by all his colleagues, Oyama's opinion is one more example of the extent to which differing explanations can be made to account for the puzzling data acquired by the biology experiments. Should Oyama's explanation turn out to be valid, it would affect more than the biology experiments. It would also help explain the nature of the magnetic particles that adhered to the magnets on the sampler head, the interactions between the atmosphere and the surface, and the early evolution of the planet. His theory begins with a simple photochemical effect in the atmosphere: the intense solar ultraviolet radiation breaks down atmospheric carbon dioxide ($CO_2$) into activated carbon monoxide ($CO$) and single atoms of oxygen ($O$). As the ultraviolet radiation continues to bombard the atmosphere, some of the carbon monoxide is further reduced to its constituents, carbon and oxygen. Some of this single-atom carbon combines with carbon monoxide to produce carbene ($C_2O$). The carbene in turn combines with carbon monoxide to form the first key element in Oyama's theory, carbon suboxide ($C_3O_2$). Oyama postulated that the carbon suboxide molecules were united to form a carbon suboxide polymer. Intriguingly, the resulting polymer has a reddish cast.

Oyama's theory is consistent with data from the three biology experiments. Looking first at the pyrolytic-release experiment, Oyama noted that the carbon-14 isotope was an important factor in explaining the results observed from this instrument. The decay of the carbon-14 isotope into nitrogen-14 released a beta particle. The resulting energy was more than sufficient to fracture carbon-carbon, carbon-hydrogen, and carbon-oxygen bonds. The breakdown would activate the red carbon suboxide polymer, allowing it to incorporate the available carbon monoxide. Heating that same polymer to about 625°C during pyrolysis would produce about four percent of the original carbon suboxide, with a carbon-14 label. This single carbon suboxide molecule (monomer) would tend to stick to the pyrolytic release experiment's organic vapor trap and with subsequent heating would be released as the critical "second peak" the specialists observed in the experiment's data. Taking this another step, Oyama reported that the presence of water vapor when the sample was exposed to the labeled atmosphere would lower the second peak.[72]

In Oyama's laboratory gas-exchange tests, the prominent release of oxygen was also less the second time. But as Oyama said, the reason was very different. In the Martian atmosphere, the same photochemical breakdown (photodissociation) that led to the formation of carbon suboxide also led to the creation of activated oxygen atoms, albeit by a different route. When these oxygen atoms struck alkaline earths (for example, oxides of magnesium or calcium), they united to form superoxides that would release oxygen upon exposure to water vapor. Oyama argued that less oxygen was released at the Utopia site than at the Chryse site because the greater amount of water vapor in the more northerly landing site had previously freed some of the oxygen in the superoxides near the surface.

In describing the reasons for the results observed in the labeled-release experiment, Oyama presented the following scenario. Hydrogen peroxide formed photochemically in the atmosphere reacted with a catalyst on the soil-grain surfaces to release oxygen, which diffused into the grains, reacting with the alkaline earths and metals to form other superoxides. Atmospheric water vapor could readily convert the superoxides to peroxides, which in turn could combine with water in the nutrient to form hydrogen peroxide, $H_2O_2$, which would oxidize the labeled components of the nutrients to release the labeled $CO_2$. John Oro of the molecular analysis team also suggested very early that the results from the gas-exchange tests and labeled release were due to the presence of peroxidelike materials in the surface of the planet. To explain the process, Oyama used the example of chemical reactions in human beings. When hydrogen peroxide ($H_2O_2$), a commonly used disinfectant, is applied to a wound, it bubbles. This, Oyama said, is caused by the presence of iron in the enzyme catalyst. When the iron combines catalytically with the hydrogen peroxide, it releases bubbles of oxygen. Oyama believed that a similar process is at work on the surface of Mars.

Having searched for possible Martian catalysts, Oyama concluded that there is one likely candidate—a form of iron oxide known as gamma $Fe_2O_3$, or maghemite. On Earth, this is usually found only around the edges of hydrothermal or magnetic activity, where the temperatures range between 300° to 400°C. The abundance of water on Earth has converted much of the maghemite into a noncatalytic form, but on Mars this material has survived

411

virtually unaltered. Oyama thinks that it probably was produced either by an episode of volcanic heating or by heating that accompanied a period of meteoritic impacts. While this probably occurred early in the planet's history, he believes that it took place after the large quantities of water others suspect once existed had disappeared. Otherwise, the maghemite would have been rendered noncatalytic, just as it has been here on Earth. This explanation is a complex one, but as Jonathan Eberhart, writing for *Science News*, has reported: "Oyama's theory will have to stand the test of time, additional data and competing theories. But it does show that looking for life on other worlds has the potential for making valuable contributions in other fields as well."[73]

That there is still disagreement over the Viking biology results has caused some hard feelings among members of the biology team. Summarizing the situation after the results were in, Jerry Soffen said that he would expect the following responses if Horowitz, Oyama, Levin, and he were asked to participate in another Mars-bound biology investigation: Horowitz would not want to participate; Viking had satisfied his curiosity on the subject. Oyama would probably take part, but he would not expect to discover life. Gil Levin still believed that life may be discovered on the Red Planet. He had started with the goal of proving that there was life on Mars, and for him it was an engineering problem: How do you prove that there is life on Mars? To some of his colleagues, this was the attitude of an engineer, not the professional skepticism of the scientist. Examining his own position, Soffen said that he had never been certain about the possible existence of life on Mars, but he had hoped that it might be found. At no time, however, had he committed himself to proving that it actually existed. Horowitz, on the other hand, had always had such strong doubts about finding life that on several occasions members of the team wondered aloud why he had remained with the group. For Soffen, disappointments aside, he would like to return to Mars and look beyond the horizon shown in the lander photos—looking not for life but for whatever was there.[74]

Biology team leader Chuck Klein also had some thoughts on the search for life. "Before we landed on Mars we had a variety of opinions, ranging from those who expected to see no life on Mars to those who expected to see a rather flourishing—maybe not terribly advanced, but at least a flourishing life on Mars." Judging from all the Viking mission's findings, there is no visible flourishing life. But Klein suggested that the scientists must look more carefully at Mars "and ask whether the sophisticated biology and the chemistry instruments have given us clues as to whether there might be some less obvious kind of life on Mars." Klein believed that they could reject their pre-Viking model of Martian microbial life, "namely the Oyama model, which says that Mars should have micro-organisms similar to large numbers of soil bacteria on this planet." At neither site was there any indication to support that kind of concept of Martian biology. That means that either there are no organisms or any existing organisms do not fit that model.

Even though two of the biology experiments gave indications that could be interpreted on first inspection as being the result of some simple organisms being present, the molecular analysis team found no detectable organic compounds in the soil samples. The absence of organics made the biology team very suspicious; the weak-to-moderate signals in the two experiments might not be due to biological processes at all. "However, the lack of organics, in and of itself, does not rule out the possibility of organisms but makes that whole idea much less attractive," said Klein. As was noted by other Viking scientists, there is evidence that the surface material of Mars contains chemicals that are highly oxidizing and could interfere with the biological tests and mimic them. "Just as a living organism can, let us say, decompose a steak by eating it and digesting it, the steak can also be decomposed by being thrown into sulfuric acid, with roughly the same end products." The equivalent to the sulfuric acid in the case of the Viking biology experiments could be an inorganic nonbiological oxidizing material. Since this kind of nonorganic material seems to be present on Mars, it could be the cause of the confusing experiment results. "We tried a few tricks on Mars to see if we could devise some experiments that might definitely rule out the possibility that the decomposition seen is due to biology. We have not been able to do that so far." Although the two landing sites were more hostile than the biologists had anticipated, Klein points out that the Viking data do not really say there is no life on Mars.

> We can certainly say that it is not rampant, but we can't be sure there isn't some scraggly form of life for which we just haven't found the right nutrients or the right location or the right incubation temperature or the right environment within which to show its presence. That's why it's going to be very difficult for me, at least, to come out and say that there is no life on Mars. I think that would not be a scientific conclusion.

Klein, for one, wanted to go back to Mars.[75]

The planetary scientists agree that Mars is a fascinating place, and Soffen believes it is significant that no one has criticized Viking or the men who brought it about because life was not found there. Philip Abelson, editor of *Science*, stated categorically in February 1965 that "we could establish for ourselves the reputation of being the greatest Simple Simons of all time" if NASA pursued the goal of looking for extraterrestrial life on Mars.[76] His editorial in *Science* in August 1976 that reported on the initial results of *Viking 1* did not repeat this complaint, however, nor did he make it in either of the two subsequent issues that dealt with the Mars findings.[77] Some writers complained that the Martian microbes had not been given a decent chance—after all, the same ultraviolet radiation that caused the various photochemical reactions postulated by Oyama could also have destroyed the organic remains of many if not all of the Martian microbes— but none faulted the space agency for having made the search.[78]

A November 1976 editorial in the *New York Times* was typical of the press reaction. Noting that Mars had gone behind the sun earlier in

413

November, interrupting for a time communications between Earth and the Viking spacecraft, the editorial suggested that the "temporary halt in the receipt of new data permits a preliminary evaluation of what has been accomplished since last summer's historic landing." It appeared that "the whole field of Martian studies has been revolutionized and provided with an abundance of new data that will take years to assimilate fully." Findings on Mars would, in turn, force a reconsideration of the hypotheses concerning the origins of life on Earth. Referring to the postulated superoxides in the Martian soil, the *Times* noted, "Now the possibility is being discussed that such a superoxide existed here on Earth in the primeval years and that it is this wierd substance that provided the oxygen that now makes Earth such a hospitable planet for human and other familiar life forms. The classic explanation that the plant life produced most of earth's free oxygen is now being re-examined." Even the experiments of Miller and Urey in the early 1950s regarding the synthesis of prebiotic molecules could be questioned in light of the Viking investigations: ". . . the data from Mars have reminded scientists that electric discharges and accompanying ultra-violet radiation can also break down and destroy complex organic molecules as well as form them. All of a sudden the conventional wisdom about the development of life on Earth seems neither so certain nor so inevitable as it did before the Viking landings last summer." Although most scientists would not agree that the results of Viking were sweeping away the foundations for the studies of the origins of life, they would agree that "the Viking experiments have already been even more fruitful than their backers expected."[79] Perhaps the basic reason that there were no serious complaints about the Viking missions was that Mars had turned out to be a far more interesting place than anyone had predicted and more exciting than generations of scientists had expected.

## OTHER RESULTS

Viking's explorations and discoveries did not stop with the search for life. The great disappointment felt by the biologists was tempered to a degree by the wealth of other findings.

### Radio Science

One group of Viking investigators who did not have any scientific instruments of their own* on the four spacecraft but whose work assisted many scientists was the radio science team led by William H. Michael of the Langley Research Center. By analyzing the radio beams sent from Viking to Earth, specialists could determine precisely where the landers touched down and certain atmospheric and ionospheric properties of Mars, as well as gather data about the surface and internal properties of the planet and

---

*An X-band downlink on the orbiters was added specifically to enhance radio science capabilities and to conduct communications experiments.

about the solar system. The team's work can be divided into three general areas, as shown in table 55.

*Table 55*
*Viking Radio Science Investigations*

---

1. Dynamical, surface, and internal properties of Mars
    Spin-axis orientation and motion
    Spin rate
    Gravity field
    Figure
    Surface dielectric constant
2. Atmospheric and ionospheric properties of Mars
    Pressure, temperature, and density-altitude profiles
    Electron-number density-altitude profiles
3. Solar system properties
    Ephemerides[a] of Mars and Earth
    Masses of Martian satellites
    Interplanetary medium
    Solar corona
    Tests of general relativity

---

[a]*Ephemerides* are tabular statements of the predicted positions of celestial bodies at regular intervals.

Investigations of the locations of the Viking landers and the dynamical properties of Mars use primarily radio tracking of the landers, with some reliance on radio tracking of the orbiters for calibration. Determination of the gravity field and atmospheric and ionospheric properties use radio tracking of the orbiters, while the solar system and surface properties investigations rely on combinations of orbiter and lander radio tracking data. On Earth, the scientists use the transmitting, receiving, and data collection facilities of NASA's Deep Space Network at the 64- and 26-meter stations in California, Australia, and Spain.[80]

Although radio science operations began during Viking's cruise to Mars when the orbiter high-gain antenna was activated and tracking data were received, this activity was mostly related to checkout procedures, with some effort devoted to data and systems calibration. More immediately useful work began after the first landing, as doppler and range data became available for the first time between Earth and a spacecraft on another planet. From the first few days of tracking, the radio science specialists were able to ascertain "the location of the lander, the radius of Mars at the landing site, and the orientation of the spin axis of Mars." Additional data from both landers led to an initial determination from Viking findings of the spin rate of the Red Planet. After analyzing signal amplitude data from the lander-orbiter relay link, Michael and his colleagues were even able to suggest that the surface material around the first lander had electrical properties similar to that for pumice or tuff, a volcanic rock.

The precision of future Mars maps will be improved considerably, especially in the 30° south to 60° north latitudes, as a result of the radio science team's work during the extended mission's low-altitude gravity survey. As the second orbiter assumed a lower orbit (about 300 kilometers), the scientists measured the effect Martian gravity had on orbiter accelerations. They noted that Olympus Mons produced a very large gravitational acceleration, while prominent, though smaller, perturbations were observed over Tharsis Montes and Elysium Planitia. Results from a bistatic radar experiment will also help specialists identify Martian features more accurately by shedding light on surface reflectivity, surface roughness, slopes at various scales, and electrical properties of the surface in regions not accessible to Earth-based radar. These surface parameters are derived "from spectral analyses of signals transmitted toward specific locations on Mars from the orbiter antennas, reflected from the surface of Mars, and received at the Earth tracking stations." Besides being useful for mapping and geological interpretations, these findings will simplify the identification of future landing sites on Mars.[81]

Other questions include confirming the Einstein theory of relativity by a time-delay test—measuring how much the spacecraft signals are slowed as they pass near the sun and how the precession rate of Mar's orbital perihelion varies. During conjunction, data were gathered for studies of the solar corona. The team was also interested in more accurately measuring the distance between Earth and Mars and in determining the masses of Martian moons Phobos and Deimos. Viking's extended mission promised to be a busy time for the radio science experimenters, as did the period immediately following actual data acquisition. It would take many years to analyze all the results.

*Physical Properties*

The physical properties team was to draw conclusions from a composite of data from other experiments, to define the physical properties of the Martian soil. Richard W. Shorthill, team leader, stated that the team had been successful in describing the characteristics of the soil. But what it encountered was unlike any soils on Earth.

At the *Viking 1* site were two kinds of surfaces to investigate, the so-called rocky flats and the sandy flats. The bulk density (the number of grams per cubic centimeter) in the rocky flats area was slightly higher than in the sandy flats. At the second landing site, the bulk density was higher than the sandy area. The team determined the properties of the Martian soil by examining photographs of the trenches dug by the surface sampler. Cohesion (how the particles stick to each other) was ascertained by taking the dimensions of the trenches and the heights of the side walls and noting the collapsed state of the walls. The cohesion exhibited in the Martian trenches was similar to that found on Earth in a trench dug in wet sand. However, since the Martian soil is so very dry, the cohesion must have been caused by the electrical properties of the soil. Adhesion (tendency of the particles to stick to other objects) was determined by observing the soil that stuck to the sides of the surface sampler head before and after it vibrated. "We actually did some laboratory accelerometer tests on the vibrator at Martin Marietta while we were still on the surface of Mars to get a calibration of the adhesive forces," remarked Shorthill. By pushing the surface sampler into the surface until the force of the action turned on a microswitch, the soil's penetration resistance could also be measured.[82]

## Magnetic Properties

The magnetic properties experiment produced some interesting data, too. This investigation revealed an abundance of magnetic particles on the Martian surface, in both the soil and the very fine dust. On Earth, the most common magnetic particles are either iron metal or iron oxides, indicating that the red coloration of Martian soil may be caused by a highly oxidized iron, which is normally nonmagnetic on Earth. Robert B. Hargraves, leader of this experiment team, noted that two kinds of iron oxide exist on Earth—magnetite and hematite. Since hematite is nonmagnetic, perhaps the red mineral on Mars is magnetite with a coating of red hematite. But Mars is not Earth. "From what we've seen from the Martian imagery, these magnetic particles themselves appear red and they appear virtually indistinguishable from the average surface material on Mars." Hargraves admits that they have no direct information with which to resolve the mystery of the magnetic red soil, but the specialists planned to continue studying supporting data from other experiments in the hope of determining its properties more accurately.[83]

## Inorganic Chemistry

In the inorganic chemistry investigation, scientists analyzed the chemical elements in the Martian soil with an x-ray fluorescence spectrometer. Lander 1 acquired five soil samples successfully, three collected during the primary mission and two during the extended mission; the second lander acquired four samples for a combined total of 620 cubic centimeters of Martian soil. Each sample was sifted through a funnel to measure the precise size of the sample and then charged with high-velocity particles from an x-ray source.

When the spectrometer was supplied with a sample, the data were sufficient to detect the presence of iron (12-16%, maximum limits), calcium (2-6%), silicon (15-30%), titanium (0.1-1%), aluminum (1.5-7%), magnesium (0-8%), sulfur (2-7%), cesium (0-2%), and potassium (0-2.5%). Lander 2 attempted to retrieve rock samples three times and failed, because what appeared to be rocks in lander images were actually small crustal particles that crumbled when disturbed. The scientists believe there are pebbles but were unable to analyze one.[84]

Benton C. Clark, deputy team leader of the inorganic analysis team, commented that the "most striking factor between the two Viking landing sites is that the soil composition [chemical] is extremely similar in both cases. This is true for all elements we can detect in the soil including the very high" sulfur content, almost 100 times greater than the amount of sulfur found in Earth or lunar soil. One specialist remarked that they would be hard pressed to find such a closely matched pair of samples at such widely divergent sites on Earth, or even on the moon. The chemists think the giant dust storms that occur approximately every two years probably have mixed up the soil very efficiently and distributed it all over the planet as a fairly uniform mixture.

Despite the similarity of the soil from the two sites, different samples from the same location did indicate some differences in soil chemistry. "In one case, we get a higher sulphur content when we pick up a little dirt clod. In other cases, when we push a rock aside and sample the surface directly beneath it, we in general get a lower iron content and a somewhat higher sulphur content." Perhaps the soil under the rock was an older soil, whereas material out in a free area may have been the result of more recent dust storms—"recent in this case meaning the last thousands to millions of years." The chemists' findings have led them to believe that the Martian soil may have been derived from rocks with a very high magnesium and iron content.[85]

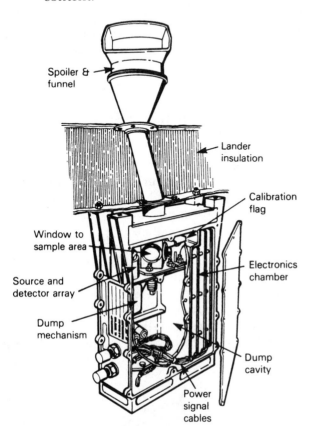

Spoiler & funnel

Lander insulation

Calibration flag

*X-ray fluorescence spectrometer*

Window to sample area

Electronics chamber

Source and detector array

Dump mechanism

Dump cavity

Power signal cables

*The Media*

Public interest may have been diminished by the failure to detect life, but many science writers continued to pursue the Viking science results. Almost weekly until the end of the prime mission in November when Mars disappeared behind the sun during conjunction, the press carried reports of scientific news from Mars. As Jerry Soffen told reporters at the second Viking science forum in August 1976, he and his colleagues were gratified by "the splendid coverage" they were getting, and he did not mean just the volume, which was considerable. The scientists had been impressed by the quality, as well:

> All of us really want to thank you and tell you how grateful we are for the remarkable clarity that has emerged as a result of this very open style that we are developing right now. . . . We have tried to each time answer your questions as clearly as we could and I know how difficult it is, as a reporter, to try to cover and clarify issues that seem to emerge one day and sometimes . . . appear to be contradictory on the next day.[86]

One example of the coverage given Viking is a series of articles in *Science News* by Jonathan Eberhart, the journal's correspondent in residence at the Jet Propulsion Laboratory during the primary mission. Respected by all his colleagues, Eberhart had a way of making understandable the complexities of science on Mars. Eberhart reported, among other accounts, efforts to move one of the rocks with the second lander's sampler arm, to find soil that had not been exposed in recent time to harsh ultraviolet radiation. As with all other maneuvers of the arm, the preparations took more than three weeks of consultations with more than a dozen specialists. The first attempt was a failure. The rock, blocking the first sample-acquisition site, refused to budge. Some persons thought that the rock might be frozen in place, but Priestley Toulmin of the inorganic analysis team argued that it was probably just the "tip-of-the-iceberg"; more of the rock was likely hidden below the surface. "Mr. Badger,"* the second candidate for displacement, was successfully moved. As the Viking lander team continued its investigations of the immediate region around the landers—pushing rocks, digging trenches, taking pictures, and measuring their findings—the science writers continued to report on the events.[87]

All the Viking mission activities prompted Gerald Soffen to comment in his Dryden lecture at the American Institute of Aeronautics and Astronautics' 16th Aerospace Science Meeting in Huntsville, Alabama, in January 1978, "How remarkable! We are performing chemical and biological experiments as though in our own laboratories. Taking pictures at will, listening for seismic shocks and making measurements of the atmosphere

---

*Henry J. Moore II named four large Martian rocks after characters—Mr. Badger, Mr. Mole, Mr. Rat, and Mr. Toad—from *Wind in the Willows* by Kenneth Grahame.

and surface. All of this from the first spacecraft ever to be landed successfully on Mars."[88]

*Management*

At the end of the primary mission in November 1976, some major changes took place within the management structure of the Viking Project Office. Several persons who had led Viking since its inception moved on to new positions. Jim Martin left NASA to become vice president of advanced programs and planning at Martin Marietta Aerospace in Bethesda, Maryland.[89] Tom Young, who had been serving both as Viking mission director and as Martin's deputy for JPL operations, took the post of director of lunar and planetary programs at NASA Headquarters.[90] For a time, Soffen maintained his position as Viking project scientist, but he was often called on to be a roving ambassador for the Mars project, traveling around the world telling scientific and lay audiences about the "real Mars" they had discovered. When Viking entered the extended mission phase in mid-December 1976, following the end of solar conjunction, however, many familiar faces still remained to complete the project. G. Calvin Broome had become project manager and mission director, and Conway Snyder, formerly orbiter scientist, first acted and then assumed full authority as project scientist.[91]

With the start of the extended mission, one phase of Mars exploration had come to an end. The goal of landing and successfully operating an unmanned scientific laboratory on the surface had been achieved, and vast archives of new and exciting information about the Red Planet had been amassed. The extended mission properly belongs to the post-Viking era, a period of evaluation and appraisal. With this initial scientific reconnaissance over, the issue facing the National Aeronautics and Space Administration was, What next? Viking, scientists hoped, was only a first step. The debate over subsequent steps would require decisions about not just exploring Mars but also how exploring Mars fitted into the overall scheme of NASA's planetary programs. One chapter closed, it was time to begin a new one.

# Epilogue

Viking was a success, both as a flight project and as a scientific investigation. Excellent hardware performance was the key to a fruitful mission. Project specifications called for a return of scientific data from the landers for a minimum of 90 days; but by the end of the primary mission on 15 November 1976, at solar conjunction, Viking lander 1 had been operating on the surface for 128 days, Viking lander 2 for 73 days. After a month-long rest while Mars disappeared from Earth's view as the planet swung behind the sun, the landers were awakened in mid-December 1976 for the extended mission, which lasted until 1 April 1978. The extended mission gave Viking scientists time to collect additional data on nearly every aspect of Mars science for which the landers had been programmed.[1]

April and May 1978 were months of transition for the Viking project. Under NASA management directives, the project was transferred to the Jet Propulsion Laboratory by the Viking staff at Langley Research Center. Old Viking hands, like G. Calvin Broome, project manager and mission director for the extended mission, left the project, and personnel from JPL took their places. Kermit Watkins, recognized for his role in preparing the Viking orbiter for flight, became project manager for what was called the continuation mission. Viking project scientists Gerald Soffen, who had accepted a new position at NASA Headquarters, was replaced by Conway Snyder.

Original plans had called for terminating the Viking mission after completion of the extended phase, but the spacecraft were functioning so satisfactorily in the spring of 1977 that the agency reconsidered the request of Viking science teams for an extension of the mission's activities. A continuation mission also received the strong endorsement of the Science Steering Group at its June 1977 meeting, but the major problem was money. Mission managers would have to reduce expenditures to a level that would continue operations without any additional funds in fiscal 1978. Once again, everyone tightened fiscal belts, and the project moved forward.[2]

Hardware problems on the Viking spacecraft began in the fall of 1977. In September, the second traveling-wave-tube amplifier on Viking lander 2 failed, and without this amplifier unit it could not communicate with Earth through the orbiter. Then a gas leak developed in the attitude control system of Viking orbiter 2, which required disabling half the control system to prevent further propellant loss. In February 1978, a more serious leak developed, losing about 22 percent of the remaining gas. A third leak in March further depleted the supply. Later that month, the flight controllers placed orbiter 2 in a roll-drift flight mode to prevent any further problems of this sort. Some atmospheric water observations were made by orbiter 2 in June and July, but on 25 July the spacecraft began to drift out of alignment with the sun, and no propellant was left to correct its attitude. At 6:01 a.m. UMT on 25 July 1978 (11:01 p.m. PDT, 24 July), orbiter 2 ceased operating during orbit 706—1049.5 days after launch from Earth.[3]

Lander 2 could communicate with Earth only via orbiter 1, while lander 1 could make direct contact. According to the continuation mission team, with these capabilities the landers could continue responding until December 1978, the start of another solar conjunction. But the scientists wanted to squeeze still more from the hardware. Mike Carr and his colleagues on the orbiter imaging team, the most vocal advocates of continuing the mission, wanted to obtain more high-resolution photographs of potential landing sites for the next Mars mission. Remembering just how harrowing the site selection and certification process had been for Viking, they argued that they needed to get as many images of the surface as the hardware would permit. In addition, they wanted to study Martian weather and atmosphere closely from January to April 1979, because this season would be similar to the one in which they had observed dust storms during 1977.[4]

Two serious limitations affected extending Viking any further than December 1978. Funds, of course, were critical, as they had always been, but also the communications loads imposed on the Deep Space Network by the Pioneer-Venus and Voyager-Jupiter missions meant that Viking could have only a limited amount of time on the air to transmit scientific information from Mars to Earth. The ability of JPL's mission control center and its Deep Space Network to squeeze the Viking transmissions into the schedule became one of the overriding factors in the continued life of Viking. In April 1979, Conway Snyder, in a memorandum to all the Viking scientists, projected that operations would come to an end in July of that year. He noted that the mission had provided the team "with a long and interesting road," and he was pleased that they had all been able to travel it together. But he also suggested that the mission might "afford us a few more surprises yet before the end."[5]

The end did not come in 1979. Viking lander 2 was shut down on 12 April 1980 after 1316.1 days on the surface. Orbiter 1 was silenced by a command from JPL on 7 August 1980, because it, too, was about out of fuel. Three of the four spacecraft were silent, but lander 1 remained active and would likely continue its transmissions to Earth for some years. Each week, the team at JPL would query the spacecraft for weather information and periodically ask for surface pictures so the specialists could monitor the Martian landscape in front of the lander for any changes.[6]

Statistical evidence of success includes 51 539 orbital images of the Red Planet and more than 4500 images from the landers. About 97 percent of the planet was photographed at a resolution of about 300 meters, while 2 percent of the planet was seen at a resolution of 25 meters or better. Together, the two landers returned more than 3 million weather reports by August 1980. Total orbital infrared observations exceeded 100 million. For generations, discussions about Mars had included such traditional topics as canals, waves of darkening, and blue clearings. But with NASA's explorations of Earth's near neighbor, man had at his disposal "a plethora of hard data about the large variety of geological features on the planet, about the composition of the surface, the atmosphere, and the polar caps, and about many aspects of Martian meteorology, including temperatures, pressures, tides, dust storms, and the abundance and transport of water vapor." Scientists, mission planners, and hardware specialists expected to spend much of the 1980s analyzing this information and preparing for another mission to Mars in the 1990s that would yield "as great a quantum jump in our understanding of this complex and fascinating planet" as did Mariner and Viking.[7]

Appendixes
Bibliographic Essay, Source Notes
Index

# Appendix A
## Orbital Relationships of Earth and Mars

The following is a brief explanation of planetary motions and, in particular, the relationships of the orbits of Earth and Mars. Mercury is the planet closest to the sun. Venus, Earth, Mars, the asteroids, Jupiter, Saturn, Neptune, and Pluto, in that order, are farther out. All move around the sun in the same direction. If the solar system were viewed from far above Earth's northern hemisphere, the planetary motion would appear to be counterclockwise. The planetary orbits lie in very nearly the same plane. The paths of the planets, if seen from the sun, would all describe the same circle, except for Mercury and Pluto, which have tilted orbits—the innermost and outermost of the sun's satellites.

The planets move in the same direction and most of them occupy a common plane, but the distance between any two planets varies considerably with time. Figure 1 illustrates the perihelia (points of closest approach to the sun) and aphelia (farthest points from the sun) of Mercury and Mars. Earth and Venus have more nearly circular orbits.

Since Earth and Mars travel around the sun in orbits of different lengths with different velocities, the distance between the planets varies constantly. About every 780 days (the actual interval ranges from 765 to 810 days), Earth overtakes its slower neighbor in their unending orbiting of the sun. (Kepler pointed out in his third law: The square of the period of a planet's complete revolution around the sun is proportional to the cube of its mean distance from the sun.) Seen from the sun, the two planets will momentarily lie along a straight line. Seen from Earth, Mars is in a

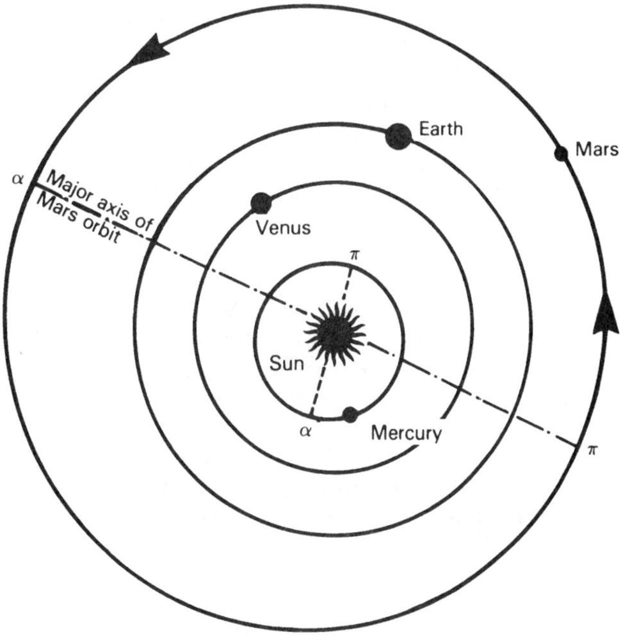

Fig. 1. Orbits of the inner planets of the solar system are drawn to scale (although the sizes of the planets are not). Orbits of Venus and Earth are nearly circular. Mercury and Mars orbits are eccentric. Perihelion points of Mercury and Mars are indicated by π and aphelion by α.

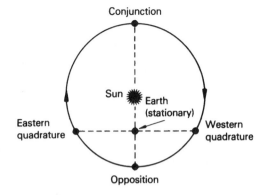

*Fig. 2. At conjunction of Mars and Earth, Mars becomes invisible in the sun's rays. Best observation period is from quadrature to quadrature. Samuel Glasstone,* The Book of Mars, *NASA SP-179, 1968.*

direction directly opposite that of the sun; Mars and the sun cannot be seen at the same time because they are on opposite sides of the Earth. This positioning, illustrated in figure 2, is called opposition. During opposition, Mars and Earth come closest together—between 55 and 102 million kilometers.

### Distances between the Two Planets

| Date | Kilometers |
|---|---|
| 30 Dec. 1960 | 90 606 067 |
| 4 Feb. 1963 | 100 101 196 |
| 9 Mar. 1965 | 99 779 328 |
| 15 Apr. 1967 | 89 801 395 |
| 31 May 1968 | 71 615 808 |
| 10 Aug. 1971 | 56 166 105 |
| 25 Oct. 1973 | 65 017 497 |
| 15 Dec. 1975 | 84 329 625 |

As Earth keeps racing ahead and Mars falls behind, there are instances when the two planets form a straight line, with the sun interposed between them. Mars disappears from Earth's view behind the disk of the sun; the planets are in conjunction. Mars is as far away from Earth as it can be—more than 350 million kilometers.

One other position in the Earth-Mars relationship is also important. When the sun, Earth, and Mars describe a right angle, Mars is said to be in quadrature. In this position, Mars does not appear as a round disk to Earth-based observers. Instead, it looks like the gibbous moon, between half- and full-moon phases. What we do not see is the night side of Mars blending with the black sky.

If an opposition takes place along the line marked 0° on figure 1, an observer on Earth would look across a shorter distance to Mars than during a 180° opposition. An opposition at Mars perihelion would offer the best opportunity for observations and spaceflight, since the distance between Earth and Mars at that point would be the shortest. Of the more recent oppositions, the one in 1924 came the closest to being at perihelion, while the 1933 opposition was almost precisely at aphelion.

For further information, see Samuel Glasstone, *The Book of Mars,* NASA SP-179 (Washington, 1968), and Willy Ley and Wernher von Braun, *The Exploration of Mars* (New York: Viking Press, 1956).

428

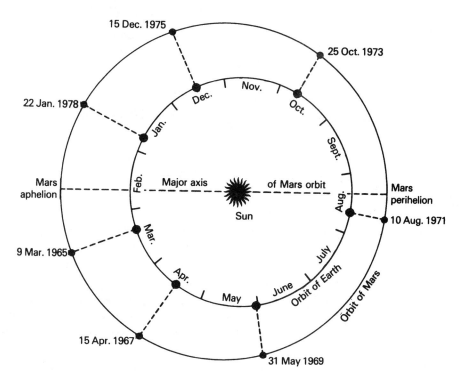

*Fig. 3. The orbits of Earth and Mars, showing the times of opposition and the separation at opposition. Samuel Glasstone,* The Book of Mars, *NASA SP-179, 1968.*

# Appendix B
## Voyager Project Highlights, 1966–1967

22 Dec. 1965     First Voyager flight deferred to 1973.

3–5 Jan. 1966     NASA management explained to spacecraft contractors rationale for changes in Voyager plans.

12 Jan. 1966     Don Hearth and Don Burcham discussed Voyager plans for next 18–24 mos.

24 Jan. 1966     First major meeting of top Space Science Office and JPL managers on revised planetary program.

11 Feb. 1966     NASA Hq. authorized JPL to extend General Electric, Boeing, and TRW phase IA study contracts in effort to keep those contractors' Voyager teams together. Work identified as task C of phase IA.

14 Feb. 1966     Planetary program review with Administrator Webb.

4 Apr. 1966     Voyager Project Guidelines revised.

7 July 1966     Viking Project Estimate 3, VPE-5, VPE-6, VPE-7, and VPE-8 presented to JPL Executive Council.

14 July 1966     VPE-5 and VPE-12 presented to Voyager Capsule Advisory Group.

19 July 1966     Voyager Capsule Advisory Group evaluated proposed missions and recommended:
        1969—Mariner-class flyby.
        1971—Mariner-class orbiter.
        1973—Voyager-class orbiter, plus soft-lander with 45 kg of landed scientific instruments.
        1975—Voyager-class orbiter, plus soft-lander with 136 kg of landed scientific instruments.
        1977—Voyager-class orbiter, plus soft-lander with automated biological laboratory.

22 July 1966     NASA-JPL management review of Voyager. VPE-5, VPE-12, and VPE-13 presented. RFP for phase B Voyager procurement for capsule system recommended for 1 Nov. 1966.

28 July 1966     Newell sent Pickering Revised Mission Guidelines letter calling for orbiter and surviving capsule in 1973. Two spacecraft with each launch vehicle in 1973 and 1975. Capsule would soft-land using retropropulsion package to slow descent.

Aug. 1966     JPL proposed Mariner Mars 1971 flyby with atmospheric probe.

| | |
|---|---|
| 14 Sept. 1966 | Voyager presentation made to Office of Space Sciences (VPE-14), including all Hq. recommendations. Two spacecraft on one Saturn V for *all* missions, 1973–1979. Capsule—2270-2720 kg. All missions soft-landing. |
| 23 Sept. 1966 | Procurement plan for capsule system, phase B, submitted to NASA Hq. by JPL. "Allowed to die." |
| 26 Sept. 1966 | Space Science Office and Office of Advanced Research and Technology discussed breaking down capsule system into delivery system and lander. Langley would work on former, JPL on latter. |
| 5 Oct. 1966 | Revised Voyager Guidelines basically approved VPE-14 with modifications. "Modifications open so broad a set of considerations as to violate VPE-14." |
| 17 Oct. 1966 | VPE-14 presented to Associate Administrator Seamans. |
| 19 Oct. 1966 | JPL and Newell's staff discussed Langley part in development of lander systems. |
| 20 Oct. 1966 | Newell and Nicks traveled to Huntsville to explore greater Marshall Space Flight Center participation in Voyager. |
| 3–4 Nov. 1966 | Hq. meeting discussed management assignments for Voyager. In addition to Space Science Office Staff, Webb, Seamans, George Mueller (OMSF), Floyd Thompson (Langley), and Wernher von Braun (MSFC) were present. |
| 18 Nov. 1966 | Revised Voyager Project Guidelines reaffirmed existing management assignments: JPL—project management and spacecraft; MSFC—launch vehicle. Langley likely to get landing systems for lander; i.e., "capsule bus." Capsule system phase B RFP rescheduled to 1 Jan. 1967. |
| 1–16 Dec. 1966 | JPL worked on several drafts of phase B procurement plan. |
| 19–20 Dec. 1966 | JPL representatives met at NASA Hq. with Space Science Office staff to discuss consolidated Voyager management plan. |
| 27 Dec. 1966 | Approved phase B procurement plan for capsule systems distributed within NASA. Not released publicly until 17 Jan. 1967. |
| 27 Jan. 1967 | Project Approval Document for phase B signed. MSFC was assigned management responsibility for both Voyager spacecraft and Saturn V launch vehicle. JPL and Langley to share responsibility for lander. |
| | Apollo 204 fire killed Virgil I. Grissom, Edward H. White II, and Roger B. Chaffee. |
| 31 Jan. 1967 | RFP for phase B issued to 36 industrial contractors. |
| 8 Feb. 1967 | Webb postponed assignment of project management to Marshall until summer 1967. |

| 16 Feb. 1967 | OSSA recommended establishment of Voyager Interim Project Office until final project management decision made. |
| 23 Feb. 1967 | Webb and Seamans approved Voyager Interim Project Office (VIPO), established Voyager as separate division within OSSA. |
| 28 Feb. 1967 | Webb advised Congress of Voyager management changes. |
| 2 Mar. 1967 | Phase B capsule proposals submitted by Grumman Aircraft Engineering Corp., Hughes Aircraft Co., Martin Marietta Corp., and McDonnell Aircraft Corp. Evaluation begun. |
| 14 Mar. 1967 | Newell distributed Voyager Guidelines. |
| 21 Mar. 1967 | Newell described revised Voyager project to House Subcommittee on Space Science and Applications. |
| 22–23 Mar. 1967 | First Voyager Management Committee meeting held in Pasadena. Committee created to coordinate all VIPO and field organization activities related to Voyager. (Met monthly thereafter.) |
| 23 Mar. 1967 | Rep. Karth and members of the House Space Science and Applications Subcommittee visited JPL. Hearth briefed them on VIPO activities. |
| 12 Apr. 1967 | Revised Project Guidelines. Surface lifetime of lander must be at least 24 hrs. |
| 5–6 May 1967 | Nicks, Hearth, Fellows, and others attended Lunar and Planetary Missions Board meeting at Stanford Univ. Board recommended Voyager surface-laboratory science be done in-house by a working group of its choice. |
| 8 May 1967 | Voyager quarterly review held at VIPO. |
| 17 May 1967 | Martin Marietta Corp. (Denver Div.) and McDonnell Aircraft Corp. (Astronautics Co.) selected by NASA for a 90-day phase B design study of landing capsule. Both companies received $500 000. Contracts dated 1 June. |
| 24 May 1967 | Newell, Naugle, and Nicks made 3½-hr. presentation on planetary program to President's Science Advisory Committee. |
| 9 June 1967 | Seamans, Newell, and others from OSSA discussed two significant issues: (1) limiting phase C procurement, fabrication of Voyager spacecraft, to phase B contractors; (2) arrangement for permanent project management assignment. Seamans established committee to study contractor question. Seamans said either Langley or Marshall could handle management of Voyager "with most factors favoring Marshall." Decision needed by end of Aug. 1967. |
| 17 June 1967 | Lunar and Planetary Missions Board examined scientific aspects of Voyager. Plans established for advisory group that would work with JPL in defining surface laboratory system. |

| 19 June 1967 | Nicks and Hearth met with Newell to review results of meeting on 17 June; all agreed to examine ways of extending expected lifetime of surface laboratory. |
| --- | --- |
| 29 June 1967 | Voyager Board of Directors meeting gave much attention to project management question. Majority favored assignment to either Langley or Marshall. |
| 5 July 1967 | Alternative budgets and programs developed by Hearth and discussed with Cortright. "Options were determined for establishing strategy in the event congressional appropriations for Voyager were less than requested." |
| 10 July 1967 | Revised Project Guidelines: 90–115-kg science subsystem must have minimum operational life of 30 days after landing. Mortar deployment of samples (as proposed for Gulliver) excluded. |
| 13–14 July 1967 | Lunar and Planetary Missions Board met at VIPO to discuss Voyager surface laboratory and experiments to be included in it. |
| 19 July 1967 | Cortright and Nicks reviewed Voyager plans, assignments, and functions of Voyager Program Office. |
| 31 July 1967 | Congressional Conference Committee reported a $42-million fiscal 1968 authorization instead of $71.5 million requested by NASA. |
| 3–4 Aug. 1967 | Fifth monthly Voyager management meeting reviewed alternative programs possible with reduced fiscal 1968 authorization. |
| 10 Aug. 1967 | Voyager announcement of flight opportunity was distributed to 5000 prospective scientific experimenters. |
| 11–12 Aug. 1967 | Lunar and Planetary Missions Board reviewed planetary program. Despite lower authorization, Board endorsed Voyager as prime means of landing large payloads on Mars. Orbital part of mission was essential. |
| 16 Aug. 1967 | NASA advised that House Appropriations Committee reported bill eliminating Voyager entirely and cutting lunar and planetary programs budget by $6.9 million. Action left funds only for Mariner 69 mission. No further projects funded. |
| 24 Aug. 1967 | NASA Public Information Office notified all centers of congressional cut-back. |
| 30 Aug. 1967 | Nicks notified all officers: "Because of a reduced FY1968 NASA Budget, it is not planned to proceed with Voyager into Phase C this fall as previously planned. All current Voyager Phase B system contracts will, however, be completed as previously planned." Voyager effectively ended on this date. |

SOURCE: Several dozen documents, among which the most helpful was Donald P. Burcham, "Listing of Voyager Important Documents & Meetings relative to Project Direction," Oct. 1967.

# Appendix C
## Summary Data from
## Mariner, Voyager, and Viking

### Abbreviations

CC—Coded commands for program update.
DC—Direct commands for switch closures.
PC—Processor commands for computer control.
QC—Qualitative commands for positioning and deflection maneuvers.
RCS—Reaction control system.
RF—Radio frequency.
RTG—Radioisotope thermoelectric generator.
TM—Telemetry.
TWTA—Traveling-wave-tube amplifier.

### MARINER A

| | |
|---|---|
| Preflight designation: | Mariner A |
| Flight designation: | Not flown. |
| Project proposed: | Study begun July 1960 at JPL. |
| Project approved: | 15 July 1960 by T. K. Glennan. |
| Launch vehicle: | Atlas-Centaur |
| Launch date: | Canceled 30 Aug. 1961 because of projected unavailability of suitable launch vehicle. |
| Program objectives: | Initial plan called for a flyby of Venus in 1962. Revised plan (February 1961) called for flights to Venus in 1962, 1964, and 1965. |
| Spacecraft shape and size: | Hexagonal frame derived from Ranger spacecraft (dimensions not available). |
| Weight: | Projected, 487–686 kg. |
| Program results: | Canceled. |
| Duration of flight to target: | Canceled. |

### MARINER B

| | |
|---|---|
| Preflight designation: | Mariner B |
| Flight designation: | Not flown. |

Project proposed:     Study begun July 1960 at JPL.

Project approved:     15 July 1960 by T. K. Glennan.

Launch vehicle:     Atlas-Centaur

Launch date:     Project deferred to Mariner-Mars 1966 on 6 May 1963.

Program objectives:     Mariner B went through a series of redefinitions:

        1. Initial plans called for an instrumented landing on Venus or Mars in 1964.

        2. In February 1961, the Venus landing was dropped from consideration.

        3. On 9 Apr. 1962, the Venus landing was again considered and the Mars landing dropped.

        4. On 14 Mar. 1963, mission changed to pre-Voyager checkout flight to Mars with lander.

        5. Mission postponed and redesignated Mariner-Mars 1966 on 6 May 1963.

Spacecraft shape and size:     Several configurations proposed; none finalized.

Weight:     Projected, 400–600 kg.

Program results:     Redesignated Mariner-Mars 1966.

Duration of flight to target:     Redesignated Mariner-Mars 1966.

## MARINER-VENUS 1962

Preflight designation:     Mariner R-1 and Mariner R-2.

Flight designation:     *Mariner 1* and *Mariner 2*

Project proposed:     Study started August 1961 at JPL and proposed to NASA Hq. by JPL 28 Aug. 1961.

Project approved:     30 Aug. 1961

Launch vehicle:     Atlas-Agena B

Launch date:     *Mariner 1*, 22 July 1962, 4:21 a.m. EST.
*Mariner 2*, 27 Aug. 1962, 1:53 a.m. EST.

Program objectives:     Launch 2 spacecraft to the near-vicinity of Venus in 1962; establish and maintain 2-way communication with the spacecraft throughout the flight; obtain interplanetary data in space and during Venus encounter; make scientific survey of planet's characteristics.

Spacecraft shape and size:     Hexagonal magnesium-frame base, 104 cm diagonally, 36 cm deep. Two solar panels attached to base span 5.05 m when deployed. Aluminum tubular superstructure, mounted atop the base, supports experiments and omnidirectional antenna. High-gain antenna mounted below base. Attitude-control jets mounted to base. Midcourse propulsion mounted in base compartment. Overall height, 3.66 m.

Weight:

| Structures/mechanical | 38.6 |
| Electrical (RF, TM, Data) | 32.2 |
| Power | 47.8 |
| Computer command | 5.1 |
| Attitude control | 22.2 |
| Pyro and cabling | 19.1 |
| Propulsion (inert) | 9.6 |
| Thermal control | 3.2 |
| Science | 18.4 |
| Expendables | 6.6 |
| Launch weight | 202.8 |

Control system:

10 $N_2$ jets
3 gyros, Earth sensor
2 primary sun sensors
2 secondary sun sensors
Sungate and sensor

Electrical power:

9800 solar cells
Panels: 152 x 76 cm (2)
Total area, 2.3 m²
148 watts at Earth
222 watts at Venus
Silver-zinc battery, 1000 watts per hr.

Telecommunications:

L-band transponder, 1-watt/3-watt output
Low-gain omnidirectional antenna
Dual low-gain turnstile/dipole antennas
High-gain parabolic antenna
Science and engineering data, 8⅓ and 33⅓ bits per sec

Propulsion:

Monopropellant hydrazine
225 newtons thrust
Rate of velocity change, 0.2 m/sec to 60 m/sec
Total impulse, 9560 newtons per sec
4-jet vane vector control

Command system:

DC—14
QC—3     @ 1 bit per sec
CC—0

Program results:

*Mariner 1*—Booster deviated from course and was destroyed by range safety officer 290 sec after launch.
*Mariner 2*—First spacecraft to scan another planet; passed within 34 762 km of Venus on 14 Dec.; made 42-min instrument survey of atmosphere and sur-

face before going into heliocentric orbit; made first comprehensive measurements of properties of solar wind. Transmissions from interplanetary experiments received until 4 Jan. 1963 from 87.4-million-km distance, establishing new communication record.

| | |
|---|---|
| Duration of flight to target: | *Mariner 1*—destroyed shortly after launch. <br> *Mariner 2*—109 days. |
| For additional information: | JPL, *Mariner-Venus 1962 Final Project Report*, NASA SP-59 (Washington, 1965). |

## MARINER-MARS 1964

| | |
|---|---|
| Preflight designation: | Mariner C <br> Mariner M (Mars) <br> Mariner C (*Mariner 3*) <br> Mariner D* (*Mariner 4*) |
| Flight designation: | *Mariner 3* and *Mariner 4* |
| Project proposed: | July–August 1962 |
| Project approved: | November 1962 (tentative); project approval document signed 1 Mar. 1963. |
| Launch vehicle: | Atlas-Agena SLV-3 <br> AA-11—*Mariner 3* <br> AA-12—*Mariner 4* |
| Launch date: | *Mariner 3*—5 Nov. 1964, 2:22 p.m. EST <br> *Mariner 4*—28 Nov. 1964, 9:22 a.m. EST |
| Program objectives: | Flyby to study surface and atmosphere of Mars, develop operational techniques, make scientific measurements of interplanetary environment, provide engineering experience in spacecraft operations during long-duration flights away from the sun. |
| Spacecraft shape and size: | Octagonal magnesium-frame base, 127 cm diagonally and 45.7 cm deep. Four solar panels attached to top of base span 6.88 m deployed (including solar-pressure-vane extensions). High-gain dish antenna mounted atop base with low-gain antenna on top of aluminum tube. Attitude-control jets mounted at solar panel tips. Midcourse propulsion mounted on side of octagon. Overall height, 2.89 m. |
| Weight: | Structures/mechanical                 49.4 <br> Electrical (RF, TM, Data)         52.6 |

---

*Mariner D was also used for a short time in the winter of 1963 to refer to an Atlas-Centaur-launched Mariner C bus with a small atmospheric capsule that was being planned for 1966.

437

| | |
|---|---:|
| Power | 61.5 |
| Computer command | 5.4 |
| Attitude control | 29.0 |
| Pyro and cabling | 15.4 |
| Propulsion (inert) | 12.8 |
| Thermal control | 6.4 |
| Science | 15.8 |
| Expendables | 12.5 |
| Launch weight | 260.8 |

**Control system:**
12 N$_2$ jets—redundant
3 gyros, Canopus sensor
Earth sensor
Mars sensor
2 primary sun sensors
2 secondary sun sensors

**Electrical power:**
28 224 solar cells
Panels: 176 x 90 cm (4)
Total area, 6.3 m$^2$
310 watts at Mars
Silver-zinc battery, 1200 watts per hr.

**Telecommunications:**
Dual, S-band, 7-watt cavity amp/10-watt TWTA transmitter, single receiver
Low-gain omnidirectional antenna
High-gain parabolic antenna
Science and engineering data, 8$^1$/$_3$ and 33$^1$/$_3$ bits per sec
Tape recorder, 5.24 million bits

**Propulsion:**
Monopropellant hydrazine
225 newtons thrust
4-jet vane vector control

**Command system:**
DC—29
QC—3     @ 1 bit per second
CC—0

**Program results:**
*Mariner 3*—Shroud failed to jettison; battery power dropped; no evidence that solar panels opened to replenish power supply; communications lost; in permanent heliocentric orbit.
*Mariner 4*—Spacecraft flew by Mars 14 July 1965, with closest approach about 9844 km; discovered densely packed lunar-style impact craters on Martian surface; ionosphere and atmosphere measured somewhat less dense than expected; carbon dioxide suggested to be major constituent in atmosphere.

**Duration of flight to target:**
*Mariner 3*—Did not reach target.
*Mariner 4*—228 days

**For additional information:**
JPL, *Mariner-Mars 1964 Final Project Report*, NASA SP-139 (Washington, 1967).

## MARINER-MARS 1966

| | |
|---|---|
| Preflight designation: | Mariner E<br>Mariner F |
| Flight designation: | Not flown. |
| Project proposed: | May 1963 |
| Project approved: | 19 Dec. 1963 |
| Launch vehicle: | Atlas-Centaur |
| Launch date: | Effectively canceled 28 July 1964; officially terminated 4 Sept. 1964. |
| Program objectives: | Mars flyby spacecraft with small atmospheric probe to replace more ambitious Mariner B. |
| Program results: | Replaced by Advanced Mariner 1969. |

## ADVANCED MARINER 1969

| | |
|---|---|
| Preflight designation: | Mariner-Mars 1969 |
| Flight designation: | Not flown. |
| Project proposed: | Initial discussions January 1964 |
| Project approved: | Project approval document signed 2 Aug. 1964. |
| Launch vehicle: | Atlas-Centaur |
| Launch date: | Canceled 20 Nov. 1964. |
| Project objectives: | Combination orbiter-lander mission designed to replace Mariner-Mars 1966 Mars flyby. |
| Program results: | Replaced by Mariner-Mars 1969 flyby mission. |
| For additional information: | JPL, "Mariner Mars 1969 Orbiter Technical Feasibility Study," EPD-250, 16 Nov. 1964; and JPL, "Mariner Mars 1969 Lander Technical Feasibility Study," EPD-261, 28 Dec. 1964. |

## MARINER-VENUS 1967

| | |
|---|---|
| Preflight designation: | Mariner E |
| Flight designation: | *Mariner 5* |
| Project proposed: | By post-Voyager 1971 deferral, 25 Dec. 1965. |
| Project approved: | 25 Dec. 1965 |
| Launch vehicle: | Atlas-Agena SLV-3<br>AA-23 |
| Launch date: | 14 June 1967, 2:01 a.m. EDT. |
| Program objectives: | Venus flyby to within 3218 km to provide data on atmosphere, radiation, and magnetic field envi- |

ronment; to return data on interplanetary environment before and after planetary encounter; to provide first exercise of turnaround ranging technique of planetary distance.

Spacecraft shape and size:

Octagonal magnesium-frame base, 127 cm diagonally and 45.7 cm deep. Four solar panels attached to top of base span 5.48 m deployed. High-gain ellipse antenna mounted atop base along with low-gain omnidirectional antenna and magnetometer supported by 2.23-m-long tube. Attitude-control jets mounted at solar panel tips. Midcourse propulsion mounted on side of octagon. Overall height, 2.89 m.

Weight:

| | |
|---|---:|
| Structures/mechanical | 49.4 |
| Electrical (RF, TM, Data) | 52.6 |
| Power | 57.4 |
| Computer command | 5.4 |
| Attitude control | 25.0 |
| Pyro and cabling | 15.4 |
| Propulsion (inert) | 12.8 |
| Thermal control | 4.5 |
| Science | 10.2 |
| Expendables | 12.2 |
| Launch weight | 244.9 |

Control system:

12 $N_2$ jets—redundant
3 gyros, Canopus tracker
2 primary sun sensors
2 secondary sun sensors
Earth sensor
Venus sensor
Venus terminator sensor

Electrical power:

17 640 solar cells
Panels: 112 x 90 cm (4)
Total area, 4.0 m²
370 watts at Earth
550 watts at Venus
Silver-zinc battery 1200 watts per hr

Telecommunications:

Dual, S-band, 6.5-watt/10.5-watt transmitter, single receiver
Low-gain omnidirectional antenna
High-gain 2-position parabolic antenna
Science and engineering data, $8^1/_3$ and $33^1/_3$ bits per sec
Tape recorder, 1 million bits

Propulsion:

Monopropellant hydrazine
225 newtons thrust
4-jet vane vector control

| Command system: | DC—29 | |
|---|---|---|
| | QC—3 | @ 1 bit per sec |
| | CC—0 | |

Program results: Spacecraft passed within 4000 km of Venus, provided data on atmospheric structure, radiation, and magnetic field; mass of Venus was further defined by processing flyby trajectory data; solar-wind interaction with Venus shown to be different from Earth interaction.

Duration of flight to target: 127 days

For additional information: JPL, *Mariner-Venus 1967 Final Project Report*, NASA SP-190 (Washington, 1971).

## MARINER-MARS 1969

Preflight designation: Mariner F and Mariner G

Flight designation: *Mariner 6* and *Mariner 7*

Project proposed: By post-Voyager 1971 deferral, 22 Dec. 1965.

Project approved: 22 Dec. 1965; project approval document signed 28 Mar. 1966.

Launch vehicle: Altas-Centaur SLV-3C
  AC20 (spacecraft 69-3)—*Mariner 6*
  AC19 (spacecraft 69-4)—*Mariner 7*

Launch date: *Mariner 6*—24 Feb. 1969, 8:29 p.m. EST
*Mariner 7*—27 Mar. 1969, 5:22 p.m. EST

Program objectives: Flyby of Mars at 3218 km to study surface and atmosphere to establish basis for future experiments in search for extraterrestrial life; develop technologies for future Mars missions. Demonstrate engineering concepts and technique required for long-duration flight away from sun.

Spacecraft shape and size: Octagonal magnesium-frame base, 138.4 cm diagonally and 45.7 cm deep. Four solar panels span 5.79 m deployed. High-gain parabolic antenna mounted atop base along with low-gain omnidirectional antenna atop 2.23-m-long tube. Attitude-control jets mounted at solar panel tips. Midcourse propulsion system mounted in base compartment. Overall height, 3.35 m.

| Weight: | Structures/mechanical | 120.7 |
|---|---|---|
| | Electrical (RF, TM, Data) | 62.1 |
| | Power | 54.9 |
| | Computer command | 10.9 |

441

|                    |       |
|--------------------|------:|
| Attitude control   | 37.2  |
| Pyro and cabling   | 35.4  |
| Propulsion (inert) | 10.9  |
| Thermal control    | 13.1  |
| Science            | 57.6  |
| Expendables        | 10.0  |
| Launch weight      | 412.8 |

Control system:
2 sets of 6 $N_2$ jets
3 gyros, Canopus tracker
2 primary sun sensors
4 secondary sun sensors

Electrical power:
17 472 solar cells
Panels: 215 x 90 cm (4)
Total area, 7.7 m²
800 watts at Earth
449 watts at Mars
Silver-zinc battery, 1200 watts per hr

Telecommunications:
Dual, S-band, 10-watt/30-watt transmitters, single receiver
Low-gain omnidirectional antenna
Engineering data, $8\frac{1}{3}$ and $33\frac{1}{3}$ bits per sec
Science data, $66\frac{2}{3}$ and 670 bits per sec
Tape recorder, 195 million bits

Propulsion:
Monopropellant hydrazine
225 newtons thrust
Total impulse 20 900 newtons per sec
4-jet vane vector control

Command system:
DC—53
QC—4      @ 1 bit per sec
CC—5

Program results:
*Mariner 6*—First Mariner launched with Atlas-Centaur; performed flyby with *Mariner 7*; acquired data on Mars with visual imager, ultraviolet spectrometer, infrared spectrometer, and temperature sensors; obtained most detailed data on Mars to date.

*Mariner 7*—Same as above; flew at a different angle from *Mariner 6*; obtained same data from different areas of the planet. Together the 2 spacecraft transmitted 143 analog pictures as they approached Mars, plus 58 photos during flyby; closeups were made of 20 percent of surface. Provided daytime and nighttime surface temperatures; confirmed presence of $CO_2$, ionized $CO_2$, CO, atomic hydrogen, and very slight traces of molecular oxygen. Confirmed ablateness estimates.

| Duration of flight to target: | Mariner 6—156 days<br>Mariner 7—133 days |
|---|---|

For additional information: JPL, *Mariner Mars 1969 Final Project Report*, JPL TR-32-1460, 3 vols. (Pasadena, 1970).

## MARINER-MARS 1971

Preflight designation:
Mariner H and Mariner I
Mission A and Mission B

Flight designation:
*Mariner 8* and *Mariner 9*

Project proposed:
November 1967 by Office of Space Science staff following cancellation of Voyager.

Project approved:
Project approval document signed 23 Aug. 1968; NASA Hq. authorized JPL to begin work on Mariner-Mars 1971 on 14 Nov. 1968

Launch vehicle:
Atlas-Centaur SLV-3C
AC-24—*Mariner 8*
AC-23—*Mariner 9*

Launch date:
*Mariner 8*—8 May 1971, 9:11 p.m. EDT.
*Mariner 9*—30 May 1971, 6:23 p.m. EDT.

Program objectives:
Orbit Mars for 90 days; provide 25–30 million bits of scientific data; take total of more than 5000 TV pictures; take scores of TV pictures of the 2 moons; map more than 70 percent of surface; study temperature and composition of surface; study composition and structure of atmosphere; determine pressure of atmosphere.

Spacecraft size and shape:
Octagonal magnesium-frame base, 138.4 cm diagonally and 45.7 cm deep. Four solar panels attached to top of the base span 6.89 m. High-gain parabolic antenna mounted on base along with low-gain omnidirectional antenna on 1.44-m-long tube. Attitude-control jets mounted on solar panel tips. Midcourse propulsion mounted on top. Overall height, 2.28 m.

Weight:

| | |
|---|---:|
| Structures/mechanical | 155.1 |
| Electrical (RF, TM, Data) | 60.8 |
| Power | 72.6 |
| Computer command | 10.4 |
| Attitude control | 39.4 |
| Pyro and cabling | 49.4 |
| Propulsion (inert) | 98.0 |
| Thermal control | 10.0 |
| Science | 63.1 |
| Expendables | 439.1 |
| Launch weight | 997.9 |

| | |
|---|---|
| Control system: | 2 sets of ACS jets, 6 jets each<br>Canopus star tracker<br>Cruise sun sensor<br>Sun gate |
| Electrical power: | 14 742 solar cells<br>Panels: 215 x 90 cm (4)<br>Total area, 7.7 m²<br>800 watts at Earth<br>500 watts at Mars<br>Nickel-cadmium battery, 20 amp-hrs. |
| Telecommunications: | Dual, S-band, 10-watt/20-watt transmitters, single receiver<br>Low-gain omnidirectional antenna<br>Medium-gain horn antenna<br>High-gain, parabolic, 2-position antenna<br>Engineering data, 8⅓ and 33⅓ bits per sec<br>Science data, 16.2 kilobits per sec maximum<br>Tape recorder, 180 million bits |
| Propulsion: | Monomethyl hydrazine and nitrogen tetroxide<br>1340 newtons thrust<br>5 restarts capability<br>Gimbaled engine ±9° |
| Command system: | DC—86<br>QC—4    @ 1 bit per sec<br>CC—5 |
| Program results: | *Mariner 8*—With ignition of Centaur main engine 265 sec after launch, the upper stage began to oscillate and subsequently tumbled end over end. At about 365 sec after launch, Centaur engine shut down, and upper stage and spacecraft fell into Atlantic about 560 km north of Puerto Rico.<br><br>*Mariner 9*—Total useful lifetime was 515 days, with 349 days in Mars orbit. Transmitted total of 7329 TV pictures of Mars and its satellites and mapped 100 percent of planet; transmitted 54 billion bits of science data (contrasting with 2 billion bits from all previous Mars flights). |
| Duration of flight to target: | *Mariner 8*—Did not reach target<br>*Mariner 9*—167 days |
| For additional information: | Chap. 10; and JPL, *Mariner Mars 1971 Project Final Report*, JPL TR-32-1550, 5 vols. (Pasadena, 1972). |

## MARINER VENUS-MERCURY 1973

| | |
|---|---|
| Preflight designation: | Mariner J |

| | |
|---|---|
| Flight designation: | *Mariner 10* |
| Project proposed: | June 1968 by the Space Science Board. |
| Project approved: | Project assigned to JPL 30 Dec. 1969. |
| Launch vehicle: | Atlas-Centaur SLV-3D/D1-A |
| Launch date: | 3 Nov. 1973, 12:45 a.m. EST. |
| Program objectives: | Flyby of Venus and encounter with Mercury (primary target) at an altitude of 1000 km; conduct exploratory investigations by obtaining measurements of environment, atmosphere, and body characteristics for both planets. Perform experiments in the interplanetary medium; obtain experience with dual planet, gravity-assist mission. |
| Spacecraft size and shape: | Octagonal magnesium-frame base, 138.4 cm deep. Two solar panels attached to top of base span 6.55 m. High-gain parabolic antenna mounted on top along with scan platform and one of the 2 low-gain omniantennas. Other low-gain antenna mounted on base. Attitude-control jets mounted on supports from octagon faces. Midcourse propulsion system mounted in base compartment. |

| | | |
|---|---|---:|
| Weight: | Structures/mechanical | 109.4 |
| | Electrical (RF, TM, Data) | 60.8 |
| | Power | 63.6 |
| | Computer command | 10.4 |
| | Attitude control | 29.9 |
| | Pyro and cabling | 30.8 |
| | Propulsion (inert) | 11.3 |
| | Thermal control | 9.5 |
| | Science | 78.2 |
| | Expendables | 29.0 |
| | Launch weight | 432.9 |

| | |
|---|---|
| Control system: | 2 sets of 6 $N_2$ jets<br>3 gyros, Canopus tracker<br>2 primary sun sensors<br>4 secondary sensors |
| Electrical power: | 19 800 solar cells<br>Panels: 215 x 120 cm (2)<br>Total area, 5.2 m²<br>500 watts on Earth<br>820 watts on Venus<br>820 watts (tilted) on Mercury<br>Nickel-cadmium battery, 20 amp-hrs |
| Telecommunications: | Dual, S-band, 10-watt/20-watt transmitter, single receiver<br>2 low-gain omnidirectional antennas |

445

High-gain parabolic antenna
Engineering data, 8⅓ and 33⅓ uncoded
Science data, 117.6 kilobits per sec maximum
Tape recorder, 180 million bits

Propulsion:

Monopropellant hydrazine
225 newtons thrust
Total impulse 20 900 newtons per sec
4-jet vane vector control

Command system:

DC—96
QC—4    @ 1 bit per sec
CC—5

Program results:

First dual-planet mission; first mission to use gravitational attraction of one planet to reach another. Venus encountered at 1:01 p.m. EDT, 5 Feb. 1974, 5800 km from surface. Some 4000 photos of Venus revealed a nearly round planet enveloped in smooth cloud layers with a slow rotational period (243 days) and only 0.05 percent of Earth's magnetic field; atmosphere mostly hydrogen, resulting from solar wind bombardment. After Venus flyby, spacecraft trajectory bent in toward sun for first exploration of Mercury. Mercury encountered at 4:47 p.m. EDT, 29 March 1974, 704 km from surface. Photos revealed intensely cratered, lunarlike surface. Atmosphere, mostly helium. High-iron-rich core makes Mercury densest planet in solar system; iron core also accounts for existence of magnetic field despite planet's extremely slow spin rate. After Mercury flyby, spacecraft entered solar orbit. Flew by Mercury again 20–23 Sept. 1974, coming within 48 069 km. Photographed sun side of planet and south polar region. Photographed total of 45 percent of Mercury's surface.

Duration of flight to
    target:

166 days to Venus

For additional information:

NASA, "Mariner Venus/Mercury: A Study in Cost Control," Nov. 1973; and James A. Dunne and Eric Burgess, *The Voyage of Mariner 10: Mission to Venus and Mercury*, NASA SP-424 (Washington, 1978).

## VOYAGER

Preflight designation:

Voyager*

---

*The name *Voyager* was later given to 2 spacecraft launched in 1977 to fly by Jupiter, Saturn, and perhaps Uranus.

Voyager 71
Voyager 73

Flight designation:            Not flown.

Project proposed:           First mentioned in spring 1960; JPL preliminary studies 1961–1962.

Project approved:           Project approval document for preliminary studies signed 21 Nov. 1962; phase I study approved 16 Dec. 1964 and revised 14 Jan. 1965 and 15 Oct. 1965.

Launch vehicle:             Initially all studies centered on use of Saturn IB–Centaur; when Saturn IB–Centaur terminated in mid-Oct. 1965, Voyager shifted to Saturn V.

Launch date:               Saturn IB–based Voyager 1971 mission canceled 22 Dec. 1965; Saturn V missions with 2 landers each for Voyager 1973 canceled 29 Aug. 1967.

Program objectives:        Originally planned as the large-weight class of spacecraft to follow the Mariner class, Voyager was scheduled to visit both Venus and Mars and release landers. Successively redefined plans called for the following missions: Venus 1967; Mars 1969, deferred in 1964 and rescheduled for 1971; Mars 1971, deferred 22 Dec. 1965 and rescheduled for 1973; Mars 1973, canceled 29 Aug. 1967.

For additional information:   Chap. 4; app. B.

## VIKING

Preflight designation:      Viking 73
Viking 75
Viking A and Viking B

Flight designation:         *Viking 1* and *Viking 2*

Project proposed:           November 1967–January 1968

Project approved:           4 Dec. 1968 by T. O. Paine; project approval document signed 8 Feb. 1969.

Launch vehicle:             Titan IIIE–Centaur
TC4—*Viking 1* (spacecraft B), VLC-1 and VO-1
TC3—*Viking 2* (spacecraft A), VLC-2 and VO-2

Launch date:               Launch deferred from 1973 to 1975:
*Viking 1*—20 Aug. 1975, 5:22 p.m. EDT
*Viking 2*—9 Sept. 1975, 2:39 p.m. EDT

Program objectives:        Soft-land on Mars; search for presence of life; compare orbital and surface data. Orbiter: deliver lander to Mars orbit; survey and select landing sites; relay data from surface to Earth; conduct orbital science

Spacecraft shape and
size:

Weight:

investigations. Lander: search for possibility of life forms; determine environment of surface.

Orbiter: Mariner-style bus; octagonal ring 45.72 cm high with alternate 139.7-cm and 50.8-cm sides. Consists of 16 modular compartments, 3 on each of the 4 long sides and 1 on each of the 4 short sides; 9.75 m across from tips of extended solar panels. Overall height, 3.29 m from lander attachment points on bottom to launch vehicle attachment points on top.

Lander: Hemispherical bioshield 360 cm diameter. Conical 70° half-angle aeroshell/heat shield 350 cm diameter. Triangular 3-leg lander configuration height 102 cm, width 284 cm (less instruments). High-gain S-band parabolic, UHF, and low-gain S-band antennas. Aeroshell, parachute, and terminal descent propulsion (18 nozzles) for deceleration.

Orbiter:

| | |
|---|---:|
| Structures and mechanisms | 267 |
| Communications | 57 |
| Data processing & storage | 45 |
| Power | 129 |
| Computer command | 18 |
| Attitude control inerts | 59 |
| Pyro & cabling | 61 |
| Propulsion inerts | 174 |
| Science instruments | 73 |
| Orbiter dry weight | 883 |
| Propellant | 1426 |
| Gas for propulsion & attitude control | 20 |
| Total expendables | 1445 |
| Orbiter launch weight | 2328 |

Lander:

| | |
|---|---:|
| Structures & mechanisms | 132 |
| Propulsion inerts | 49 |
| Pyro and cabling | 43 |
| Thermal control | 36 |
| Guidance and control | 79 |
| Power | 103 |
| Communications and telemetry | 57 |
| Science instruments | 91 |
| Lander dry weight | 590 |
| Residual propellants at landing | 22 |
| Total VL weight at landing | 612 |

| | |
|---|---|
| Propellant at launch | 73 |
| Lander launch weight | 663 |

Viking lander capsule:
Aerodecelerator

| | |
|---|---|
| Structures | 46 |
| Parachute & mortar | 56 |
| Thermal control | 4 |
| Miscellany | 3 |
| Total | 109 |

Aeroshell

| | |
|---|---|
| Structures | 120 |
| Propulsion inerts | 29 |
| Thermal control | 7 |
| Cabling | 7 |
| Science instruments | 9 |
| Miscellany | 9 |
| Total dry weight | 181 |
| Propellants | 88 |
| Total loaded weight | 269 |

Bioshield base

| | |
|---|---|
| Structures | 45 |
| Thermal control | 10 |
| Power | 15 |
| Miscellany | 4 |
| Total | 74 |

Bioshield cap

| | |
|---|---|
| Structures | 47 |
| Thermal control | 3 |
| Miscellany | 4 |
| Total | 54 |
| Loaded capsule weight | 1168 |

| | |
|---|---|
| Orbiter | 2328 |
| Lander and capsule | 1185 |
| Capsule mounting adapter | 14 |
| Total launch weight | 3527 |

Control system:

Orbiter:
2 sets of $N_2$ ACS jets, 6 jets each
Canopus star tracker
Acquisition sun sensor
Cruise sun sensor
Sun gate
6 gyros, 2 accelerometers

Lander:
Inertial control
4 gyros
Aerodecelerator

449

Radar altimeter
Terminal descent and landing radar

Electrical power:

Orbiter:
34 800 solar cells
Panels: 157 x 123 cm (8)
Total area, 15.4 m²
1400 watts on Earth
620 watts on Mars
2 nickel-cadmium, 30-amp-hr batteries

Lander:
Radioisotope thermal generator
2 RTG units, 90 watts
4 nickel-cadmium, 8-amp-hr batteries

Telecommunications:

Orbiter:
S-band, 20-watt transmitter, 2 10- and 20-watt
TWTAs
High-gain antenna, 2-axis steerable
Low-gain antenna, fixed
Engineering data, 8⅓ and 33⅓ kilobits per sec
Science data, 1, 2, 4, 8, and 16 kilobits per sec
2 tape recorders, 128 x 10⁷ bits
Relay radio, 381 MHz

Lander:
S-band, 20-watt transmitter, 2 20-watt TWTAs
High-gain antenna, 2-axis steerable
Low-gain antenna, fixed
Engineering data, 8⅓ and 33⅓ kilobits per sec
Science data, 250, 500, and 1000 bits per sec
Tape recorder, 4 x 10⁷ bits
Relay radio, 381 MHz, 30 watts, 4 and 16 kilobits
per sec

Propulsion:

Orbiter:
Monomethyl hydrazine and nitrogen tetroxide
1323 newtons thrust
Rate-of-velocity-change ($\Delta$ V) capability 1480
meters per second
Gimbaled engine ±9°

Lander:
RCS/deorbit: monomethyl hydrazine, 35 newtons
thrust, 12 nozzles, $\Delta$ V 180 meters per sec
Terminal descent: monomethyl hydrazine, 2650
newtons maximum thrust, 3 (18-nozzle) engines

Command system:

Orbiter:
DC—171
CC— 40    @ 4 bits per sec
PC—  4 operator words

Lander:
  DC—25     @ 4 bits per sec
  6000-word (maximum) for command instructions

Program results:          See chaps. 11 and 12.

Duration of flight to     *Viking 1*—304.1 days
  target:                 *Viking 2*—332.7 days

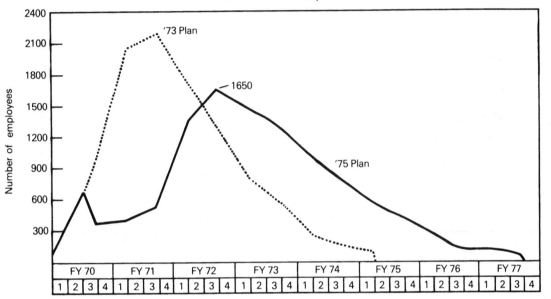

**Martin Marietta Manpower Plan**

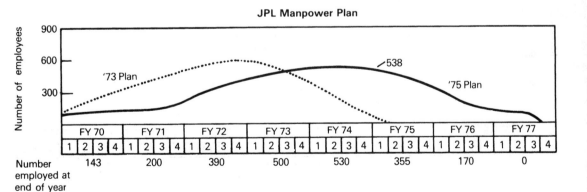

**JPL Manpower Plan**

*Shifting from a 1973 to a 1975 Viking launch brought changes in manpower levels. Martin Marietta personnel numbers were significantly lower for the 1975 mission (1650 vs. about 2200), although JPL figures were not significantly affected (538 vs. about 600). From Langley Research Center, Viking Proj. Off., "Viking Project, Resource Planning Report, July 1970," 30 July 1970.*

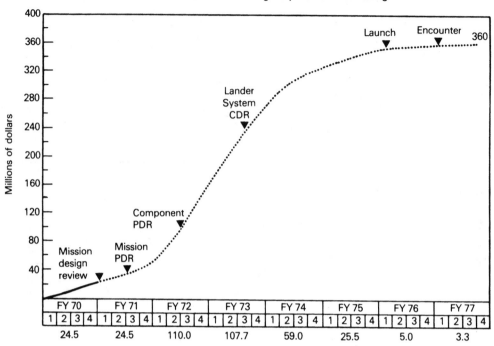

**Martin Marietta Funding Requirement for Viking**

| | FY 70 | FY 71 | FY 72 | FY 73 | FY 74 | FY 75 | FY 76 | FY 77 |
|---|---|---|---|---|---|---|---|---|
| | 24.5 | 24.5 | 110.0 | 107.7 | 59.0 | 25.5 | 5.0 | 3.3 |

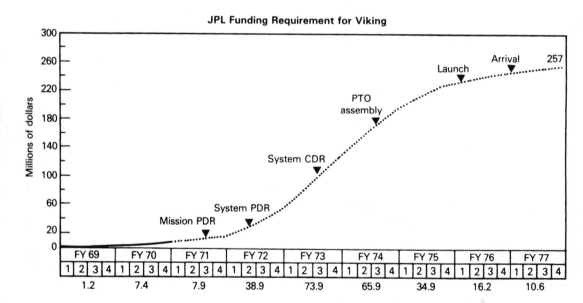

**JPL Funding Requirement for Viking**

| | FY 69 | FY 70 | FY 71 | FY 72 | FY 73 | FY 74 | FY 75 | FY 76 | FY 77 |
|---|---|---|---|---|---|---|---|---|---|
| | 1.2 | 7.4 | 7.9 | 38.9 | 73.9 | 65.9 | 34.9 | 16.2 | 10.6 |

*The graphs illustrate the major milestones for the Viking lander and orbiter and the amount of money required for each phase of the project. From Langley Research Center, Viking Proj. Off., "Viking 75 Project, Resource Planning Report, July 1970," 30 July 1970. PDR = preliminary design review. CDR = critical design review. PTO = proof-test orbiter.*

# Appendix D
# Mars Experiments, Science Teams,
# and Investigators

## SPACE SCIENCE BOARD'S PRINCIPAL RECOMMENDATIONS FOR PLANETARY INVESTIGATION, 1968–1975

1. We recommend that the planetary exploration program be presented, not in terms of a single goal, but rather in terms of the contribution that exploration can make to a broad range of scientific disciplines (page 3).

2. We recommend that a substantially increased fraction of the total NASA budget be devoted to unmanned planetary exploration (page 3).

3. (a) We recommend that duplicate missions for a particular opportunity be undertaken only when a clear gain in scientific information will result from such double launches (page 4).

(b) We recommend that NASA initiate now a program of Pioneer/IMP-class spinning spacecraft to orbit Venus and Mars at every opportunity and for exploratory missions to other targets (page 5).

(c) We recommend the following larger missions to Mars: A Mariner orbiter mission in 1971, and a Mariner-type orbiter and lander mission, based on a Titan-Centaur, in 1973 (page 5).

(d) We accord next priorities (in descending order) to a Mariner-class Venus-Mercury fly-by in 1973 or 1975, a multiple drop-sonde mission to Venus in 1975, and a major lander on Mars, perhaps in 1975 (page 6).

4. (a, b) Rather than attempt to define in detail payloads to be carried aboard high priority missions, we have selected several sample payloads (page 6).

(c) We recommend that with regard to Mars and Venus, NASA continually reassess, in the light of current knowledge of the planets, its program, methods, and mathematical model for meeting the internationally agreed objectives on planetary quarantine (page 11).

5. (a) We recommend strongly that NASA support radar astronomy as an integral part of its planetary program. In particular, we recommend that NASA fund the development and operation of a major new radar observatory to be used primarily for planetary investigation (page 12).

(b) We recommend that NASA planetary program planning be closely coordinated with Earth-orbital telescopes being designed for the 1970's and with the infrared aircraft telescopes now under construction (page 13).

(c) We recommend that the NASA program of ground-based optical planetary astronomy continue to receive strong support and that opportunities for planetary astronomical investigations be increased by:

---

Space Science Board, *Planetary Exploration, 1968–1975*, recommendations of June 1968 study, published August 1968.

(1) Construction of an intermediate sized optical telescope in the Southern Hemisphere

(2) Construction of an infrared telescope employing a very large collecting area and permitting interferometric measurements at a dry site

(3) Development of new infrared devices, including improved detectors and high resolution interferometers (page 14)

(d) We recommend that steps be taken to facilitate the analysis by qualified investigators of the data secured by the photographic planetary patrol (page 14).

6. (a) We recommend that NASA openly solicit participation in all future planetary missions by the issuance of flight opportunity announcements with adequate time for response from the scientific community (page 15).

(b) We recommend that NASA develop a summer institute program expressly designed to introduce interested scientists and engineers to the science, technology, and administration of the planetary program (page 15).

7. We recommend that those resources currently intended for support of manned planetary programs be reallocated to programs for instrumented investigation of the planets (page 16).

8. We recommend a coordinated effort involving representatives of NASA, the Department of State, and the National Academy of Sciences, for the purpose of contacting knowledgeable Soviet scientists in an informal way with regard to the possibility of joint planetary exploration (page 16).


## Excerpt from
## Viking "Biology Science Instrument Teams Report"

*Introduction*

This document is an attempt to synthesize several types of biological investigations in a manner to permit their performance in an integrated package. It is subject to modification which may result either from engineering evaluations or from scientific considerations as a result of Mariner VI and VII results.

The examination of the Martian surface for living organisms is based upon the following approaches: visual imagery, atmospheric analysis, chemical composition of the surface, biochemical activity, and enumeration of active particles. In this context visual images represent a high risk–high gain observation, in the sense that the detection of what is unmistakenly a living organism would be highly conclusive, while the absence of such an observation provides the biologist with no direct information on the presence or absence of living organisms, though it contributes to his understanding of the environment. . . .

After describing several types of measurements that can be conducted, an integrated instrument is proposed which combines several of the measurements described, and the directions in which the capability of such an instrument could be expanded should weight and power considerations make such an expansion possible is [*sic*] indicated.

*General Scientific Objectives*

1. At the time of the first examination of the surface of Mars the structure of the experiments must be based on an unavoidable minimum of geocentric assumptions. It is

From Langley Research Center, Viking Proj. Off., "Viking Lander Science Instrument Teams Report," M73-112-0, draft, July 1969.

assumed that, should living organisms exist on Mars, their biochemistry is based on carbon and water. While alternative assumptions are experimentally approachable, they are inappropriate for a first mission.

2. Measurements must be carried out more than twice. . . .

3. It is more important to repeat experiments in time rather than in space. That is, if experimental capability is severely restricted, it is preferable to study a single sample site repeatedly, hopefully in the course of seasonal variation, rather than examining many sample sites a single time within a short period.

4. It is more important to examine a single sample by different principles, rather than carry out a sample examination with refined variations of a single principle. . . .

5. The sensitivity of an observation should be directed primarily at the detection of *any* life. The characterization of such life is at present of secondary importance.

6. In general experimental conditions should be close to the conditions of the Martian environment, except that a variation in water content is contemplated, on the assumption that water may be one of the most important limiting factors for life on Mars.

7. The observations chosen should complement each other in such a manner as to confirm results and minimize ambiguities.

8. Stress has been laid on the formation or fixation of carbon dioxide. This stress results partially from the present view of the composition of the Martian atmosphere, and partly from the geocentric assumptions made. Each experiment which describes the fixation or evolution of carbon dioxide is intended to include at the same time carbon monoxide in the same proportion in which it exists in the Martian atmosphere. . . .

9. The integrated package must withstand terminal sterilization. . . .

*Sampling*

The biological interest is centered on samples taken from approximately the top three centimeters. The instrument is designed to examine a mixture of the top three centimeters. In the event that the sampling device is capable of reaching down several tens of centimeters, it is required that the top three centimeters be examined separately. . . .

The type of sample most useful for biological examination is loosely divided soil. . . . Particles up to two millimeters in size are ideal for the investigations. . . .

## VIKING SCIENCE TEAMS

| Team | As Announced 25 Feb. 1969 | Selected 15 Dec. 1969 | As of Summer 1976 |
|---|---|---|---|
| Active biology | Norman H. Horowitz, California Institute of Technology | Horowitz | Horowitz |
| | Joshua Lederberg, Stanford University | Lederberg | Lederberg |
| | Gilbert V. Levin, Biospherics Research, Inc. | Levin | Levin |
| | Vance I. Oyama, Ames Research Center | Oyama | Oyama |
| | Alexander Rich, MIT | Rich | Rich |

## *Viking Science Teams, Continued*

| Team | Feb. 1969 | Dec. 1969 | Summer 1976 |
| --- | --- | --- | --- |
| | Wolf V. Vishniac, University of Rochester | Vishniac | Harold P. Klein, Ames Research Center (team leader) |
| Lander imagery | Alan B. Binder, Illinois Inst. of Technology Research Inst. | Binder | Binder, Science Applications Institute |
| | Elliot C. Morris, U.S. Geological Survey | Morris | Morris |
| | Thomas A. Mutch, Brown University | Mutch | Mutch (team leader) |
| | Carl Sagan, Cornell University | Sagan | Sagan |
| | | Friedrich Huck, Langley Research Center | Huck |
| | | Elliott C. Levinthal, Stanford University | Levinthal |
| | | Andrew T. Young, JPL | James A. Pollack, Ames Research Center |
| Surface sampler–pyrolyzer–gas analysis (team name changed to molecular analysis, Dec. 1969) | Duwayne M. Anderson, U.S. Army Terrestrial Science Center | Anderson | Anderson |
| | Klaus Biemann, MIT | Biemann | Biemann (team leader) |
| | Melvin Calvin, University of California, Berkeley | | |
| | Leslie E. Orgel, Salk Institute | Orgel | Orgel |
| | John Oró, University of Houston | Oró | Oró, Ames Research Center |
| | Tobias Owen, Illinois Institute of Technology Research Institute | Owen | Owen, State University of New York, Stony Brook |
| | Garson P. Shulman, JPL | Shulman | |
| | Priestley Toulmin, U.S. Geological Survey, Reston | Toulmin | Toulmin |

## Viking Science Teams, Continued

| Team | Feb. 1969 | Dec. 1969 | Summer 1976 |
|---|---|---|---|
| | Harold C. Urey, University of California, San Diego | Urey | Urey |
| | | | Alfred O. C. Nier, University of Minnesota |
| Entry science | Alfred O. C. Nier, University of Minnesota | Nier | Nier (team leader) |
| | Alvin Seiff, Ames Research Center | Seiff | Seiff |
| | Nelson W. Spencer, Goddard Space Flight Center | Spencer | Spencer |
| | | William B. Hanson, University of Texas, Dallas | Hanson |
| | | Michael B. McElroy, Harvard University | McElroy |
| Meteorology | Seymour L. Hess, Florida State University | Hess | Hess (team leader) |
| | Conway B. Leovy University of Washington | Leovy | Leovy |
| | Jack A. Ryan, McDonnell Douglas Astronautics, Western Division | Ryan | Ryan |
| | | Robert M. Henry, Langley Research Center | Henry |
| | | | James E. Tillman, University of Washington |
| Radio science | Dan L. Cain, JPL | Cain | Cain |
| | Von R. Eshelman, Stanford University | G. Levy, JPL | C. T. Stelzried, JPL |
| | Mario D. Grossi, Raytheon Company | Grossi | Grossi |
| | William H. Michael, Langley Research Center | Michael | Michael (team leader) |

## *Viking Science Teams, Continued*

| Team | Feb. 1969 | Dec. 1969 | Summer 1976 |
|------|-----------|-----------|-------------|
| | | Leonard G. Tyler, Stanford University | Tyler |
| | | | Joseph Brenkle, JPL |
| | | | John G. Davies, University of Manchester |
| | | Gunnar Fjeldbo, JPL | Fjeldbo |
| | | Irwin I. Shapiro, MIT | Shapiro |
| | | | Robert H. Tolson, Langley Research Center |
| Ultraviolet photometer (experiment not flown) | Charles A. Barth, University of Colorado | | |
| | Charles W. Hord, University of Colorado | | |
| | Jeffrey B. Pearce, University of Colorado | | |
| Seismology | Don L. Anderson, California Institute of Technology | Anderson | Anderson (team leader) |
| | Robert L. Kovach, Stanford University | Kovach | Kovach |
| | Gary V. Latham, Columbia University | Latham | Latham, University of Texas Medical Branch, Galveston |
| | Frank Press, MIT | | |
| | George H. Sutton, University of Hawaii | Sutton | Sutton |
| | N. Nafi Toksoz, MIT | Toksoz | Toksoz |
| Physical properties | | Richard W. Shorthill, University of Utah Research Institute | Shorthill (team leader) |
| | | Robert E. Hutton, TRW Systems Group | Hutton |

## Viking Science Teams, Continued

| Team | Feb. 1969 | Dec. 1969 | Summer 1976 |
|---|---|---|---|
| | | Henry J. Moore II, U.S. Geological Survey, Menlo Park | Moore |
| | | Ronald Scott, California Institute of Technology | Scott |
| Magnetic properties | | | Robert B. Hargraves, Princeton University (team leader) |
| Inorganic chemistry | | | Priestley Toulmin III, U.S. Geological Survey, Reston (team leader) |
| | | | Alex K. Baird, Pomona College |
| | | | Benton C. Clark, Martin Marietta |
| | | | Klaus Keil, University of New Mexico |
| | | | Harry J. Rose, U.S. Geological Survey, Reston |

## EXPERIMENT INVESTIGATORS

| Scientist | Experiment | Mission | | | |
|---|---|---|---|---|---|
| | | Mariner 1964 | Mariner 1969 | Mariner 1971 | Viking 1975 |
| Alexander, W. M.* | Cosmic-dust detector | X | | | |
| Allen, R. D. | Television | X | | | |
| Anderson, D. L. * | Seismology | | | | X |
| Anderson, D. M. | Molecular analysis | | | | X |
| Anderson, H. R. | Ionization-chamber/ particle-flux detector | X | | | |
| Anderson, J. D.* | Celestial mechanics | | X | X | |
| Arthur, D. | Television | | | X | |

## Experiment Investigators, Continued

| Scientist | Experiment | Mission | | | |
|---|---|---|---|---|---|
| | | Mariner 1964 | Mariner 1969 | Mariner 1971 | Viking 1975 |
| Baird, A. K. | Inorganic chemistry | | | | X |
| Barth, C. A.* | Ultraviolet spectrometer | | X | X | |
| Batson, R. | Television | | | X | |
| Baum, W. A. | Orbiter imaging | | | | X |
| Berg, O. E. | Cosmic-dust detector | X | | | |
| Biemann, K.* | Molecular analysis | | | | X |
| Binder, A. B. | Lander imaging | | | | X |
| Blasius, K. R. | Orbiter imaging | | | | X |
| Borgeson, W. | Television | | | X | |
| Born, G. | Radio science | | | | X |
| Brenkle, J. P. | Radio science | | | | X |
| Bridge, H. S.* | Solar-plasma probe | X | | | |
| Briggs, G. | Television; orbiter imaging | | | X | X |
| Burke, T. | Infrared interferometer spectrometer | | | X | |
| Cain, D. L. | Occultation; S-band occultation; radio science | X | | X | X |
| Carr, M. H.* | Television; orbiter imaging | | | X | X |
| Chase, S. C. | Infrared radiometer; thermal mapping | | X | X | X |
| Clark, B. C. | Inorganic chemistry | | | | X |
| Coleman, P. J., Jr. | Helium magnetometer | X | | | |
| Conrath, B. | Infrared interferometer spectrometer | | | X | |
| Cutts, J. | Television; orbiter imaging | | | X | X |
| Davies, D. W. | Water-vapor mapping | | | | X |
| Davies, J. G. | Radio science | | | | X |
| Davies, M. E. | Television | | X | X | |
| Davis, L., Jr. | Helium magnetometer | X | | | |
| Drake, F. | Occultation | X | | | |
| Duennebier, F. | Seismology | | | | X |
| Duxbury, T. C. | Orbiter imaging | | | | X |
| Eshelman, V. R. | Occultation | X | | | |
| Farmer, C. B.* | Water-vapor mapping | | | | X |
| Fastie, W. G. | Ultraviolet spectrometer | | X | | |
| Fjeldbo, G. | Occultation; S-band occultation; radio science | X | X | X | X |
| Frank, L. A. | Radiation detector | X | | | |

## Experiment Investigators, Continued

| Scientist | Experiment | Mission | | | |
|---|---|---|---|---|---|
| | | Mariner 1964 | Mariner 1969 | Mariner 1971 | Viking 1975 |
| Gause, K. | Ultraviolet spectrometer | | X | | |
| Greeley, R. | Orbiter imaging | | | | X |
| Grossi, M. D. | Radio science | | | | X |
| Guest, J. E. | Orbiter imaging | | | | X |
| Hanel, R.* | Infrared interferometer spectrometer | | | X | |
| Hanson, W. B. | Entry science | | | | X |
| Hargraves, R. B.* | Magnetic properties | | | | X |
| Hartman, W. | Television | | | X | |
| Henry, R. M. | Meteorology | | | | X |
| Herr, K. C. | Infrared spectrometer | | X | | |
| Herriman, A. G. | Television | | X | | |
| Hess, S. L.* | Meteorology | | | | X |
| Hord, C. B. | Ultraviolet spectrometer | | X | X | |
| Horowitz, N. H. | Television; biology | | X | | X |
| Hovis, W. | Infrared interferometer spectrometer | | | X | |
| Howard, K. A. | Orbiter imaging | | | | X |
| Huck, F. O. | Lander imaging | | | | X |
| Hutton, R. E. | Physical properties | | | | X |
| Jones, D. E. | Helium magnetometer | X | | | |
| Keil, K. | Inorganic chemistry | | | | X |
| Keiffer, H. H.* | Infrared radiometer; thermal mapping | | | X | X |
| Kelley, K. K. | Ultraviolet spectrometer | | X | | |
| Klein, H. P.* | Biology | | | | X |
| Kliore, A. J.* | Occultation; S-band occultation | X | X | X | |
| Kovach, R. L. | Seismology | | | | X |
| Krimijis, S. M. | Radiation detector | X | | | |
| Kunde, V. | Infrared interferometer spectrometer | | | X | |
| Lane, A. | Ultraviolet spectrometer | | | X | |
| Laporte, D. | Water-vapor mapping | | | | X |
| Latham, G. V. | Seismology | | | | X |
| Lazarus, A. | Solar-plasma probe | X | | | |
| Lederberg, J. | Television; biology | | | X | X |
| Leighton, R. B.* | Television | X | X | | |

## Experiment Investigators, Continued

| Scientist | Experiment | Mariner 1964 | Mariner 1969 | Mariner 1971 | Viking 1975 |
|---|---|---|---|---|---|
| Leovy, C. B. | Television; meteorology | | X | X | X |
| Levin, G. V. | Infrared interferometer spectrometer; biology | | | X | X |
| Levinthal, E. | Television; lander imaging | | | X | X |
| Levy, G. S. | Occultation | X | | | |
| Liebes, S., Jr. | Lander imaging | | | | X |
| Lorell, J.* | Celestial mechanics | | | X | |
| Lowman, P. | Infrared interferometer spectrometer | | | X | |
| McCauley, J. | Television | | | X | |
| McCracken, C. W. | Cosmic-dust detector | X | | | |
| McElroy, M. B. | Entry science | | | | X |
| Mackey, E. F. | Ultraviolet spectrometer | | X | | |
| Martin, W. L. | Celestial mechanics | | X | X | |
| Masursky, H.* | Television; orbiter imaging | | | X | X |
| Michael, W. H.* | Radio science | | | | X |
| Milton, D. | Television | | | X | |
| Miner, E. D. | Infrared radiometer; thermal mapping | | | X | X |
| Moore, H. J., II | Physical properties | | | | X |
| Morris, E. C. | Lander imaging | | | | X |
| Munch, G. | Infrared radiometer; thermal mapping | | X | X | X |
| Murray, B. C. | Television | X | X | X | |
| Mutch, T. A.* | Lander imaging | | | | X |
| Neher, H. V.* | Ionization-chamber/ particle flux detector | X | | | |
| Neugebauer, G.* | Infrared radiometer; thermal mapping | | X | X | X |
| Nier, A. O. C.* | Molecular analysis; entry science | | | | X |
| O'Gallagher, J. | Cosmic-ray telescope | X | | | |
| Orgel, L. E. | Molecular analysis | | | | X |
| Oró, J. | Molecular analysis | | | | X |
| Owen, T. | Molecular analysis | | | | X |
| Oyama, V. I. | Biology | | | | X |
| Pearce, J. B. | Ultraviolet spectrometer | | X | | |
| Pearl, J. | Infrared interferometer spectrometer | | | X | |

## Experiment Investigators, Continued

| Scientist | Experiment | Mariner 1964 | Mariner 1969 | Mariner 1971 | Viking 1975 |
|---|---|:---:|:---:|:---:|:---:|
| Pimentel, G. C.* | Infrared spectrometer | | X | | |
| Pollack, J. | Television; lander imaging | | | X | X |
| Prabhakara, C. | Infrared interferometer spectrometer | | | X | |
| Rasool, S. I. | S-band occultation | | X | X | |
| Reasenberg, R. | Celestial mechanics | | | | X |
| Rich, A. | Biology | | | | X |
| Rose, H. J. | Inorganic chemistry | | | | X |
| Ruehle, R. | Ultraviolet spectrometer | | X | | |
| Ryan, J. A. | Meteorology | | | | X |
| Sagan, C. | Television; lander imaging | | | X | X |
| Schlachman, B. | Infrared interferometer spectrometer | | | X | |
| Scott, R. F. | Physical properties | | | | X |
| Secretan, L. | Cosmic-dust detector | X | | | |
| Seidel, B. | S-band occultation | | | X | |
| Seiff, A. | Entry science | | | | X |
| Shapiro, I. I. | Celestial mechanics; radio science | | | X | X |
| Sharp, R. P. | Television | X | X | X | |
| Shipley, E. | Television | | | X | |
| Shorthill, R.* | Physical properties | | | | X |
| Simpson, J. A.* | Cosmic-ray telescope | X | | | |
| Sjoren, W. | Celestial mechanics | | | X | |
| Sloan, R. K. | Television | X | X | | |
| Smith, B. A. | Television; orbiter imaging | | X | X | X |
| Smith, E. J. | Helium magnetometer | X | | | |
| Snyder, C. W. | Solar-plasma probe | X | | | |
| Soderblom, L. A. | Television; orbiter imaging | | | X | X |
| Spencer, N. W. | Entry science | | | | X |
| Stelzried, C. T. | Radio science | | | | X |
| Stewart, I. | Ultraviolet spectrometer | | | X | |
| Sutton, G. | Seismology | | | | X |
| Tillman, J. E. | Meteorology | | | | X |
| Toköz, N. | Seismology | | | | X |
| Tolson, R. H. | Radio science | | | | X |
| Toulmin, P., III* | Molecular analysis; inorganic chemistry | | | | X |

## Experiment Investigators, Continued

| Scientist | Experiment | Mission | | | |
|---|---|---|---|---|---|
| | | Mariner 1964 | Mariner 1969 | Mariner 1971 | Viking 1975 |
| Tyler, G. L. | Radio science | | | | X |
| Urey, H. C. | Molecular analysis | | | | X |
| Van Allen, J. A.* | Radiation detector | X | | | |
| Vaucouleurs, G. de | Television | | | X | |
| Veverka, J. | Television; orbiter imaging | | | X | X |
| Wellman, J. B. | Orbiter imaging | | | | X |
| Wildey, R. | Television | | | X | |
| Wilhelms, D. | Television | | | X | |
| Wilshusen, F. C. | Ultraviolet spectrometer | | X | | |
| Young, A. | Television | | X | X | |
| | | | | | |
| Total experimenters per project | | 29 | 29 | 55 | 80 |

Grand total for all projects—193

    1 scientist worked on all 4 projects
    9 scientists worked on 3 projects
    24 scientists worked on 2 projects

    2 scientists worked on Mariner 1964 and Mariner 1969 exclusively
    3 scientists worked on Mariner 1964, Mariner 1969, and Mariner 1971 exclusively
    7 scientists worked on Mariner 1969 and Mariner 1971 exclusively
    5 scientists worked on Mariner 1969, Mariner 1971, and Viking 1975 exclusively
    14 scientists worked on Mariner 1971 and Viking 1975 exclusively

*Principal investigator.

# Appendix E
## Launch Vehicles for Mars Missions

Atlas-Agena launch vehicles—used by NASA during the 1960s to launch a variety of payloads to Earth orbit, the moon, and the near planets—sent Mariner spacecraft (200-260 kilograms) on their way to Venus or Mars. The Agena upper stage, developed by Lockheed Missiles and Space Co. for the Air Force, was capable of restarting its engines, thus permitting the spacecraft to be positioned more precisely. Paired with the Atlas booster, Agena B was used in 1962 for the first two Mariner flights (the Atlas stage malfunctioned during the *Mariner 1* launch). Atlas–Agena D, with an improved upper stage that could accept a greater variety of payloads, launched the next three Mariners in 1964 and 1967.

Advanced mission planners of the early 1960s had based their planetary exploration schedules on the early availability of the high-energy Centaur upper stage. Centaur, a liquid-hydrogen-fueled stage developed for NASA by General Dynamics/Convair, did not go into service until 1966, however. It was 1969 before Atlas-Centaur sent two 400-kilogram Mariners flying by Mars. In May 1971, a Centaur failure led to the destruction of the next Mariner spacecraft. A second attempt weeks later saw *Mariner 9* (990 kilograms) off on its journey to Mars. In November 1973, Atlas-Centaur boosted *Mariner 10* (500 kilograms) to an interplanetary (Venus and Mercury) trajectory.

Centaur also had a role in the Viking Mars landing project. Mated with a Titan IIIE two-stage vehicle, the improved Centaur could boost the 3500-kilogram, two-part Viking spacecraft to the Red Planet. NASA had used another model of the Air Force Titan, the Titan II, for the manned Gemini program, 1965–1966. The Titan IIIE, a modified version of the Titan IIID used by the Air Force since 1971 as a satellite launcher and made by Martin Marietta Corporation, was a powerful, versatile vehicle. NASA's first test launch of Titan IIIE–Centaur in February 1974 ended in failure, however, because of the malfunction of a Centaur component. *Helios 1*, a German satellite, was successfully launched by NASA with Titan-Centaur the following December. The bulbous launch vehicle with its two powerful strap-on booster rockets performed equally well in 1975 for Viking.

## *Atlas–Agena B Characteristics*

| Characteristic | 1st Stage, Atlas | 2d Stage, Agena B | Total with Adapter |
|---|---|---|---|
| Height (m) | 21.9 | 7.2 | 30.6 |
| Diameter (m) | 3 | 1.5 | |
| Launch weight (kg) | 117 780 | 7022 | 124 802 |
| Propulsion system | | | |
|   Powerplant | MA-5 propulsion system | Bell XLR-81-Ba-9 (model 8081; upgraded to 8096) | |
|   Thrust (kilonewtons) | 1600 | 71.2 | 1670 |
|   Propellant[a] | LOX/RP-1 | IRFNA/UDMH | |
| Payload capacity (kg) | | | 2627 to Earth orbit 340 to escape trajectory 204 to Mars or Venus |
| Origin | Uprated Atlas–Agena A | | |
| Contractors | Consolidated Vultee Aircraft Corp. (prime). North American Aviation, Inc. (engines). | Lockheed Missiles and Space Co. (prime). Bell Aerospace (engine). | |
| Program use | Mariner, Ranger, and OGO. | | |
| Remarks | Capable of engine restart. | | |

[a]LOX/RP-1 = liquid oxygen and modified kerosene.
IRFNA/UDMH = inhibited red-fuming nitric acid and unsymmetrical dimethylhydrazine.

## *Atlas–Agena D Characteristics*

| Characteristic | 1st Stage, Atlas | 2d Stage, Agena B | Total with Adapter |
|---|---|---|---|
| Height (m) | 21.9<br>23.2 (SLV-3C) | 7.2 | 30.6<br>32.1<br>(w/SLV-3C) |
| Diameter (m) | 3 | 1.5 | |
| Launch weight (kg) | 117 780<br>128 879 (SLV-3C) | 7248 | 125 028<br>136 127<br>(w/SLV-3C) |
| Propulsion system | | | |
|   Powerplant | MA-5<br>propulsion<br>system | Bell<br>XLR-81-Ba-9<br>(model 8247;<br>upgraded to 8533) | |
|   Thrust (kilonewtons) | 1600<br>1750 (SLV-3C) | 71.2 | 1670<br>1820<br>(w/SLV-3C) |
|   Propellant[a] | LOX/RP-1 | $N_2O_4$/UDMH | |
| Payload capacity (kg) | | | 2718 to Earth<br>orbit<br>385 to escape<br>trajectory<br>250 to Mars<br>or Venus |
| Origin | Uprated Atlas–Agena B | | |
| Contractors | Consolidated Vultee<br>Aircraft Corp. (prime).<br>North American<br>Aviation, Inc. (engines). | Lockheed Missiles<br>and Space Co. (prime).<br>Bell Aerospace (engine). | |
| Program use | Mariner, OAO, Lunar Orbiter, and ATS. | | |
| Remarks | The Agena D stage could accept a greater variety of payloads than<br>could the Agena B model. | | |

[a]LOX/RP-1 = liquid oxygen and modified kerosene.
$N_2O_4$/UDMH = nitrogen tetroxide and unsymmetrical dimethylhydrazine.

## *Atlas–Centaur Characteristics*

| Characteristic | 1st Stage, Atlas SLV-3C/ Atlas SLV-3D | 2d Stage, Centaur D-1A | Total |
|---|---|---|---|
| Height (m) | 23.2 | 13<br>14.6 w/payload fairing | 34 |
| Diameter (m) | 3.05 | 3.05 | |
| Launch weight (kg) | 128 879 | 17 145 | 146 024 |
| Propulsion system | | | |
|   Powerplant | MA-5 propulsion system | Pratt & Whitney (2) RL-10 | |
|   Thrust (kilonewtons) | 1700/1900 | 133.4 | 1890/2050 |
|   Propellant[a] | LOX/RP-1 | LOX/LH$_2$ | |
|   Payload capacity (kg) | | | 3857/4536 to Earth orbit 1225/1882 to synchronous orbit 815/907 to Venus or Mars |
| Origin | Air Force ICBM | General Dynamics studies for a high-energy second stage; development supported by NASA | |
| Contractors | General Dynamics/ Convair (formerly Consolidated Vultee Aircraft Corp.) (prime). North American Aviation (engines). | General Dynamics/Convair | |
| Program use | Surveyor, ATS, OAO, Mariner, Intelsat, Pioneer. | | |
| Remarks | Centaur, the first U.S. launch vehicle to use liquid hydrogen as a propellant, was originally scheduled for operations in the early 1960s for Mars and Venus probes. Because of delays in the vehicle's development, however, it was not ready until 1966. One of the serious problems with the stage's development was hydrogen loss; heat transfer between the oxygen and hydrogen fuel tanks caused the liquid hydrogen to evaporate.. | | |

[a]LOX/RP-1 = liquid oxygen and modified kerosene.
LOX/LH$_2$ = liquid oxygen and liquid hydrogen.

## Titan IIIE – Centaur Characteristics

| Characteristic | Stage 0 Solid-Fueled Rocket Motors (2) | 1st Stage Titan | 2d Stage Titan | 3d Stage, Centaur D-1T | Centaur Standard Shroud | Total |
|---|---|---|---|---|---|---|
| Height (m) | 25.9 | 22.2 | 7.1 | 9.7 | (17.7) | 48.8 |
| Diameter (m) | 3.05 | 3.05 | 3.05 | 3.05 | 4.3 | |
| Launch weight (kg) | 226 800 | 123 830 | 33 112 | 17 700 | 3092 | 631 334 |
| Propulsion system | | | | | | |
|   Powerplant | United Technology 1205 | Aerojet YLR87-AJ-11 | Aerojet YLR91-AJ-11 | Pratt & Whitney (2) RL-10A-3-3 | | |
|   Thrust (kilonewtons) | 10 680 (combined) | 2310 | 449.2 | 133.4 | | 13 550 |
|   Burn time (sec) | 110 | 150 | 208 | 450 | | 918 |
|   Propellant[a] | powdered aluminum/ ammonium perchlorate | $N_2H_4$- UDMH/ $N_2O_4$ | $N_2H_4$- UDMH/ $N_2O_4$ | $LH_2$/LOX | | |
| Payload capacity (kg) | | | | 15 000 to Earth orbit 3 000 to synchronous ( 3 400 to Mars | | |
| Origin | Air Force Titan IIID modified to NASA requirements | | | NASA design | | |
| Contractors | Chemical Systems Div., United Technologies | Martin Marietta Corp. | | General Dynamics/ Convair | | |
| Program manager | R. A. Mattson, NASA Hq. | | | | | |
| Project manager | Andrew J. Stofan, Lewis Research Center | | | | | |
| Program use | | Viking, Voyager | | | | |
| Remarks | In this configuration, the Centaur upper stage replaced the standard Titan third stage, called the transtage; Centaur was capable of restarting its two engines, a desirable characteristic for planetary missions. During Centaur's coast phase, 14 small hydrogen peroxide thrusters controlled attitude. When the two five-segment solid-fueled rocket motors, together known as "stage 0," were jettisoned, the Titan first stage ignited. These strap-on motors provided more than four times the thrust of the Atlas booster at liftoff. | | | | | |

[a]$N_2H_4$-UDMH/$N_2O_4$ = hydrazine–unsymmetrical dimethylhydrazine and nitrogen tetroxide.
$LH_2$/LOX = liquid hydrogen and liquid oxygen.

# Appendix F
## Major Viking Contractors and Subcontractors

| Organization | Location | Responsibility |
|---|---|---|
| Bendix Aerospace Systems Division | Ann Arbor, Mich. | Bendix was responsible to Martin Marietta for two of the entry science instruments—the upper atmospheric mass spectrometer and the retarding potential analyzer—and one Martian surface instrument—the seismometer. Design, fabrication, assembly, and testing were conducted at the Ann Arbor facilities. |
| Celesco Industries | Costa Mesa, Calif. | Celesco Industries built the surface-sampler arm, housing, and drive mechanism that picked up the surface samples and delivered them to the lander instruments. Celesco acted as a subcontractor to Martin Marietta. |
| Goodyear Aerospace Corporation | Akron, Ohio | Goodyear was responsible to Martin Marietta for the decelerator system used on the lander. Goodyear personnel designed, built, and tested the decelerator system and managed subtier suppliers and subcontractors. |
| Honeywell Aerospace Division | St. Petersburg, Fla. | Under contract to Martin Marietta, Honeywell designed, manufactured, and tested the lander guidance, control, and sequencing computer and data-storage memory. |
| Itek Corporation, Optical Systems Division | Lexington, Mass. | Itek was responsible to Martin Marietta for all aspects of the lander camera system. Itek produced and tested the cameras and their supporting Earth-based ground reconstruction sets. Itek also provided the computer software necessary to operate and control the cameras and to drive the ground reconstruction equipment in reconstructing the photographs. |
| Jet Propulsion Laboratory | Pasadena, Calif. | JPL was responsible to NASA for the orbiter and the mission control center (Space Flight Operations Facility). JPL also operated the Deep Space Network. |

| Organization | Location | Responsibility |
|---|---|---|
| Litton Industries, Guidance and Control Systems Division | Woodland Hills, Calif. | Litton contracted with NASA for the production and integration of the design technology used in the gas chromatograph–mass spectrometer. |
| Martin Marietta Aerospace | Denver, Colo. | Martin Marietta was responsible to NASA's Langley Research Center for the overall integration of the Viking project and was prime contractor for the lander and its subsystems (designing, testing, and building the lander and managing all lander subcontractors). Martin Marietta also designed and built the photo sensor array for the Viking cameras, the temperature transducers, and the x-ray fluorescence spectrometer. In addition, Martin Marietta built the Titan IIIE launch vehicles used in the project. |
| RCA Astro-Electronics Division | Princeton, N.J. | As a subcontractor to Martin Marietta, RCA designed, built, and tested the lander communications subsystem, including an ultrahigh-frequency radio transmitter, an antenna for beaming signals to the orbiter, an S-band antenna for broadcasting directly to Earth, and an S-band low-gain unit to receive direct commands from Earth. |
| Rocket Research Corporation | Redmond, Wash. | Rocket Research, under contract to Martin Marietta, was responsible for developing and manufacturing the throttleable, monopropellant-hydrazine, landing engines and the control and deorbit engines. |
| Sheldahl, Inc. | Northfield, Minn. | For Martin Marietta, Sheldahl designed and built four load-carrying high-altitude balloons, which were used for the balloon launch-decelerator test program for the lander, and the disk-gap-band parachute used as part of the decelerator system. Sheldahl also fabricated the bioshields used to encapsulate the lander and the lander leg covers. |
| Space and Missile Systems Organization (SAMSO) | Los Angeles, Calif. | SAMSO was the U.S. government agency responsible for developing the Titan III launch vehicle. The SAMSO 6555 Aerospace Test Wing at Cape Canaveral Air Force Station managed the Ti- |

471

| Organization | Location | Responsibility |
|---|---|---|
| | | tan launch facility and supported NASA in launching the Titan III Centaur. |
| Teledyne Ryan Aeronautical | San Diego, Calif. | Teledyne Ryan subcontracted with Martin Marietta as designer, tester, and builder of the radar altimeter and the terminal-descent and landing radar used on the lander. |
| TRW Inc. | Redondo Beach, Calif. | As a subcontractor to Martin Marietta, TRW built the biology and meteorology instruments carried on the lander. |

# Appendix G
# Organization Charts

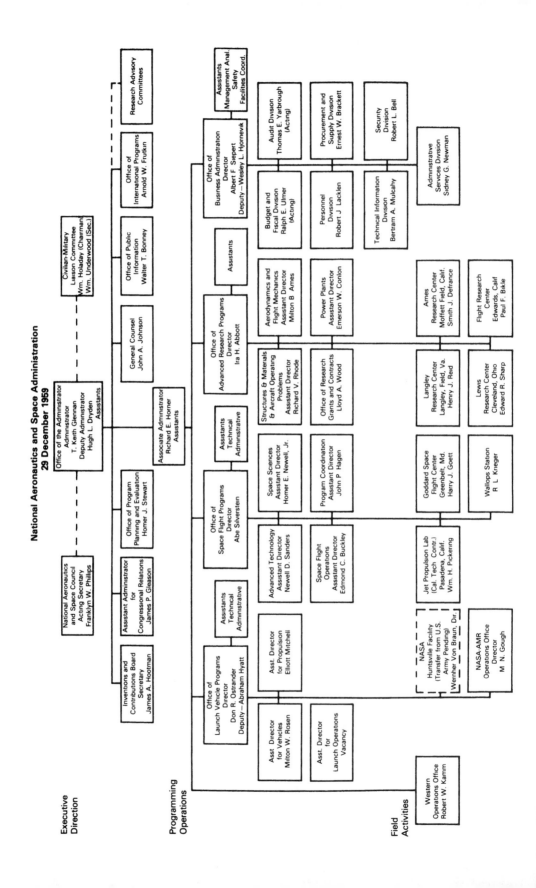

National Aeronautics and Space Administration
29 December 1959

Executive Direction

Programming

Operations

Field Activities

Office of the Administrator
Administrator
T. Keith Glennan
Deputy Administrator
Hugh L. Dryden
Assistants

National Aeronautics and Space Council
Acting Secretary
Franklyn W. Phillips

Civilian-Military Liaison Committee
Wm. Holaday (Chairman)
Wm. Underwood (Sec.)

Research Advisory Committees

Office of International Programs
Arnold W. Frutkin

Office of Public Information
Walter T. Bonney

General Counsel
John A. Johnson

Assistant Administrator for Congressional Relations
James P. Gleeson

Office of Program Planning and Evaluation
Homer J. Stewart

Inventions and Contributions Board
Secretary
James A. Hootman

Associate Administrator
Richard E. Horner
Assistants

Office of Advanced Research Programs
Director
Ira H. Abbott

Assistants
Technical
Administrative

Office of Space Flight Programs
Director
Abe Silverstein

Assistants
Technical
Administrative

Office of Launch Vehicle Programs
Director
Don R. Ostrander
Deputy — Abraham Hyatt

Office of Business Administration
Director
Albert F. Siepert
Deputy — Wesley L. Hjornevik

Assistants
Management Anal.
Safety
Facilities Coord.

Assistants

Aerodynamics and Flight Mechanics
Assistant Director
Milton B Ames

Power Plants
Assistant Director
Emerson W. Conlon

Structures & Materials & Aircraft Operating Problems
Assistant Director
Richard V. Rhode

Office of Research Grants and Contracts
Lloyd A. Wood

Space Sciences
Assistant Director
Homer E. Newell, Jr.

Program Coordination
Assistant Director
John P Hagen

Advanced Technology
Assistant Director
Newell D. Sanders

Space Flight Operations
Assistant Director
Edmond C. Buckley

Asst. Director for Propulsion
Elliott Mitchell

Asst. Director for Vehicles
Milton W. Rosen

Asst. Director for Launch Operations
Vacancy

NASA
Huntsville Facility
(Transfer from U.S. Army Pending)
Wernher Von Braun, Dir.

Jet Propulsion Lab
(Cal. Tech Contr.)
Pasadena, Calif.
Wm. H. Pickering

Goddard Space Flight Center
Greenbelt, Md.
Harry J. Goett

Wallops Station
R L Krieger

NASA-AMR
Operations Office
Director
M. N. Gough

Audit Division
Thomas E. Yarbrough
(Acting)

Procurement and Supply Division
Ernest W. Brackett

Security Division
Robert L. Bell

Budget and Fiscal Division
Ralph E. Ulmer
(Acting)

Personnel Division
Robert J Lacklen

Technical Information Division
Bertram A. Mulcahy

Administrative Services Division
Sidney G. Newman

Ames Research Center
Moffett Field, Calif.
Smith J. DeFrance

Langley Research Center
Langley, Field, Va.
Henry J. Reid

Lewis Research Center
Cleveland, Ohio
Edward R. Sharp

Flight Research Center
Edwards, Calif
Paul F. Bikle

Western Operations Office
Robert W. Kamm

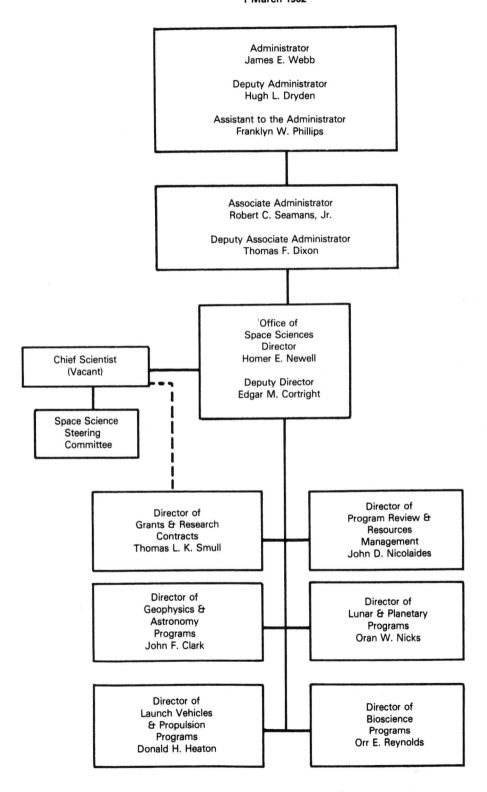

**National Aeronautics and Space Administration
1 March 1962**

Administrator
James E. Webb

Deputy Administrator
Hugh L. Dryden

Assistant to the Administrator
Franklyn W. Phillips

Associate Administrator
Robert C. Seamans, Jr.

Deputy Associate Administrator
Thomas F. Dixon

'Office of
Space Sciences
Director
Homer E. Newell

Deputy Director
Edgar M. Cortright

Chief Scientist
(Vacant)

Space Science
Steering
Committee

Director of
Grants & Research
Contracts
Thomas L. K. Smull

Director of
Program Review &
Resources
Management
John D. Nicolaides

Director of
Geophysics &
Astronomy
Programs
John F. Clark

Director of
Lunar & Planetary
Programs
Oran W. Nicks

Director of
Launch Vehicles
& Propulsion
Programs
Donald H. Heaton

Director of
Bioscience
Programs
Orr E. Reynolds

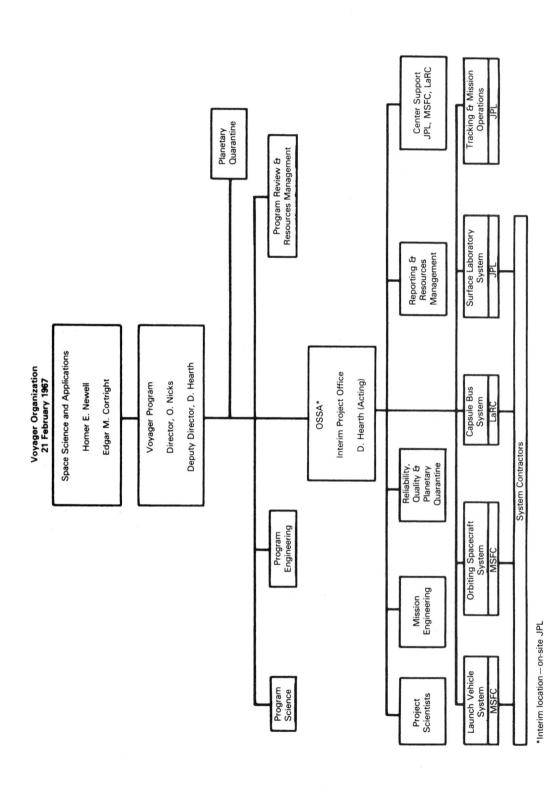

Voyager Organization
21 February 1967

Space Science and Applications
Homer E. Newell
Edgar M. Cortright

Voyager Program
Director, O. Nicks
Deputy Director, D. Hearth

Planetary Quarantine

Program Review & Resources Management

Program Science

Program Engineering

OSSA*
Interim Project Office
D. Hearth (Acting)

Project Scientists

Mission Engineering

Reliability, Quality & Planetary Quarantine

Reporting & Resources Management

Center Support
JPL, MSFC, LaRC

Launch Vehicle System
MSFC

Orbiting Spacecraft System
MSFC

Capsule Bus System
LaRC

Surface Laboratory System
JPL

Tracking & Mission Operations
JPL

System Contractors

*Interim location—on-site JPL

**Voyager Management at Langley Before and After
Establishment of the Voyager Capsule Bus Management Office**

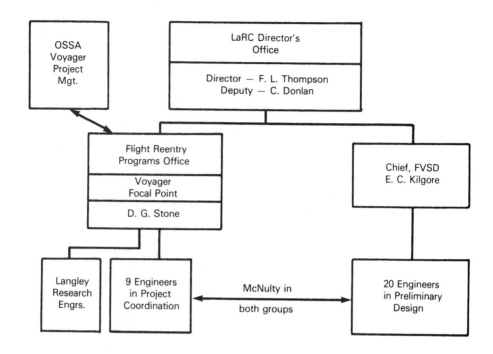

OSSA Voyager Project Mgt.

LaRC Director's Office

Director — F. L. Thompson
Deputy — C. Donlan

Flight Reentry Programs Office

Voyager Focal Point

D. G. Stone

Chief, FVSD
E. C. Kilgore

Langley Research Engrs.

9 Engineers in Project Coordination

McNulty in both groups

20 Engineers in Preliminary Design

**Organization January 1967.**

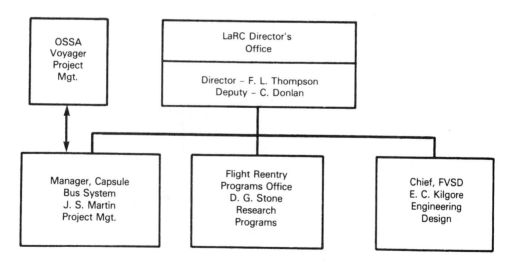

OSSA Voyager Project Mgt.

LaRC Director's Office

Director – F. L. Thompson
Deputy – C. Donlan

Manager, Capsule Bus System
J. S. Martin
Project Mgt.

Flight Reentry Programs Office
D. G. Stone
Research Programs

Chief, FVSD
E. C. Kilgore
Engineering Design

**Organization June 1967.**

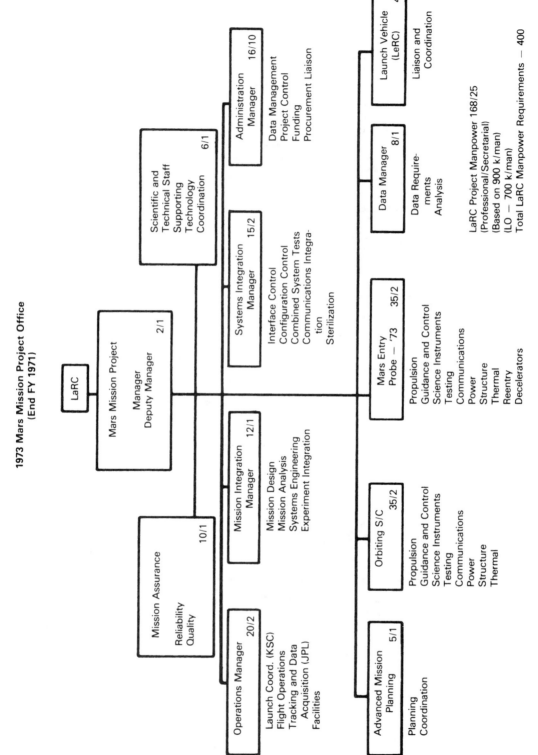

**1973 Mars Mission Project Office**
(End FY 1971)

LaRC

**Mars Mission Project**
Manager
Deputy Manager    2/1

**Scientific and Technical Staff**
Supporting Technology Coordination    6/1

**Mission Assurance**    10/1

Reliability
Quality

**Mission Integration Manager**    12/1

Mission Design
Mission Analysis
Systems Engineering
Experiment Integration

**Systems Integration Manager**    15/2

Interface Control
Configuration Control
Combined System Tests
Communications Integration
Sterilization

**Administration Manager**    16/10

Data Management
Project Control
Funding
Procurement Liaison

**Operations Manager**    20/2

Launch Coord. (KSC)
Flight Operations
Tracking and Data Acquisition (JPL)
Facilities

**Advanced Mission Planning**    5/1

Planning
Coordination

**Orbiting S/C**    35/2

Propulsion
Guidance and Control
Science Instruments
Testing
Communications
Power
Structure
Thermal

**Mars Entry Probe — '73**    35/2

Propulsion
Guidance and Control
Science Instruments
Testing
Communications
Power
Structure
Thermal
Reentry
Decelerators

**Data Manager**    8/1

Data Requirements Analysis

**Launch Vehicle (LeRC)**    4/1

Liaison and Coordination

LaRC Project Manpower 168/25
(Professional/Secretarial)
(Based on 900 k/man)
(LO — 700 k/man)
Total LaRC Manpower Requirements — 400

January 11, 1968

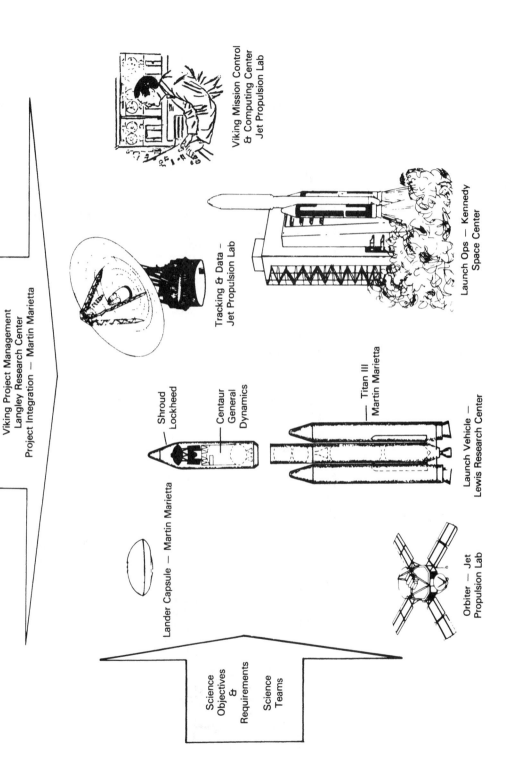

**Viking Project Responsibilities**

Viking Project Management
Langley Research Center
Project Integration — Martin Marietta

Viking Mission Control
& Computing Center
Jet Propulsion Lab

Launch Ops — Kennedy
Space Center

Tracking & Data –
Jet Propulsion Lab

Shroud
Lockheed

Centaur
General Dynamics

Titan III
Martin Marietta

Lander Capsule — Martin Marietta

Launch Vehicle —
Lewis Research Center

Orbiter — Jet
Propulsion Lab

Science
Objectives
&
Requirements

Science
Teams

**Viking Orbiter
Project Office**

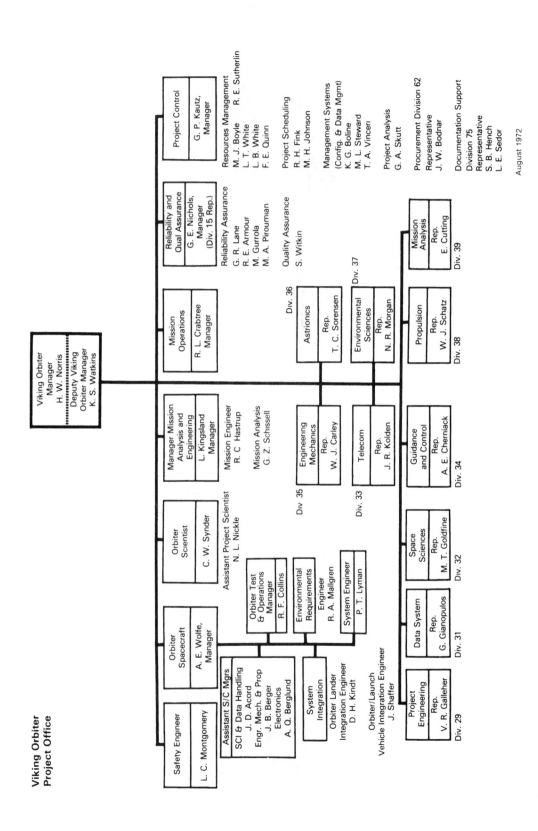

August 1972

# Bibliographic Essay

This essay is designed to serve as a guide to the sources used in preparing this history, rather than as an inclusive catalog. The source notes are the main guide to the materials used, but some discussion about research techniques and the items cited in the source notes will enhance the reader's understanding of how the history was prepared. We also hope this note will be useful to subsequent researchers. Much of this book was written very close in time to the events described, and the subject deserves further study.

From the standpoint of sources, the book can be divided into two parts—chapters 1 through 4 and chapters 5 through the Epilogue. In the former, we relied heavily on traditional sources familiar to the historical researcher: books, periodicals, newspapers, and occasional documents from the National Aeronautics and Space Administration (NASA), the Space Science Board (SSB) of the National Academy of Sciences, and related organizations. In the second part of the book we have used, for the most part, internal NASA documentation: memorandums, letters, telexes, reports, weekly activity reports, minutes of meetings. This NASA paperwork represented the collective product of the Viking Program Office at NASA Headquarters, the Viking Project Office at the Langley Research Center (LaRC), the Viking Orbiter Office at the Jet Propulsion Laboratory (JPL), and the industrial contractors. Similar documents for the Mariner and Voyager projects were also used. Specific comments for each of the two parts follow.

## Pre-Viking

Scientific interest in the Red Planet, always very keen, increased with the coming of the space age. The literature on Mars is ever growing. Researchers interested in the literature that has evolved since 1958 should consult the following bibliographic aids:

NASA RECON (a computerized system for access to aerospace literature). RECON permits a quick review of the technical periodicals and report literature related to NASA engineering and science projects. The NASA publication *STAR* (Scientific and Technical Aerospace Reports) and the American Institute of Astronautics and Aeronautics' *International Aerospace Abstracts* duplicate in a printed version much of the bibliographical information contained in RECON.

*Icarus,* international journal of solar system studies, 1962–present. *Icarus* is the single most important journal for scientific studies related to Mars. Many of the issues discussed in the planning for NASA flights to Mars were first debated in the pages of this journal founded by Carl Sagan. Over the years since 1962, most of the major findings of Martian research were reported in *Icarus,* including the results of Soviet investigations.

Scientific and Technical Information Division. *Extraterrestrial Life: A Bibliography, 1952–1964.* NASA SP-7015. Washington, September 1964. This annotated bibliography contains 183 citations.

Shneour, Elie A., and Ottesen, Eric A., comps. *Extraterrestrial Life: An Anthology and Bibliography.* NAS publication 1296A. Washington: National Academy of Sciences, 1966. This collection of readings and bibliographical entries was prepared to accompany the following publication.

Pittendrigh, Colin S.; Vishniac, Wolf; and Pearman, J. P. T., eds. *Biology and the Exploration of Mars: Report of a Study Held under the Auspices of the Space Science Board, National Academy of Sciences–National Research Council, 1964–1965.* NAS publication 1296. Washington: National Academy of Sciences, 1966. Part 3, the bibliography *(Extraterrestrial Life: An Anthology and Bibliography),* contains more than 2000 selected references to published literature through mid-1964, with an addendum of papers published through the latter part of 1965.

Magnolia, L. R.; Gogin, S. A.; and Turley, J. A. *Exobiology: A Bibliography.* Research bibliography 52. TRW STL Technical Library: Redondo Beach, Calif., October 1964. The report contains 400 annotated citations with indexes to authors, subjects, serials, and Defense Technical Information Center (AD) and NASA (N) accession numbers. It can be retrieved on the NASA/RECON system as document N 65-19834.

Jet Propulsion Laboratory. *Mariner Mars 1964 Bibliography.* Project Document 67, Rev. 1. Pasadena, 7 November 1966. Intended primarily for internal use at JPL, the bibliography covers internal reports and other documentation for the period 1962–1966.

————. *JPL Bibliography of Voyager Spacecraft Related Documents.* Pasadena, 8 February 1967. This bibliography lists NASA, JPL, and contractor documents for 1964 and 1967.

Magnolia, L. R. and Gogin, S. A. *Manned Mars Missions: A Bibliography.* Research Bibliography 53. TRW STL Technical Laboratories: Redondo Beach, Calif., April 1965. The bibliography contains 348 annotated references to manned Mars flyby and stopover missions, unmanned preparatory missions, and Earth-based studies of Mars for the years 1955 to January 1965. Supplemental data on manned Mars missions can be found in the NASA/RECON system.

Magnolia, L. R. *The Planet Mars: A Selected Bibliography.* TRW Systems Group Special Literature Survey 61, 20 April 1973.

General publications that were useful in the preparation of chapters 1 through 4 include:

Berkner, Lloyd V., and Odishaw, Hugh, eds. *Science in Space*. New York, Toronto, London: McGraw-Hill, 1961. The first overview of the field of space science, this book is an essential starting point for students of American scientific activities in space.

Blum, Harold F. *Time's Arrow and Evolution*. Princeton: Princeton University Press, 1951.

Glasstone, Samuel. *The Book of Mars*. NASA SP-179. Washington, 1968.

Hall, R. Cargill. *Lunar Impact: A History of Project Ranger*. NASA SP-4210. Washington, 1977.

_____. *Project Ranger: A Chronology*. JPL/HR-2. Pasadena: Jet Propulsion Laboratory, 1971.

Hoyt, William Graves. *Lowell and Mars*. Tucson: University of Arizona Press, 1976.

Jet Propulsion Laboratory. *Exploration of the Moon, the Planets, and Interplanetary Space*, ed. Albert R. Hibbs. JPL report 30-1. Pasadena: Jet Propulsion Laboratory, 1959.

Young, Richard S.; Painter, Robert B.; and Johnson, Richard D. *An Analysis of the Extraterrestrial Life Detection Problem*. NASA SP-75. Washington, 1965.

Ley, Willy, and Bonestell, Chesley. *The Conquest of Space*. New York: Viking Press, 1949.

_____, and von Braun, Wernher. *The Exploration of Mars*. New York: Viking Press, 1956.

Logsdon, John M. *The Decision to Go to the Moon: Project Apollo and the National Interest*. Cambridge, Mass., and London: MIT Press, 1970.

National Academy of Sciences–National Research Council. *A Review of Space Research: The Report of the Summer Study Conducted under the Auspices of the National Academy of Sciences at the State University of Iowa, Iowa City, Iowa, June 17–August 10, 1962*. NAS–NRC publication 1079. Washington: National Academy of Sciences, 1962.

Newell, Homer E. *Beyond the Atmosphere: Early Years of Space Science*. NASA SP-4211. Washington, 1980.

Phillips, Charles R. *The Planetary Quarantine Program: Origins and Achievements*. NASA SP-4902. Washington, 1974.

Rosholt, Robert L. *An Administrative History of NASA, 1958–1963*. NASA SP-4101. Washington, 1966.

von Braun, Wernher. *The Mars Project*. Urbana, Ill.: University of Illinois Press, 1953.

The following document collections for chapters 1 through 3 are held by the National Archives and Records Service:

Federal Records Center, Suitland, Md. (Washington, DC 20409).

NASA, Office of the Administrator, Secretariat.

Meetings, NASA participation (National Academy of Sciences/SSB et al.) 1958–1960: box 11, RG 255, accession 66-A-184.

Meetings of Space Science Board, beginning to 1963: box 63, RG 255, accession 67-A-601.

Committees for NAS/Space Science Board et al., 1958–1969: box 17, RG 255, accession 72-A-3070.

NASA, Office of Space Sciences (Office of Space Science and Applications, 1963–1971). Lunar and Planetary Programs Office.

Chronological files:
1962–1964: box 50, RG 255, accession 74-A-663.
1965–1971: box 51, RG 225, accession 74-A-663.

Mariner C, R, B files (reports, correspondence, etc.), 1964: boxes 3–8, RG 255, accession 65-A-836 (destroyed August 1973, and not available for reference).

Voyager Phase IA reports and evaluations prepared by JPL June 1965 to September 1965: box 1, RG 255, accession 66-A-1089.

Voyager spacecraft final technical reports, July 1965: boxes 1–6, RG 225, accession 66-A-578.

Proposals for Voyager spacecraft system, February 1965: boxes 1–2, RG 255, accession 66-A-1155.

Proposals for Voyager spacecraft system, January 1966: boxes 1–5, RG 255, accession 67-A-785.

Voyager spacecraft phase B contractor reports, July 1965: boxes 1–5, RG 255, accession 67-A-202.

Reports on Voyager studies, August 1965 to November 1967: boxes 1–17, RG 255, Accession 68-A-6256.

Reports on Voyager capsule phase B studies, August 1967: boxes 1–4, RG 255, accession 69-A-3065.

Federal Records Center, Bell, CA 90201.

NASA, Voyager Project Office, Pasadena.

Closeout records of Voyager Project Office, 1967 and earlier: boxes 35341 through 35364, RG 255, accession 68-A-746.

Jet Propulsion Laboratory.

Voyager history files, 1962–1967: box 1, RG 255, accession 70-A-779.

## THE VIKING ERA

The raw materials for the narrative of the evolution of the Viking Project were found in the files of the Viking Project Office at Langley Research Center. The files were maintained by the General Electric Corporation as part of a documentation support contract with that office. All correspondence and reports were received by the General Electric personnel stationed at Langley, who indexed the documents chronologically, coded them for easy retrieval, and microfilmed them on a Kodak Recordak microfilm-cassette format. The cassettes are stored (as of 1982) in the Langley Research Center Technical Library. All the paper copies of the documents were disposed of in 1978, except some groups that the authors of this history selected for their use. The papers they used in preparation of chapters 5 through Epilogue were turned over to the NASA History Office archivist in the summer of 1982. These documents ultimately will become a part of the Historian's Source Files deposited at the Washington National Records Center by the NASA Headquarters History Office.

In addition to the primary NASA documentation, which the source notes cite, a number of more conventional publications were repeatedly useful. These include:

Adelson, H. E., et al. *The Viking Lander Biology Instrument.* TRW Systems Group Report 21020-6003-RU-00, August 1975. The report describes the operation of each experiment of the biology instrument and the conversion of the experimental concepts to space hardware.

American Geophysical Union. *Scientific Results of the Viking Project.* Washington, 1977. Reprints of articles from *Journal of Geophysical Research* 82 (30 Sept. 1975).

Biemann, Hans-Peter. *The Vikings of '76.* Hans-Peter Biemann, 1977.

Burgess, Eric. *To the Red Planet.* Irvington, N.Y.: Columbia University Press, 1978.

Collins, Stewart A. *The Mariner 6 and 7 Pictures of Mars.* NASA SP-263. Washington, 1971.

French, Bevan M. *Mars: The Viking Discoveries.* NASA EP-146. Washington, 1977.

Goodell, Rae. *The Visible Scientists.* Boston, Toronto: Little, Brown, 1975.

Hartmann, William K., and Odell Raper. *The New Mars: The Discoveries of Mariner 9.* NASA SP-337. Washington, 1974.

Ley, Willy. *Mariner IV to Mars.* New York: New American Library, 1966.

Martin Marietta Corp. *The Viking Mission to Mars.* Denver, 1975.

———, Denver Division. "Viking Lander 'as Built' Performance Capabilities." June 1976. This report was written for readers with a technical background and some familiarity with the Viking mission. Its intent is to describe the "as-built" capabilities of the landers and compare them with capabilities in the environments imposed during separation, entry, descent, and landing. Subsystem components whose performance was essential to mission success through landing are discussed. Particular emphasis is given to items that required performance margins because of uncertainties in the knowledge of Mars at the time.

Martin Marietta Corp. *Viking: Mars Expedition 1976.* Denver, 1978.

Morgenthaler, George W., ed. *Exploration of Mars, Proceedings of the American Astronautical Society Symposium on the Exploration of Mars.* Vol. 15, *Advances in the Astronautical Sciences.* Denver, 1963.

Phillips, Charles R. *The Planetary Quarantine Program: Origins and Achievements, 1956–1973.* NASA SP-4902. Washington, 1974.

Scientific and Technical Information Office, NASA. *Mariner-Venus 1962 Final Project Report.* NASA SP-59. Washington, 1965.

_____. *Spacecraft Sterilization Technology, Beckman Auditorium, Pasadena, California, November 16–18, 1965.* NASA SP-108. Washington, 1966.

_____. *Mariner-Mars 1964 Final Project Report.* NASA SP-139. Washington, 1967.

_____. *Mariner-Mars 1969: A Preliminary Report.* NASA SP-225. Washington. 1969.

_____. *Mariner-Venus 1967 Final Project Report.* NASA SP-190. Washington, 1971.

_____. *Mars as Viewed by Mariner 9: A Pictorial Presentation by the Mariner 9 Television Team and the Planetology Program Principal Investigators.* NASA SP-329. Washington, 1974.

_____. *Viking 1: Early Results.* NASA SP-408. Washington, 1976.

Viking Project Office, Langley Research Center. "Viking Flight Team Organization and Staffing." 23 June 1976.

_____. "Viking Personnel Directory." July 1976.

_____, and Viking Mission Operations, Jet Propulsion Laboratory. "Viking Project Mission to Mars, Viking-1 Mission Timeline." Rev. 1, 7 June 1976.

Viking Lander Imaging Team. *The Martian Landscape.* NASA SP-425. Washington, 1978.

Washburn, Mark. *Mars at Last!* New York: Putnam, 1977.

Werber, Morton. *Objectives and Models of the Planetary Quarantine Program.* NASA SP-344. Washington, 1975.

Some postmission documents will be essential reading for future scholars interested in the history of the Viking Project:

Holmberg, Neil A.; Faust, Robert P.; and Holt, H. Milton. *Viking '75 Spacecraft Design and Test Summary.* NASA RP-1027. Washington, 1980. Vol. 1, *Lander Design;* vol. 2, *Orbiter Design;* vol. 3, *Engineering and Test Summary.*

Goddard Space Flight Center, National Space Science Data Center/World Data Center A for Rockets and Satellites, *Catalog of Viking Mission Data,* Robert W. Vostreys, ed. NSSDC report no. 78–01. Greenbelt, Md., May 1981. This document catalogs available scientific data acquired by the Viking science teams. It is the starting point for anyone wishing to use these materials.

Tucker, Robert. *Viking Lander Imaging Investigation: Picture Catalog of Primary Mission Experiment Data Record.* NASA RP-1007. Washington,

February 1978. A general reference for imaging data from Viking, the volume presents results of procedures applied to the imaging data to produce an organized record as complete and as error-free as possible. It contains all images returned by the two Viking landers during the primary mission. Skyline drawings display the outlines of each image.

# Errata

*On p. 235*, lines 21–22, the statement that the Antarctic dry valleys were considered to be sterile before 1972 is incorrect, as reference 70, this chapter, shows. The same error occurs on p. 236, line 20.

*On p. 235*, lines 25–27 the statement that Norman Horowitz, Roy E. Cameron, and Jerry S. Hubbard failed to detect microorganisms in the soil of the dry valleys of Antarctica in five years of research is incorrect. The cited reference (ref. 70) shows that these authors detected bacteria in 86 percent of soil samples examined; 14 percent were sterile.

# Source Notes

## Introduction

1. This introductory section is based on notes taken by the authors during the mission and tape recordings of communications audio circuits. See Jet Propulsion Laboratory (JPL), "Viking Status Report," 2:00 a.m. PDT, 20 July 1976; and Roy Calvin, "Viking: Intellect and Ingenuity Triumphant," *Martin Marietta Today*, no. 3 (1976): 3.
2. Calvin, "Viking: Intellect and Ingenuity Triumphant," pp. 3–4; and "Viking Status Report," 4:15 a.m. PDT, 20 July 1976.
3. Calvin, "Viking: Intellect and Ingenuity Triumphant," p. 4.

## Chapter 1

1. A recent treatment of Lowell's work is William Graves Hoyt, *Lowell and Mars* (Tucson: Univ. of Arizona Press, 1976). See Brian W. Aldiss, *Billion Year Spree: The True History of Science Fiction* (Garden City, N.Y.: Schocken Books, 1973); and Robert M. Philmus, *Into the Unknown: The Evolution of Science Fiction from Francis Godwin to H.G. Wells* (Berkeley, Los Angeles: Univ. of Calif. Press, 1970).
2. Robert Philmus and David Y. Hughes, eds., *H. G. Wells: Early Writings in Science and Science Fiction* (Berkeley, Los Angeles: Univ. of Calif. Press, 1975), pp. 175–78.
3. Willy Ley and Chesley Bonestell, *The Conquest of Space* (New York: Viking Press, 1949), pp. 105–15; and Alfred Wallace, *Is Mars Habitable?* (London: Macmillan and Co., 1907). Wallace (1823–1913), best known for his independent anticipation of Darwin's work on the origin of species, developed a cogent and amazingly accurate description of the climatic conditions on Mars. He argued that the average temperature would be about -40°C and that few life forms could survive such temperatures. He also predicted that the surface would be cratered and that the polar caps consisted of frozen carbon dioxide.
4. Ernst Stuhlinger et al., eds., *Astronautical Engineering and Science: From Peenemünde to Planetary Space, Honoring the Fiftieth Birthday of Wernher von Braun* (New York, Toronto, London: Mc Graw-Hill, 1963), p. 371; and Wernher von Braun, "Das Marsprojekt: Studie einer interplanetarischen Expedition," *Weltraumfahrt*, special issue, 1952, trans. as *The Mars Project* (Urbana, Ill.: Univ. of Illinois Press, 1953).
5. Von Braun, *The Mars Project*, pp. 2–5, 65–66, 75–76.
6. Ibid, p. 78.
7. See von Braun, "Crossing the Last Frontier," *Collier's* 22 (March 1952): 22–29, one of a series of articles; von Braun et al., *Across the Space Frontier*, ed. Cornelius Ryan (New York: Viking Press, 1952); and von Braun, F[red] L. Whipple, and W[illy] Ley, *Conquest of the Moon*, ed. Ryan (New York: Viking Press, 1953).
8. Von Braun, with Ryan, "Can We Get to Mars," *Collier's* 30 (April 1954): 22–29.
9. Ibid., p. 23. Von Braun's plans for a Mars venture were elaborated in Ley and von Braun, *The Exploration of Mars* (New York: Viking Press, 1956).
10. House Select Committee on Astronautics and Space Exploration, *The Next Ten Years in Space, 1959–1969*, staff rpt., H. doc. 115, 86th Cong., 1st sess. (henceforth 86/1), 2 Jan. 1959, p. 211.
11. Army Ballistic Missile Agency, Development Operations Div., "Proposal: A National Integrated Missile and Space Vehicle Development Program," rpt. D-R-37, 10 Dec. 1957, JSC History Archives.
12. Loyd S. Swenson, Jr., James M. Grimwood, and Charles C. Alexander, *This New Ocean: A History of Project Mercury*, NASA SP-4201 (Washington, 1966), pp. 75–76.
13. NACA, Special Committee on Space Technology, Working Group on Vehicular Program, "A National Integrated Missile and Space Vehicle Development Program," 18 July 1958, JSC History Archives.

14. Robert L. Rosholt, *An Administrative History of NASA, 1958-1963*, NASA SP-4101 (Washington, 1966), pp. 19-70.

15. Eli Ginzberg et al., *Economic Impact of Large Public Programs: The NASA Experience* (Salt Lake City: Olympus, 1976), p. 83.

16. William D. Metz, "Report on Astronomy: A New Golden Age," *Science* 177 (21 July 1972): 247.

17. Ginzberg et al., *Economic Impact*, pp. 81-86ff; J. Tuzo Wilson, *I.G.Y.: The Year of New Moons* (New York: Knopf, 1961), pp. 9-10; and Homer E. Newell interview by Edward C. Ezell, 25 May 1977.

18. Leo Goldberg, "Research with Solar Satellites," *Astrophysical Journal* 191 (1 July 1974): 1-37.

19. Ginzberg et al., *Economic Impact*, p. 90.

20. Singer's role in the IGY satellite project is detailed in Homer E. Newell's *Beyond the Atmosphere: Early Years of Space Science*, NASA SP-4211, 1980; and Constance McLaughlin Green and Milton Lomask, *Vanguard: A History*, NASA SP-4202 (Washington, 1970).

21. Green and Lomask, *Vanguard*, p. 23.

22. Newell discusses the satellite race in chap. 7 of his *Beyond the Atmosphere*.

23. Charles M. Atkins, "NASA and the Space Science Board of the National Academy of Sciences," NASA HHN-62, Sept. 1966, p. 13; Hugh Odishaw to Detlev Bronk, 24 Dec. 1957, "Space Science Board" file, NASA History Off.; and Lloyd V. Berkner and Odishaw, eds., *Science in Space* (New York, Toronto, London: McGraw-Hill, 1961), p. 429. Unless otherwise stated, all unpublished Space Science Board materials are in the NASA History Off. Archives.

24. Newell, *Beyond the Atmosphere*, chap. 8.

25. Ibid.

26. Ibid.

27. Space Science Board, "Minutes of the Second Meeting," 19 July 1958, pp. 14-15.

28. Douglas Aircraft Co., Inc., Missile and Space Systems Div., "The Thor History," Douglas rpt. SM-41860, May 1963, pp. 1-6; and Space Technology Laboratories, "Interplanetary Probes, Tracking, and Communications," proposal 26-17, 27 June 1958, pp. 1-8.

29. Space Science Board ad hoc Committee on Interplanetary Probes and Space Stations, "Minutes," 13 Sept. 1958, pp. 2-3.

30. Odishaw to T. Keith Glennan, 2 Dec. 1958, with encl., Odishaw to Admin. NASA, Dir. NSF, and Dir. ARPA, "Memorandum Report," 1 Dec. 1958, p. 16.

31. R. Cargill Hall, *Lunar Impact: A History of Project Ranger*, NASA SP-4210 (Washington, 1977), p. 15; Harold C. Urey, *The Planets: Their Origin and Development* (New Haven: Yale Univ. Press, 1952); and Robert Jastrow, *Red Giants and White Dwarfs: The Evolution of Stars, Planets and Life* (New York, Evanston, London: Harper and Row, 1967), pp. 3-4.

32. William H. Pickering to James R. Killian, 9 July 1958, NASA History Off. Archives; and Pickering, "The Jet Propulsion Laboratory and the U.S. Space Program," draft [ca. 1958], box 15, White House Off., Off. of Science and Technology, Eisenhower Papers, Dwight D. Eisenhower Library, Abilene, Kans.

33. R. Cargill Hall, *Project Ranger: A Chronology*, JPL/HR-2 (Pasadena, 1971), pp. 14, 46-47, 56, 60; and Clarence R. Gates, "Mars Proposal," JPL, 19 Sept. 1958.

34. John F. Froehlich, "Minutes of Meeting on N.A.S.A. Space Program of October 27, 1958," memo, 28 Oct. 1958, JPL Library Archives.

35. Froehlich, "Minutes of Meeting No. 2 on N.A.S.A. Space Program Held October 28, 1958," memo, 29 Oct. 1958, JPL Library Archives.

36. Froehlich, "Minutes of Meeting No. 3 on N.A.S.A. Space Program Held 29 October 1958," memo, 30 Oct. 1958, JPL Library Archives.

37. V. C. Larsen, Jr., to Abe Silverstein, 7 Nov. 1958, with encl., JPL, "Proposal for Space Flight Program Study," JPLHF-2-260, 7 Nov. 1958; Homer J. Stewart interview, no. 13, 20 Sept. 1972, by J. H. Wilson, JPL Library Archives; Ralph Green to G. W. Green, TWX, 18 Nov. 1958; Hall, *Project Ranger*, p. 73; and J. D. McKenney, "Minutes of Meeting of the NASA Program Study Committee with the Senior Staff, Held Dec. 12, 1958," memo, 15 Dec. 1958, roll 211-2, JPL Vellum Collection (JPLVC).

38. Newell, "Conference Report," memo for record, 18 Dec. 1958.

39. McKenney, "Minutes of the Meeting of the NASA Program Study Committee and the JPL Senior Staff with Dr. Silverstein, Dr. Stewart, Dr. Newell, and Mr. Rosen of NASA . . . January 12-13, 1959," memo, 16 Jan. 1959; and "Meeting of JPL," 12 Jan. 1959, in Newell, "Conference Notes," notebook, box 28, Newell papers.

40. Newell to Silverstein, "Hiring of Zavasky, Schilling, Berning, Mange, and Bordean," 11 Dec. 1958, with encl., "Objectives of Space Science Program," 11 Dec. 1958. The disciplinary approach

continued to be reflected in the documents prepared by Newell's office for several years. See NASA, Off. of Asst. Dir. for Space Sciences, "The NASA Space Sciences Program," 16 April 1959; and NASA, Off. of Space Flight Programs, "Space Flight Programs," 1 Apr. 1960.

41. McKenney, "Minutes of the Meeting," 16 Jan. 1959.
42. Glennan to Hugh L. Dryden, 24 Dec. 1958; and Glennan to Homer J. Stewart, 16 Jan. 1959.
43. Rosholt, *Administrative History*, pp. 130-31, details some of the agency's reluctance to release even the confidential version. See JPL, *Exploration of the Moon, the Planets, and Interplanetary Space*, ed. Albert R. Hibbs, JPL report 30-1 (Pasadena, 1959); NASA, "Minutes of Meeting of Research Steering Committee on Manned Space Flight," 25-26 May 1959; and NASA, Off. of Program Planning and Evaluation, "The Ten Year Plan of the National Aeronautics and Space Administration," 18 Dec. 1959.

# Chapter 2

1. Edgar M. Cortright to Ad Hoc Committee to Name Space Science Projects and Objects, "Revised Suggestions for Lunar and Planetary Program," 18 May 1960.
2. Homer E. Newell, *Beyond the Atmosphere: Early Years of Space Science*, NASA SP-4211, 1980.
3. R. Cargill Hall, *Lunar Impact: A History of Project Ranger*, NASA SP-4210 (Washington, 1977), pp. 6-10; Henry L. Richter, Jr., ed., *Instruments and Spacecraft, October 1957-March 1965*, NASA SP-3028 (Washington, 1966), pp. 31-32, 34-35, 45-46, 81, 123-24, 146; and Peter L. Smolders, *Soviets in Space*, trans. Marian Powell (Guildford, London: Lutterworth Press, 1973), pp. 220-21.
4. Senate Committee on Aeronautical and Space Sciences, *NASA Authorization for Fiscal Year 1960*, hearings on S. 1582, 86th Cong., 1st sess. (henceforth 86/1), pt. 1, 7-10 Apr. 1959, p. 26; General Dynamics Corp., Convair (Astronautics) Div., "Vega, Final Technical Report, Contract NASW-45," report AP60-0294, 25 May 1960; and Jet Propulsion Laboratory, *The Vega Program*, report 30-6 (Pasadena, 1959).
5. Comptroller General of the U.S., "Report to the Committee on Science and Astronautics, House of Representatives, Review of Cancelled Atlas-Vega Launch Vehicle Development, December 1958-December 1959," report B133308, Apr. 1960, pp. 4, 10-12, 29-30, 32-37. See also Homer J. Stewart to T. Keith Glennan, 28 Sept. 1959; Stewart to Glennan, 14 Oct. 1959, box 7, accession 70A5793, Record Group 255, Washington National Records Ctr., Suitland, MD; Paul Means, "Vega-Agena-B Mix-up Cost Missions," *Missiles and Rockets* 6(20 June 1960): 19-20; Evert Clark, "Vega Study Shows Early NASA Problems," *Aviation Week & Space Technology* 72 (27 June 1960): 62-68; and George Kistiakowsky, *A Scientist at the White House: The Private Diary of President Eisenhower's Special Assistant for Science and Technology* (Cambridge: Harvard Univ. Press, 1976), p. 129.
6. Stewart to Glennan, "Consideration of Vega, Agena B and Centaur," 2 Oct. 1959; and R. Cargill Hall, *Project Ranger: A Chronology*, JPL/HR-2 (Pasadena, 1971), p. 127.
7. Hall, *Lunar Impact*, p. 23; Newell, "Trip Report for the Visit to Jet Propulsion Laboratory on 28 December 1959 by Homer E. Newell, Jr., Newell Sanders, J. A. Crocker, Morton J. Stroller," memo for record, 30 Dec. 1959; and Newell notes, "Meeting at JPL, 28 Dec 59, 0900."
8. Richard E. Horner to William H. Pickering, 16 Dec. 1959. The story of ABMA's transfer to NASA is outlined in Robert L. Rosholt, *An Administrative History of NASA, 1958-1963*, NASA SP-4101 (Washington, 1966), pp. 45-47, 107-14.
9. Horner to Pickering, 16 Dec. 1959; and Rosholt, *Administrative History*, pp. 115-16.
10. Horner to Pickering, 16 Dec. 1959.
11. Robert Jastrow to Newell, "Report on December 1 Meeting of the Lunar Science Group," 11 Dec. 1959; and Abe Silverstein to Pickering, 21 Dec. 1959, roll 614-167, JPLVC. See also Pickering, interoffice memo no. 16, 16 Dec. 1969; and J. D. Bruker, memo, Jan. 1959, roll 211-4, JPLVC.
12. Newell, "Trip Report," memo for record, 30 Dec. 1959.
13. Ibid.; and Pickering to Silverstein, 29 Dec. 1959.
14. Senate Committee on Aeronautical and Space Sciences, *NASA Scientific and Technical Programs*, hearings, 87/1, 28 Feb., 1 Mar. 1961, pp. 254-65.
15. House Committee on Science and Astronautics, *Centaur Program*, hearings before Subcommittee on Space Sciences, 87/2, 15 & 18 May 1962, p. 63.
16. Ibid., pp. 4-5, 63-66; John L. Sloop, *Liquid Hydrogen as a Propulsion Fuel, 1945-1959*, NASA SP-4404 (Washington, 1978), chap. 10; and Horner to Cmdr., Air Research and Development Command, 17 Aug. 1959.
17. Silverstein to Ira H. Abbott, "Establishment of Centaur Project Technical Team," 17 Nov. 1959.

18. Stewart to Glennan, "Considerations of Vega, Agena B and Centaur," 2 Oct. 1959; Senate Committee on Aeronautical and Space Sciences, *NASA Authorization for Fiscal Year 1960*, pp. 24–35; Sloop, *Liquid Hydrogen*, chap. 12; and House Committee on Science and Astronautics, *Centaur Program*, pp. 7–8.

19. Robert J. Parks, "Presentation of the Planetary Program," in JPL, "Planetary Program Briefing Summary, July 8, 1960, Revised as Presented," 14 July 1960, pp. 1–2.

20. Ibid., p. 3.

21. Ibid., pp. 4, 20–22.

22. NASA, "Planetary Program Chronology," encl. to "Planetary Program Briefing—October 17, 1962."

23. JPL, "Mariner B Study Report," TM 33-34, March 1961, p. 7; and JPL, "Mariner B Capsule Study, Mars 1964 Mission," Engineering Planning Document 79, 20 Apr. 1962.

24. Ibid.

25. Massachusetts Institute of Technology, Instrumentation Laboratory, "A Recoverable Interplanetary Space Probe," rpt. R-235, 4 vols., 1 July 1959.

26. JPL, "Mariner B Study Report," pp. 8, 12.

27. NASA's outlook on Centaur in the spring of 1960 can be followed in: Horner to Herbert F. York (draft retyped), 7 March 1960; Horner to Bernard A. Schriever, 18 March 1960, with encls.; and Alfred J. Gardner, "Minutes of First Meeting of Centaur Working Committee, Headquarters, NASA, Washington," 4 May 1960.

28. Don R. Ostrander to Glennan, "Centaur Program; Report of Explosion on Vertical Test Stand at Pratt and Whitney," 17 Nov. 1960.

29. House Committee on Science and Astronautics, *Centaur Program*, p. 8.

30. Cortright to Thomas F. Dixon, "Recommendations on the Centaur Program," 17 Jan. 1961. The Able M design is described in Space Technology Laboratories, Inc., "Able Mars Feasibility Study (NASW-246)," 24 July 1961.

31. NASA, "Planetary Program Chronology."

32. Oran W. Nicks to Edward C. Ezell, 6 June 1977. For Mariner R experiments, see: Charles P. Sonett to Jack James, 16 Oct. 1961; Paul J. Coleman to Silverstein, "Notification for Mariner-A Experimenters of Project Cancellation," 17 Oct. 1961; Sonett to James, 26 Oct. 1961; and Coleman to Newell, "Experiments and Experimenters for the 1962 Mariner-R Mission to Venus," 5 Dec. 1961.

33. NASA, "Planetary Program Chronology"; and *Mariner-Venus 1962 Final Project Report*, NASA SP-59 (Washington, 1965), pp. 11–12.

34. Hall, *Lunar Impact*, p. 160.

35. Hall, *Project Ranger*, p. 340; Hall, *Lunar Impact*, pp. 160–77; Ronald J. Ostraw, "Mariner II's Space Triumph Is Result of $13-Million Battle against Odds," *Washington Post*, 17 Dec. 1962; and Jonathan Spivak, "Mariner Probe of Venus Called a Success; Power Failure Cuts Use of Relay Satellite," *Wall Street Journal*, 17 Dec. 1962.

36. NASA, *Mariner-Venus 1962 Final Project Report*, pp. 12–15, 87–120; John E. Naugle, *Unmanned Space Flight* (New York: Holt, Rinehart & Winston, 1965), p. 133; and Marcia Neugebaur et al., "The Mission of Mariner II: Preliminary Observations," *Science* 138 (7 Dec. 1962): 1095–100.

37. NASA, "Planetary Program Chronology"; and Fred D. Kochendorfer to Chairman, Space Sciences Steering Committee, "Scientific Experiments for the 1964 Mariner C Mission to Mars," 18 Mar. 1963.

38. NASA, "Membership of Space Sciences Steering Committee and Subcommittees," circular 125, 7 Feb. 1961; and NASA, "Space Science and Applications Steering Committee," NASA Management Instruction 1152.51, 5 Nov. 1973. See also Space Science Board, "Minutes of the First Meeting, Committee on Chemistry of Space and Exploration of Moon and Planets of the Space Science Board," 13 Nov. 1958; and Space Science Board, "Ad Hoc Committee on Chemistry of Space and Exploration of the Moon and Planets," interim rpt., 8 Jan. 1959.

39. Newell to Hugh Odishaw, 10 Mar. 1961.

40. Odishaw to Newell, 31 Mar. 1961, with encl., "Mariner B Consideration Planetary Atmospheres Study Group," 31 Mar. 1961.

41. The best single source on the Apollo decision is John M. Logsdon, *The Decision to Go to the Moon: Project Apollo and the National Interest* (Cambridge, Mass., and London: MIT Press, 1970).

42. Hall, *Lunar Impact*, p. 120.

43. Rosholt, *Administrative History*, pp. 197–239; and Hall, *Lunar Impact*, pp. 121–22. The changes

included: 1. All of the field centers, including JPL and the new Manned Spacecraft Center to be established in Houston, reported directly to Associate Administrator Robert C. Seamans, Jr., on all matters concerning institutional operations. 2. Headquarters program offices for Advanced Research Programs (Ira H. Abbott), Space Flight Programs (Abe Silverstein), Launch Vehicle Programs (Don Ostrander), and Life Science Programs (Charles Roadman) were abolished. 3. Those offices were replaced by four new ones—Advanced Research and Technology (Abbott), Space Sciences (Newell), Manned Space Flight (D. Brainerd Holmes, brought in from RCA), and Applications (vacant). An agency-wide support office for Tracking and Data Acquisition (Edmond Buckley) was also established. These program offices were to supervise the projects run by the field centers in their areas of responsibility.

44. The Mariner B experiment selection process can be followed in the following representative documents: Newell, memo, 7 Apr. 1961; Sonett to Pickering, 25 Apr. 1961; Roger C. Moore to Silverstein, "List of Tentative Experiments and Experimenters Recommended for 1964 Mariner B Mission to Mars (P-70)," 18 Oct. 1961; Silverstein to Pickering, 19 Oct. 1961; Moore and Coleman to Sonett, TWX, 18 Oct. 1961; Moore to Silverstein, "Research Proposal for Mariner B Instrumentation Support," 25 Oct. 1961; Coleman to Thomas L. K. Smull, "Disposition of Mariner B Proposals," 2 Nov. 1961; Coleman to Smull, "Additional Rejected Mariner-B Proposals," 29 Nov. 1961; Sonett to Parks, 13 Dec. 1961; Moore to Newell, "Revised List of Experimenters Recommended for 1964 Mariner Mission to Mars (P70 and P71)," 3 May 1962; Newell to Pickering, 4 May 1962; and Newell to Harry J. Goett, "1964 Mariner B Mission to Mars," 4 May 1962, with encls.
45. Sonett to Nicks, "Problems Requiring Clarification and/or Decisions," 19 Feb. 1962; and Newell to Pickering, 4 May 1962.
46. Newell to Robert E. Bourdeau, 4 May 1962.
47. Kochendorfer to Parks, 10 May 1962; and Newell to Goett, "Proposed Project Development Plan for Planetary Capsule," 14 May 1962.
48. Coleman to Nicks, "Status of the Scientific Payloads of the Mariner B Bus and Capsule (as of June 1, 1962)," 18 June 1962; Coleman to Newell," A Radio Propagation Experiment for the Mariner-B Mission," 17 Aug. 1962; and Newell to Goett, "Electron Experiment for the Mariner-B Program," 5 Sept. 1962.
49. NASA, "Planetary Program Chronology"; and NASA, Off. of Space Sciences, Mariner Project, "Administrator's Progress Report," 31 Aug. 1962.
50. Glenn A. Reiff, "Mariner for Mars 1964 Meeting," memo, 17 Sept. 1962.
51. Parks to Nicks, 20 Sept. 1962.
52. Ibid.
53. D. L. Forsythe to Newell, "Program Management for Light and Medium Launch Vehicles," 18 Sept. 1962. Also see Erasmus H. Kloman, "Surveyor and Lunar Orbiter; Case Studies of Project Management," 30 June 1970, p. II-5, prepared under contract for the Off. of Space Sciences, MS. in NASA History Off. Archives.
54. Wernher von Braun to Newell, 20 Sept. 1962; and Newell to von Braun, 29 Mar. 1961, in which Centaur payload needs were clearly delineated.
55. Sparks to Newell, 13 Sept. 1962, encl.
56. Ibid.
57. Ibid.
58. Sparks to Newell, 2 Oct. 1962; Thomas F. Dixon to Robert C. Seamans, Jr., "Implementation of Centaur Transfer," 4 Oct. 1962; von Braun to Div. Directors and Off. Chiefs, "Centaur Transfer," 16 Oct. 1962; Seamans to Silverstein and von Braun, "Centaur Development Project," 24 Oct. 1962; J. R. Dempsey to "Full Supervision" distr., "Centaur Reorganization," General Dynamics/Astronautics notice 19, 9 Nov. 1962; Kloman, "Surveyor and Lunar Orbiter," pp. II-9, II-10; and Philip Geddes, "Centaur: How It Was Put Back on Track," *Aerospace Management* 7 (Apr. 1964): 24–29.
59. Newell to Sparks, 23 Oct. 1962.
60. Donald P. Hearth, "Voyager/Mariner B Discussions, 17 December 1962," memo for record, 3 Jan. 1963.
61. Newell to Goett, "Study of Capsule Development for Mariner B," 21 Dec. 1962; and Nicks to Parks, 21 Dec. 1962.
62. Kochendorfer to Nicks, "Redirection of Mariner B," 4 Mar. 1963; and Kochendorfer to Nicks, "Redirection of Mariner B, Addendum I," 14 Mar. 1963.
63. Newell interview by Edward C. Ezell, 25 May 1977; and Kennedy Space Center, *A Summary of Major NASA Launchings*, KSC Historical Rpt. 1, pp. B-6, B-10.

## Chapter 3

1. Charles Darwin letter in Melvin Calvin, "The Origin of Life on Earth and Elsewhere," *Annals of Internal Medicine* 54 (May 1961): 956.
2. Ibid, pp. 956-57.
3. Stanley L. Miller, "Production of Amino Acids under Possible Primitive Earth Conditions," *Science* 117 (15 May 1953): 528-29.
4. Shirley Thomas, *Men of Space: Profiles of Scientists Who Probe for Life in Space*, 6 (Philadelphia, New York: Chilton Co., 1963): 249.
5. Miller and Harold C. Urey, "Organic Compound Synthesis on the Primitive Earth," *Science* 130 (31 July 1959): 245-46.
6. Thomas, *Men of Space*, 6: 250.
7. Ibid., p. 251.
8. Miller and Urey, "Organic Compound Synthesis on the Primitive Earth," p. 251.
9. Joshua Lederberg, "A View of Genetics," *Stanford Medical Bulletin* 17 (Aug. 1959): 123. Lederberg cites A. J. Kluyver and C. B. van Niel, *The Microbe's Contribution to Biology* (Cambridge, Mass., 1956), as the basic overview of comparative biochemistry. "A View of Genetics" was also published in *Science* 131 (29 Jan. 1960): 269-76.
10. Lederberg, "Exobiology: Approaches to Life beyond the Earth," *Science* 132 (12 Aug. 1960): 398. This article later appeared as chap. 9 in Lloyd V. Berkner and Hugh Odishaw, eds., *Science in Space* (New York: McGraw-Hill, 1961), pp. 407-25, and was an adaptation of a talk given at the 1st International Space Science Symposium, Nice, 13 Jan. 1960, and published in H. K. Kallman Bijl, *Space Research: Proceedings of the First International Space Science Symposium (Nice, January, 11-16, 1960)* (Amsterdam: North-Hollard Publishing Co., 1960), pp. 1153-70.
11. Lederberg, "Memo on Initial Planetary Quarantine," 2 Jan. 1976, from Lederberg's personal files.
12. Lederberg, "Lunar Biology?" Dec. 1957, letter to colleague from personal files; Lederberg, "Cosmic Microbiology," Jan. 1958, letter to colleague from personal files; Lederberg to Hugh L. Dryden, 4 Feb. 1958; and Lederberg and Dean B. Cowie, "Moon Dust," *Science* 127 (27 June 1958): 1473-75.
13. National Academy of Sciences, "Minutes of the Council Meeting," 8 Feb. 1958; NAS, "Addendum to Minutes of the Council of the National Academy of Sciences," 8 Feb. 1958; NAS-National Research Council, *A Review of Space Research: The Report of the Summer Study Conducted under the Auspices of the National Academy of Sciences at the State University of Iowa, Iowa City, Iowa, June 17-August 10, 1962* (Washington, 1962), p. 10-11; and Stewart Alsop, "Race for the Moon," *Washington Post*, 21 Feb. 1958.
14. NAS-NRC, *Review of Space Research*, pp. 10-12, 10-13; "Development of International Efforts to Avoid Contamination of Extraterrestrial Bodies," *Science* 128 (17 Oct. 1958): 887-89; and Charles R. Phillips, *The Planetary Quarantine Program: Origins and Achievements, 1956-1973*, NASA SP-4902 (Washington, 1974), pp. 3-7, 9-11.
15. Space Science Board, "Minutes of the Fifth Meeting," 7-9 May 1959.
16. T. Keith Glennan to Odishaw, 13 Oct. 1959.
17. R. Cargill Hall, *Project Ranger: A Chronology*, JPL/HR-2 (Pasadena, 1971), p. 121; and Phillips, *Planetary Quarantine Program*, p. 10.
18. NASA, "Report of National Aeronautics and Space Administration Bioscience Committee," 25 Jan. 1960.
19. Ibid. Harold Urey also pressed for the early flight of life detection devices to Mars; see Urey to Homer E. Newell, Jr., 29 Mar. 1961.
20. Lederberg interview by Edward C. Ezell, 23 Aug., 1977.
21. Harold F. Blum, *Time's Arrow and Evolution* (Princeton, N.J.: Princeton Univ. Press, 1951).
22. Gerald A. Soffen interview by Ezell, 17 Nov. 1978 and 23 Dec. 1978.
23. Newell to Harry H. Hess, 20 Feb. 1964.
24. Richard W. Porter to Hess, 26 Feb. 1964. Porter, a staunch supporter of the U.S space program, noted, "Although I personally consider the objective to be very important, I believe there would be ample scientific reason to explore the planets even if we were sure that no evidence of extraterrestrial life would be found."
25. Lederberg interview, 23 Aug. 1977.
26. For an understanding of both the 1964 Summer Study and the state of scientific knowledge on the eve of the first Mariner flights, the report of the sessions is essential reading: Colin S. Pittendrigh, Wolf Vishniac, and J. P. T. Pearman, eds., *Biology and the Exploration of Mars: Report of a Study Held under the Auspices of the Space Science Board, National Academy of Sciences-National*

*Research Council, 1964-1965*, NAS pub. 1296 (Washington, 1966; the quoted passage is on p. 5); and Elie A. Shneour and Eric A. Ottesen, comps., *Extraterrestrial Life: An Anthology and Bibliography*, NAS pub. 1296A (Washington, 1966).

27. Pittendrigh, Vishniac, and Pearman, *Biology and the Exploration of Mars*, p. 6.
28. Ibid., p. 7. William M. Sinton, "Further Evidence of Vegetation on Mars," *Science* 130 (6 Nov. 1969): 1234-37, argued for the presence of life on Mars but disputed the green color, believing it to be "a complementary hue produced by the bright orange colors of the deserts."
29. Pittendrigh, Vishniac, and Pearman, *Biology and the Exploration of Mars*, p. 7; pt. 5 of this book is devoted to "Some Extrapolations and Speculations," including a "Model of Martian Ecology," pp. 229-42, by Vishniac, K. C. Atwood, R. M. Bock, Hans Gaffron, T. H. Jukes, A. D. McLaren, Carl Sagan, and Hyron Spinrod.
30. Ibid., p. 8.
31. Ibid., p. 8.
32. Ibid., p. 10.
33. Ibid., p. 12.
34. Robert F. Fellows to Abe Silverstein, "Supporting Information re Proposed Contract for the Development of an Instrument for the Detection of Microorganisms on Other Planets," 25 Mar. 1959, from Newell reading file, box 11.
35. Thomas, *Men of Space*, 6: 275; and Hall, *Project Ranger*, pp. 78-79. Vishniac, with his Latvian parents and sister, escaped Germany in 1939 and came to America.
36. Thomas, *Men of Space*, 6: 276.
37. Richard S. Young, Robert B. Painter, and Richard D. Johnson, *An Analysis of the Extraterrestrial Life Detection Problem*, NASA SP-75 (Washington, 1965), p. 4.
38. Ibid.
39. F. Jacob and J. Monod, *Biological Organization at the Cellular and Supercellular Level*, ed. R. J. C. Harris (New York: Academic Press, 1963), p. 1.
40. H. J. Muller, "Genetic Nucleic Acid: Key Material in the Origin of Life," *Perspectives in Biology and Medicine* 5 (Autumn, 1961): 11.
41. Young, Painter, and Johnson, *An Analysis of the Life Detection Problem*, p. 6.
42. Pittendrigh, Vishniac, and Pearman, *Biology and the Exploration of Mars*, p. 12. Lederberg presented an early description of an automated microscope system in "Exobiology: Experimental Approaches to Life beyond the Earth," in Berkner and Odishaw, *Science in Space*, pp. 420-21.
43. Young, Painter, and Johnson, *Analysis of the Life Detection Problem*, p. 6.
44. Pittendrigh, Vishniac, and Pearman, *Biology and the Exploration of Mars*, p. 13.
45. Freeman H. Quimby, ed., *Concepts for Detection of Extraterrestrial Life*, NASA SP-56 (Washington, 1964), with supplementary materials. See also NASA, *Extraterrestrial Life: A Bibliography*, pt. 1: *Report Literature*. A Selected listing of annotated references to unclassified scientific and technical reports, 1952-1964, NASA SP-7015 (Washington, 1964).
46. "Minutes of the Working Group Meeting of the Exobiology Summer Study," 15 June 1964, from Lederberg's personal files.
47. Ames Research Center, Life Detection Team, "A Survey of Life-Detection Experiments for Mars," TMX-54946, Aug. 1963, pp. 29-43, N66-29419; Harold D. Watkins, "Ames Study Supports Mars Life Theory," *Aviation Week & Space Technology* (18 Nov. 1963): 61, 65-66, 71; and Young to Ezell, "Life Sciences Work at Ames," 20 June 1978.
48. Quimby, *Concepts for Detection of Extraterrestrial Life*, p. 11.
49. Ibid., p. 13; and Soffen, "Extraterrestrial Optical Microscopy," *Applied Optics* 8 (July 1969): 1341-47.
50. Ames Life Detection Team, "Survey of Life-Detection Experiments," p. 15.
51. Lederberg interview, 23 Aug. 1977, by Ezell.
52. Vishniac, "Extraterrestrial Microbiology," *Aerospace Medicine* 31 (Aug. 1960): 678-80, reprinted in Shneour and Ottesen, comps., *Extraterrestrial Life: An Anthology and Bibliography*, pp. 282-84.
53. Ames Life-Detection Team, "Survey of Life-Detection Experiments," p. 14.
54. Lederberg, "Multivator: Proposal for Mariner B Experiment-Capsule," submitted in Lederberg, "Cytochemical Studies of Planetary Microorganisms—Explorations in Exobiology," NsG 81-60, 1 Apr. 1960 to 31 Mar. 1961 (N102 172).
55. Ames Life-Detection Team, "Survey of Life-Detection Experiments," p. 12; and E. Levinthal, L. Hundley, and Lederberg, "Multivator—A Biochemical Laboratory for Martian Experiments," in M. Florkin and A. Dolfuss, eds., *Life Sciences and Space Research II* (Amsterdam: North-Holland Publishing Co.; New York: John Wiley & Sons; 1964), pp. 112-23.

56. Ames Life-Detection Team, "Survey of Life-Detection Experiments," pp. 12-13.

57. Gilbert V. Levin interview by Ezell, 31 Aug. 1977.

58. Ames Life-Detection Team, "Survey of Life-Detection Experiments," pp. 9-10; Levin and A. Wendell Carriker, "Life on Mars," *Nucleonics* 20 (Oct. 1962): 71-72; Levin et al., " 'Gulliver'—A Quest for Life on Mars," *Science* 138 (12 Oct. 1962): 114-21; Levin, "Rapid Microbiological Determinations with Radioisotopes," *Advances in Applied Microbiology* 5 (1963): 95-133; and Levin et al., "'Gulliver,' an Experiment for Extraterrestrial Life Detection and Analysis," in Florkin and Dolfuss, *Life Sciences and Space Research II*, pp. 124-32. See also Ira Blei and J. W. Liskowitz, "Review of Concepts and Investigations for the Use of Optical Rotation as a Means of Detecting Extraterrestrial Life," in Florkin, ed., *Life Sciences and Space Research III* (Amsterdam: North-Holland Publishing Co.; New York: John Wiley & Sons; 1965), pp. 86-94.

59. Pittendrigh, Vishniac, and Pearman, *Biology and the Exploration of Mars*, pp. 12-17.

60. Fred D. Kochendorfer to Newell, "Selection of Experiments for Mariner C," 8 Feb. 1963, Lunar and Planetary Programs (SL) chron file.

61. NASA, *Mariner-Mars 1964: Final Project Report*, NASA SP-139 (Washington, 1967), prepared under contract by JPL, provides the best summary of this project from conception to conclusion.

62. Ibid., p. 131; NASA, JPL, *Report from Mars: Mariner IV, 1964-1965*, EP-39 (Pasadena, 1966), p. 21; and Oran W. Nicks, memo for record, "Mariner '64 Bulletin No. 2," 18 Nov. 1964.

63. "Mariner III Agena Shroud," 2 March 1965, SL chron file.

64. NASA, *Mariner-Mars 1964*, pp. 132-54.

65. Robert B. Leighton to Glen A. Reiff, 11 Jan. 1965.

66. William Hines, "U.S. Plans Double-Barrel Mars Shot," *Evening Star* (Washington), 14 Aug. 1964.

67. A sample of *Mariner 4* and related articles includes: "First of 2 Mars Shots Due Next Week," *Evening Star* (Washington), 29 Oct. 1964; Ronald Kotulak, "Space Probes to Mars Set within Month: Seek to Learn if Life Exists," *Chicago Tribune*, 2 Nov. 1964; "Search for Life in Space: Mars Called Prime Target," *New York Herald Tribune*, 17 Nov. 1964; Howard Simons, "Early Search for Mars Life Urged," *Washington Post*, 17 Nov. 1964; J. Allen Hynek, "Probe May Clear Age-Old Mars Mysteries," *Indianapolis Star*, 20 Nov. 1964; "Journey to July; Mariner IV," *Newsweek* 64 (7 Dec. 1964): 67; Walter Sullivan, "Panel Finds Mars Life Likely and Urges Exploration by U.S.," *New York Times*, 27 April 1965; Simons, "Scientists Feel Mars Has Life; Seek Probes," *Washington Post*, 27 April 1965; Simons, "Get Ready for a Peek at the Red Planet," *Washington Post*, 11 July 1965; and "Searching for Life: Mars and a Magnificent Flying Machine," *National Observer*, 12 July 1965.

68. Sullivan, "First U.S. Rocket Aimed for Mars Is Ready for Launching Tomorrow," *New York Times*, 3 Nov. 1964.

69. David Hoffman, "Mariner Takes First Mars Closeups Today," *New York Herald Tribune*, 14 July 1964.

70. NASA, "Initial Scientific Interpretation of Mariner IV Photography," news release 65-249, 29 July 1965.

71. Leighton et al., "Mariner IV Photography of Mars: Initial Results," *Science* 149 (6 Aug. 1965): 627-30, reprinted in Shneour and Ottesen, *Extraterrestrial Life*, pp. 307-12; and Richard K. Sloan, "Scientific Results of Mariner Missions to Mars and Venus," in Virginia Polytechnic Institute, *Proceedings of the Conference on the Exploration of Mars and Venus*, August 23-27, 1965 (Blacksburg, Va., 1965), pp. IX-1 to IX-37.

72. John W. Finnery, "Biologist Backs Space Plan Foes," *New York Times*, 9 June 1963; Senate Committee on Aeronautical and Space Sciences, *Scientists' Testimony on Space Goals*, hearings, 88th Cong., 1st sess. (henceforth 88/1), 10-11 June 1963, pp. 3 ff.; and Thomas, *Men of Space*, 6, pp. 1-27.

73. John Barbour, "Scientist Abelson Raps Race for Man on Moon," *Evening Star*, (Washington), 2 Sept. 1963.

74. Karl Abraham, "Scientist Attacks NASA's Proposed Project to Search for Life on Planet Mars," *Evening Bulletin* (Philadelphia), 28 Dec. 1964.

75. Philip H. Abelson, "The Martian Environment," *Science* 147 (12 Feb. 1965): 683; and Evert Clark, "Scientist Decries Mars Life Search," *New York Times*, 13 Feb. 1965. For Abelson's earlier involvement, see Thomas, *Men of Space*, 6, pp. 1-27; and Abelson to Clark T. Randt, 14 Oct. 1960.

76. Rae Goodell, *The Visible Scientists* (Boston, Toronto: Little Brown & Co., 1977), discusses the general issue of scientists as public promoters and critics of science and the social, economic, and political implications of conducting science in a democratic society.

77. "The Dead Planet," *New York Times*, 30 July 1965.

78. Pittendrigh, Vishniac, and Pearman, *Biology and the Exploration of Mars*, p. 19; and NAS–National Research Council, Space Science Board, *Biology and the Exploration of Mars: Summary and*

*Conclusions of a Study Supported by the National Aeronautics and Space Administration* (Washington, 1965). Academy President Frederick Sietz sent this report to NASA Administrator James E. Webb 16 April 1965.

79. Pittendrigh, Vishniac, and Pearman, *Biology and the Exploration of Mars*, p. 20.
80. Ibid., pp. 20–21.

# Chapter 4

1. Edgar M. Cortright to Ad Hoc Committee to Name Space Science Projects and Objects, "Revised Suggestions for Lunar and Planetary Program," 18 May 1960; and Cortright, "Exploration of the Moon, Planets, and Interplanetary Space," in NASA, *NASA-Industry Program Plans Conference, July 28–29, 1960* (Washington, 1960), p. 71.
2. NASA, *Fourth Semiannual Report to Congress, April 1, 1960 through September 30, 1960* (Washington, 1960), p. 66.
3. Raymond A. Bauer and Richard F. Meyer et al., "NASA Planning and Decision Making, Final Report," vol. 1, rpt. X70-90256, Harvard University, Graduate School of Business Administration, ca. 1970, p. III-4; JPL, "Advanced Planetary Spacecraft Study Report," vol. 1, Engineering Planning Document (EPD) 139, 28 Dec. 1962, p. I-1; JPL, "Study of Mars and Venus Orbiter Missions Launched by the 3-Stage *Saturn* C-1B Vehicle," vol. 2, pts. 1 and 2, EPD-132, 1 Mar. 1963; and JPL, "Study of Mars and Venus Orbiter Missions Launched by the 3-Stage Saturn C-1B Vehicle," vol. 3, pts. 1 and 2, EPD-139, 31 Dec. 1963.
4. Donald P. Hearth to Oran W. Nicks, "Voyager Review at JPL, November 1962," 19 Nov. 1962. Unless noted otherwise, all letters, memos, and telexes are in the chronological files of Hearth's office (NASA Hq. code SL), boxes 50 and 51, accession 255-74-663, record group 255, Federal Records Ctr., Washington. In Nicks to Hearth, 4 June 1962, Hearth is offered the job of Staff Engineer, Advanced Programs, Lunar and Planetary Programs Office.
5. Hugh L. Dryden to James E. Webb, "Current Status of Certain Programs," 31 July 1962.
6. Hearth to Nicks, "Proposed Industry Studies on Voyager in FY63," 19 Nov. 1962.
7. Forrest Hogg to Hearth, 22 Jan. 1963; E. F. Wilcomb to Hearth, 23 Jan. 1963; George N. Mangurian to Hearth, 31 Jan. 1963; R. L. Winters to Hearth, 28 Jan. 1963; Robert W. Menzel to Hearth, 11 Feb. 1963; and Hearth, "OSS Review—January 17, 1963 Advanced Programs Status Report," 17 Jan. 1963.
8. Hearth, "OSS Review," 17 Jan. 1963; and Nicks to Homer E. Newell, Jr., "Proposed Industry Studies on Voyager," 30 Jan. 1963.
9. Hearth, "Presentation for Administrator's Review—9 Feb. 63," n.d.
10. Nicks to Newell, "Proposed Industry Studies on Voyager," 30 Jan. 1963.
11. Carl M. Grey, "Request for Proposal No. 10-929, Voyager Design Studies," memo, 5 Mar. 1963, with encl.
12. Ibid. The schedule and process for evaluating the proposals is outlined in Hearth, "Industry Voyager Studies," memo, 6 Mar. 1963; Hearth to Grey, "Method for Technical Evaluation of Proposals in Response to RFP No. 10-929," 12 Mar. 1963; NASA, "NASA Simplifies Names of Saturn Launch Vehicles," news release 63-26, 7 Feb. 1963; and General Electric Co., Missile and Space Div., Valley Forge Space Technology Center, "Voyager Design Study," vol. 1, "Design Summary," doc. 635D801, 15 Oct. 1963, p. 1-1 (N65-32171).
13. Grey, "Request for Proposal Number 10-929, Voyager Design Studies."
14. Hearth to Grey, "Selection of Contractors for Voyager Studies," 5 Apr. 1963; Grey to Hearth, "Technical Evaluation of Proposals Received in Response to RFP No. 10-929," 4 Apr. 1963; Hearth, note of conversation with Grey and R. Wallace Lord, 4 April 1963; and Hearth and Andrew Edwards, Jr., to Grey, "Technical Evaluation of Proposals Received in Response to RFP No. 10-929," 2 Apr. 1963.
15. House Committee on Science and Astronautics, *1964 NASA Authorization*, hearings before Subcommittee on Space Sciences and Advanced Research and Technology, 88th Cong., 1st sess. (henceforth 88/1), pt. 3(a), Mar.–May 1963, pp. 1620–22; and "Voyager—A $700 Million Program," *Space Daily*, 13 Mar. 1963.
16. Charles A. Taylor to Hearth, 7 May 1963.
17. Hearth to William P. Neacy, 28 June 1963; and Hearth to Taylor, 28 June 1963.
18. Hearth to Nicks, "FY Funding for Voyager," 26 June 1963.
19. Bauer and Meyer, "NASA Planning and Decision Making," 1: III-4 to III-5.
20. Ibid., pp. III-5 to III-6.
21. Lewis D. Kaplan, Guido Munch, and Hyron Spinrad, "An Analysis of the Spectrum of Mars,"

*Astrophysical Journal* 139 (1 Jan. 1964): 1-15; and Alvin Seiff and David E. Reese, Jr., "Defining Mars' Atmosphere—A Goal for the Early Missions,"*Astronautics & Aeronautics* 3 (Feb. 1965): 16-21.

22. Nicks to Newell, "Changes to AVCO and GE Voyager Studies," 9 Aug. 1963; and Senate Committee on Aeronautical and Space Sciences, *NASA Authorization for Fiscal Year 1965*, hearings, 88/2, pt. 1, *Scientific and Technical Programs*, 4 Mar. 1964, pp. 131-33.

23. Nicks, "Planetary Program," in NASA, "Briefing for the Administrator on Possible Expansion of Lunar and Planetary Programs," 2 Dec. 1963; Nicks to Milton Rosen,"Centaur Success," 29 Nov. 1963; Newell to William H. Pickering, TWX, 19 Dec. 1963; and Urner Liddel to Chairman, Space Sciences Steering Committee, "Mars 1966 Mission," 31 Dec. 1963.

24. Hearth to Nicks, "JPL's Plan to Suspend Their Voyager Study," 8 Aug. 1963.

25. Ibid.

26. Hearth, "Mariner Plan and Voyager Concepts," in NASA, "Briefing for the Administrator on Possible Expansion of Lunar and Planetary Programs," 2 Dec. 1963, pp. 34-36.

27. GE Missile and Space Div., "Design Summary," p. 2-1.

28. Ibid., pp. 5-9 to 5-15; and AVCO Corp., Research and Advanced Development Div., "Voyager Design Studies," vol. 1, "Summary," 15 Oct. 1963, pp. 67-74.

29. Hearth, "Mariner Plan and Voyager Concepts," p. 41.

30. Hearth, "OSSA Review—February 24, 1964: Advanced Studies Status Report," n.d.; and Liddel, "OSSA Review, February 24, 1964, Mars 1966 Mission," n.d.

31. Nicks to Liddel et al., "Line Assignments for Mariner 66 Mission Definition," 6 Jan. 1964.

32. Seiff, "Call from Mr. Donald Easter, Office of Lunar and Planetary Programs, NASA Headquarters to Set Meeting for Potential 1966 Mars Mission Experimenters," telephone memo, 7 Jan. 1964; Liddel, "OSSA Review, January 23, 1964, Mariner 66 Mission Definition," n.d.; Seiff, "Call by Mr. John Dimeff and Writer to Mr. Joseph Koukol, Chief, Communication System Development Branch, JPL, concerning Capsule to Bus Communication System for 1966 Mars Mission," telephone memo, 5 Feb. 1964; Liddel to Donald P. Burcham, TWX, 20 Feb. 1964; Liddel, "OSSA Review, February 24, 1964, Mars 1966 Mission," n.d.; and Nicks to Newell, "JPL Manpower Requirements," 13 March 1964.

33. NASA, "NASA Statement on Appropriation Bill," news release 64-221, ca. 30 Aug. 1964. For a general discussion of NASA budgets from 1964 to 1969, see NASA, *Preliminary History of the National Aeronautics and Space Administration during the Administration of President Lyndon B. Johnson, November 1963-January 1969* (Washington, 1969).

34. Hearth to Nicks, "Mariner Missions to Mars in 1969," 18 Mar. 1964.

35. Ibid.; Hearth to Nicks, "Call from Dr. Newell Relative to Mariner '64 Occultation Experiment," 11 May 1964; and Nicks to Chairman, Space Sciences Steering Committee, "Mars Atmosphere Experiment for Mariner '64," 5 June 1964, with encl.

36. Hearth to Nicks, "Mariner Missions to Mars in 1969."

37. House Committee on Science and Astronautics, *Investigation of Project Ranger*, hearings before Subcommittee on NASA Oversight, 88/2, April-May 1964, pp. 211-46. See also R. Cargill Hall, *Lunar Impact: A History of Project Ranger*, NASA SP-4210 (Washington, 1977), pp. 240-55.

38. Robert C. Seamans, Jr., to NASA Management Committee, "Outline for Discussion—Probable Contributing Factors to Project Schedule Slippage," 17 June 1964.

39. For the story of Mariner-Mars 69, consult: Donald P. Burcham, "Listing of Voyager Important Documents & Meetings Relative to Project Direction," 14 Dec. 1964, rev. 14 March 1967 and Oct. 1967; Newell to Pickering, 7 Aug. 1964; Newell to Pickering, TWX, 9 Sept. 1964; Newell to Seamans, "Voyager Program and the Fiscal Year 1966 Budget," 25 Sept. 1965; Hearth, "Discussion of Planetary Programs with the Associate Administrator, September 23, 1964," memo for record, 30 Sept. 1964; Hearth, "Current Actions and Guidelines on the Post-1964 Mars Program," memo for record, 13 Oct. 1964; Nicks to Alvin R. Luedecke, 19 Oct. 1964; Hearth, "OSSA Review of November 3, 1964: Mariner Mars 69," 5 Nov. 1964; Newell to Pickering, 7 Jan. 1965; Lockheed Missiles & Space Co., "Final Report, Mariner Mars 1969 Orbiter Study Contract No. JPL 950877," report M-29-64-1, 4 Oct. 1964 (N65-21458); JPL, "Mariner Mars 1969 Orbiter Technical Feasibility Study," EPD-250, 16 Nov. 1964 (N65-27875); and JPL, "Mariner Mars 1969 Lander Technical Feasibility Study," EPD-261, 28 Dec. 1964.

40. Seamans to Donald F. Hornig, 14 Dec. 1964; and James O. Spriggs to Newell, "Voyager 1969," 7 Dec. 1964.

41. Hearth to Nicks, "Proposed Headquarters Organization for Voyager," 14 Dec. 1964; Hearth to Pickering, TWX, 1 Jan. 1965; Newell to Pickering, 7 Jan. 1965; and Nicks to staff, "SL Organization" 29 Jan. 1965.

42. NASA, "NASA Asks Proposals on Mars Voyager Design Definition," news release 65-15, 15 Jan. 1965.

43. Cortright to Pickering, TWX, 1 Jan. 1965; Hearth, "Mr. Hilburn's Suggestion Relative to Submission by Voyager Phase IA Proposers of Fully Expected Contracts," memo for record, 8 Jan. 1965; George J. Vecchietti to Seamans, "Voyager Procurement Plan," 12 Jan. 1965; Cortright to Pickering, TWX, 13 Jan. 1965; Hearth, "Status of Voyager RFP," memo, 25 Jan. 1965; Newell to Paul Ross, 28 Jan. 1965; Earl D. Hilburn to Newell, "Voyager Procurement Plan," 8 Feb. 1965; Hearth to Carl Schreiber, "Contractor Performance Information on Voyager Proposers," 16 Feb. 1965; Hearth to Nicks, "Status of Voyager Proposals and Evaluation," 24 Feb. 1965; Hearth to C. W. Cole, 10 Mar. 1965; NASA, "NASA Selects Three Firms for Voyager Design," news release 65-135, 22 Apr. 1965; and Hilburn to Newell, "Fixed-Price Contract for Phase IA of the Procurement of a Voyager Spacecraft System by the Jet Propulsion Laboratory," 26 Apr. 1965.

44. William J. Normyle, "Space Sciences Receive Budget Reprieve," *Aviation Week & Space Technology* 82 (18 Jan. 1965): 23-24; NASA, "Background Material and NASA FY 1966 Budget Briefing, 23 January 1965," news release, 25 Jan. 1965; John W. Finney, "99.7 Billion Budget Puts Emphasis on Broader Aid for Schools and Welfare," *New York Times*, 26 Jan. 1965; "Still in the Space Race," *New York Times*, 30 Jan. 1965; George C. Wilson, "Johnson Maps Strong Aerospace Efforts," *Aviation Week & Space Technology* 82 (1 Feb. 1965): 16-18; and Robert Hotz, "The Aerospace Budget," *Aviation Week & Space Technology* 82 (1 Feb. 1965): 11.

45. House Committee on Science and Astronautics, *1966 NASA Authorization*, hearings, 89/1, pt. 1, Feb. and Apr. 1965, pp. 43-44, 278, and pt. 2, Mar. 1965, p. 947; and House Committee on Science and Astronautics, *1966 NASA Authorization*, hearings before Subcommittee on Space Science and Applications, 89/1, pt. 3, Mar. 1965, pp. 154-55.

46. Arvydas J. Kliore et al., "The Mariner 4 Occultation Experiment," *Astronautics & Aeronautics* 3 (July 1965): 72-80.

47. Nicks to Newell, "Voyager Capsule Coordination," 23 Mar. 1965; Newell to Pickering, 31 Mar. 1965; Newell to Floyd L. Thompson, "Voyager Capsule Coordination Group," 7 Apr. 1965; Newell to Wernher von Braun, "Meeting on Martian Capsule Technology," 28 Apr. 1965; and Hearth, "Meeting on Martian Capsule Technology, May 11, 1965," memo, 30 Apr. 1965.

48. Newell, "Establishment of the Voyager Capsule Advisory Group," memo, 14 July 1965.

49. Robert Lindsey, "Mars Atmospheric Probe Proposed," *Missiles and Rockets* 16 (8 Feb. 1965): 13; Rex Pay, "Experts at Odds over Mars Goals," *Missiles and Rockets* 16 (22 Feb. 1965): 39; Michael Yaffe, "Martian Atmosphere Experiment Urged for 1969 Voyager Vehicle," *Aviation Week & Space Technology* 82 (8 Feb. 1965): 61-62; Temple W. Neumann, "The Automated Biological Laboratory," in George W. Morgenthaler and Robert G. Morra, eds., *Unmanned Exploration of the Solar System*, Advances in the Astronautical Sciences, vol. 19 (North Hollywood: American Astronautical Society, 1965), pp. 237-51; and Seiff and Reese, "Defining Mars' Atmosphere—A Goal for the Early Missions," pp. 16-20.

50. Bruce C. Murray, 'A Martian Horror Story: Requirements vs. Capabilities for the Photographic Exploration of Mars," in Morgenthaler and Morra, *Unmanned Exploration of the Solar System*, pp. 153-73. The volume contains several other articles pertaining to Mars exploration. The Cal Tech-JPL Planetary Exploration Study Group studied the Mars exploration program 1964-1965 in an effort to influence NASA policy. A collection of documents is available as "B. C. Murray-NASA-CIT Voyager Dialogue, 1964-1965," JPLHF 3-481.

51. Pay, "Experts at Odds over Mars Goals;" and Irving Stone, "Mariner Data May Limit Voyager Payload," *Aviation Week & Space Technology* 83 (2 Aug. 1965): 55, 57, 60.

52. John E. Naugle to Newell, Nicks, and Hearth, "Interim Report of the Voyager Capsule Advisory Group," 20 Aug. 1965; and Paul Tarver, memo, "Summary Minutes of Meeting No. 1," 27 July 1965.

53. Newell to Pickering, 10 Sept. 1965; and Hearth, "Meeting with Dr. Seamans, Mr. Hilburn, Mr. Rieke on September 3, 1965," memo for record, 14 Sept. 1965.

54. NASA, "NASA Considers Saturn V for Voyager Program," news release 65-328, 14 Oct. 1965; "Major Reorganization Upcoming in Voyager Program," *Missile/Space Daily* 15 (1 Oct. 1965): 151-52; Hal Taylor, "Voyager Program Facing Reorganization," *Missiles and Rockets* 17 (4 Oct. 1965): 14; Howard Simons, "NASA Weighs Major Changes in Voyager's Mission to Mars," *Washington Post*, 9 Oct. 1965; and Seamans to Newell, 8 Oct. 1965.

55. Donald E. Fink, "NASA Reorients Lunar, Planet Missions," *Aviation Week & Space Technology* 83 (27 Dec. 1965): 20; Taylor, "Seamans Crystallizes Voyager Plans," *Missiles & Rockets* 17 (25 Oct. 1965): 17; Fink, "NASA Revamping Voyager Development," *Aviation Week & Space Technology* 83 (1 Nov. 1965): 20; Stone, "Atmosphere Data to Alter Voyager Design," *Aviation Week & Space*

*Technology* 83 (22 Nov. 1965): 66–67, 69; "War to Limit NASA Budget, Teague Says," *Houston Post*, 30 Aug. 1965; Hearth, "Revisions to Planetary Program (Discussions with NASA)," 17 Dec. 1965; and Nicks, "Discussion with Mr. Fink of 'Aviation Week,' on Oct. 7, 1965," memo for record, 14 Oct. 1965.

56. General Electric, Missile and Spacecraft Div., Spacecraft Dept., "Voyager Spacecraft System Study (Phase I—Titan IIIC Launch Vehicle), Final Report," vol. 1, "Summary," doc. 64SD933, 7 Aug. 1964 (N65-24896); and General Electric Co., "Voyager Spacecraft System Study (Phase II, Saturn V Launch Vehicle), Final Report," vol. 1, "Summary," doc. 64SD4376, 9 Dec. 1964 (N66-15824).

57. Newell interview by Edward C. Ezell, 11 Jan. 1978; Cortright to Seamans, "Launch Vehicles for Voyager," 11 Jan. 1965; and V. L. Johnson to Rosen, "Proposed Statement on Launch Vehicle for Voyager," 21 Jan. 1965.

58. Hearth, "Presentation for Administrator's Review," 9 Feb. 1965; and "Saturn V May Boost Two Voyagers," *Missiles and Rockets* 17 (18 Oct. 1965): 15.

59. Seamans to E. C. Welsh, 2 Dec. 1965.

60. Newell interview, 11 Jan. 1978.

61. Bauer and Meyer, "NASA Planning and Decision Making," vol. 1, pp. III-11 to III-12.

62. Ibid., p. III-14. See also Burcham, "Voyager Notes," 23 Jan. 1966, JPL History File (JPLHF) 3-270, JPL Library, for detailed JPL comments on changes in the Voyager project.

63. NASA, "NASA Defers Voyager: Schedules Three Mariners," news release 65-389, 22 Dec. 1965; JPL Off. of Public Information, "NASA Defers Voyager; Schedules Three New Mariners," 22 Dec. 1965, JPLHF 3-165, JPL Library; and Newell to Luedecke, TWX, 22 Dec. 1965.

64. Fink, "Fund Bite Halts Voyager Bus Work: Lander Capsule Design to Continue," *Aviation Week & Space Technology* 84 (3 Jan. 1966): 24; and Hearth, "Voyager Plans," memo for record, 14 Jan. 1966.

65. Bauer and Meyer, "NASA Planning and Decision Making," vol. 1, p. III-16; Newell interview, 11 Jan. 1978; James E. Webb to Clinton P. Anderson, 2 May 1966; and House Committee on Science and Astronautics, *1967 NASA Authorization*, hearings before Subcommittee on Space Science and Applications, 89/2, pt. 3, Feb.–Mar. 1966, pp. 22–23, 223–24.

66. Bauer and Meyer, "NASA Planning and Decision Making," vol. 1, pp. III-16 to III-17; [Hearth], "Webb Review, Planetary Program," 14 Feb. 1966; and [Hearth], "Answers to Nasty Questions, Planetary Programs," 15 Feb. 1966.

67. JPL, "Voyager Presentation to Executive Council," 7 July 1966; JPL, "Voyager Project Study—Backup Data," 7 July 1966; JPL, "Voyager Project Study—Backup Data," 12 July 1966; JPL, "Voyager Mars Mission Science Models," 19 July 1966; JPL, "Voyager Project Study Presentation to NASA—Backup Data," 21 July 1966; JPL, "Voyager Project Study—Presentation to NASA," 14 Sept. 1966; and Newell to Pickering, 5 Oct. 1966.

68. JPL, "Voyager Project Study—Presentation to NASA (Seamans)," 12 Oct. 1966; and Burcham, "Listing of Voyager Important Documents & Meetings Relative to Project Direction," 14 Dec. 1966, and revisions.

69. [Hearth] "Voyager Project Management Chronology," 9 June 1967.

70. Hearth, "Chronology of Voyager Management Decisions," memo for record, 11 Jan. 1967; Hearth, "Chronology of Voyager Management," memo for record, 12 Jan. 1967; Hearth to David G. Stone and Burcham, TWX, 13 Jan. 1967; NASA, "Memorandum of Agreement between OSSA and OMSF on the Voyager Project," 9 June 1967; and Newell to Webb and Seamans, "Voyager Management Decisions," 21 Feb. 1967.

71. Newell to Webb and Seamans, "Voyager Management Decisions"; and NASA, "NASA Organizes Management for Voyager Project," news release 67-40, 28 Feb. 1967.

72. House Committee on Science and Astronautics, *1968 NASA Authorization*, hearings before Subcommittee on Space Science and Applications, 90/1, pt. 3, Mar.–Apr. 1967, pp. 539–40; and Bauer and Meyer, "NASA Planning and Decision Making," vol. 1, p. III-27. The lengthy discussions of the impact of managing Voyager on the size of the JPL staff are documented in JPL, "File on Manpower-Management Negotiations with OSSA, Oct.–Dec. 1966," n.d., JPLHF 3-271, JPL Library.

73. Bauer and Meyers, "NASA Planning and Decision Making," vol. 1, p. III-27.

74. William Hines, "Apollo Fire May Delay Mars Probe till 1979," *Washington Evening Star*, 3 Feb. 1967. See also Jane Van Nimmen and Leonard C. Bruno with Robert L. Rosholt, *NASA Historical Data Book, 1958–1968*, vol. 1, *NASA Resources*, NASA SP-4012 (Washington, 1976), p. 118.

75. Senate Committee on Aeronautical and Space Sciences, *NASA Authorization for Fiscal Year 1968*, hearings, 90/1, pt. 1, Apr. 1967, p. 30.

76. House Committee on Science and Astronautics, *1968 NASA Authorization*, hearings before the Subcommittee on Space Science and Applications, 90/1, pt. 3, Mar.–Apr. 1967, p. 1061.

77. Senate Committee on Aeronautical and Space Sciences, *NASA Authorization for Fiscal Year 1968*, 90/1, rpt. 353, 23 June 1967, p. 35.

78. The fate of the NASA budget authorization for fiscal 1968 can be followed in the *Congressional Record*; NASA, "Chronological History, Fiscal Year 1968 Budget Submission," 8 Nov. 1967; and the press. For the press, see "Budget Looks to Era beyond Moon-Landing," *Chicago Tribune*, 25 Jan. 1967; "Mars Can Wait," *Chicago Tribune*, 5 Mar. 1967; Philip H. Abelson, "The Future Space Program," *Science* 155 (17 Mar. 1967): 1367; "Space Budget Request Kept Nearly Intact by House Unit, but It Faces Tougher Tests," *Wall Street Journal*, 17 May 1967; "Space Agency's Budget Faces House Challenges," *Washington Evening Star*, 22 June 1967; J. V. Reistrup, "House NASA Budget Cut Doubled by Senate Group," *Washington Post*, 24 June 1967; Hines, "NASA Meets the Budget Ax," *Sunday Star*, 25 June 1967; Jerry Kluttz, "NASA Budget Cut $23 Million," *Washington Post*, 27 June 1967; "House and Senate Cut NASA Request," *New York Times*, 29 June 1967; "House and Senate Approve Separate Space Budgets," *Wall Street Journal*, 29 June 1967; "Space Funds Cut Deeply by House, Senate," *Aviation Week & Space Technology* 87 (3 July 1967): 28; Luther J. Carter, "Space Budget: Congress Is in a Critical, Cutting Mood," *Science* 157 (14 July 1967): 170-73; and Edward C. Welsh, "Selling the Space Program," *Aviation Week & Space Technology* 87 (24 July 1967): 11.

79. Hotz, "The Turbulent Summer," *Aviation Week & Space Technology* 87 (31 July 1967): 11; and David Tarr, "Civil Rights," in Congressional Quarterly Service, *Congress and the Nation*, vol. 2, *1965-1968* (Washington, 1969), pp. 354-55.

80. Senate Committee on Appropriations, *National Aeronautics and Space Administration Appropriations for Fiscal Year 1968*, hearings before the Subcommittee, 90/1, [26 July] 1967, pp. 73-77; Katherine Johnsen, "Webb Refuses to Choose Program for Cuts," *Aviation Week & Space Technology* 87 (31 July 1967): 20; Louis Harris, "Space Programs Losing Support," *Washington Post*, 31 July 1967; and " 'Our Priorities Are out of Balance,' " *Washington Post*, 31 July 1967.

81. MSC contracting officer to prospective contractors, "Request for Proposal No. BG721-28-7-528P . . . Planetary Surface Sample Return Probe Study for Manned Mars/Venus Reconnaissance/Retrieval Missions," 3 Aug. 1967.

82. L. R. Magnolia and S. A. Gogin compiled *Manned Mars Missions: A Bibliography*, research bibliography 35, for TRW Space Technology Laboratories (Redondo Beach, Calif., 1965), containing 348 entries. Several hundred more recent references can be obtained through NASA's RECON computerized information retrieval system. See also Robert B. Merrifield, "A Historical Note on the Genesis of Manned Interplanetary Flight," paper XVE 3, Joint National Meeting, American Astronautical Society (15th Annual) and Operations Research Society (35th National), 17-20 June 1969.

83. Normyle, "Priority Shift Blocks Space Plans," *Aviation Week & Space Technology* 87 (11 Sept. 1967): 27.

84. NASA Public Affairs Off., to field centers, TWX, 24 Aug. 1967; Hotz, "New Era for NASA," *Aviation Week & Space Technology* 87 (7 Aug. 1967): 17; Carter, "Space: 1971 Mariner Mission Knifed by Budget Cutters," *Science* 157 (11 Aug. 1967): 658-60; "Fund Cuts Redirect Space Plans," *Aviation Week & Space Technology* 87 (21 Aug. 1967): 23; "NASA Budget Cut $500 Million by House," *Washington Post*, 23 Aug. 1967; "House Votes Fund for Space Agency," *New York Times*, 23 Aug. 1967; and "White House Stand Blocks NASA Budget Restoration," *Aviation Week & Space Technology* 87 (28 Aug. 1967): 32.

85. Thomas O'Toole, "Some of Cut Funds Regained by NASA," *Washington Post*, 4 Oct. 1967; Reistrup, "Senate Boosts NASA Funds," *Washington Post*, 7 Oct. 1967; and Albert Sehlstedt, Jr., "Conferees Vote 4.5 Billion to NASA, None for Voyager," *Baltimore Sun*, 26 Oct. 1967.

86. "People," *Electronics* 40 (4 Sept. 1967): 8, 10.

# Chapter 5

1. Unless otherwise noted, this section is based on James F. McNulty, "The Defining of Mars Project Viking," unpublished typescript [ca. 1976], pp. 27 ff.

2. Ibid., p. 32.

3. Leonard Roberts, "Entry into Planetary Atmospheres," *Astronautics & Aeronautics* 2 (Oct. 1964): 22-29. Also see Roberts, "Probe and Lander Design Problems," *Proceedings of the Conference on the Exploration of Mars and Venus, August 23-27, 1965*, Virginia Polytechnic Institute Engineering Extension Series Circular 5 (Blacksburg, Va., 1965), pp. xvi-1 to xvi-10. For background information on studies related to Mars missions, see the following Langley Working Papers (LWP): Lawrence D. Guy, "Tension-Shell Configurations for Low-Density Entry Vehicles,"

LWP-51, 4 Nov. 1964; Perry A. Newman, "Tables of Thermodynamic Properties of the Six Proposed Mars Atmospheres, VM-1 to 6," LWP-149, 14 Sept. 1965: William M. Adams, Jr., and J. W. Young, "Study of an Unmanned Mars Mission Beginning with Separation of a Lander Capsule from a Fly-by Bus and Ending Near the Surface of Mars," LWP-205, 6 May 1966; Richard N. Green and John F. Newcomb, "A Parametric Analysis of Orbital Geometry for Mars Voyager," LWP-248, 22 July 1966; Richard J. Bendura and Charles H. Whitlock, "A Dispersion and Motion Analysis of the Balloon Launched Phase of the Planetary Entry Parachute Project," LWP-290, 1 Oct. 1966; L. D. Guy and M. S. Anderson, "Technology Programs in Supersonic Decelerators," LWP-307, Oct. 1966; staff, Flight Reentry Programs Off., Langley Research Ctr., "Guidelines for Phase B LRC Voyager Capsule Bus Mission Mode Study," LWP-366, 15 Feb. 1967; Vernon L. Alley, Jr., and Raye C. Mathis, "An Analysis and Computing Program for Three-body Parachute Deployment Dynamics with Specific Applications to the PEPP (Balloon) Program," LWP-398, 20 April 1967; J. L. Humble and W. W. Fernald, "Voyager/Mars Aerodynamic Decelerator Trajectories," LWP-401, 25 April 1967; and Langley Research Ctr. Inhouse Study Team, "Voyager Capsule Bus System Baseline and Mission Mode Description—1973 Mission on Saturn V," LWP-478, 28 Sept. 1967.

4. McNulty, "Defining Project Viking," pp. 44–47, describes the details of the statement of work.

5. "Voyager Components Must Withstand 293F," *Aviation Week & Space Technology* 82 (26 April 1965): 100; Hal Taylor, "JPL to Manage Voyager Lander," *Missiles and Rockets* 16 (3 May 1965): 14; and "AVCO Will Perform Two Separate Studies . . .," *Astronautics & Aeronautics* 3 (June 1965): 108. Also see AVCO, Research and Development Div., "Mars Probe Final Oral Presentation to Langley Research Center, . . . Contract No. NAS1-5224," 1 March 1966, Viking Project Off. Files (VPOF). Unless otherwise noted, letters, memos, telexes, and related documents are from VPOF or Code SL chron. files.

6. McNulty, "Defining Project Viking," p. 75.

7. Parachute deployments at high speeds were tested using test vehicles launched by balloons and Honest John–Nike rockets. The full-scale tests lifted a 4.6-meter simulated entry body to about 40 000 meters, the point at which Earth's atmosphere approached the 10-millibar pressure of Mars. At that altitude, the shell would be released. Twelve small rockets would accelerate it to mach 1.2. The 26-meter test chute would then be deployed and, with an instrument package, would subsequently be separated from the shell. John C. McFall of Langley was NASA project manager for the Planetary Entry Parachute Project. The balloons were launched by the Air Force Cambridge Research Laboratory at the balloon launch facility, Holloman Air Force Base, N. Mex., and Walker Air Force Base, N. Mex. The balloons were fabricated by G. T. Schjeldahl Co., Northfield, Minn. Overall cost of the project was far below that projected for similar tests using the Little Joe II rocket. Honest John–Nikes were launched at the White Sands Missile Range, New Mexico. Consult the following documents: H. Lee Dickinson to Richard T. Mittauer, "AFCRL to Launch Largest High Altitude Balloon Made for NASA's Voyager Program," 14 Apr. 1966; NASA, "NASA to Explore Use of Parachutes for Mars Landing," news release 66-90, 27 Apr. 1966; "Launch of Huge Balloon Delayed," *Washington Evening Star*, 22 June 1966; Walter Sullivan, "Giant Balloon Is Lofted in Test for Mars Landing," *New York Times*, 19 July 1966; NASA, "Martin to Build Planetary Entry Parachute Units," news release 66-229, 25 Aug. 1966; NASA, "First Planetary Parachute Test Planned Aug. 29," news release 66-225, 26 Aug. 1966; NASA, "Parachute Test in New Mexico Complete Success," news release 66-241, Sept. 1966; NASA Off. of Advanced Research and Technology (OART), "Post-Launch Preliminary Report for Small Flight Project, . . . Planetary Entry Parachutes," 7 Sept., 18 Nov., and 21 Nov. 1966; "Chuting for Mars," *Newsweek* 68 (12 Sept. 1966): 59; NASA, "Parachute Entry Experiment Fails in WSMR Launch [rocket flight]," news release 66-292, 10 Nov. 1966; NASA, "Soft-Lander Parachute Test a Success," news release 67-120, 10 May 1967; NASA, "Parachute Tests Reach Halfway Point," news release 67-162, 20 June 1967; NASA, "Large Balloon to Launch NASA Parachute Test," news release 67-170, 6 July 1967; and "Voyager Parachute Test Completed at White Sands," *Space Business Daily*, 23 Oct. 1967.

8. McNulty, "Defining Project Viking," p. 79.

9. Homer E. Newell, Jr., to William Pickering, 14 July 1965.

10. Donald P. Burcham to Roberts, 29 June 1965.

11. McNulty, "Defining Project Viking," pp. 80–82.

12. Floyd L. Thompson to R. W. May, Jr., "Redirection of NASA Contract NAS1-5224, 'Comparative Studies of Conceptual Design and Qualification Procedures for a Mars Probe/Lander,'" 18 Nov. 1965.

13. McNulty, "Defining Project Viking," pp. 87–88; "Impact of Voyager upon the OART Program," ca. 22 Oct. 1965; and Donald E. Fink, "Fund Bite Halts Voyager Bus Work; Lander Capsule Design

to Continue," *Aviation Week & Space Technology* 84 (3 Jan. 1966): 24.

14. McNulty, "Defining Project Viking," pp. 96–97; and AVCO, "Mars Probe Final Oral Presentation," 1 Mar. 1966.

15. McNulty, "Defining Project Viking," pp. 114–16; and McNulty, Daniel B. Snow, and Roberts, "Modal and Conceptual Design Comparisons for the Voyager Capsule," Langley Working Paper-326, 2 Dec. 1966.

16. McNulty, "Defining Project Viking," pp. 118–26; and Donald P. Hearth, "Chronology of Voyager Management Activities and Decisions since September 1966," memo for record, 7 Feb. 1967.

17. McNulty, "Defining Project Viking," pp. 137–38; [NASA], "Biographical Data—James S. Martin, Jr.," n.d.; and Bruce K. Byers, *Destination Moon: A History of the Lunar Orbiter Program*, NASA TMX-3487 (Washington, 1977), pp. 51, 208–10.

18. McNulty, "Defining Project Viking," pp. 137–40.

19. McNulty, "Minutes of the Twenty-fifth Meeting of the Planetary Missions Technology Steering Committee, 1st Session, September 6, 1967, . . . [and] 2nd Session, September 11, 1967," n.d.; Oran W. Nicks to Charles J. Donlan, TWX, 29 Aug. 1967; Nicks to H. Julian Allen, 31 Aug. 1967; and Nicks to Mac C. Adams, "Assistance from NASA Centers in Planning a Planetary Program," 1 Sept. 1967.

20. Eugene S. Love to Clifford H. Nelson, "Comments on In-House Feasibility Study Items in Hand-out of 8/31/67 relating to near Planet Exploration," 1 Sept. 1967.

21. Ibid.

22. Hearth, "Minutes, Voyager Management Committee, September 7, 1967," [8 Sept. 1967].

23. McNulty, "Minutes of the Twenty-fifth Meeting"; and Clifford H. Nelson, "In-House Feasibility Studies—Planetary Exploration Program," rev. 8 Sept. 1967. Also see Anthony J. Calio to R. J. Parks, E. C. Draley, D. G. Newby, and Roberts, TWX, 11 Sept. 1967.

24. James S. Martin, Jr., to Donlan, "OSSA Proposed Planetary Programs," 5 Oct. 1967; and "Planetary Program Extension FY 1968-1969 Program Issues and Options," 9 Oct. 1967.

25. "Planetary Program Extension," 9 Oct. 1967.

26. Ibid.; Hearth to Nicks, "Work Assignments," 18 Oct. 1967; Nicks to Thompson and Martin, TWX, 19 Oct. 1967; Eugene Love to Martin, "Comments on Role of Planetary Missions Technology Steering Committee," 19 Oct. 1967; Martin, "Planetary Projects—Implementation Information," memo, 30 Oct. 1967, with encl., C. W. McKee and R. N. Parker, "Proposed Project Organization and Procurement Implementation Plans for Selected Planetary Programs," 25 Oct. 1967; Hearth to Earle J. Sample and Parks, TWX, 30 Oct. 1967; and Nicks to Webb, "Planetary Program Discussion with the Congress," 7 Nov. 1967.

27. Senate Committee on Aeronautical and Space Sciences, *NASA's Proposed Operating Plan for Fiscal Year 1968*, hearing, 90th Cong., 1st sess. (henceforth 90/1), 8 Nov. 1967, p. 16; and Luther J. Carter, "Planetary Exploration: How to Get by the Budget-Cutters," *Science* 157 (24 Nov. 1967): 1025–28.

28. Carter, "Planetary Exploration"; William J. Normyle, "NASA Pushes Planetary Program," *Aviation Week & Space Technology* 87 (27 Nov. 1967): 16–17: and Normyle, "Planetary Program Support Seen Lacking," *Aviation Week & Space Technology* 87 (11 Dec. 1967): 32–33.

29. "NASA FY 1969 Budget Reclama of BOB Tentative Allowance," 30 Nov. 1967; and [Hearth], "Reclama to FY 1969 Bureau of the Budget Mark," 1 Dec. 1967.

30. "Planetary Exploration Program: Collection of Comments, Policy Statements, etc. (Excerpts from NASA Press Conf. on FY69 Budget, Jan. 29, 1968)," n.d.

31. Ibid.; and Newell to Hugh Odishaw, 10 Mar. 1961.

32. Naugle to Donlan, TWX, 9 Feb. 1968.

33. McNulty, "Defining Project Viking," pp. 164–65; "Fourteen Bids in on New Mars Mission Studies," *Space Business Daily*, 18 Mar. 1968; Martin, "Procurement Planning—Mars '73 Mission," memo, 13 Feb. 1968; Charles W. Cole to Martin, 17 Feb. 1968; McNulty, "Minutes of the Twenty-eighth Meeting of the Planetary Missions Technology Steering Committee," 21 Feb. 1968; A. J. Kullas to John Naugle, 12 Feb. 1968; Hearth to Naugle, "Letter from Martin Marietta Corporation Dated February 12, 1968," 23 Feb. 1968; Martin to LOPO staff, "Mars '73 Mission—Coordination Meetings with JPL," 27 Feb. 1968; Israel Taback, "Coordination Meeting with JPL Personnel regarding Tasks Required for Defining the 1973 Mars Mission," memo for record, 4 Mar. 1968; William J. Boyer to Martin, "Analysis of Proposed Assignment of the Operations System for the Mars '73 Project to the Jet Propulsion Laboratory," 19 Mar. 1968; E. B. Lightner, "Minutes of Mars '73 Mission Planning Meeting, March 26, 1968," 27 Mar. 1968; Martin, "Oral Presentations for Mars '73 Study Contracts," memo, 4 Apr. 1968; Martin to LOPO staff, "Mars 1973 Mission Definition Report No. M73-101-0," 8 Apr. 1968, with encl.; Pickering to Donlan, 17 Apr. 1968;

Langley Research Ctr., "Mars '73 Mission Status: Material Discussed with OSSA/SL on April 17, 1968," n.d.; and Martin, "Mars '73 Statement of Work," memo, 13 June 1968, with encl.

34. Pickering to Donlan, 17 April 1968.

35. Draley, "Langley Research Center Management Proposal for 1973 Mars Mission," memo, 18 Apr. 1968.

36. [Langley Research Ctr.], briefing charts, 20 Apr. 1968.

37. "NASA-LRC/JPL Management Agreement for Advanced Planetary Mission Technology Mars Lander-Mission Study," Aug. 1968; Martin to Edgar M. Cortright, "Meeting with JPL Representatives to Discuss Management Options for the Mars '73 Mission, May 21, 1968," 22 May 1968; Cortright to Draley, "Mars '73 Management," 21 May 1968; Langley Research Ctr., Announcement 29-68, "Change in Organization in the Office of Assistant Director for Flight Projects," 7 June 1968; Martin to Draley, "Mars 73 Mission Design Steering Committee," 10 June 1968; and A. Gustaferro, "Minutes, Mars '73 Mission Design Steering Committee June 5th, 1968, Jet Propulsion Laboratory, Pasadena, California," [11 June 1968].

38. "NASA Hopes to Remain in Orbit by Selling Congress on Mars, Venus, Mercury Probes," *Wall Street Journal*, 26 Jan. 1968; NASA, "Background Material, NASA FY 1969 Budget Briefing," news release, 29 Jan. 1968.

39. Philip M. Boffey, "LBJ's New Budget: Another Tight Year for Research and Development," *Science* 159 (2 Feb. 1968): 509; William Leavitt, "A Dreary Season for NASA," *Air Force and Space Digest* 51 (Feb. 1968): 72-76; "Physicist Says Space Budgets Cuts 'Kill' Scientific Exploration," *New York Times*, 24 Apr. 1968; and Joel A. Strasser, "Tight Budget Spurs New Look at Low-Cost Planetary Plans," *Aerospace Technology* 21 (6 May 1968): 22-25.

40. NASA Off. of Admin., Budget Operations Div., "Chronological History Fiscal Year 1969 Budget Submission," 14 Oct. 1968; Carter, "Space Budget: Down 20 Percent in 1 Year—at Least," *Science* 160 (10 May 1968): 634; "President Limits NASA to $4.008 Billion Appropriation," *Space Business Daily*, 20 May 1968; Jerry Kluttz, "NASA, Glamor Gone, Becomes Pet Target for Economy Cuts," *Washington Post*, 27 May 1968; Richard D. Lyons, "NASA Will Drop 1,600 Men and Cut Projects," *New York Times*, 9 Aug. 1968; John B. Campbell, "Is NASA Viable?" *Space Astronautics* 52 (July 1968): 45; Victor Cohn, "U.S. Science Is Feeling Budget Pinch," *Washington Post*, 4 Aug. 1968; Neal Stanford, "Budget Cuts About U.S. Space Plans," *Christian Science Monitor*, 12 Aug. 1968; "War's Effect on U.S. Budget to Continue for Years after Peace, Humphrey Is Told," *Wall Street Journal*, 14 Aug. 1968; and Executive Off. of the President, Bureau of the Budget, "Summer Review of the 1969 Budget," Sept. 1968.

41. NASA, "NASA Interim Operating Plan," news release 68-141, 8 Aug. 1968.

42. Hearth to Naugle, "Background on Titan Mars '73 for Congressional Hearings on the Interim Operating Plan," [ca. Sept. 1968]; Martin to Draley, "Mars 73 Mission and Related Items Discussed during Telephone Conversation with Mr. Don Hearth on June 28, 1968," 28 June 1968; and Martin, "FY69 Planetary Program Information," memo for record, 26 June 1968.

43. Harry H. Hess to Webb, 3 Nov. 1967; and Peter L. Smolders, *Soviets in Space*, trans. Marian Powell (Guildford, London: Lutterworth Press, 1973), pp. 227-28.

44. Hess to Webb, 3 Nov. 1967, and 14 Dec. 1967. See Webb to Hess, undated letter not sent; Jim Long to Nicks, 7 Feb. 1968; Nicks to Hearth, 12 Feb. 1968; and Naugle to Webb, "Comments on Letters of Dr. Hess from the Space Science Board (SSB) concerning NASA Programs," 16 Feb. 1968.

45. National Academy of Sciences-National Research Council, *Planetary Exploration 1968-1975: Report of a Study by the Space Science Board*, (Washington, 1968); "Space Science Board Recommends Expanded Planetary Effort," *Space Business Daily*, 15 Aug. 1968; "Academy Renews Its Stand against Manned Exploration," *Space Business Daily*, 16 Aug. 1968; "Space Science Board Emphasizes Mars Surface Studies," *Space Business Daily*, 19 Aug. 1968. More detailed comments on relations between NASA and the scientific community are found in Raymond A. Bauer, Richard F. Meyer, et al., "NASA Planning and Decision Making, Final Report," vol. 1, rpt. X70-90256, Harvard Univ. Grad. School of Business Admin. [ca. 1970], pp. III-37 to III-39.

46. Webb to Norman F. Ramsey, 14 Jan. 1966. See "NASA Ad Hoc Science Advisory Committee Report to the Administrator," 15 Aug. 1966, pp. 27-30, OSSA files, for a list of issues that Webb asked Ramsey's committee to address itself to. For a general overview of the Lunar and Planetary Missions Board's history see Barry Rutizer, "The Lunar and Planetary Missions Board," HHN-138, 30 Aug. 1976, NASA History Off. Archives.

47. "NASA Ad Hoc Science Advisory Committee Report," pp. 1-25.

48. For discussion of the reaction, see Rutizer, "Lunar and Planetary Missions Board," p. 12. See also Newell, "Interim Response to the Report of the Ad Hoc Science Advisory Committee," 7 June 1967, OSSA files.

49. Rutizer, "Lunar and Planetary Missions Board," pp. 13–14.
50. Ibid., p. 16.
51. Ibid., p. 19 and note 72.
52. John W. Findlay to Newell, 11 Oct. 1967; Newell to Findlay, 24 Oct. 1967; and Findlay to Webb, 2 Nov. 1967.
53. Wolf Vishniac to Findlay, 7 Nov. 1967; Lester Lees to Findlay, 10 Nov. 1967; Gordon J. F. MacDonald, 13 Nov. 1967; Findlay to Webb, 11 Dec. 1967; Webb to Findlay, n.d.; and Jesse L. Greenstein to Findlay, 15 Dec. 1967, personal files of Findlay and OSSA files. See also Carter, "Planetary Exploration: How to Get by the Budget-Cutters," p. 1026, for a summary of the NASA–Mission Board disagreement. Carter's article in *Science* (note 27) reflects one disgruntled member of the board.
54. A. Thomas Young to Martin, "Mars Panel Meeting Summary," 14 June 1968; and George C. Pimentel to Martin, 5 June 1968, with encl.
55. Taback and Young to Martin, "Mars Panel Meeting, June 28, 1968," 2 July 1968; Young to Martin, "Report on Lunar and Planetary Missions Board Meeting," 19 July 1968; Young to G. C. Broome and J. C. Moorman, "Candidate Objectives for Mars '73 Mission," 19 July 1968; Young to Martin, "Report on Lunar and Planetary Missions Board Meeting, September 5 and 6, 1968," 10 Sept. 1968; Hearth to Naugle, "Solicitation to the Scientific Community for Participation in the Development of Instruments for Mars Landers," 10 Sept. 1968; and Findlay to Newell, 17 Sept. 1968 and 11 Oct. 1968.
56. "Mars Panel Report," 5 Oct. 1968, with encl., Young, "Science Critique Meeting, October 3, 1968," memo for record, 18 Oct. 1968; and Young, "Lunar and Planetary Missions Board Meeting, October 4–5, 1968," memo for record, 18 Oct. 1968. For discussion of the launch vehicle and the lander decision, see Gustaferro, "Minutes of July 29, 1968, Briefing on Mars '73 Mission Planning Status to Mr. Edgar M. Cortright," memo for record, 31 July 1968; Martin to R. N. Conway, "Mars '73 Research and Development POP 68-2 Funding Requirements for OSSA," 13 Aug. 1968; G. W. Brewer to Martin, "Request for SRT Support of a Study for Assessment of Technical and Cost Impact of a Mars '73 Hard Lander," 20 Aug. 1968; C. F. Mohl, "Minutes of the 3d Orbiter Design Team Meeting Held 29 August 1968," memo, 30 Aug. 1968; Gustaferro, "Minutes of August 29, 1968, Briefing on Titan Mars '73 Mission Planning Status to Mr. Edgar M. Cortright," memo for record, 3 Sept. 1968; Cortright to Naugle, "Titan/Mars '73 FY 1969 Project Approval Document (Code 84-840-815)," 7 Oct. 1968; M. G. Dietl, "Minutes of the 7th Orbiter Design Team Meeting Held on 10 October 1968," memo, 11 Oct. 1968; and Hearth to Naugle, "Solicitation to the Scientific Community for Participation in the Development of Instruments for Mars Landers," 10 Sept. 1968, with encl., "Solicitation for Participation in the Development of Instruments for Mars Landers," draft, 3 Sept. 1968.
57. Boeing Co., "Study of Powered Spacecraft for Mars Missions," D2-140028-5, 2 vols., Sept. 1968, 68N34279; General Electric Co., Missile and Space Div., "Titan/Mars Hard Lander," 68SD7041, 2 vols., 6 Jan. 1969, 69N19165-6; General Electric Co., Missile and Space Div., "Mars Hard Lander Capsule Study," 68SD952-1-8, 4 vols., 31 July 1968, 68N35980-7; General Electric Co., Missile and Space Div., "Final Report Direct Versus Orbital Entry for Mars Mission," 68SD4293, 3 vols., 1 Aug. 1968, privately funded rpt.; Hughes Aircraft Co., Space Systems Div., "Mars Spinning Support Module Study," NASl-8638, contractor rpt. CR66732-1, CR66732-2, 2 vols., 31 Jan. 1969, 69N19945-6; Martin Co., "Study of Direct Versus Orbital Entry for Mars Missions," NASl-7976, 6 vols., Aug. 1968, 68N31831-6; McDonnell Douglas Corp., McDonnell Astronautics Co., "Soft Lander Mars Lander Capsule Study (Entry from Orbit), Final Report" [prelim. draft], G346, 1 July 1968; McDonnell Douglas Corp., McDonnell Astronautics Co., "Soft Lander," NASl-7977, 9 pts., Sept. 1968, 68N34050-8; Martin Co., "Study of a Soft Lander/Support Module for Mars Missions," NASl-7976, vols. 1–3, 7, Jan. 1969, 69N15600-2, 68N37340.
58. Young, "Titan Mars '73 Mission Mode Meetings Summary," memo for record, 14 Nov. 1968, with encl., "Summary Titan Mars '73 Mission Mode Meeting, November 8-9, 1968, at Langley Research Center," n.d.; Langley Research Ctr., "Titan Mars '73 Mission Mode Briefing, November 7-8, 1968," 7 Nov. 1968; and W. I. Watson, "Viking Project Phase B Report," M63-110-0 [Circa Nov. 1968].
59. Young, "Titan Mars '73 Mission Mode Meetings Summary," 14 Nov. 1968.
60. Naugle interview by Ezell, 17 Apr. 1978. See also Hearth to Parks, TWX, 12 Nov. 1968; Naugle, "Review of Titan Mars 73 Options," memo, 12 Nov. 1968; Naugle to Thomas O. Paine, "Reprogramming of FY 1969 Funds for Titan Mars 1973," 13 Nov. 1968; Hearth, memo, 18 Nov. 1968, with encl., "Actions Required That Have Been Deferred Pending Resolution of FY70 Budget," 18 Nov. 1968; Martin to multiple addressees, TWX, "ASPO Review of the Draft Copy of the JPL Task

Order Dated November 11, 1968," 25 Nov. 1968; Julian Scheer to Paine, 2 Dec. 1968; NASA, "Scientific Payloads for Mars '73," news release 68-207, 5 Dec. 1968; Naugle to Scheer, "Titan Mars 1973 Orbiter/Lander," 22 Nov. 1968; and Naugle, "Project Viking," 23 Dec. 1968.

61. Robert F. Allnutt to Joseph E. Karth, 4 Dec. 1968.

62. Bauer and Meyer, "NASA Planning and Decision Making," vol. 1, p. III-45.

63. Allnutt to Karth, 4 Dec. 1968.

64. Langley Research Ctr, Announcement 60-68, "Establishment of Interim Viking Project Office," 6 Dec. 1968

# Chapter 6

1. Allen E. Wolfe and Henry W. Norris, "The Viking Orbiter and Its Mariner Inheritance," in *Proceedings of the Twelfth Space Congress: Technology Today and Tomorrow* (Cocoa Beach, Fla., 1975), pp. 6-17 and 6-27.

2. The initial project is described in Jet Propulsion Laboratory, "Mariner Mars 1969 Orbiter Technical Feasibility Study," EPD-250, 16 Nov. 1964, N65-27875; and JPL, "Mariner Mars 1969 Lander Technical Feasibility Study," EPD-261, 28 Dec. 1964. See also NASA, "NASA Defers Voyager: Schedules Three New Mariners," news release 65-389, 22 Dec. 1965; and Homer E. Newell to Alvin R. Luedecke, TWX, 22 Dec. 1965.

3. JPL, *Mariner Mars 1969 Final Project Report: Development, Design and Test*, vol. 1, JPL TR 32-1460 (Pasadena, 1976), p. 199.

4. Ibid., pp. 7-8; and William R. Corliss, "A History of the Deep Space Network," NASA CR-15195, Nov. 1976, p. 199.

5. JPL, *Mariner Mars 1969 Final Project Report*, 1: 9.

6. James H. Wilson, *Two over Mars: Mariner VI and Mariner VII, February to August 1969*, NASA EP-90 (Pasadena: JPL, 1971), pp. 11-12.

7. Ibid., p. 13; and JPL, *Mariner Mars 1969 Final Project Report*, 1: 12-13.

8. NASA, "Background Material NASA FY 1969 Budget Briefing," news release, 29 Jan. 1968; NASA, "NASA Interim Operating Plan," news release 68-141, 8 Aug. 1968; and NASA Off. of Administration, Budget Operations Div., "Chronological History Fiscal Year 1969 Budget Submission," 14 Oct. 1968.

9. NASA, "JPL Building Mariners," news release 68-196, 14 Nov. 1968.

10. Oran W. Nicks, "Applying Surveyor and Lunar Orbiter Techniques to Mars," paper, American Institute of Aeronautics and Astronautics meeting, Washington, D.C., 5 Dec. 1968.

11. Robert A. Schmitz to A. Thomas Young, "Wave of Darkening," 20 May 1969, with encl., Schmitz, "Martian Seasonal Darkening," 9 May 1969; James S. Martin, Jr., "Martian Wave of Darkening Phenomena," memo, 27 May 1969; Duwayne M. Anderson to Martin, 4 June 1969; Melvin Calvin, 5 June 1969; Schmitz to Young, "Martian Wave of Darkening Phenomena relative to the Mars 1973 Mission," 31 July 1969; and Gerald A. Soffen, "Information on the Martian Wave of Darkening Requested at June 10-11, 1970 Science Steering Group Meeting," memo, 26 June 1970.

12. Schmitz to Young, "Position of Martian Polar Caps during the Period January–August 1974," 26 Aug. 1969.

13. JPL, *Mariner Mars 1971 Project Final Report: Project Development through Launch and Trajectory Correction Maneuver*, vol. 1, JPL TR 32-1550 (Pasadena, 1973), pp. 2-3, N73-20855; and E. S. Travis, *Mariner Mars 1971 Mission Specification and Plan: Mission Specification*, vol. 1, PD-610-16, rev. A, change 1 (Pasadena: JPL, 1971).

14. Nicks, "Applying Surveyor and Lunar Orbiter Techniques to Mars."

15. JPL, *Mariner Mars 1971 Project Final Report*, 1: 3.

16. Ibid., pp. 3-4; and Dan Schneiderman et al., "The First Mars Orbiters," *Astronautics & Aeronautics* 8 (April 1970): 65.

17. Casper F. Mohl telephone interview by Edward C. Ezell, 24 Oct. 1978; Mohl, "Minutes of the 1st M73 Orbiter Design Team Held 8 August 1968," memo, 12 Aug. 1968; and Mohl, "Minutes of the 2nd Orbiter Design Team Meeting Held 22 August 1968," memo, 23 Aug. 1968.

18. Mohl, "Minutes of the 2nd Orbiter Design Team Meeting"; Mohl, "Minutes of the 3rd Orbiter Design Team Meeting Held 29 August 1968," memo, 30 Aug. 1968; Martin G. Dietl, "Minutes of the 7th Orbiter Design Team Meeting Held 10 October 1968," memo, 11 Oct. 1968; and Dietl, "Minutes of the 9th TM'73 Out-of-Orbit Design Team Meeting Held 21 November 1968," memo, 4 Dec. 1968.

19. JPL, "Titan Mars 1973 Orbiter: Option I Minimum Orbiter Study," 9 Oct. 1968; and JPL, "Titan Mars 1973 Orbiter Option II," 30 Oct. 1968.

20. Charles W. Cole to Martin and Donald P. Hearth, TWX, 12 Nov. 1968; and Dietl, "Minutes of the 10th TM'73 Out-of-Orbit Design Team Meeting Held 5 December 1968," memo, 6 Dec. 1968.
21. Cole to Martin, "Viking Baseline Orbiter Conceptual Design Description Document," 7 Feb. 1969; and JPL, "Viking Project Orbiter System Visual Presentation," 13-14 Feb. 1969.
22. L. I. Mirowitz to Martin, "Viking Orbiter Presentation, February 14, 1969" [21 Feb. 1969].
23. A. J. Kullas to Martin, 21 Feb. and 28 Feb. 1969; and T. T. Yamauchi to Martin, "Informal Comments on Orbiter-Lander Interfaces," 19 Mar. 1969.
24. See enclosures to S. R. Schofield, "Minutes of the 17th Viking 1973 Orbiter Design Team Meeting Held 20 March 1969," memo, 24 Mar. 1969.
25. JPL news release, 17 April 1969; and JPL, "Henry W. Norris, Biographical Sketch," n.d.
26. JPL news release, 17 April 1969; and "Conway W. Snyder, Biographical Sketch," 1 June 1976.
27. Norris, "Viking Project Meeting Minutes," memo, 1 Apr. 1969.
28. Cole to Martin, "JPL Resource Requirements for Viking Project," 10 Feb. 1969.
29. Schofield, "Minutes of the 17th Viking 1973 Orbiter Design Team Meeting"; Norris, "Viking Project Meeting Minutes, April 9, 1969," memo, 11 Apr. 1969; idem for April 16, dated 21 Apr. 1969; William H. Pickering to Edgar M. Cortright, 21 Apr. 1969; Norris, "Viking Orbiter Project Staff Meeting, Minutes of May 7, 1969," memo, 9 May 1969; idem for May 14, dated 19 May 1969, and for May 21, dated 26 May 1969.
30. Norris, "Viking Orbiter Staff Meeting, Minutes of June 4, 1969," memo, 9 June 1969; Rolf C. Hastrup and Richard Case, "Minutes of 26th Viking 1973 Orbiter Design Team Meeting Held 12 June 1969," memo, 17 June 1969; Donald H. Kindt to Raymond S. Wiltshire, "Action Items 3, 5 and 8 from the Spacecraft Interface and Integration Working Group Meeting of June 10 and 11," 23 June 1969; and Joseph Shaffer, Jr., to Wolfe, "Action Item 11 (Possible Reduction of VLV/VO Dynamic Envelope) from Spacecraft Interface and Integration Working Group Meeting of 10-11 June 1969," 24 June 1969.
31. Norris, "Viking Orbiter Project Staff Meeting, Minutes of August 27, 1969," memo, 29 Aug. 1969.
32. Martin to Viking Project Management Council (VPMC), "Establishment of VPMC," 10 Mar. 1969; and Martin, "Transmittal of Minutes of the Viking Project Management Council," memo, 29 Sept. 1969, with encl., Fred W. Bowen, Jr., "Minutes, Viking Project Management Council Meeting, Denver, Colorado, August 18-19, 1969," n.d.
33. Norris, "Viking Orbiter Project Staff Meeting, Minutes of August 20, 1969," memo, 22 Aug. 1969.
34. Langley Research Ctr., Viking Project Off. (VPO), "Viking Project Mission Definition No. 2," M73-112-0, Aug. 1969. Following is a summary of the revisions of this document through Aug. 1969:

| Revision | Date | Description |
|---|---|---|
| 1 | 5/22/68 | Table of Contents and Introduction added. Other sections amplified, including minor modifications. Lander lifetime requirements changed. |
| 2 | 8/12/68 | With exceptions of sections 1.0, 2.0, and 3.0, document completely revised in accordance with new guidelines. |
| 3 (Mission Definition No. 1) | 1/21/69 | Document almost completely revised to reflect selection of mission mode for Viking Project. Cover and title sheet revised to reflect establishment of Viking Project Office. |
| 4 (Mission Definition No. 2) | 6/23/69 | Sections 3, 4, 6, and 12 revised to reflect action of Viking Science Steering Group. Section 5 added. Section 10 revised to include 3 85-foot dish antennas. Section 2 and 11 updated to reflect events since last revision. Table of Contents and Preface revised accordingly. |
| 5 (Mission Definition No. 2) | 8/11/69 | Sections 1 and 2, minor word changes. Section 3 revised to include last listed orbiter objective and necessary discussion of objective. Section 4 revised to include orbiter science requirements. Sections 4 and 5 revised to include modifications of entry and lander science requirements based on July Science Steering Group meeting. Section 6 changed landing zone to 20°S-30°N. Date for Mission Definition no. 2 changed in section 11. |

35. VPO, "Viking Project Definition No. 2," p. 18; and VPO, "Viking Lander Science Instrument Teams Report," M73-112-0, Aug. 1969.

36. Norris, "Viking Orbiter Project Staff Meeting, Minutes of August 27, 1969," memo, 29 Aug. 1969.

37. VPO, "Viking Project Definition No. 2," pp. 18-19; and Hastrup, "Minutes of 7th Orbiter Mission Design Team Meeting of 2 September 1969," memo, 3 Sept. 1969.

38. NASA, "Progress Report—Mariner Mars 1969 Mission (Mariner 6 and Mariner 7)," transcript of press conference, 11 Sept. 1969.

39. Robert B. Leighton, Norman H. Horowitz, Bruce C. Murray, Robert P. Sharp, Alan H. Herriman, A. Thomas Young, Bradford A. Smith, Merton E. Davies, and Conway B. Levoy, "Mariner 6 and 7 Television Pictures: Preliminary Analysis," *Science* 166 (3 Oct. 1969): 49-67 (also idem, "Television Observations from Mariners 6 and 7," *Mariner Mars 1969: A Preliminary Report*, NASA SP-225 (Washington: 1969), pp. 37-82). See also Leighton et al., "Mariner 6 Television Pictures: First Report," *Science* 165 (15 Aug. 1969): 684-90.

40. Leighton et al., "Mariner 6 and 7 Television Pictures."

41. NASA, "Progress Report."

42. Ibid.

43. Leighton et al., "Mariner 6 and 7 Television Pictures," pp. 66-67.

44. R. A. Schmitz, "Minutes, Viking Science Steering Group Meeting in Washington, D.C., on September 11, 1969."

45. JPL, "NASA Viking Orbiter Science Briefing," 12 Sept. 1969.

46. Norris, "Viking Orbiter Project Staff Meeting, Minutes of Sept. 10, 1969," memo, 12 Sept. 1969.

47. Martin Marietta Corp., Denver Div., "Viking Project Quarterly Review Held October 7 & 8, 1969 at Langley Research Center: Presentation Material," PM-3700005, Oct. 1969.

48. Harper E. Van Ness to Martin, "The Center Directors' Meeting on October 9, 1969," 15 Oct. 1969; Van Ness to Martin, "Viking Project Management Council Meeting on October 9, 1969," 15 Oct. 1969; and Bowen, "Minutes, Viking Project Management Council Meeting Langley Research Center, October 9, 1969" [21 Oct. 1969].

49. Walter Jakobowski, "LPMB Meeting on September 13, 1969," memo, 19 Sept. 1969, with encl., "Attachment B [resolution adopted by Lunar and Planetary Missions Board, 13 Sept. 1969]."

50. Nicks to John E. Naugle, "Discussion with Dr. Paine on the Viking Program, May 29, 1969," 4 June 1969; and Jakobowski, "Briefing to Dr. Paine on May 28, 1969: Viking Funding Estimates," memo for record, 2 June 1969.

51. André J. Meyer, Jr., notebooks on advanced planning, 31 July 1969, JSC History Archives.

52. Homer E. Newell, "Conference Report of Space Task Group Meeting of March 22," 25 Mar. 1969; NASA, "America's Next Decade in Space," draft, 9 July 1969, pp. i-iii; NASA, *America's Next Decades in Space: A Report for the Space Task Group*, Sept. 1969; and Hearth to Newell, "Study of Manned Planetary Program Alternatives," 8 Apr. 1969. The language in NASA's draft report for the Space Task Group said," ". . . we recommend that the United States begin preparing for a manned expedition to Mars at an early date." In the published version, the sentence was changed to, "Manned expeditions to Mars could begin as early as 1981." A series of articles and editorials was submitted by Rep. Ryan to *Congressional Record*, 13 Aug. 1969: "After Apollo—Mars?" *New York Times*, 18 July 1969; John A. Hamilton, "Meanwhile, back on Earth," *New York Times*, 28 July 1969; and "Poll Finds Public Cool to Mars Trip—Opinions Split by Age, with Young Adults in Favor," *New York Times*, 7 Aug. 1969. See also Clinton P. Anderson, "Future Space Goals," *Congressional Record*, 29 July 1969, p. S8739.

53. Space Task Group, *The Post-Apollo Space Program: Directions for the Future* (Washington, 1969); "Nixon Backs Mars Flight but Rejects All-Out Drive," *New York Times*, 16 Sept. 1969; and William Leavitt, "A Muted Martian Manifesto," *Air Force/Space Digest* 52 (Nov. 1969): 51-54.

54. Naugle to Cortright, "The Cost of the Viking Project," 26 Aug. 1969.

55. House Committee on Science and Astronautics, *Supplemental Review—NASA—OSSA Projects*, hearings before Subcommittee on Space Science and Applications, 91st Cong., 1st sess. (henceforth 91/1), 16 Oct. 1969, pp. 6-7, 11-12, 20; Thomas O. Paine to Clinton P. Anderson, 31 Dec. 1969; and Robert P. Allnutt to Henry M. Jackson, 12 Jan. 1970. See also Jakobowski, "Briefing to Dr. Paine on May 28, 1969," memo for record; and Thomas Campbell to William E. Lilly and Joseph F. Malaga, "Review of Viking Project Costs," 12 Aug. 1969.

56. Paine to Robert P. Mayo, 5 Dec. 1969; Hearth to Naugle, "Headquarters Controls on the Viking Program," 26 Nov. 1969; Naugle to Nicks, "Importance of Controlling Viking Costs," 8 Dec. 1969; George M. Low to Naugle, "Control of Viking Costs," 19 Dec. 1969; Low to Naugle, "Questions on the Viking Program," 19 Dec. 1969; Nicks to Hearth, "Viking Program Management," 24 Dec. 1969; and Naugle to Low, "The Viking Program," 30 Dec. 1969.

57. NASA Off. of Administration, Budget Operations Div., "Chronological History Fiscal Year 1968 Budget Submission," 8 Nov. 1967; idem for FY 1969, 14 Oct. 1968; for FY 1970, 5 Dec. 1969; for FY 1971, 11 June 1971; and "NASA FY 1971 Budget Reclama of BoB Tentative Allowance" [17 Nov. 1969].

58. Naugle, "Decision to Reschedule Viking to 1975," memo for record, 4 Jan. 1970; Victor Cohn, "Scientists Cite Social Needs; Cut in Space Program Urged," *Washington Post*, 29 Dec. 1969; and Walter Orr Roberts, "After the Moon, the Earth," *Science* 167 (2 Jan. 1970): 11–16.

59. Paine statement at press conference, release, 13 Jan. 1970; and NASA, "NASA Future Plans Press Conference," transcript, 13 Jan. 1970.

60. Naugle, TWX, 13 Jan. 1970.

61. Norris, "Viking Orbiter Project Staff Meeting, Minutes of January 13 and 14, 1970," memo, 19 Jan. 1970.

62. Ibid.

63. Low to Naugle, "Priority of Viking," 6 Feb. 1970; Naugle to Cortright, TWX, 10 Feb. 1970; and Norris, "Viking Orbiter Staff Meeting, Minutes of February 18, 1970," memo, 20 Feb. 1970.

64. Paine to Philip Handler, 24 Dec. 1969.

65. Naugle to Paine, Low, et al., 18 Mar. 1970, with encl., "Viking Review Panel Statement," draft, Mar. 1970; and Handler to Paine, 31 Mar. 1970, with encl., "Viking Review Panel Statement," 24 Mar. 1970.

66. Joseph R. Goudy to Martin, "Weekly Report," 9 Feb. 1973.

67. Norris, "Viking Orbiter Project Staff Meeting, Minutes of February 4, 1970," memo, 6 Feb. 1970; idem for March 4, dated 6 Mar. 1970; and Bowen, "Minutes, Viking Project Management Council Meeting, Martin Marietta Corporation, Denver, Colorado, February 26, 1970" [11 Mar. 1970].

68. See R. Cargill Hall, *Lunar Impact: A History of Project Ranger*, NASA SP-4210 (Washington, 1977), pp. 34–37, for comments on early NASA reactions to the JPL organizational structure.

69. Norris, "Review Meetings with Viking Project Manager," memo, 25 Mar. 1970.

70. Norris, "Viking Orbiter Project Staff Meeting, Minutes of April 8, 1970," memo, 13 Apr. 1970.

71. Van Ness to Cortright and Nicks, "Viking Weekly Highlights Report," 28 Oct. 1971.

72. Van Ness to Cortright and Nicks, "Weekly Report, Week of July 9, 1973," 19 July 1973; and Van Ness to Martin, "Summary of Executive Session following the Orbiter CDR," 30 July 1973.

73. The progress of the orbiter can be followed in Norris, "Viking Orbiter Project Staff Meeting Minutes," 1972–1973; Goudy, "Weekly Report," 1972–1973; JPL, "Viking 75 Orbiter System Monthly Progress Report," 1972–1973; VPO, "Viking 75 Project Quarterly Review, Presentation Material," PM-370233, 15–16 Feb. 1972; VPO, "Viking 75 Project Quarterly Review, Presentation Material," PM-3720276, 14–15 June 1972; VPO, "Viking Project Status Presentation to Dr. George M. Low . . . and Dr. John E. Naugle . . .," 23 Apr. 1973; VPO, "Viking Project Status Presentation to . . . Low . . . and . . . Naugle . . .," 11 May 1973; and Jack E. Harris to Martin, "Jet Propulsion Laboratory (JPL) Status Summary," 7 Feb. 1973.

74. Goudy to Martin, "Weekly Report," 10, 24, and 31 Aug. and 14 Sept. 1973; JPL, "Viking 75 Orbiter System Monthly Progress Report, August 1973," 31 Aug. 1973.

75. Goudy to Martin, "Weekly Report," 26 Oct. and 9 Nov. 1973 with encl., R. Glaser and M. Trummel to R. Malgren, "VO'75 DTM Pyro Shock Data," 8 Nov. 1973; and E. L. Leppert, B. K. Wada, and R. Miyakawa, *Modal Test of the Viking Orbiter*, JPL TM-33-688 (Pasadena, 1974).

76. JPL, "Viking 75 Orbiter System Monthly Progress Report, December 1973" [1 Jan. 1974] and "January 1974," [Feb. 1974]; Goudy to Martin, "Weekly Report," 5 Apr. and 3 May 1974.

77. Goudy to Martin, "Weekly Report," 10, 17, and 24 May 1974; and JPL, "Viking 75 Orbiter System Monthly Progress Report, May 1974" [June 1974].

78. JPL, "Viking 75 Orbiter System Monthly Progress Report, June 1974" [June 1974]; idem for July and Aug. 1974; and Goudy to Martin, "Weekly Report," 5, 12, 19, and 26 July and 2 Aug. 1974.

79. Goudy to Martin, "Weekly Report," 27 Sept. 1974; and JPL, "Viking 75 Orbiter System Monthly Progress Report, September 1974" [Sept. 1974].

80. Goudy to Martin, "Weekly Report," 4 Oct. 1974.

81. Goudy to Martin, "Weekly Report," 10 and 31 Jan. 1975; and JPL, "Viking 75 Orbiter System Monthly Progress Report, February 1975" [Feb. 1975].

82. Carl D. Newby telephone interview by Ezell, 8 July 1979.

# Chapter 7

1. "The Authors of the 1973 Viking Voyage to Mars," *Astronautics & Aeronautics* 7 (Nov. 1969): 59; and Gerald A. Soffen interview by Edward C. Ezell, 17 Nov. 1978.

2. James S. Martin to Eugene C. Draley, "Weekly Reports," 19 and 30 Aug., 16 and 30 Sept., 14 Oct., 8 Nov., and 16 Dec. 1968.

3. Martin, TWX, "Establishment of Science Instrument Working Group (SIWG)," 27 Aug. 1968; and C. H. Stembridge, "Minutes, Meeting of Science Instrument Working Group, September 4, 1968," 5 Sept. 1968.

4. Martin to [Langley Research Center Deputy Director], "Mars '73 Mission—Project Scientist," 27 Feb. 1968; and Soffen interview, 17 Nov. 1978.

5. Edgar M. Cortright to William H. Pickering, "Request for JPL Support for Mars 73 Mission Planning," 30 Aug. 1968; Angelo Guastaferro, "Report of September 12, 1968, Briefing on Titan/Mars '73 Mission Planning Status to Mr. Edgar M. Cortright," memo for record, 16 Sept. 1968; and Soffen interview, 17 Nov. 1968.

6. Soffen interview, 17 Nov. 1968.

7. "Authors of the 1973 Viking Voyage to Mars," p. 59.

8. Martin to Draley, "Weekly Reports," 19 Aug. 1968.

9. Soffen, "Mars '73 Science Instrument Team Selection," memo, 3 Dec. 1968, with encl., "Selection Criteria for Team Membership," n.d.

10. Ibid.; Viking Lander Imaging Team, The Martian Landscape, NASA SP-425 (Washington, 1978), p. 7; Maynard D. MacFarlane, "Digital Pictures Fifty Years Ago," Proceedings of the IEEE 60 (July 1972): 768-70; and Rafael C. Gonzalez and Paul Wintz, Digital Image Processing (Reading, Mass., 1977), pp. 1-11.

11. Klaus Biemann, "Detection and Identification of Biologically Significant Compounds by Mars Spectrometry," in M. Florkin, ed., Life Sciences and Space Research III, a Session of the Fifth International Space Science Symposium, Florence, 12-16 May 1964 (Amsterdam, 1965), pp. 77-85; and Biemann to Monte D. Wright, "'The Exploration of Mars, 1958-1978' by Edward and Linda Ezell," 13 Apr. 1978.

12. Soffen, "Mars '73 Science Instrument Team Selection."

13. John E. Naugle, "Solicitation for Participation in the Development of Instruments for Mars Landers," 27 Sept. 1968; Langley Research Ctr., Viking Project Off. (hereafter VPO), "Viking Science Management Plan," M73-105-1, 21 Aug. 1969, pp. 6-7, 21; and Milton A. Mitz to Space Science and Applications Steering Committee, "Scientists Selection for the 1973 Viking Mars Lander," 22 Jan. 1969.

14. The conflict between scientists and engineers can be studied in: R. Cargill Hall, Lunar Impact: A History of Project Ranger, NASA SP-4210 (Washington, 1977), pp. 57, 74-76, 223-28; and W. David Compton and Charles D. Benson, "History of Skylab," in process, of which chap. 3 deals with science in manned spaceflight before Skylab and pp. 115-16 the different outlooks of scientists and engineers.

15. Martin to Joshua Lederberg, 11 Feb. 1969; and file copy of Martin to Lederberg, 11 Feb. 1969, with 38 other addressees listed.

16. Martin, "Meeting of Science Instrument Teams for the Viking Project," memo, 12 Feb. 1969, with encl.; "Minutes, Viking Project Science Steering Group Meeting," 21 Feb. 1969; Martin to Lederberg, 5 Mar. 1969; and "Minutes of Viking Biology Instrument Team Meeting," 19-20 Feb. 1969.

17. NASA, "Viking Scientific Teams," news release 69-31, 25 Feb. 1969. Possible conflict of interest on the part of nongovernment scientists whose companies might subsequently bid on instrument contracts (e.g., Jack A. Ryan, Gilbert Levin, and Mario D. Grossi) was examined by persons at NASA Headquarters, who decided that it was not a real concern. See George J. Vecchietti to Donald P. Hearth, "Selection of Scientists to Participate in the Development of Instruments for the Viking Project," 12 Mar. 1969.

18. VPO, "Minutes, Viking Project Science Steering Group Meeting," 28-29 Apr. 1969.

19. Evolution of the "science definition" is described in "Minutes, Viking VPO, Project Science Steering Group Meeting, May 27-28, 1969," 13 June 1969.

20. VPO, "Minutes, Viking Project Science Steering Group Meeting, June 17-18, 1969," 11 July 1969.

21. Donald G. Rea to Naugle, "Plan for Viking Investigator Selection," 22 Apr. 1969; Walter Jakobowski to Martin, "Plans for Viking Instrument Selection," 3 July 1969; VPO, "Viking Science Management Plan," pp. 27-35, includes Naugle, "Opportunities for Participation in Space Flight Investigations Memorandum Change 23, Viking 1973 Mars Missions," memo change 23 in NHB 8030.1A, 15 July 1969; and VPO, "Viking Lander Science Instrument Teams Report," M73-112-0, draft, July 1969.

22. VPO, "Minutes, Viking Project Science Steering Group Meeting, July 17-18, 1969," 12 Aug. 1969; Walter Jakobowski, "Viking Science Review," memo, 15 Aug. 1969; and Hearth to Naugle, "Viking Science Constraints," 11 Sept. 1969.

510

23. Mitz to Soffen, "Selection of Scientists for Viking 1973 Mars Mission," 16 Dec. 1969, with encl.; Mitz to Martin, "Project Support for Viking Science Proposal Evaluation by the Subcommittees of the Space Science and Applications Steering Committee," 10 Oct. 1969; Miltz, "Record of Meeting Held on September 28, 1969, on Evaluation of Viking Proposals," memo, 13 Oct. 1969; and Israel Taback, memo, "Assessment of Science Proposals," 16 Oct. 1969.

24. Martin to Cortright, "Final VPO Assessment of Proposed Science Payload for Viking 73," 8 Dec. 1969.

25. Hearth to Naugle, "Viking Science Constraints," 11 Sept. 1969.

26. Martin to Cortright, "Final VPO Assessment of Proposed Science Payload for Viking 73," 8 Dec. 1969.

27. Richard S. Young to Lederberg, 24 Feb. 1969.

28. Ibid.

29. Ibid.

30. Thomas O. Paine, "Selection of Contractor for Viking Lander System and Project Integration, Langley Research Center" [26 Aug. 1969].

31. NASA, "Viking Contract Award," news release 69-82, 29 May 1969.

32. David B. Ahearn telephone interview by Ezell, 22 Nov. 1978; and NASA Audit Div., "Report on Audit of Viking Program Project Initiation to Contractor Selection, Langley Research Center, Hampton, Virginia," rpt. LR-DU: 33-70, 18 Dec. 1969, p. 17.

33. Ahearn telephone interview, 22 Nov. 1978.

34. VPO, "Viking 75 Project Planning Status Report, Presentation Material," 3 Feb. 1970.

35. Harper E. Van Ness to Draley, "Weekly Reports," 19 Jan.; 2, 9, 13, and 20 Feb.; 2 and 24 Mar.; 28 Apr.; 5 May; and 13 July 1970; Clarence A. Robins, Jr., to Martin, "Weekly Report for Week of January 5, 1970," 19 Jan. 1970; idem for "Week of January 12, 1970," 26 Jan. 1970; "Week of January 19, 1970," 27 Jan. 1970; and VPO, "Viking 75 Project, Resources Planning Report, July 1970," 30 July 1970.

36. Martin to Ahearn, "Proposed Contracts for Science Team Leaders," 10 Dec. 1969; Martin to Ahearn, "Proposed Contracts for Science Team Principal Investigator," 11 Dec. 1969; Martin to Ahearn, "Procurement of Services from Other Government Agencies," 11 Dec. 1969; Guastaferro to Frank W. McCabe, "NASA Contract NAS1-9000, Provisions for VPO and Science Team Review and Comment on Proposals Received in Response to Request for Proposal Associated with WB S 4.0, Science Instruments," 30 Dec. 1969; Soffen, letter, 2 Jan. 1970; Van Ness to Draley, "Weekly Reports," 9 and 19 Jan. 1970.

37. Soffen, "Viking 1975," memo, 21 Jan. 1970.

38. Guastaferro to McCabe, "NASA Contract NAS1-9000, Technical Discussions of Scientific Instrument Specifications prior to Subcontract Negotiations between Martin Marietta Corporation and Selected Vendor," 19 Feb. 1970.

39. Martin Marietta, "Viking Science Newsletter," no. 1, 3 Apr. 1970, pp. 1-2.

40. Soffen, "Science Review for the Viking Project," memo, 11 Mar. 1970; Martin, "Minutes of Special Viking Science Meetings," memo, 7 May 1970; Andrew T. Young to Soffen, 20 Apr. 1970; and "Viking Science Newsletter," no. 2, 24 Apr. 1970.

41. Martin to Charles W. Coles, 28 Aug. 1968; VPO, "Viking Project Viking Lander Science Insrument Teams Report," 30 July 1969, pp. 46-57.

42. Dale R. Rushneck et al., "Viking Gas Chromatograph-Mass Spectrometer," *Review of Scientific Instruments* 49 (June 1978): 817-34.

43. VPO, "Viking Science Activities," no. 6, 19 June 1970; no. 7, 9 July 1970; no. 8, 31 July 1970; and no. 9, 21 Aug. 1970.

44. G. Calvin Broome to Martin, "VPO/JPL Management Meeting, Inputs relative to GCMS Management Problems," 3 Sept. 1970.

45. Van Ness to Cortright and Oran W. Nicks, "Viking Project 'Highlights' Report," 16 Feb. 1971.

46. Ibid.

47. Robins to Martin, "Activity Report for Week of March 1, 1971," 8 Mar. 1971; and VPO, "Minutes, Viking Science Steering Group Meeting, 2-3 March 1971," 19 Mar. 1971.

48. Van Ness to Cortright and Nicks, "Viking Project 'Highlights' Report," 8 Mar. and 20 July 1971.

49. Van Ness to Cortright and Nicks, "Viking Project 'Highlights' Report," 22 Mar. 1971.

50. VPO, "Minutes, Viking Science Steering Group Meeting, June 2-3, 1971," 24 June 1971; and Joseph C. Moorman, "Preliminary Report: Gas Chromatograph Mass Spectrometer (GCMS) Requirements Review Panel," 21 May 1971.

51. VPO, "Viking Science Symposium, April 19-20, 1971," 19 Apr. 1971; Joel S. Levine, "Notes on the Viking Science Symposium on Mars Held at the NASA Langley Research Center, April 19-20, 1971," memo, 4 May 1971; and VPO, "Viking Science Activities," no. 23, 5 May 1971.

52. Harold P. Klein to Martin, 23 Sept. 1971.
53. Don L. Anderson to Soffen, 24 Sept. 1971. Other reactions can be found in M. Nafi Toksoz to Anderson, 22 Sept. 1971; Richard W. Shorthill to Soffen, 28 Sept. 1971; Hugh Kieffer to Soffen, 24 Sept. 1971; Robert B. Hargraves to Soffen, 28 Sept. 1971; Welcome W. Bender to Soffen, 27 Sept. 1971; and William H. Michael, Jr., to Soffen, 1 Oct. 1971.
54. VPO, "Minutes, Viking Science Steering Group Meetings," 19 Sept. and 6-7 Oct. 1971.
55. Van Ness to Cortright and Nicks, "Viking Weekly Highlights Report," 12, 20, and 28 Oct. and 3 Nov. 1971; Martin to John J. Paulson, 26 Oct. 1971; Paulson to Martin, "Viking Gas Chromatograph Mass Spectrometer (GCMS) Objectives for the Period December 1971 through May 1972," 5 Nov. 1971; and Martin to Robert J. Parks, 16 Dec. 1971.
56. Alfred O. C. Nier to Soffen, 20 Sept. 1971.
57. VPO, "Minutes, Viking Science Steering Group Meeting, February 16-18, 1972," 3 Apr. 1972; Van Ness to Cortright and Nicks, "Viking Weekly Highlights Report, Week of January 24, 1972," 3 Feb. 1972; idem for "Week of February 2. 1972," 2 Mar. 1972; "Week of February 28, 1972," 10 Mar. 1972; "Weeks of March 6 and 13, 1972," 22 Mar. 1972; Naugle to George M. Low, "Status of the Gas Chromatograph Mass Spectrometer (GCMS)," 8 Mar. 1972; and Naugle to Pickering, TWX, 10 Mar. 1972.
58. Fred S. Brown, "The Biology Instrument for the Viking Mars Mission," *Review of Scientific Instruments* 49 (Feb. 1978): 139-82; TRW Systems Group, "Hybrid Life Detection Instrument Design Review Data Package," 3 vols., 30 June 1969; and Bendix Aerospace Systems Div., "Viking Integrated Life Detector Instrument Final Report," BSR 2797, 2 vols., Dec. 1969.
59. Loyal G. Goff, "Viking Biology Instrument Narrative History," 8 Nov. 1974.
60. Klein, Lederberg, and Alexander Rich, "Biological Experiments: The Viking Mars Lander," *Icarus* 16 (Feb. 1972): 139-46.
61. Rodney A. Mills to Jakobowski, "Visit to TRW regarding Biology Instrument," 22 Feb. 1972.
62. Martin, "Viking '75 Malfunction Protection Policy," Viking Project Directive 6, 1 July 1971.
63. Van Ness to Cortright and Nicks, "Viking Weekly Highlights Report, Week of 4 October 1971," 12 Oct. 1971; and TRW, "Viking Lander Biology Instrument Preliminary Design Review," 2 vols., 4-6 Oct. 1971.
64. Langley Research Ctr. "Viking Project Manager's Presentation to Physics Panel of SPAC," 22 Nov. 1971; VPO, "Viking Science Activities," no. 32, 3 Dec. 1971; and Van Ness to Cortright and Nicks, "Viking Weekly Highlights Report, Week of January 31, 1972," 9 Feb. 1972.
65. Jakobowski, "Viking Instruments Status Report No. 8," memo, 3 Feb. 1972; "Minutes, Viking Science Steering Group," 16-18 Feb. 1971; Young, "Memo of Telephone Conversation with Dr. N. Horowitz . . . Re: Deletion of Vishniac Experiment (Light Scattering) from Viking '75 Payload," memo for record, 13 Mar. 1972; Young to Naugle, "Letter—Telephone Dictated on 15 March from Dr. Lederberg and Dr. Rich concerning the Deletion of the Light Scattering Experiment from Viking Biology Package," 15 Mar. 1972; Lederberg to Young, 15 Mar. 1972; Naugle to Wolf V. Vishniac, 17 Mar. 1972; and Lederberg interview by Ezell, 23 Aug. 1977.
66. Klein to Naugle, 22 Mar. 1972; and VPO, "Minutes of the Viking Biology Team Meeting, March 19, 1972," 30 Mar. 1972.
67. Vishniac to Naugle, 6 Apr. 1972.
68. Naugle to Vishniac, 20 Apr. 1972; and Naugle to Klein, 20 Apr. 1972.
69. Vishniac to Soffen, 16 Mar. 1973.
70. Horowitz, Roy E. Cameron, and Jerry S. Hubbard, "Microbiology of the Dry Valleys of Antarctica," *Science* 176 (21 Apr. 1972): 242-45.
71. Terry Dillman, "Death in Frozen Wasteland," *Times-Union* (Rochester), 12 Dec. 1973; "Mars Landing Instrument to Be Tested," *Houston Chronicle*, 12 Dec. 1971; "Mars Instrument Test in Antarctica Hikes Confidence," *Chicago Sun-Times*, 30 Mar. 1972; and Lederberg interview, 23 Aug. 1977.
72. Martin to Cortright, "Meeting at TRW on July 20, 1973, to Discuss Status of V'75 Biology Instrument—Schedule, Cost, Management," 20 July 1973.
73. Cortright to George Solomon, 15 Oct. 1973.
74. Martin to Walter O. Lowrie, "Top Ten Problem," 26 Oct. 1973; Cortright to Solomon, 31 Oct. 1973; Cortright to Richard D. DeLauer, 14 Nov. 1973; and DeLauer to Cortright, 4 Dec. 1973.
75. Loyal G. Goff to Jakobowski, "Viking Biology Instrument," 1 Feb. 1974; and Broome to Martin, "Management Interface Problems between VPO/MMA Resident Team at TRW," 7 Feb. 1974.
76. Martin to Lowrie and Norris, "Viking Contingency Planning," 21 June 1974; Cortright to DeLauer, 22 July 1974; Martin to Lowrie, 17 Dec. 1974; Broome to Martin, "Residual Top Ten Problems," 28 May 1975; Martin to Lowrie, 6 June 1975; and Jakobowski telephone interview by Ezell, 11 Dec. 1978.

77. VPO, "Viking Science Activities, no. 83," 14 Mar. 1975.
78. Jakobowski telephone interview, 11 Dec. 1978.

## Chapter 8

1. John D. Goodlette, "Challenge of the Viking Mars Lander System," in *Proceedings of the Twelfth Space Congress: Technology Today and Tomorrow* (Cocoa Beach, Fla.: Canaveral Council of Technical Societies, 1975), p. 6-3. Unless otherwise noted, the first part of this chapter is based on this article.

2. James S. Martin to Henry Norris, "Viking Top Ten Problems," 4 May 1970; identical letters were also sent to John J. Paulson and Albert J. Kullas. See also Harper Van Ness to Edgar M. Cortright and Oran W. Nicks, "Viking Project 'Highlights' Report, Week of December 14, 1970," 21 Dec. 1970; Van Ness, "Procedures for 'Statusizing' Viking Project Top Ten Problems," memo, 3 Aug. 1971; Langley Research Ctr., Viking Project Office (hereafter VPO), "Viking Top Ten Problems," Viking Project Directive 7, 4 Oct. 1971; and Martin to Cortright, "Viking Top Ten Problems," 20 Apr. 1972.

3. Cortright to John E. Naugle, 18 Jan. 1973, with encls. describing problems and suggested materials for letters; Cortright to Naugle, 6 Feb. 1973; and James C. Fletcher to George A. Roberts, 2 Feb. 1973; similar letters were sent by Fletcher to Russell D. O'Neal, Hudson Drake, Harry J. Gray, Barry J. Shillito, Stephen F. Keating, and Frauldin A. Lindsay.

4. Gray to Fletcher, 14 Feb. 1973; O'Neal to Fletcher, 19 Feb. 1973; Fletcher to O'Neal, 8 Mar. 1973; Lindsay to Fletcher, TWX, 6 Mar. 1973; Frank J. Madden to Cortright, 20 Mar. 1973; Fletcher to Lindsay, 4 Apr. 1973; Fletcher to Roberts, 16 Mar. 1973; Roberts to Fletcher, 10 and 27 Apr. 1973; Shillito to Fletcher, 2 Mar. and 17 May 1973; Martin to Shillito, 20 Aug. 1973; Fletcher to Charles B. Thornton, 7 Mar. 1973; Thornton to Fletcher, 15 Mar. 1973; and Fred W. O'Green to Fletcher, 17 Apr. 1973.

5. [VPO], "Development History of the Viking Lander Computer," 26 Nov. 1974.

6. Fletcher to John W. Anderson, 8 Mar. 1973.

7. Anderson to Fletcher, 22 Mar. 1973; George M. Low to Anderson, 9 Apr. 1973; Martin to Cortright, "GCSC Computer Development at Honeywell," 16 Aug. 1973; Sherwood L. Butler to Nolan I. Jones, "Guidance and Control and Sequencing Computer (GCSC)," 3 Oct. 1973; Martin to Walter O. Lowrie, "Computer (GCSC) Contingency Plans," 16 Oct. 1973; Low to Fletcher, "Status of the Viking Lander Computer," 29 Jan. 1974; Robert S. Kraemer to Low and Noel W. Hinners, "Status of the Viking Lander Computer," 29 July 1974; Hinners to Low, "Status of the Viking Lander Computer," 14 Aug. 1974.

8. [VPO], "Development History of the Viking Lander Computer," 26 Nov. 1974.

9. Cortright to Low, 30 Oct. 1973.

10. [VPO], "Development History of the Viking Lander Computer," 26 Nov. 1974.

11. Lowrie to Martin, "Computer," datafax, 15 Jan. 1975.

12. Unless otherwise noted, this section is based on Henry Caruss, "Testing the Viking Lander," *Environmental Sciences* 20 (Mar.–Apr. 1977): 11–17.

13. Stanley Barrett, "The Development of Sine Wave Vibration Test Requirements for Viking Lander Capsule Components," in Institute of Environmental Sciences, comp., *Cost Effectiveness in the Environmental Sciences: Proceedings of the Twentieth Annual Meeting* (Mount Prospect, Ill., 1974), pp. 77–82; and R. E. Snyder et al., "Specification and Correlation of the Sine Vibration for Viking 75," paper, Society of Automotive Engineers, National Aerospace Engineering and Manufacturing Meeting, San Diego, 1–3 Oct. 1974.

14. A. F. Leondis, "Viking Dynamic Simulator Vibration Testing and Analysis Modeling," *Shock and Vibration Bulletin*, no. 45, pt. 3, n.d., pp. 103–13; and H. N. McGregor, "Simulation of Viking Spacecraft Acoustic Environment," in *Cost Effectiveness in the Environmental Sciences*, pp. 183–85.

15. Stanley Barrett, "The Development of Pyro Shock Test Requirements for Viking Lander Capsule Components," in Institute of Environmental Sciences, comp., *Proceedings, 21st Annual Technical Meeting*, 2 (Mount Prospect, Ill., 1975): 5–10.

16. R. P. Parrish, Jr., "Performance and Operating Characteristics of a 4.48-M Diameter Solar Simulator for Viking Space Simulation Tests," Goddard Space Flight Center, comp., *8th Conference on Space Simulators* (Greenbelt, Md., 1975), pp. 409–17; T. Buna, "Special Techniques of the Viking Lander Capsule Thermal Vacuum Test Program," in *8th Conference on Space Simulators*, pp. 419–33; T. R. Tracey, Theodore F. Morey, and David N. Gorman, "Thermal Design of the Viking Lander Capsule," paper, AIAA 12th Aerospace Sciences Meeting,

Washington, D.C., 30 Jan.-1 Feb. 1974; Tracey and Buna, "Thermal Testing of the Viking Lander Capsule System," paper, Intersociety Conference on Environmental Systems, 29 July-1 Aug. 1974 (place not known); Morey and Gorman, "Development of the Viking Mars Lander Thermal Control Subsystem Design," paper, AIAA and ASME Thermophysics and Heat Transfer Conference, Boston, 15-17 July 1974; Buna and T. C. Shupert, "Cost Effectiveness as Applied to the Viking Lander Systems-Level Thermal Development Test Program," in *Cost Effectiveness in the Environmental Sciences*, pp. 133-37; and Gorman and Morey, "Thermal Design, Analysis, and Testing of a Full-Size Planetary Lander Model," paper, AIAA Thermophysics Conference, San Francisco, 16-18 June 1969.

17. NASA, "Viking Parachute Tests Scheduled," news release 72-118, 8 June 1973; "Balloon Launching Delayed," *New York Times*, 11 June 1973; NASA White Sands Test Facility, "NASA News," news release, 26 July 1972; Buna and H. H. Battley, "Thermal Design and Performance of the Viking Balloon-Launched Decelerator Test Vehicle (BLDTV)," paper, AIAA and ASME Thermophysics and Heat Transfer Conference, Boston, 15-17 July 1974; James L. Raper, Frederick C. Michel, and Reginald R. Lundstrom, "The Viking Parachute Qualification Test Technique," paper, AIAA Fourth Aerodynamic Decelerator Systems Conference, Palm Springs, 21-23 May 1973; Sy Steinberg et al., "Development of the Viking Parachute Configuration by Wind Tunnel Investigation," idem; N. N. Murrow, "Development Flight Tests of the Viking Decelerator System," idem; Richard D. Moog et al., "Qualification Flight Tests of the Viking Decelerator System," idem; Clarence L. Gillis, "The Viking Decelerator System—An Overview," idem; and Jesse D. Timmons, "Viking Balloon Launched Decelerator Tests," IAF paper 76-155, 27th International Astronautical Federation, Anaheim, Calif., 10-16 Oct. 1976.

18. Buna and J. R. Ratliff, "Operation of a Large Thermal Vacuum Chamber at Martian Pressure Levels," in J. C. Richmond, ed., *Space Simulation: Proceedings of a Conference Held at NBS, Gaithersburg, Maryland, September 14-16, 1970*, National Bureau of Standards Special Publication 336 (Washington, 1970), pp. 725-48; and Buna, "Thermal Testing under Simulated Martian Environment," paper, AIAA Third Thermophysics Conference, Los Angeles, 24-26 June 1968.

19. Frank W. McCabe to David B. Ahearn, "Contract NAS1-9000, End-to-End Systems Level Science Testing," 18 Oct. 1973; Arlen F. Carter, "Science 'End-to-End' Test (SEET) Objectives, Requirements, and Success Criteria," memo, 21 Jan. 1974, with draft, "Science End-to-End Test Objectives," 17 Jan. 1974; and Carter, "Science End-to-End Testing (SEET), Test Requirements Outline," memo, 24 Jan. 1974, with encl.

20. Craig Covault, "Mars Lander Proof Vehicle Passes Tests," *Aviation Week & Space Technology* 101 (4 Nov. 1974): 44-46.

21. VPO, "Viking Science End to End Test (SEET) Bulletin No. 1," 26 Aug. 1974; No. 3, 6 Sept. 1974; No. 6, 16 Sept. 1974; No. 7, 17 Sept. 1974; No. 8, 18 Sept. 1974; No. 9, 19 Sept. 1974; No. 10, 20 Sept. 1974; No. 11, 22 Sept. 1974; and No. 13, 24 Sept. 1974.

22. Welcome W. Bender to Gerald A. Soffen, 26 Oct. 1971; "Summary Minutes: Ad Hoc Subcommittee to Evaluate Viking Inorganic Instruments—The Alpha Backscatter (ABS) and X-ray Fluorescence (XRF) Spectrometer," 4 Jan. 1972; Kraemer to Naugle, "Viking Science Payload Changes," 14 Mar. 1972; Naugle to Anthony L. Turkevich, 17 Mar. 1972; Stephen E. Dwornik to Walter Jakobowski, "Viking Inorganic Analysis Experiment," 4 Apr. 1972; Fletcher to Turkovich, 11 May 1972; and Naugle to Fletcher, "Inorganic Analysis Experiment for Viking," 8 May 1972.

23. Priestley Toulmin III to Soffen, 9 Sept. 1974.

24. Toulmin to Martin, 7 Oct. 1974.

25. G. Calvin Broome to Martin, "Considerations relative to Post-Launch End to End Biology Test on PTC Lander," 30 June 1975; VPO, "Viking Science End to End Test (SEET) Bulletin No. 15," 9 Oct. 1974; Carter to Soffen, "Preparation of the Unknown Sample for GCMS Analysis during Biology Performance Verification Test," 7 Feb. 1975; VPO, "Viking BIO PV/GCMS End to End Test Bulletin No. 1," 12 Feb. 1975; No. 3, 19 Feb. 1975; and No. 4, 27 Feb. 1975.

26. H. E. Adelson et al., "The Viking Lander Biology Instrument," TRW report 21020-6003-RU-00, Aug. 1975, pp. II-1 to II-3.

27. Carter to Harold P. Klein [29 Sept. 1975]; and Martin to Hinners, 7 Oct. 1975.

28. Klein to Martin, 17 Nov. 1975; and Jakobowski to Hinners, "Viking Biology Instrument Testing Program," 1 Dec. 1975.

29. Martin to Klein, 9 Dec. 1975.

30. Martin to Hinners, "Closure of All Open Items relative to Further Prelanding Biology Testing," 9 Dec. 1975; and Hinners to Naugle, "Viking Biology Testing," 30 Dec. 1975.

31. NASA, "NASA Reorganization and Key Personnel Appointments," special announcement, 5 Mar. 1974; and NASA, "Naugle Named Associate Administrator at NASA," news release, 75-297, 21 Nov. 1975.

32. VPO, "Monthly Financial Report, January/February/March," 3 Apr. 1974.
33. Rocco A. Petrone to Hinners, "Viking Management Plan," 16 Oct. 1974; VPO, "Viking Project Status Presentation to Dr. Rocco A. Petrone, Associate Aministrator," 23 May 1974; VPO, "Viking Project Cost Status Presentation, September 5, 1974, to Dr. Noel W. Hinners, Associate Administrator for Office of Space Science," 5 Sept. 1974; and VPO, "Viking Project Program Plan and Cost Management Presentation, October 22, 1974, to Dr. Noel W. Hinners, Associate Administrator for Office of Space Science," 22 Oct. 1974.
34. Langley Research Center, "New Director at NASA Langley," news release 75-19, 11 July 1975; and NASA, "Appointment of New Director of JPL," announcement, key personnel change, 23 June 1975.
35. House Committee on Science and Astronautics, *Viking Project*, hearings before Subcommittee on Space Science and Applications, 93d Cong., 2d sess. (henceforth 93/2), 1974, passim.
36. VPO, "Minutes of the Viking Science Steering Group Meeting," 7 Oct. 1974; Hinners to Cortright, "Viking Budget Guidelines," 3 Oct. 1974; and Covault "Orbiter, Backup Lander Cut from Viking," *Aviation Week & Space Technology* 101 (28 Oct. 1974): 18-19.
37. VPO, "Mission Operations Status Bulletin," no. 3, 25 Apr. 1975; and Hinners to Jakobowski, 17 Sept. 1974.
38. VPO, "Mission Operations Status Bulletin," no. 4, 9 May 1975.
39. VPO, "Mission Operations Status Bulletin," no. 5, 26 May 1975; and W. R. Durrett, "Lightning— Apollo to Shuttle . . . Case Histories and Spacecraft Protection," in Canaveral Council of Technical Societies, comp., *Technology for the New Horizon: Proceedings of the Thirteenth Space Congress, Cocoa Beach, Florida, April 7-9, 1976 (Cocoa Beach, 1976), pp. 4-27 to 4-32.
40. "Mission Operations Status Bulletin," no. 9, 29 July 1975.

# Chapter 9

1. Langley Research Ctr., Viking Project Office (VPO), "Minutes, Viking Project Science Steering Group Meeting at Langley Research Center," 21 Feb. 1969, p. 2. Mars orbital imaging had been a topic of discussion for a number of years. For an early view of the site selection problem, see Paul R. Swan and Carl Sagan, "Martian Landing Sites for the Voyager Mission," *Journal of Spacecraft and Rockets* 2 (Jan.-Feb. 1965): 18-25. See also J. W. Kiefer, "The Influence of Visual Imaging Experiments on Unmanned Mars Spacecraft Missions," paper, 104th Technical Conference of the Society of Motion Picture and Television Engineers, 10-15 Nov. 1968; this 16-page paper was prepared by McDonnell Douglas Corp. under contract 952 000 for JPL as part of its studies for Project Voyager.
2. VPO, "Minutes, Viking Project Science Steering Group Meeting at Stanford University," 20-21 Mar. 1969, p. 3.
3. James S. Martin, "Landing Site Group," memo, 20 Aug. 1970.
4. VPO, "Minutes, Viking '75 Project Landing Site Group Meeting at MIT," 2 Sept. 1970, pp. 1-5.
5. John E. Naugle, "Decision to Reschedule Viking to 1975," memo for record, 4 Jan. 1970; and Victor Cohn, "Scientists Cite Social Needs; Cut in Space Program Urged," *Washington Post*, 29 Dec. 1969.
6. VPO, "Minutes, Viking '75 Project Landing Site Group Meeting at NASA Ames Research Center," 26-27 Oct. 1970, p. 5.
7. VPO, "Minutes, Viking '75 Project Science Steering Group Meeting at Jet Propulsion Laboratory," 10-11 June 1970, pp. 5-6; and "Viking Science Activities," 7 July 1970, p. 2.
8. Orbiter Imaging Team, "An Assessment of Viking Orbital Imaging and a Comparison of the Viking and MM71 Camera Systems" [Oct. 1970], encl. to Michael H. Carr to Martin, "Orbiter Imaging Review," 13 Oct. 1970; and Harold Masursky telephone interview, 21 Oct. 1976, by Edward C. Ezell.
9. Orbiter Imaging Team, "An Assessment of Viking Orbital Imaging."
10. VPO, "Minutes, Landing Site Group Meeting," 26-27 Oct. 1970, pp. 5-6; and VPO, "Minutes, Viking Science Steering Group Meeting," 28-29 Oct. 1970, p. 3.
11. VPO, "Minutes, Viking Science Steering Group Meeting," 19 Sept. 1971 and 6-7 Oct. 1971; and Masursky telephone interview.
12. VPO, "Minutes, Landing Site Group Meeting," 26-27 Oct. 1970, pp. 6-9; Robert S. Kramer, "Coordination of Mars Investigation Programs," memo, 25 May 1970; Earl W. Glahn to Dan Schneiderman, 28 Jan. 1971; and Steve Z. Gunter to John Newcomb, Edward Hinson, and Jim Hardy, 20 June 1973, with encl., N. R. Haynes et al., "Mariner Mars 1971 Adaptive Mission Planning," AIAA paper 72-944, AIAA/AAS Astrodynamics Conference, Palo Alto, Calif., 11-12 Sept. 1972.

515

13. Sagan, "Provisional Criteria for Viking Landing Site Selection," n.d., encl. to "Minutes, Viking Landing Site Group Meeting," 2-3 Dec. 1970; and Glahn to Kraemer, "Radar Studies of Mars for Topographic Data," 30 Mar. 1970.

14. Sagan, "Provisional Criteria for Viking Landing Site Selection"; and Alan Binder, "Viking Lander Site Criteria," n.d., encl. to Binder to A. Thomas Young, 23 Sept. 1970.

15. Sagan, "Provisional Criteria for Viking Landing Site Selection."

16. VPO, "Minutes, Viking Science Steering Group Meeting," 15-16 Dec. 1970.

17. "Mariner Mars 1971 Project Office/Viking Project Office Memorandum of Agreement for Viking Program for Viking Participation in Mariner '71 Mission Operations," 2 Feb. 1971, encl. to VPO, "Minutes, Fourth Viking '75 Project Landing Site Group Meeting at NASA, Langley Research Center," 21 Apr. 1971.

18. VPO, "Minutes, Landing Site Group Meeting," 21 Apr. 1971; Martin, "Membership to the Viking Data Analysis Team for Participation in the Mariner Mars '71 Mission Operations," memo, 19 July 1971; and Angelo Guastaferro to Frank W. McCabe, 17 Aug. 1971.

19. JPL, Mariner Mars 1971 Status Bulletin 2, "Mariner 8 Lost at Sea," 10 May 1971, and Bull. 3, "Mariner I Launch Scheduled 29 May '71," 25 May 1971; George M. Low to James C. Fletcher, "Mariner I Launch Readiness," 27 May 1971; JPL, Mariner Mars 1971 Status Bull. 4, "Mariner 9—Right on!" 1 June 1971; Naugle to Fletcher, "Mariner 9 Post Launch #1, " 2 June 1971; and JPL, Mariner Mars 1971 Status Bull. 17, "Mariner 9 MOI—How Sweet It Is!" 15 Nov. 1971.

20. Unless otherwise noted, the information on the *Mariner 9* mission is taken from William K. Hartmann and Odell Raper, *The New Mars: The Discoveries of Mariner 9*, NASA SP-337 (Washington, 1974). See also the following press accounts: Eric Burgess, "First Maps Produced of Martian Weather System," *Christian Science Monitor*, 24 June 1971; John Noble Wilford, "Thick Dust Shown on Mars," *New York Times*, 16 Nov. 1971; Walter Sullivan, "Mars: The Strange Storms of the Red Planet," *New York Times*, 21 Nov. 1971; and Sullivan, "Mars: Threat of a 'Major Scientific Disaster,' " *New York Times*, 12 Dec. 1971.

21. Hartmann and Raper, *The New Mars*, pp. 39-42; Peter Smolders, *Soviets in Space* (New York: Taplinger Publishing Co., 1974), pp. 230-35; and VPO, "Minutes, Viking Science Steering Group Meeting," 8-9 Dec. 1971, pp. 4-5, in which C. W. Snyder summarized what the Soviets had told Sagan and what they had released to the press: "1. The Soviet Mars 2 and 3 spacecraft were identical. 2. The Mars orbit inclination for both spacecraft was 40 degrees. 3. There was some difficulty with one spacecraft and it might not orbit." From this, Snyder hypothesized the following: "1. Soviet Mars 3 had trouble in achieving orbit as evidenced by its 11-day orbit period. 2. The lander from Soviet Mars 2 said by the Russians to have landed a pennant on the surface had crashed. 3. The lander from Soviet Mars 3 had landed successfully but transmitted (through its orbiter) only a part of a picture." See Charles F. Capen and Leonard J. Martin, "The Developing Stages of the Martian Yellow Storm of 1971," *Lowell Observatory Bulletin* 7 [Jan. 1972]: 211-16; JPL, Mariner Mars 1971 Status Bull. 20, "Mariner 9 TV Pictures," 10 Dec. 1971, and Bull. 22, "Dust Storm Dying," 12 Jan. 1972; and William A. Baum to Robert A. Schmitz, 10 Feb. 1972.

22. V. V. Prokof'eva and V. A. Fenchak, "O zatukhanii global'noi pylevoi buri 1971 g. na Marse" [Subsiding of the 1971 global dust storm on Mars], *Astronomicheskii Vestnik* 9 (Oct.-Dec. 1975): 201-09, trans. as "Dying Down of the 1971 Global Storm on Mars." *Solar System Research* 9 (April 1976): 165-71; Peter J. Gierasch, "Martian Dust Storms," *Reviews of Geophysics and Space Physics* 12 (Nov. 1975): 730-34; Barney J. Conrath, "Thermal Structures of the Martian Atmosphere during the Dissipation of the Dust Storm of 1971," *Icarus* 24 (1975): 36-46; and Baum, "Results of Current Mars Studies at the IAU Planetary Research Center," in A. Woszgzyk and C. Iwaniszewska, eds., *Exploration of the Planetary System* (Dordrecht, Holland, Boston, 1974), pp. 241-51. See also A. B. Whitehead, "Reports on Russian Mission," memo, 27 Sept. 1972, with encl., trans. of 25 Aug. 1972 *Pravda* article, "Initial Results of Mars Probe Analyzed"; Young, "Transmittal of USSR Technical Papers relating to the Mars 2 and 3 Missions," memo, 26 June 1973, with encls., V. I. Moroz, "Rabochaya model' atmosfery Marsa" [Working model of the atmosphere of Mars], NASA TT F-14 906, Apr. 1973; M. K. Rozhdestvenskiy and V. I. Shkirina, "Mars-3: Otsenka parametrov atmosfery v meste posadki sa" [Mars 3 estimate of parameters of the atmosphere at the landing point of the spacecraft], NASA TT F-14 907, Apr. 1973; Moroz, "Orbital'nyye apparaty Mars-2 i Mars-3; Rezul'taty issledovaniy poverkhnosti i atmosfery Marsa" [Mars 2 and Mars 3 orbital spacecraft; Results of studies of the surface and atmosphere of Mars], NASA TT F-14 908, Apr. 1973; and Gerald A. Soffen to Viking Scientists, "Pravda Article by Dr. I. Koval," 15 Aug. 1973.

23. VPO, "Minutes, Viking Science Steering Group Meeting," 7-8 Dec. 1971; "Minutes of Viking Data Analysis Team Meeting at JPL," 16 Dec. 1971; Schmitz, "Viking Data Analysis Report Number 1," memo, 28 Jan. 1972; Baum, "Where Will the Martian Dust Be When Viking Arrives," paper, AAS

Div. of Planetary Sciences Meeting, Kona, HI, 20-24 Mar. 1972; JPL, Mariner Mars 1971 Status Bull. 23, "Pits, Spots, Cracks and Rilles," 14 Jan. 1972 and Bull. 24, "Unique Martian Landform," 20 Jan. 1972.

24. VPO, "Minutes, Viking Science Steering Group Meeting," 16-18 Feb. 1972.

25. NASA, "News Conference on Mariner 9," news release, 2 Feb. 1972. For an evaluation of the *Mariner 6* and 7 television science team work, see special issue, *Journal of Geophysical Research* 76 (10 Jan. 1971): 293-472, which includes Robert B. Leighton and Bruce C. Murray, "One Year's Process and Interpretation—An Overview," 293-96, Murray et al., "The Surface of Mars: 1. Cratered Terrains," 313-30, Robert P. Sharp et al., "The Surface of Mars: 2. Uncratered Terrains," 331-42, James A. Cutts et al., "The Surface of Mars: 3. Light and Dark Markings," 343-56; Sharp et al., "The Surface of Mars: 4. South Polar Cap," 357-68. See also J. A. Stallkamp, A. G. Herriman, and the Mariner Mars 1969 Experimenters, *Mariner Mars 1969 Final Project Report: Scientific Investigations*, JPL Technical Rpt. 32-1460, vol. 3 (Pasadena, 1971); and Whitehead to MM 71 Investigators, "Bruce Murray Seminar," 2 Oct. 1972, with encl., "A New View of Mars," 28 Sept. 1972.

26. VPO, "Minutes, Viking Science Steering Group Meeting," 16-18 Feb. 1972; VPO, "Viking Science Activities," no. 29, 23 Sept. 1971, p. 5; and Schmitz, "Viking Data Analysis Report Number 1."

27. Soffen to Science Steering Group, "Meetings during the Week of February 14, 1972," 9 Feb. 1972.

28. VPO, "Minutes, Viking Science Steering Group Meeting," 16-18 Feb. 1972, pp. 6, 8-10.

29. C. Howard Robins, Jr., "Review of M75-140-0, Viking '75 Project Landing Site Selection Plan," memo, 10 Feb. 1972, with draft plan encl.

30. VPO, "Coordination Copy, M75-141-0, Viking 75 Project Landing Site Selection Plan," n.d.; Masursky to Martin, 28 July 1971; Harold F. Hipsher, "Mapping Coverage of Mars with Mariner Mars '71 (MM'71)," memo for record, 4 June 1971; Martin to Vincent McKelvey, 3 Sept. 1971; Martin to McKelvey, "Mars Mapping Support for Viking Project," 18 Oct. 1971; Guastaferro to John P. Campbell Jr., "Mars Mapping Support for Viking, Purchase Request 06.000.155," 12 Nov. 1971; Masursky to George Recant, "Viking Cartography," 3 Nov. 1971; Masursky to N. J. Trask, "Viking Cartography," 29 Oct. 1971; Masursky to Robins, 8 May 1972; and U.S. Geological Survey, Branch of Astrogeologic Studies, "Proposal to Conduct Geologic Mapping in Support of the Viking '75 Landers," 8 May 1972.

31. Thomas A. Mutch to Schmitz, 1 June 1972, with encl., Mutch, "VDAT Terrain Analysis," 30 May 1972; VPO, "Minutes, Viking Data Analysis Team Meeting, Jet Propulsion Lab," 15 March 1972; Schmitz to Viking Data Analysis Team Members, "VDAT Final Report on Participation in the Mariner Mars '71 Mission," 16 May 1972; and Schmitz, "Viking Data Analysis Team Report on MM'71 Participation," memo, 13 July 1972, with encl., "Viking Data Analysis Team Report," prelim., M75-144-0, n.d.

32. VPO, "Minutes, Viking 75 Project Landing Site Working Group Meeting at Jet Propulsion Laboratory," 25 Apr. 1972.

33. Recant, "Rating of Viking Target Areas to Be Photographed by Mariner 9," memo, 30 May 1972.

34. Robins to Landing Site Working Group, "Revision to Near-Term Selection Activities Schedule," 25 May 1972.

35. Recant, "Rating of Viking Target Areas."

36. Schmitz to Martin, "Status of Planning for Coverage of Viking Targets in the Mariner 9 Extended Mission," 6 June 1972.

37. Masursky et al. to Schmitz, "Additional Viking Targeting Sites for Extended Mariner 9 Mission," 9 June 1972.

38. VPO, "Viking Science Activities," no. 42, 23 June 1972, p. 4; Mutch to Naugle, 10 July 1972; and VPO, "Minutes, Viking Science Steering Group Meeting," 13 June 1972.

39. Young, "Transmittal of Mariner 9 Data on Viking Target Sites," memo, 19 July 1972; George P. Wood to Martin, "Some Inferences from Martian Crater Trails," 13 July 1972; Young, "August 4-5 Meeting," memo, 13 July 1972; and Kraemer to Naugle, "Viking Landing Sites," 31 July 1972.

40. Young, handwritten notes from 4 Aug. 1972 meeting.

41. VPO, "Minutes, Viking 75 Project Landing Site Working Group Meeting at Langley Research Center," 4-5 Aug. 1972; and Young, "Regions of Interest for Landing Sites," memo, 15 Aug. 1972.

42. Henry W. Norris to Martin, " 'North Polar Region' Landing Site Impact," 22 Sept. 1972; Joshua Lederberg to Robins [26 June 1972], with encl., Lederberg to James D. Porter, "LSWG" [26 June 1972], both items handwritten.

43. Young to Landing Site Working Group, " 'Polar Region' Landing Site Study," 25 Aug. 1972; and Young to Martin, 17 Aug. 1972.

44. VPO, "Minutes, Viking 75 Project Landing Site Working Group Meeting at Langley Research

Center," 28 Sept. 1972; and "Viking Project Science Steering Group Minutes for Meeting Held September 29, 1972 at the Langley Research Center," 29 Sept. 1972.

45. Young to Landing Site Working Group, "Candidate Landing Site Locations," TWX, 19 Oct. 1972; Masursky to Robert H. Steinbacher, 19 Oct. 1972; Young to Sagan, 20 Oct. 1972; Young to Porter, 24 Oct. 1972; Young to L. Kingsland, Jr., 24 Oct. 1972; and Leonard V. Clark to Young, "LSWG Comments regarding Hal Masursky's Suggested Changes to Candidate Landing Site Locations," 25 Oct. 1972.

46. Naugle to Fletcher, "Mariner Mars '71 Programs, Assessment of Extended Mission; Post Launch Mission Operation Report No. S-819-71-01/02," 4 Dec. 1972; Low to Naugle, "Viking Landing Site," 13 Nov. 1972; VPO, "Landing Site Selection Status and 'North Polar Region' Landing Site Study," presentation to Fletcher, 22 Nov. 1972; and Martin to Kraemer, "Viking Landing Sites," 29 Nov. 1972.

47. VPO, "Minutes, Viking 75 Project Landing Site Working Group Meeting at Kennedy Space Center," 4-5 Dec. 1972; VPO, "Viking Project Science Steering Group Minutes for Meeting Held at Martin Marietta Aerospace Orlando Facility," 7 Dec. 1972; and Soffen and Young to Science Steering Group and Landing Site Working Group, "Joint Meeting in December 1972," 10 Nov. 1972.

48. Young, "Candidate 'North Polar Region' Sites," memo, 15 Dec. 1972, and "Candidate Backup 'A' Site," memo, 26 Dec. 1972; Don L. Anderson to Soffen, 21 Dec. 1972; and Alfred O. C. Nier to Soffen, 22 Dec. 1972.

49. Michael H. Carr to Soffen, 26 Dec. 1972.

50. Richard W. Shorthill to Soffen, 5 Jan 1972; Seymour L. Hess to Soffen, 13 Feb. 1973; and Hess to Soffen, 22 Dec. 1972.

51. Harold P. Klein to Soffen, 9 Jan. 1973.

52. Mutch to Soffen, 10 Jan. 1973.

53. Robert B. Hargraves to Soffen, 10 Jan. 1973; Hugh H. Kieffer to Soffen, 10 Jan. 1973; Kieffer to Young, 11 Jan. 1973; C. Barney Farmer to Soffen, 11 Jan. 1973; and Klaus Biemann to Soffen, 12 Jan. 1973.

54. Sagan to Young, 12 Jan. 1973; and Sagan to Martin, 3 Feb. 1973.

55. VPO, "Minutes, Viking 75 Project Landing Site Working Group Meeting at Langley Research Center," 8 Feb. 1973; and Soffen and Young to Science Steering Group and Landing Site Working Group, "Viking Landing Sites," 12 Feb. 1973.

56. Naugle to Fletcher, "Viking Landing Sites in Polar Latitudes," 1973; and Naugle to Fletcher, "Viking Landing Sites in Polar Latitudes," 7 Mar. 1973.

57. Naugle to Fletcher, "Viking Landing Sites in Polar Latitudes" [6 March 1973].

58. VPO, "Minutes, Viking 75 Project Landing Site Working Group Meeting at Langley Research Center," 2-3 Apr. 1973.

59. Wolf V. Vishniac to Lederberg, 1 Mar. 1973. Discussion of water, biology, and temperature can be followed in: Leighton, "The Richtmyer Memorial Lecture: A Physicist Looks at Mars," *American Journal of Physics* 40 (Nov. 1972): 1569-75; Sagan to Farmer, 13 Feb. 1973; Michael C. Malin to Soffen, 20 Feb. 1973; Soffen, "Ad Hoc Meeting on February 28, 1973," TWX, 21 Feb. 1973; Conway B. Leovy to Farmer, 22 Feb. 1973; Sagan to Farmer, 26 Feb. 1973; Masursky, "Statement of Mariner 9 Volatiles Working Group on the Probable Distribution of Liquid Water on Mars," draft, 27 Feb. 1973; Lederberg to Sagan, 2 Mar. 1973; Leslie E. Orgel to Soffen, 5 Mar. 1973; Lederberg to Farmer, 6 Mar. 1973; Klein to Walter Jakobowski, 8 Mar. 1973; Leovy, "Exchange of Water Vapor between the Atmosphere and Surface of Mars," *Icarus* 18 (1973): 120-25; Lederberg to Klein, 12 Mar. 1972; Mutch to Soffen, 4 Apr. 1973; and Norman L. Crabill to Young, "Viking 75 Mission B Site Selection Process," 6 Apr. 1973.

60. NASA, "News Conference on Mariner 9," news release, 2 Feb. 1972, p. 29.

61. Lederberg to Klein, 12 Mar. 1973; VPO, "Minutes, Landing Site Working Group," 2-3 Apr. 1973; and Soffen to Martin, "Landing Site Recommendation," 3 Apr. 1973.

62. Young interview by Ezell, 9 Sept. 1976.

63. Masursky telephone interview by Ezell, 15 Nov. 1976.

64. NASA, "News Conference on Viking Landing Site Selection," news release, 7 May 1973, and "Viking Landing Sites," news release 73-91, 2 May 1973. Press reactions included: Marvin Miles, "1st Viking Spacecraft Targeted to Land on Mars July 4, 1976," *Los Angeles Times*, 8 May 1973; Harold M. Schmeck, Jr., "2 Mars Sites Chosen for Unmanned Landings in '76," *New York Times*, 8 May 1973; and Vern Haugland, "Mars Probe Aims at Canyon Mouth," *Washington Post*, 8 May 1973.

# Chapter 10

1. Langley Research Ctr., Viking Project Office (VPO), "Minutes, Viking Science Steering Group Meeting Held at Viking Project Office, Langley Research Center," 16-17 Aug. 1973.

2. VPO, "Minutes, Viking Science Steering Group Meeting," 13-14 Feb. 1974, p. 2; and Robert C. Blanchard, "Draft Copy of Landing Site Certification Strategy," memo, 4 Sept. 1973, with encl.

3. "Viking Site Planners Expect Data from New Soviet Mars Probes," *Aerospace Daily*, 14 May 1973, p. 79; Kenneth W. Gatland, "Russians Renew Search for Life On Mars," *Christian Science Monitor*, 19 Feb. 1974; "More Martian Vapor Discovered," *Philadelphia Inquirer*, 15 Mar. 1974: Christopher S. Wren, "A Soviet Capsule Descends to Mars," *New York Times*, 15 Mar. 1974; and "Mars Probe by Soviets Is Failure," *Washington Post*, 15 Mar. 1974.

4. S. S. Sokolov et al., "Funktsionirovaniye spuskayemogo apparata AMS 'Mars-6' v atmosfere Mars" [Operation of descent module of the Mars 6 automatic interplanetary station in the Martian atmosphere], *Kosmicheskiye Issledovaniya* 13 (Jan.-Feb. 1975): 9-15, trans. as NASA TT F-16334; V. G. Istomin et al., "Eksperiment po izmereniyu sostava atmosfery Marsa no spuskayemom apparate kosmicheskoy stantsiya 'Mars-6' " [Experiment measuring composition of Martian atmosphere on board the descent module of the Mars 6 space station], *Kosmicheskiye Issledonaniya* 13 (Jan.-Feb. 1975): 16-20, trans. as NASA TTF-16335; R. B. Zezin et al., "Analiz rel'yefnykh usloviy v rayone posadki SA AMS 'Mars-6' " [Analysis of terrain conditions in the landing site of the Mars 6 automatic interplanetary station], *Kosmicheskiye Issledonaniya* 13 (Jan.-Feb. 1975): 99-107, trans. as NASA TT F-16349; "A Valuable Contribution to Planetology," *Pravda*, 17 Mar. 1974, p. 3 [interview of R. Z. Sagdeyev, director of the Institute of Cosmic Research, Soviet Academy of Sciences]; E. Kodin and D. Pipko, "From Hypothesis to Facts—Experiment 'Earth-Mars,'" *Socialist Industry*, 17 Mar. 1974; and "Mars Probes Reveal Signs of Life Supporting Environment," Moscow Tass broadcast in English, 25 Mar. 1974, as reported in Foreign Broadcast Information Service, *Daily Report: Soviet Union*, 26 Mar. 1974, p. U1.

5. VPO, "Minutes, Viking Science Steering Group Meeting Held at the Jet Propulsion Laboratory," 9 Apr. 1974, p. 2; idem at Langley Research Center, 20 June 1974, p. 1; idem at Martin Marietta Corp., 13 Aug. 1974, p. 4; Edgar M. Cortright to John E. Naugle, "Soviet/U.S. Discussions on Mars Landing Sites," 7 May 1971, with encl., Tobias Owen to A. Thomas Young, 2 Apr. 1971; Robert S. Kraemer to Naugle and Vincent L. Johnson, "Preparation for US/USSR Joint Working Group on Space Research," 9 July 1971; James S. Martin to Ichtiaque Rasool, "USSR Mars Data," 14 Aug. 1972; Martin to Henry W. Norris, 27 Mar. 1973, with encl., "Protocol, a Working Session of Soviet and American Scientists for the Discussion of Questions concerning Explorations of Mars and Venus," n.d.: Rasool to N. J. Trask and Harold Masursky, "Maps from Mariner 9 Data for USSR Mission to Mars in 1973," 17 July 1973; Masursky to Young, 17 June 1974, with encl., Masursky to Rasool, "Meeting in the USSR on June 4-8, 1974," 17 June 1974; and Masursky, "USSR Mars Coverage," memo, 4 Sept. 1975.

6. The nature of the early discussions about the use of radar and the interpretation of the data collected can be followed in: Lincoln Laboratory, MIT, "Final Report, Radar Studies of Mars," 15 Jan. 1970; Robert A. Schmitz, "Radar Measurements of Martian Topography," memo, 17 Feb. 1970; Martin to Walter Jakobowski, "Earth-Based Radar and Mariner '71 Data," 17 Mar. 1970; Jakobowski to Earl W. Glahn, "Radar Studies of Mars for Topographic Data," 24 Mar. 1970; Schmitz to Jakobowski, "Viking Requirement for Radar Measurement," 26 Mar. 1970; Glahn to Jakobowski, "Radar Studies of Mars for Topographic Data," 30 Mar. 1970; Martin to Jakobowski, "Viking Requirement for Radar Measurements on Mars," 31 Mar. and idem 20 Apr. 1970; Cortright to Naugle, "Viking Requirement for Radar Measurements of Mars," 20 May 1970; R. Goldstein to Schmitz, "Mars Radar Observing Schedule, 1971," 5 June 1970; William E. Brunk, "Funding of Radar Astronomy at the MIT Haystack Antenna for the Fiscal Year 1971," memo, 10 June 1970; Naugle to Cortright, "Viking Requirements for Radar Measurements of Mars," 16 June 1970; Schmitz to Young, "Contour Map of Mars," 24 Aug. 1970; N. A. Renzetti to Norman Pozinsky, "NASA Support Plan for Mars Planetary Radar Observations for Viking," 17 May 1971, with encl., "DSN Support Plan for Mars Planetary Radar Operations in Support of the Viking Project," 12 Apr. 1971; Schmitz to Gordon H. Pettengill, 6 Mar. 1972, with encl., R. E. Hutton to Schmitz, "Analysis of 1971 Haystack Radar Data," 25 Feb. 1972; Hutton to Schmitz, "Analysis of 1967 and 1969 Haystack Radar Data," 9 Mar. 1972; Pettengill to Schmitz, 13 Mar. 1972; Schmitz to C. H. Robins, Jr., "1971 Goldstone Radar Measurements of Mars Elevations," 1 June 1972; Leo D. Staton 1972; to Viking Project Off., "Comments concerning the Validity of Radar Results for Dielectric Constants on Mars," 28 Nov. 1972; Young, "Coordination of Earth-Based Radar

Measurements of Mars," memo, 12 Jan. 1973, with encl., N. N. Krupeniyo and N. Ya. Shapirovskaya, "Dielectric Permeability and Density of the Substance of the Surface Layer of Mars," n.d., NASA TTF-14369; Pettengill to Carl Sagan, 15 Feb. 1973; Staton to Viking Project Off., "Supplementary Notes and Recommendations concerning the Validity of Radar Results for Dielectric Constants on Mars," 21 Feb. 1973; Martin to Doug J. Mudgway, "Mars Radar Observations," 16 Mar. 1973; G. Leonard Tyler to Masursky, 5 Dec. 1973; and Tyler to Young, 6 Oct. 1975.

7. Sagan to Martin, 3 Feb. 1973.

8. Martin to Cortright, 23 Mar. 1973; Martin to Sagan, 27 Mar. 1973; Young to Rolf B. Dyce, 13 Mar. 1973, with encl., "Minutes, Conference on Earth-based Radar Measurements of Mars for Viking—Arecibo, Puerto Rico," 21-23 Feb. 1973; and Tyler, "Proposal to the National Aeronautics and Space Administration . . . for Analysis of Radar Data from Mars," Apr. 1973, proposal RL-24-73. The progress of Tyler's work can be followed in Tyler to Young, "Monthly Progress Report for Viking Mars Radar Study," 4 Dec. 1973, 4 Jan. 1974, 4 Feb. 1974, 5 Mar. 1974, 8 Apr. 1974, 1 May 1974, 4 June 1974, 3 July 1974, 7 Aug. 1974, and 12 Sept. 1974.

9. VPO, "Landing Site Working Group Minutes," 4-5 Nov. 1974; and U.S. Geological Survey, Center for Astrogeology, Masursky and Mary H. Strobell, "Geological Maps and Terrain Analysis Data for Viking Mars '75 Landing Sites Considered in February and April 1973," Interagency Rpt.: Astrogeology 60, Dec. 1975, p. 3.

10. VPO, "Minutes, Meeting, Landing Site Working Group Subcommittee, Stanford University," 12 Dec. 1974; VPO, "Minutes, 'C' Site Subcommittee for the Landing Site Working Group," 6 Feb. 1975; Norman L. Crabill, "Minutes for the February 6, 1975, Meeting of the 'C' Site Subcommittee of the Landing Site Working Group at the Jet Propulsion Laboratory," memo, 14 Mar. 1975; and VPO, "Viking Science Activities," no. 82, 21 Feb. 1975, p. 4.

11. VPO, "Landing Site Working Group Minutes," 24 Feb. 1975, p. 4; VPO, "Viking Science Activities," no. 83, 14 Mar. 1975; Henry J. Moore to Young, 17 Dec. 1974; Crabill to Tyler, 14 Feb. 1975; and Dick Simpson to Crabill, "Proposed Arecibo Mars Observations," 21 Mar. 1975.

12. U. M. Lovelace, "Site A-1 Certification Procedure Review," memo, 27 May 1975, with encl., SA&MP Directorate, Landing Site Staff, "Site A-1 Certification Procedure," draft, SAM-1-14, 23 May 1975.

13. W. B. Green, "Evaluation of Viking Site Certification Test I," memo, 14 July 1975; Pat M. Bridges to Masursky, "Analysis of Viking Test Site Data," 10 July 1975; and VPO, "Viking Science Activities," no. 88, 25 July 1975, p. 4.

14. Young to M. S. Johnson, "A-1 Time Line," 30 Sept. 1975; J. F. Newcomb to James D. Porter, "Changes to the Mission 'A' Certification Timeline," 22 Oct. 1975; Newcomb, "Site Certification Timeline for Site A1—Revised October 21, 1975," memo, 24 Oct. 1975; Crabill, "B-1 Timeline Problems," memo, 9 Jan. 1976; Crabill to B. Gentry Lee, "B-1 Site Certification Timeline Review, January 13, 1976," 19 Jan. 1976; VPO, "Viking Science Activities," no. 93, 23 Jan. 1976, p. 4; Crabill to Viking Flight Team, "Change in Date of Preliminary and Firm Commitment Decision Dates in A-1 Site Certification Timeline," 2 Feb. 1976; and Crabill, "Minutes of SAMP Director's Review of B1 Timeline," memo, 11 Feb. 1976.

15. VPO, "Launch and Mission Operations Status Bulletin," no. 10, 20 Aug. 1975, and no. 11, 20 Aug. 1975; W. O. Lowrie and H. Wright, "Daily KSC Status (FAX), Dated August 11, 1975," memo, 11 Aug. 1975; idem (daily) for 12-15 and 18-21 Aug. 1975; and Martin and Young, "Viking to Mars: Profile of a Space Expedition," Astronautics & Aeronautics 14 (Nov. 1976): 31-32.

16. Lowrie and Wright, "Daily KSC Status (FAX), Dated August 22, 1975," memo, 22 Aug. 1975, and idem (daily) for 25-29 Aug. and 2-5 Sept. 1975; SL/Viking Prog. Mgr. to A/Administrator, "Viking Mission Post Launch Report #1," 28 Aug. 1975; idem, #2, 16 Sept. 1975.

17. VPO, "Launch and Mission Operations Status Bulletin," no. 12, 10 Sept. 1975; "Mission Operations Status Bulletin," no. 13, 10 Nov. 1975, and no. 15, 9 Jan. 1976.

18. Martin and Young, "Viking to Mars," p. 32.

19. VPO, "Viking Science Activities," no. 91, 21 Nov. 1975, and no. 93, 23 Jan. 1976; Crabill, "SAMPD-1 Test Minutes," memo, 12 Feb. 1976; and Lee interview by Edward C. Ezell, 12 Sept. 1976.

20. Young interview by Ezell, 9 Sept. 1976.

21. Martin interview by Ezell, 9 Sept. 1976; Young interview; and Martin and Young, "Viking to Mars," p. 33.

22. Martin and Young, "Viking to Mars," p. 34.

23. Crabill, "LSS #1 Minutes," memo, 17 June 1976; "LSS #2 Meeting Minutes," 18 June 1976; "LSS #3 Meeting Minutes," 21 June 1976; and "LSS #4 Meeting Minutes," 22 June 1976.

24. Crabill, "LSS #5 Meeting Minutes," memo, 23 June 1976; and Viking press briefing, 23 June 1976 [authors' notes from briefings held at JPL, as well as material distributed to the press].
25. Mike Carr interview, 8 Sept. 1976, and Gerald A. Soffen interview, 11 Sept. 1976, both by Ezell.
26. Masursky interview by Ezell, 11 Sept. 1976.
27. Crabill, "LSS #6 Meeting Minutes," memo, 24 June 1976; Carr interview; Lee interview; and Viking press briefing, 6 July 1976.
28. Viking press briefing, 24 June 1976.
29. Crabill, "LSS #7 Meeting Minutes," memo, 25 June 1976.
30. Viking press breifing, 25 June 1976.
31. Crabill, "LSS #8 Meeting Minutes," 26 June 1976.
32. Crabill, "LSS #9 Meeting Minutes," 28 June 1976.
33. Nick Panagakos interview, 30 June 1976, and Robert Shafer interview, 2 July 1976, both by Ezell.
34. Viking press briefings, 27 and 28 June 1976.
35. Masursky interview by Ezell, 10 Sept. 1976.
36. Lee interview.
37. Viking press briefing, 29 June 1976; and Crabill, "LSS #12 Meeting Minutes," 30 June 1976.
38. Crabill, "#13 LSS Meeting Minutes," 5 July 1976.
39. Crabill, "Minutes of Project Manager's Landing Site Meeting, July 1, 1976, 8:00-10:00 a.m.," memo, 1 July 1976.
40. Viking press briefing, 1 July 1976; and Martin interview.
41. Lee interview; and Masursky interview, 10 Sept. 1976.
42. Crabill, "#14 LSS Meeting Minutes," memo, 5 July 1976; "LSS #15 Meeting Minutes," 6 July 1976; "#16 LSS Meeting Minutes," 6 July 1976; and "#17 LSS Minutes," 7 July 1976.
43. Crabill, "A1NW Preliminary Radar Assessment Meeting Minutes, July 7, 1976—8:00-10:00 a.m.," memo, 8 July 1976; Crabill, "A-1 Preliminary Radar Assessment, June 6, 1976," memo, 9 June 1976; and Crabill "A1/A2 Radar Assessment, June 18, 1976," memo, 20 June 1976.
44. Viking press briefing, 7 July 1976.
45. Crabill, "LSS Meeting #18," memo, 16 Aug. 1976; and Carr interview.
46. VPO, "Viking Mission Status Report" [7 July 1976].
47. Viking press briefing, 8 July 1976; and Crabill, "A1/A2 Radar Assessment," memo, 20 June 1976.
48. Viking press briefing, 8 July 1976.
49. Crabill, "#19 LSS Minutes," memo, 10 July 1976; "#20 LSS Meeting Minutes," 11 July 1976; and "#21 LSS Meeting Minutes," 12 July 1976.
50. Crabill, "#22 LSS Meeting Minutes," 14 July 1976; Carr interview, and Masursky and Crabill, "The Viking Landing Sites: Selection and Certification," Science 193 (27 Aug. 1976): 809-12.
51. Carr, interview.
52. Crabill, "#23 LSS Meeting Minutes," 18 July 1976; and Crabill to Lee, "C Site Latitudes," 20 May 1976.
53. Masursky and Crabill, "The Viking Landing Sites," p. 810.
54. Crabill, "#24 LSS Meeting Minutes," 19 July 1976, and "#25 LSS Meeting Minutes," 20 July 1976.
55. Crabill, "#29 LSS Meeting Minutes," 24 July 1976.
56. Masursky interview, 10 Sept. 1976.
57. Crabill, "#31 LSS Meeting Minutes," 27 July 1976.
58. Masursky and Crabill, "Search for the Viking 2 Landing Site," Science 194 (1 Oct. 1976): 66.
59. Viking press briefing, 25 July 1976; Crabill, "#33 LSS Meeting Minutes," 28 July 1976; and Masursky interview, 10 Sept. 1976.
60. Masursky and Crabill, "Search for the Viking 2 Landing Site," p. 66; Crabill, "#39 LSS Meeting Minutes," 13 Aug. 1976; and Masursky interview, 10 Sept. 1976.
61. Crabill, "#42 LSS Meeting Minutes," 18 Aug. 1976; and Viking press briefing, 18 Aug. 1976.
62. Masursky interview, 10 Sept. 1976.
63. Lee interview; Crabill interview by Ezell, 13 Sept. 1976; and Masursky interview, 10 Sept. 1976.
64. Crabill, "#42 LSS Meeting," 18 Aug. 1976.
65. Masursky interview, 10 Sept. 1976.
66. Crabill, "#43 LSS Meeting Minutes," 19 Aug. 1976; "Minutes of 44th LSS Meeting," 20 Aug. 1976; and "Minutes of 45th LSS Meeting," 23 Aug. 1976; and Masursky interview, 10 Sept. 1976.
67. Crabill, "Minutes of 46th LSS Meeting," 28 Aug. 1976.
68. Masursky and Crabill, "Search for the Viking 2 Landing Site," p. 68.
69. Ibid., and Crabill, "Minutes of 46th LSS Meeting," 28 Aug. 1976.
70. Viking press briefing, 27 Aug. and 3 Sept. 1976.
71. Tape recording of communications audio circuits made by J. S. Kukowski; Martin and

Young, "Viking to Mars," pp. 44–45; JPL, "Mission Status Bulletin," no. 41, 3–4 Sept. 1976; and Viking press briefing, 4 Sept. 1976.

72. Martin and Young, "Viking to Mars," p. 45.
73. Martin interview.
74. Young interview.
75. Soffen interview.
76. Masursky interview, 10 Sept. 1976.
77. Soffen interview.
78. Lee interview.
79. Carr interview; and Carr et al., "Preliminary Results from the Viking Orbiter Imaging Experiment," *Science* 193 (27 Aug. 1976): 766–76.
80. Masursky interview, 10 Sept. 1976.
81. Masursky telephone interview by Ezell, 15 Nov. 1976.
82. Sagan interview by Ezell, 13 Sept. 1976.

# Chapter 11

1. Conway B. Leovy, "The Atmosphere of Mars," *Scientific American* 237 (July 1977): 34.
2. Ibid.
3. Martin Marietta Corp., *Viking Mars Expedition, 1976* (Denver, 1978), p. 30.
4. Michael H. Carr et al., "Some Martian Volcanic Features as Viewed from the Viking Orbiters," *Journal of Geophysical Research* 82 (30 Sept. 1977): 3985. This issue contains 53 articles about Viking; it was also published as American Geophysical union, comp., *Scientific Results of the Viking Project* (Washington, 1977).
5. Carr, "Some Martian Features," pp. 4003–11; and Gerald A. Soffen, "Mars and the Remarkable Viking Results," paper, American Institute of Aeronautics and Astronautics 16th Aerospace Sciences Meeting, Huntsville, Ala., 16–18 Jan. 1978, AIAA paper 78-191.
6. Martin Marietta Corp., *Viking Mars Expedition, 1976*, p. 30.
7. Ibid., p. 31.
8. Soffen, "Mars and the Remarkable Viking Results."
9. Harold Masursky et al., "Classification and Time of Formation of Martian Channels Based on Viking Data," *Journal of Geophysical Research* 82 (30 Sept. 1977): 4016–38.
10. Carr and Gerald O. Schaber, "Martian Permafrost Features," *Journal of Geophysical Research* 82 (30 Sept. 1977): 4039–54; and Martin Marietta Corp., *Viking Mars Expedition, 1976*, p. 31.
11. Carr et al., "Martian Impact Craters and Emplacement of Ejecta by Surface Flow," *Journal of Geophysical Research* 82 (30 Sept. 1977): 4055–65.
12. Karl R. Blasius et al., "Geology of the Valles Marineris: First Analysis of Imaging from Viking 1 Orbiter Primary Mission," *Journal of Geophysical Research* 82 (30 Sept. 1977): 4090.
13. Carr to James S. Martin, "October Imaging Review," 13 Oct. 1970, with encls.
14. Soffen, "Mars and the Remarkable Viking Results."
15. Ronald Greeley et al., "Geology of Chryse Planitia," *Journal of Geophysical Research* 82 (30 Sept. 1977): 4109.
16. John E. Guest, P. E. Butterworth, and Greeley, "Geological Observations in the Cydonia Region of Mars from Viking," *Journal of Geophysical Research* 82 (30 Sept. 1977): 4119.
17. James A. Cutts et al., "North Polar Region of Mars: Imaging Results from Viking 2," *Science* 194 (17 Dec. 1976): 1329–37.
18. Hugh H. Kieffer et al., "Infrared Thermal Mapping of the Martian Surface and Atmosphere: First Results," *Science* 193 (27 Aug. 1976): 780–86; and Kieffer et al., "Thermal and Albedo Mapping of Mars during the Viking Primary Mission," *Journal of Geophysical Research* 82 (30 Sept. 1977): 4257–62.
19. Martin Marietta Corp., *Viking Mars Expedition, 1976*, p. 27.
20. Ibid.
21. Ibid., pp. 28–29.
22. Crofton B. Farmer et al., "Mars: Water Vapor Observations from the Viking Orbiters," *Journal of Geophysical Research* 82 (30 Sept. 1977): 4225–48; Leon Kosofsky, *Viking 1: Early Results*, NASA SP-408 (Washington, 1976), pp. 18–19; and Soffen, "Mars and the Remarkable Viking Results."
23. Martin Marietta Corp., *Viking Mars Expedition, 1976*, p. 25.
24. Soffen, "Mars and the Remarkable Viking Results."
25. Kosofsky, *Viking 1: Early Results*, pp. 25–26; and Martin Marietta Corp., *Viking 75 Project: Viking*

*Lander System Primary Mission Performance Report*, by G. C. Cooley and J. G. Lewis, NASA CR-145148 (Denver, 1977), pp. V-11 to V-12, contract NAS1-9000.

26. Kosofsky, *Viking 1: Early Results*, p. 26; W. B. Hanson, S. Santani, and D. R. Zuccaro, "The Martian Ionosphere as Observed by the Viking Retarding Potential Analyzers," *Journal of Geophysical Research* 82 (30 Sept. 1977): 4351-63; and A. Tan and S. T. Wu, "Post-Viking Martian Ionospheric Model," *Acta Astronomica* 28, 2d issue (1978): 187-94.

27. Soffen, "Mars and the Remarkable Viking Results"; and A.O. Nier and M. B. McElroy, "Structure of the Neutral Upper Atmosphere of Mars," *Science* 194 (17 Dec. 1976): 1298-300.

28. Martin Marietta Corp., *Viking Mars Expedition, 1976*, p. 39; Mark Washburn, *Mars at Last!* (New York, 1977), p. 230; and JPL, Viking Science Forum, 28 July 1976, transcript.

29. Kosofsky, *Viking 1: Early Results*, pp. 27-28; Nier and McElroy, "Composition and Structure of Mars' Upper Atmosphere: Results from the Neutral Mass Spectrometers on Viking 1 and 2," *Journal of Geophysical Research* 82 (30 Sept. 1977): 4341-49; and Klaus Biemann to Monte D. Wright, " 'The Exploration of Mars, 1958-1978' by Edward and Linda Ezell," 13 Apr. 1979.

30. JPL, Viking Science Symposium, 28 July 1976, transcript; and Washburn, *Mars at Last!*, p. 229.

31. Soffen, "Mars and the Remarkable Viking Results."

32. Kosofsky, *Viking 1: Early Results*, p. 29; Alvin Seiff and Donn B. Kirk, "Structure of the Atmosphere of Mars in Summer at Mid-Latitudes," *Journal of Geophysical Research* 82 (30 Sept. 1977): 4364-78; and Seiff and Kirk, "Structure of Mars' Atmosphere up to 100 Kilometers from the Entry Measurements of Viking 2," *Science* 194 (17 Dec. 1976): 1300-03.

33. Viking Lander Imaging Team. *The Martian Landscape*, NASA SP-425 (Washington, 1978), p. 23.

34. Notes taken by the authors during the mission, 20 July 1976; Washburn, *Mars at Last!*, pp. 219-20; and "Mission Status Bulletin," no. 35, 20 July 1976.

35. Viking Lander Imaging Team, *Martian Landscape*, p. 27.

36. JPL, "Viking Status Report," 10:00 a.m. PDT, 21 July 1976 [reports were issued several times a day, as events took place, to help keep the press informed].

37. JPL "Viking Status Report," 11:30 a.m. PDT, 22 July 1976; and Viking press briefing, 21 July 1976.

38. Viking Lander Imaging Team, *Martian Landscape*, p. 27; Viking press briefing, 26 July 1976; JPL, "Viking Lander Color Imaging," press handout, 26 July 1976; and VPO, Viking Project Bulletin Special Edition, "The Colors of Mars," Dec. 1976.

39. Soffen interview by Edward C. Ezell, 23 Dec. 1978.

40. Martin Marietta Corp., *Viking Mars Expedition, 1976*, pp. 34-36.

41. Soffen interview.

42. Seymour L. Hess, "Weather at Chryse Planitia on Sols 2 and 3 of the Viking Era." 24 July 1976; Hess, "Weather at Chryse on Sols 0 and 1," 21 July 1976; and Hess, "Weather Report for Chryse Planitia on Sols 9 & 10," meteorology team press handout, 1 Aug. 1976.

43. Martin Marietta Corp., *Viking Mars Expedition 1976*, pp. 22-24.

44. Martin Marietta Corp., *Viking Lander System Primary Mission Performance Report*, pp. VI-21 to VI-22.

45. Martin Marietta Corp., *Viking Mars Expedition, 1976*, pp. 48-49.

46. Washburn, *Mars at Last!*, p. 223.

47. Viking press briefing, 24 July 1976.

48. Martin Marietta Corp., *Viking Lander System Primary Mission Performance Report*, p. VI-7.

49. Ibid.

50. JPL, "Viking Status Report," 6:30 a.m. PDT, 22 July 1976.

51. Viking press briefing, 22 July 1976; and notes taken by the authors during the mission, 22 July 1976.

52. JPL, "Viking Status Report," 1:00 p.m. PDT, 25 July 1976; JPL, "Mission Status Bulletin," no. 36, 27 July 1976; notes taken by the authors during the mission, 25 July 1976; Cary R. Spitzer, "Unlimbering Viking's Scoop," *IEEE Spectrum* 13 (Nov. 1976): 92-93; and Martin Marietta Corp., *Viking Lander System Primary Mission Performance Report*, pp. VI-24 to VI-25.

53. Narrative is based heavily on Washburn, *Mars at Last!*, pp. 225-27. See also JPL, "Mission Status Bulletin," supplement, 10 Sept. 1976.

54. Martin Marietta Corp., *Viking Lander System Primary Mission Performance Report*, pp. VI-25 to VI-26; and Viking press briefing, 29 July 1976.

55. Martin Marietta Corp., *Viking Lander System Primary Mission Performance Report*, pp. VI-26 to VI-27.

56. Viking press briefing, 31 July 1976.

57. Ibid.

58. Ibid.

59. Washburn, *Mars at Last!*, pp. 249–50.
60. Viking press briefing, 7 Aug. 1976.
61. JPL, Viking Science Forum, 10 Aug. 1976, transcript.
62. Norman H. Horowitz, "The Search for Life on Mars," *Scientific American* 237 (Nov. 1977): 57–58.
63. Ibid., pp. 58–60.
64. Ibid., pp. 60–61.
65. Timothy Ferris, "The Odyssey and the Ecstasy," *Rolling Stone*, 7 Apr. 1977.
66. Viking press briefing, 7 Aug. 1976.
67. Viking press briefing, 13 Aug. 1976.
68. Martin Marietta Corp., *Viking Mars Expedition, 1976*, pp. 20–21.
69. Soffen interview; and Joshua Lederberg and Carl Sagan, "Microenvironments for Life on Mars," *Proceedings of the National Academy of Sciences* 48 (Sept. 1962): 1473–75.
70. Horowitz, "The Search for Life on Mars," p. 61.
71. David L. Chandler, "Life on Mars," *Atlantic* 242 (June 1977): 34.
72. "One Man's Mars: No Martians," *Science News* 111 (5 Mar. 1977): 149.
73. Ibid., p. 150.
74. Soffen interview.
75. Martin Marietta Corp., *Viking Mars Expedition, 1976*, pp. 18–19.
76. Philip H. Abelson, "The Martian Environment," *Science* 147 (12 Feb. 1965): 683.
77. Abelson, "Viking 1," *Science* 193 (27 Aug. 1976): 723.
78. Chandler, "Life on Mars," pp. 29–36; and Ferris, "The Odyssey and the Ecstasy."
79. "The Mars Experiment," *New York Times*, 18 Nov. 1976; and "Life on Mars? . . ." *New York Times*, 20 Sept. 1976.
80. William H. Michael et al., "Viking Radio Science Investigations," *Journal of Geophysical Research* 82 (30 Sept. 1977): 4293–95. See also "Viking: Riches in a Radio Beam," *Science News* 110 (20 Nov. 1976): 325–26.
81. Michael, "Viking Extended Mission Radio Science Results," unpublished paper, n.d.
82. Martin Marietta Corp., *Viking Mars Expedition, 1976*, pp. 42–43.
83. Ibid., p. 44.
84. Ibid., pp. 45–47.
85. Ibid.
86. JPL, Viking Science Forum, 10 Aug. 1976, transcript.
87. "Mars Soil Similar at Two Viking Sites," *Science News* 110 (16 Oct. 1976): 245–46; JPL, "Mission Status Bulletin," no. 44, 29 Sept. 1976, and no. 45, 21 Oct. 1976.
88. Soffen, "Mars and the Remarkable Viking Results."
89. Langley Research Ctr., "Viking Project Manager Leaving NASA," news release, 76-37, 5 Nov. 1976.
90. Noel W. Hinners, "Personnel Changes," NASA Special Announcement, 5 Nov. 1976.
91. "Biographical Data, G. Calvin Broome," n.d., in NASA History Office biography files.

# Epilogue

1. Conway W. Snyder, "The Extended Mission of Viking," *Journal of Geophysical Research* 84 (30 Dec. 1979): 7917.
2. Snyder to All Viking Scientists, "The Viking Mission Status Plans," 9 Apr. 1979; and A. Thomas Young to Noel Hinners, "Future of the Viking Program," 21 Nov. 1977.
3. Snyder, "The Extended Mission of Viking," p. 7930.
4. Snyder to scientists, 9 Apr. 1979; and Mike Carr to Snyder, "Imaging from VO-1 during the December 1, 1978 to April 1, 1979 Time Period," 29 Aug. 1978.
5. Snyder to scientists, 9 Apr. 1979.
6. NASA, "Transmitter Switched Off on Viking Orbiter 1," news release 80-129, 8 Aug. 1980; and JPL, "Viking Facts" [4 Aug. 1980].
7. Snyder, "The Planet as Seen at the End of the Viking Mission," *Journal of Geophysical Research* 84 (30 Dec. 1979): 8487, 8515.

# Index

# The Authors

Edward Clinton Ezell, born in Indianapolis, Indiana (1939), received his A.B. from Butler University (1961); M.A. from the University of Delaware (1963), where he was a Hagley Fellow; and Ph.D. in the history of science and technology from Case Institute of Technology, Cleveland (1969). He taught at North Carolina State University, Raleigh, and Sangamon State University, Springfield, Illinois, before contracting with the National Aeronautics and Space Administration to write *The Partnership* (Linda N. Ezell, coauthor, 1978), a history of the Apollo-Soyuz Test Project. In 1980, Ezell became the historian at NASA's Johnson Space Center, Houston, Texas. Also active in the field of military technology, Ezell had recently written *Handguns of the World* (1981), a companion volume to the 12th edition of *Small Arms of the World* (in press) (Harrisburg, Pa.: Stackpole Books). He is now a curator of military history at the Smithsonian Institution's National Museum of American History in Washington.

Linda Neuman Ezell, born in Fulton County, Illinois (1951), graduated from Sangamon State University in 1974. Under contract to NASA Headquarters, she has also written *NASA Historical Data Book, 1958-1968*, vol. 2, *Programs and Major Projects* (in press) and is working on another volume that will describe NASA programs during the agency's second decade. Ms. Ezell, a part-time graduate student at George Washington University in Washington, is also an emergency medical technician and firefighter for a volunteer fire department in northern Virginia.

# List of NASA Photograph Numbers